E. C Rice

Tables for Calculating Excavation and Embankment of Regular and Irregular Cross Sections

Third Edition

E. C Rice

Tables for Calculating Excavation and Embankment of Regular and Irregular Cross Sections
Third Edition

ISBN/EAN: 9783744689977

Printed in Europe, USA, Canada, Australia, Japan

Cover: Foto ©berggeist007 / pixelio.de

More available books at **www.hansebooks.com**

TABLES

FOR CALCULATING

EXCAVATION AND EMBANKMENT

OF

REGULAR AND IRREGULAR CROSS SECTIONS.

BY

E. C. RICE, C. E.

THIRD EDITION.

ST. LOUIS, MO.
R. P. STUDLEY & CO., PRINTERS AND PUBLISHERS, 221 NORTH MAIN STREET.
1880.

THE following Tables for calculating Excavations, Embankments and Culverts have, at our request, been furnished us for publication by E. C. RICE, Chief Engineer of the St. Louis and Southeastern Railway, and the Cairo and Vincennes Railroad, and formerly Chief Engineer of the Ohio and Mississippi Railway, the St. Louis, Vandalia and Terre Haute, and other Railroads.

R. P. STUDLEY & CO.

ST. LOUIS, MO., 1873.

CUBIC YARDS.

AREAS.

INDEX TO TABLES—*Continued.*

AREAS OF TRIANGLES.

EXPLANATIONS.

FIG. 1.

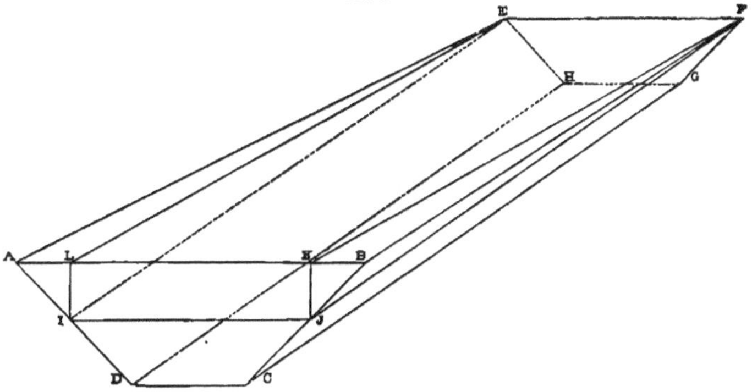

FIG. 1.—Let d represent the difference of the end heights, s the ratio of the slope to 1, and l the length of the prismoid.

Draw AB parallel to DC, and EF to HG, and make IJDC equal to EFHG, LK equal to EF, and AL equal to KB. The contents of the prisms IJDC, EFHG and LKIJEF equals the sum of the areas EFGH, IJDC, LKIJ, by one-half l, and the contents of the pyramids ALIE and KBJF equals the sum of the areas of their bases by one-third l. The result obtained by multiplying the sum of the end areas of the prismoid by one-half the length is therefore too large by the difference between a half and a third of the product of the bases of the pyramids by their length, or $\frac{sd^2l}{6}$.

FIG. 2.

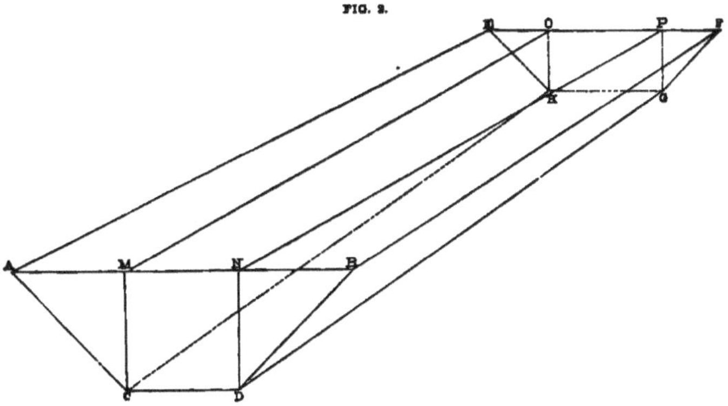

If, in calculating the solid contents of a prismoid, we multiply the middle area by the length, we shall obtain a result too small by $\frac{sd^2l}{12}$, when d represents the difference of the end heights, s the ratio of the slope to 1, and l the length of the prismoid.

Fig. 2.—Let ND=MC=H, PG=OH=h, CD=HG=b, GD=HC=l, NB=MA=sH, PF=OE=sh.

It is evident that the prismoid is composed of the prismoid MNDCOPHG, and the equal frustra of pyramids NDBPGF and MACOEH.

The solid contents of the prismoid MNDCOPGH is $bl\,(\frac{H+h}{2}.)$ The solid contents of the frustra is $\frac{l}{3}\,(sH^2+sh^2+sHh)=(4H^2+4Hh+4h^2)\,\frac{sl}{12}=(3H^2+6Hh+3h^2+H^2-2Hh+h^2)\,\frac{sl}{12}=(H+h)^2\frac{sl}{4}+(H-h)^2\frac{sl}{12}$.

The contents of the whole prismoid is $bl\times\frac{H+h}{2}+\frac{(H+h)^2sl}{4}+(H-h)^2\frac{sl}{12}=[2b+s(H+h)]\,(H+h)\,\frac{l}{4}+(H-h)^2\frac{sl}{12}$.

The quantity obtained by multiplying the middle area by the length is $[2b+s\,(H+h)]\,(H+h)\,\frac{l}{4}$, which is therefore too small by $(H-h)^2\frac{sl}{12}=\frac{sd^2l}{12}$.

Again, the result obtained by multiplying the middle area of the prismoid (Fig. 1) by the length, is too small by the difference between a third and a fourth of the product of the bases of the pyramids by their length, or $\frac{sd^2l}{12}$, as before shown; for while the contents of the prisms will be correctly given, the contents of the two pyramids, by this method, is $\frac{sd^2l}{4}$, when their true contents is $\frac{sd^2l}{3}$, and $\frac{sd^2l}{3}-\frac{sd^2l}{4}=\frac{sd^2l}{12}$.

The contents of any prism is equal to the area of the base multiplied by the height.

$ah=(a+4a+a)\,\frac{h}{6}$=the sum of the end areas and four times the middle area by one-sixth of the height, which is the Prismoidal Formula.

The contents of any pyramid equals the area of the base multiplied by one-third the perpendicular height.

$\frac{ah}{3}=(a+a+o)\frac{h}{6}$=the sum of the end areas, and four times the middle area, by one-sixth of the height.

FIG. 3.

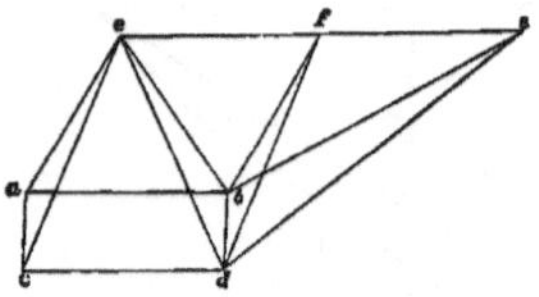

Let $ab=cd=ef=e^1f$.

h=the perpendicular height.

a=area of the base $abcd$.

The contents of $abcdef=\frac{ah}{2}=\quad(a+2a+o)\frac{h}{6}$.

The contents of $abcde=\quad(a+\;a+o)\frac{h}{6}$.

The contents of $bdef=bde^1f=\quad(o+\;a+o)\frac{h}{6}$.

hence the wedge may be calculated by the Prismoidal Formula.

Frustra of pyramids may be divided into wedges and prisms, and hence their contents may be calculated by the Prismoidal Formula.

Whatever the shape of a prismoid, intercepted between two parallel cross sections, it may be divided into prisms, pyramids, wedges, or frustra of pyramids, to all of which, and therefore to the prismoid, the Prismoidal Formula may be correctly applied.

FIG. 4.

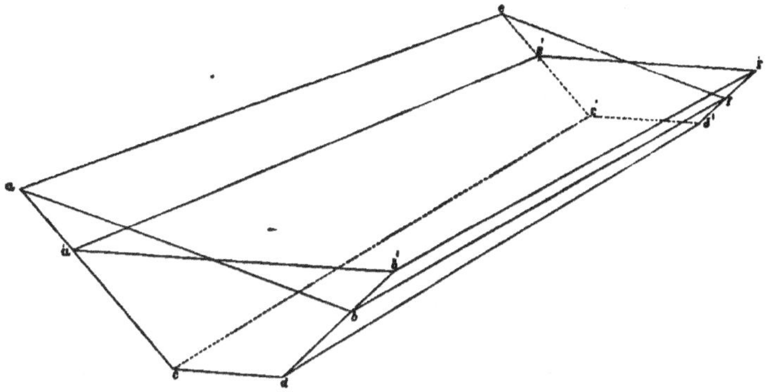

FIG. 4.—Make the area of the level cross section a^1b^1cd = the area of the cross section $abcd$, and the area of the level cross section $e^1f^1c^1d^1$ = the area of the cross section efc^1d^1: then the contents of the two prismoids are equal.

The following Tables give the contents of prismoids in cubic yards, obtained by multiplying the middle area by the length, together with the correction, in cubic yards, to add for the difference of the end heights, to obtain the true quantity:

Height.	Area in Feet.

Height.	$\frac{ml}{27}$ Cubic Yards.	$\frac{md^2l}{12 \times 27}$ Correction.

To CALCULATE ANY PRISMOID BY THE TABLES.—If the cross section is level, in the column of heights, find half the sum of the two end heights of the prismoid, and set down the quantity found opposite in the column of Cubic Yards; then, in the same column of heights, find the difference of those end heights, and set down the quantity found opposite in column of corrections. The sum of these two quantities is the correct quantity required.

Example First.—If we have a prismoid of level cross section, base 20 feet, slope 1 1-2 to 1, length 40 feet, and end heights of 50.22 and 10.12 feet, then $\frac{50.22+10.12}{2}$=30.17=mean height, and 50.22−10.12=40.10=difference of end heights. By Table No. 12 of cubic yards.

In column of cubic yards, opposite height 30.1, is found	7263.00
In column of difference, between heights 30.1 and 30.2, is found 40.9, which multiplied by .7 gives -	28.63
In column of corrections, opposite 40.10 is found	744.40
Total, ...	8036.03=

the number of cubic yards in a prismoid 100 feet in length, which, multiplied by $\frac{40}{100}$, gives 3214.412 cubic yards in the prismoid 40 feet in length.

If the cross section of the ground is irregular, first find the area of each end cross section of the prismoid; then in the column of areas find these areas, and opposite, in the column of heights, will be found the heights of equivalent level cross sections. With the heights thus obtained proceed as before.

Example Second.—If we have a prismoid of irregular cross section, base 20 feet, slope 1 1-2 to 1, length 100 feet. with end areas of 226.42 and 8465.64 square feet. By Table No. 12 of Areas.

In column of heights, opposite area 225.9, is found - - - - - - - - - - - - -	7.3
In column of difference, between areas 225.9 and 230.1, is found 4.2, by which divide .52, which gives	.012
Total equivalent height, - - - - - - - - -	7.312
Opposite area 8453.5 is found in column of heights - - - - - - - - - - - - -	68.7
In column of difference, between 8453.5 and 8476.2, is found 22.6, by which divide 12.14, which gives	.054
Total equivalent height, - - - - - - - - - - - - - - - -	68.754

With these equivalent heights, proceed with Table 12 as in Example First.

Cross-sections for calculations of quantities must be taken near enough to each other that the end sections of each solid may be as near similar as practicable; for unless the surface of a solid is a plane and the end sections are similar, the quantities obtained by the Tables will be a *very little too small*, though occasionally the irregularities of the surface will tend to counterbalance this deficiency.

It is not practicable to obtain the *exact* quantity of a solid of excavation or embankment, by any feasible method, on account of the usually warped and irregular surface of the ground.

To make ample allowance for any deficiency on account of the warped condition and irregularity of the surface of the solid in *Example Second*, call the equivalent heights 7.35 and 68.80 in lieu of 7.312 and 68.754, and call the half sum of these heights 38.1.

Field and office notes should be kept in such a manner that any engineer can understand them without explanation or study, and each one of an engineer department should have the same method of keeping notes.

All irregular cross-sections, and many of the regular cross-sections, should be platted, because it is the best way to show the solid and the different kinds of material, and will facilitate making calculations of areas and of quantities.

A center line plug should be driven even with the surface of the ground at each regular station and wheresoever a cross-section is required, and the heights of all these plugs should be taken and tested when the location of the line is made.

Intermediate centers may usually be set without the use of the transit, and where practicable they should be set at some multiple of five or ten feet, for convenience in calculating the cubic yards.

Cross-section levels can usually be taken quickest, and with the least liability to error, with a rod ten feet long and spirit level.

The heights should be taken at some multiple of five or ten feet, when practicable, for convenience in platting and calculating areas, and they should be taken far enough each side of the center line to allow for any change in grade, or in width of road-bed, or in slope, which may possibly be required.

The heights should be made up with reference to the grade line.

The cross-sections should be platted consecutively, and there should be upon the same sheets columns for Areas, Equivalent Heights, Average Heights, Cubic Yards, etc.

The Tables of Areas of Triangles will be found useful in calculating the areas of irregular cross-sections.

The Tables of Culverts will be found useful in making approximate estimates.

E. C. RICE.

TABLE No. 1.

EXCAVATION AND EMBANKMENT.

CUBIC YARDS.

Prismoids, 100 feet long. Breadth of Base, 15 feet. Slopes, 3 and 2 Horizontal to 1 Perpendicular.

Height.	Cub. Yds.	Diff.	Cor.	Height.	Cub. Yds.	Diff.	Cor.	Height.	Cub. Yds.	Diff.	Cor.	Height.	Cub. Yds.	Diff.	Cor.	Height.	Cub. Yds.	Diff.	Cor.
0.0	0.0	0.0	0.0	8.0	1037.0	20.3	49.4	16.0	3259.3	35.1	197.5	24.0	6666.7	49.9	444.4	32.0	11259.3	64.7	790.1
1	5.6	5.6	0.0	1	1057.5	20.5	50.6	1	3294.5	35.3	200.0	1	6716.8	50.1	448.2	1	11324.2	64.9	795.1
2	11.5	5.8	0.0	2	1078.1	20.6	51.9	2	3330.0	35.5	202.5	2	6767.0	50.3	451.9	2	11389.3	65.1	800.0
3	17.5	6.0	0.1	3	1099.0	20.8	53.2	3	3365.6	35.6	205.0	3	6817.5	50.5	455.6	3	11454.5	65.3	805.0
4	23.7	6.2	0.1	4	1120.0	21.0	54.4	4	3401.5	35.8	207.5	4	6868.1	50.6	459.4	4	11520.0	65.5	810.0
5	30.1	6.4	0.2	5	1141.2	21.2	55.7	5	3437.5	36.0	210.1	5	6919.0	50.8	463.2	5	11585.6	65.6	815.0
6	36.7	6.6	0.3	6	1162.6	21.4	57.1	6	3473.7	36.2	212.6	6	6970.0	51.0	466.9	6	11651.5	65.8	820.0
7	43.4	6.8	0.4	7	1184.2	21.6	58.4	7	3510.1	36.4	215.2	7	7021.2	51.2	470.7	7	11717.5	66.0	825.1
8	50.4	6.9	0.5	8	1205.9	21.8	59.8	8	3546.7	36.6	217.8	8	7072.6	51.4	474.6	8	11783.7	66.2	830.1
9	57.5	7.1	0.6	9	1227.9	21.9	61.1	9	3583.4	36.8	220.4	9	7124.2	51.6	478.4	9	11850.1	66.4	835.2
1.0	64.8	7.3	0.8	9.0	1250.0	22.1	62.5	17.0	3620.4	36.9	223.0	25.0	7175.9	51.8	482.3	33.0	11916.7	66.6	840.3
1	72.3	7.5	0.9	1	1272.3	22.3	63.9	1	3657.5	37.1	225.6	1	7227.9	51.9	486.1	1	11983.4	66.8	845.4
2	80.0	7.7	1.1	2	1294.8	22.5	65.3	2	3694.8	37.3	228.3	2	7280.0	52.1	490.0	2	12050.4	66.9	850.5
3	87.9	7.9	1.3	3	1317.5	22.7	66.7	3	3732.3	37.5	230.9	3	7332.3	52.3	493.9	3	12117.5	67.1	855.6
4	95.9	8.1	1.5	4	1340.4	22.9	68.2	4	3770.0	37.7	233.6	4	7384.8	52.5	497.8	4	12184.8	67.3	860.8
5	104.2	8.2	1.7	5	1363.4	23.1	69.6	5	3807.9	37.9	236.3	5	7437.5	52.7	501.7	5	12252.3	67.5	865.9
6	112.6	8.4	2.0	6	1386.7	23.2	71.1	6	3845.9	38.1	239.0	6	7490.4	52.9	505.7	6	12320.0	67.7	871.1
7	121.2	8.6	2.2	7	1410.1	23.4	72.6	7	3884.2	38.2	241.7	7	7543.4	53.1	509.6	7	12387.9	67.9	876.3
8	130.0	8.8	2.5	8	1433.7	23.6	74.1	8	3922.6	38.4	244.5	8	7596.7	53.2	513.6	8	12455.9	68.1	881.5
9	139.0	9.0	2.8	9	1457.5	23.8	75.6	9	3961.2	38.6	247.2	9	7650.1	53.4	517.6	9	12524.2	68.2	886.7
2.0	148.1	9.2	3.1	10.0	1481.5	24.0	77.2	18.0	4000.0	38.8	250.0	26.0	7703.7	53.6	521.6	34.0	12592.6	68.4	892.0
1	157.5	9.4	3.4	1	1505.6	24.2	78.7	1	4039.0	39.0	252.8	1	7757.5	53.8	525.6	1	12661.2	68.6	897.2
2	167.0	9.5	3.7	2	1530.0	24.4	80.3	2	4078.1	39.2	255.6	2	7811.5	54.0	529.7	2	12730.0	68.8	902.5
3	176.8	9.7	4.1	3	1554.5	24.5	81.9	3	4117.5	39.4	258.4	3	7865.6	54.2	533.7	3	12799.0	69.0	907.8
4	186.7	9.9	4.4	4	1579.3	24.7	83.5	4	4157.0	39.5	261.2	4	7920.0	54.4	537.8	4	12868.1	69.2	913.1
5	196.8	10.1	4.8	5	1604.2	24.9	85.1	5	4196.8	39.7	264.1	5	7974.5	54.5	541.9	5	12937.5	69.4	918.4
6	207.0	10.3	5.2	6	1629.3	25.1	86.7	6	4236.7	39.9	266.9	6	8029.3	54.7	546.0	6	13007.0	69.5	923.7
7	217.5	10.5	5.6	7	1654.5	25.3	88.3	7	4276.8	40.1	269.8	7	8084.2	54.9	550.1	7	13076.8	69.7	929.1
8	228.1	10.6	6.0	8	1680.0	25.5	90.0	8	4317.0	40.3	272.7	8	8139.3	55.1	554.2	8	13146.7	69.9	934.4
9	239.0	10.8	6.5	9	1705.6	25.6	91.7	9	4357.5	40.5	275.6	9	8194.5	55.3	558.3	9	13216.8	70.1	939.8
3.0	250.0	11.0	6.9	11.0	1731.5	25.8	93.4	19.0	4398.1	40.6	278.5	27.0	8250.0	55.5	562.5	35.0	13287.0	70.3	945.2
1	261.2	11.2	7.4	1	1757.5	26.0	95.1	1	4439.0	40.8	281.5	1	8305.6	55.6	566.7	1	13357.5	70.5	950.6
2	272.6	11.4	7.9	2	1783.7	26.2	96.8	2	4480.0	41.0	284.4	2	8361.5	55.8	570.9	2	13428.1	70.6	956.0
3	284.2	11.6	8.4	3	1810.1	26.4	98.5	3	4521.2	41.2	287.4	3	8417.5	56.0	575.1	3	13499.0	70.8	961.5
4	295.9	11.8	8.9	4	1836.7	26.6	100.3	4	4562.6	41.4	290.4	4	8473.7	56.2	579.3	4	13570.0	71.0	966.9
5	307.9	11.9	9.5	5	1863.4	26.8	102.0	5	4604.2	41.6	293.4	5	8530.1	56.4	583.5	5	13641.2	71.2	972.4
6	320.0	12.1	10.0	6	1890.4	26.9	103.8	6	4645.9	41.8	296.4	6	8586.7	56.6	587.8	6	13712.6	71.4	977.9
7	332.3	12.3	10.6	7	1917.5	27.1	105.6	7	4687.9	41.9	299.5	7	8643.4	56.8	592.0	7	13784.2	71.6	983.4
8	344.8	12.5	11.1	8	1944.8	27.3	107.4	8	4730.0	42.1	302.5	8	8700.4	56.9	596.3	8	13855.9	71.8	988.9
9	357.5	12.7	11.7	9	1972.3	27.5	109.3	9	4772.3	42.3	305.6	9	8757.5	57.1	600.6	9	13927.9	71.9	994.5
4.0	370.4	12.9	12.3	12.0	2000.0	27.7	111.1	20.0	4814.8	42.5	308.6	28.0	8814.8	57.3	604.9	36.0	14000.0	72.1	1000.0
1	383.4	13.1	13.0	1	2027.9	27.9	113.0	1	4857.5	42.7	311.7	1	8872.3	57.5	609.3	1	14072.3	72.3	1005.6
2	396.7	13.2	13.6	2	2055.9	28.1	114.8	2	4900.4	42.9	314.8	2	8930.0	57.7	613.6	2	14144.8	72.5	1011.1
3	410.1	13.4	14.3	3	2084.2	28.2	116.7	3	4943.4	43.1	318.0	3	8987.9	57.9	618.0	3	14217.5	72.7	1016.7
4	423.7	13.6	14.9	4	2112.6	28.4	118.6	4	4986.7	43.2	321.1	4	9045.9	58.1	622.3	4	14290.4	72.9	1022.3
5	437.5	13.8	15.6	5	2141.2	28.6	120.6	5	5030.1	43.4	324.3	5	9104.2	58.2	626.7	5	14363.4	73.1	1028.0
6	451.5	14.0	16.3	6	2170.0	28.8	122.5	6	5073.7	43.6	327.4	6	9162.6	58.4	631.1	6	14436.7	73.2	1033.6
7	465.6	14.2	17.0	7	2199.0	29.0	124.5	7	5117.5	43.8	330.6	7	9221.2	58.6	635.6	7	14510.1	73.4	1039.3
8	480.0	14.4	17.8	8	2228.1	29.2	126.4	8	5161.5	44.0	333.8	8	9280.0	58.8	640.0	8	14583.7	73.6	1044.9
9	494.5	14.5	18.5	9	2257.5	29.4	128.4	9	5205.6	44.2	337.0	9	9339.0	59.0	644.5	9	14657.5	73.8	1050.6
5.0	509.3	14.7	19.3	13.0	2287.0	29.5	130.4	21.0	5250.0	44.4	340.3	29.0	9398.1	59.2	648.9	37.0	14731.5	74.0	1056.3
1	524.2	14.9	20.1	1	2316.8	29.7	132.4	1	5294.5	44.5	343.5	1	9457.5	59.4	653.4	1	14805.6	74.2	1062.0
2	539.3	15.1	20.9	2	2346.7	29.9	134.4	2	5339.3	44.7	346.8	2	9517.0	59.5	657.9	2	14880.0	74.4	1067.8
3	554.5	15.3	21.7	3	2376.8	30.1	136.5	3	5384.2	44.9	350.1	3	9576.8	59.7	662.4	3	14954.5	74.5	1073.5
4	570.0	15.5	22.5	4	2407.0	30.3	138.5	4	5429.3	45.1	353.4	4	9636.7	59.9	666.9	4	15029.3	74.7	1079.3
5	585.6	15.6	23.3	5	2437.5	30.5	140.6	5	5474.5	45.3	356.7	5	9696.8	60.1	671.5	5	15104.2	74.9	1085.1
6	601.5	15.8	24.2	6	2468.1	30.6	142.7	6	5520.0	45.5	360.0	6	9757.0	60.3	676.0	6	15179.3	75.1	1090.9
7	617.5	16.0	25.1	7	2499.0	30.8	144.8	7	5565.6	45.6	363.3	7	9817.5	60.5	680.6	7	15254.5	75.3	1096.7
8	633.7	16.2	26.0	8	2530.0	31.0	146.9	8	5611.5	45.8	366.7	8	9878.1	60.6	685.2	8	15330.0	75.5	1102.5
9	650.1	16.4	26.9	9	2561.2	31.2	149.1	9	5657.5	46.0	370.1	9	9939.0	60.8	689.8	9	15405.6	75.6	1108.3
6.0	666.7	16.6	27.8	14.0	2592.6	31.4	151.2	22.0	5703.7	46.2	373.5	30.0	10000.0	61.0	694.4	38.0	15481.5	75.8	1114.2
1	683.4	16.8	28.7	1	2624.2	31.6	153.4	1	5750.1	46.4	376.9	1	10061.2	61.2	699.1	1	15557.5	76.0	1120.1
2	700.4	16.9	29.7	2	2655.9	31.8	155.6	2	5796.7	46.6	380.3	2	10122.6	61.4	703.7	2	15633.7	76.2	1126.0
3	717.5	17.1	30.6	3	2687.9	31.9	157.8	3	5843.4	46.8	383.7	3	10184.2	61.6	708.4	3	15710.1	76.4	1131.9
4	734.8	17.3	31.6	4	2720.0	32.1	160.0	4	5890.4	46.9	387.2	4	10245.9	61.8	713.1	4	15786.7	76.6	1137.8
5	752.3	17.5	32.6	5	2752.3	32.3	162.2	5	5937.5	47.1	390.6	5	10307.9	61.9	717.8	5	15863.4	76.8	1143.7
6	770.0	17.7	33.6	6	2784.8	32.5	164.5	6	5984.8	47.3	394.1	6	10370.0	62.1	722.5	6	15940.4	76.9	1149.7
7	787.9	17.9	34.6	7	2817.5	32.7	166.7	7	6032.3	47.5	397.6	7	10432.3	62.3	727.2	7	16017.5	77.1	1155.6
8	805.9	18.1	35.7	8	2850.4	32.9	169.0	8	6080.0	47.7	401.1	8	10494.8	62.5	732.0	8	16094.8	77.3	1161.6
9	824.2	18.2	36.7	9	2883.4	33.1	171.3	9	6127.9	47.9	404.6	9	10557.5	62.7	736.7	9	16172.3	77.5	1167.6
7.0	842.6	18.4	37.8	15.0	2916.7	33.2	173.6	23.0	6175.9	48.1	408.2	31.0	10620.4	62.9	741.5	39.0	16250.0	77.7	1173.6
1	861.2	18.6	38.9	1	2950.1	33.4	175.9	1	6224.2	48.2	411.7	1	10683.4	63.1	746.3	1	16327.9	77.9	1179.6
2	880.0	18.8	40.0	2	2983.7	33.6	178.3	2	6272.6	48.4	415.3	2	10746.7	63.2	751.1	2	16405.9	78.1	1185.7
3	899.0	19.0	41.1	3	3017.5	33.8	180.6	3	6321.2	48.6	418.9	3	10810.1	63.4	755.9	3	16484.2	78.2	1191.7
4	918.1	19.2	42.3	4	3051.5	34.0	183.0	4	6370.0	48.8	422.5	4	10873.7	63.6	760.8	4	16562.6	78.4	1197.8
5	937.5	19.4	43.4	5	3085.6	34.2	185.4	5	6419.0	49.0	426.1	5	10937.5	63.8	765.6	5	16641.2	78.6	1203.9
6	957.0	19.5	44.6	6	3120.0	34.4	187.8	6	6468.1	49.2	429.8	6	11001.5	64.0	770.5	6	16720.0	78.8	1210.0
7	976.8	19.7	45.7	7	3154.5	34.5	190.2	7	6517.5	49.4	433.4	7	11065.6	64.2	775.4	7	16799.0	79.0	1216.1
8	996.7	19.9	46.9	8	3189.3	34.7	192.6	8	6567.0	49.5	437.1	8	11130.0	64.4	780.3	8	16878.1	79.2	1222.3
9	1016.8	20.1	48.2	9	3224.2	34.9	195.1	9	6616.8	49.7	440.7	9	11194.5	64.5	785.2	9	16957.5	79.4	1228.4

TABLE No. 2.

EXCAVATION AND EMBANKMENT.

CUBIC YARDS.

Prismoids, 100 Feet Long Breadth of Base, 15 Feet. Slopes, 2 and 1 1-2 Horizontal to 1 Perpendicular.

Height.	Cub. Yds.	Diff.	Cor.	Height.	Cub. Yds.	Diff.	Cor.	Height.	Cub. Yds.	Diff.	Cor.	Height.	Cub. Yds.	Diff.	Cor.	Height.	Cub. Yds.	Diff.	Cor.
0.0	0.0	0.0	0.0	8.0	859.3	15.9	34.6	16.0	2548.1	26.2	138.3	24.0	5066.7	36.6	311.1	32.0	8414.8	47.0	553.1
1	5.6	5.6	0.0	1	875.3	16.0	35.4	1	2574.5	26.4	140.0	1	5103.4	36.7	313.7	1	8461.9	47.1	556.5
2	11.4	5.8	0.0	2	891.4	16.1	36.3	2	2601.0	26.5	141.8	2	5140.3	36.9	316.3	2	8509.1	47.2	560.0
3	17.3	5.9	0.0	3	907.6	16.3	37.2	3	2627.6	26.6	143.5	3	5177.3	37.0	318.9	3	8556.5	47.4	563.5
4	23.3	6.0	0.1	4	924.0	16.4	38.1	4	2654.4	26.8	145.3	4	5214.4	37.1	321.6	4	8604.0	47.5	567.0
5	29.4	6.1	0.1	5	940.5	16.5	39.0	5	2681.3	26.9	147.0	5	5251.6	37.3	324.2	5	8651.6	47.6	570.5
6	35.7	6.3	0.2	6	957.1	16.6	39.9	6	2708.3	27.0	148.8	6	5289.0	37.4	326.9	6	8699.4	47.8	574.0
7	42.1	6.4	0.3	7	973.9	16.8	40.9	7	2735.4	27.1	150.6	7	5326.5	37.5	329.5	7	8747.3	47.9	577.5
8	48.6	6.5	0.3	8	990.8	16.9	41.8	8	2762.7	27.3	152.4	8	5364.1	37.6	332.2	8	8795.3	48.0	581.1
9	55.3	6.7	0.4	9	1007.8	17.0	42.8	9	2790.1	27.4	154.3	9	5401.9	37.8	334.9	9	8843.4	48.1	584.6
1.0	62.0	6.8	0.5	9.0	1025.0	17.2	43.8	17.0	2817.6	27.5	156.1	25.0	5439.8	37.9	337.6	33.0	8891.7	48.3	588.2
1	69.0	6.9	0.7	1	1042 3	17.3	44.7	1	2845.3	27.7	157.9	1	5477.8	38.0	340.3	1	8940.1	48.4	591.8
2	76.0	7.0	0.8	2	1059.7	17.4	45.7	2	2873.0	27.8	159.8	2	5516.0	38.2	343.0	2	8988.6	48.5	595.3
3	83.2	7.2	0.9	3	1077.3	17.5	46.7	3	2901.0	27.9	161.7	3	5554.3	38.3	345.7	3	9037.3	48.7	598.9
4	90.5	7.3	1.1	4	1094.9	17.7	47.7	4	2929.0	28.0	163.5	4	5592.7	38.4	348.5	4	9086.0	48.8	602.5
5	97.9	7.4	1.2	5	1112.7	17.8	48.7	5	2957.2	28.2	165.4	5	5631.3	38.5	351.2	5	9135.0	48.9	606.2
6	105.5	7.6	1.4	6	1130.7	17.9	49.8	6	2985.5	28.3	167.3	6	5669.9	38.7	354.0	6	9184.0	49.0	609.8
7	113.2	7.7	1.6	7	1148.7	18.1	50.8	7	3013.9	28.4	169.2	7	5708.7	38.8	356.7	7	9233.2	49.2	613.4
8	121.0	7.8	1.8	8	1166.9	18.2	51.9	8	3042.5	28.6	171.1	8	5747.7	38.9	359.5	8	9282.5	49.3	617.1
9	129.0	8.0	1.9	9	1185.3	18.3	52.9	9	3071.2	28.7	173.1	9	5786.7	39.1	362.3	9	9331.9	49.4	620.7
2.0	137.0	8.1	2.2	10.0	1203.7	18.5	54.0	18.0	3100.0	28.8	175.0	26.0	5825.9	39.2	365.1	34.0	9381.5	49.6	624.4
1	145.3	8.2	2.4	1	1222.3	18.6	55.1	1	3129.0	29.0	176.9	1	5865.3	39.3	367.9	1	9431.2	49.7	628.1
2	153.6	8.3	2.6	2	1241.0	18.7	56.2	2	3158.0	29.1	178.9	2	5904.7	39.5	370.8	2	9481.0	49.8	631.8
3	162.1	8.5	2.9	3	1259.8	18.8	57.3	3	3187.3	29.2	180.9	3	5944.3	39.6	373.6	3	9531.0	50.0	635.4
4	170.7	8.6	3.1	4	1278.8	19.0	58.4	4	3216.6	29.3	182.9	4	5984.0	39.7	376.4	4	9581.0	50.1	639.2
5	179.4	8.7	3.4	5	1297.9	19.1	59.5	5	3246.1	29.5	184.9	5	6023.8	39.8	379.3	5	9631.3	50.2	642.9
6	188.3	8.9	3.7	6	1317.1	19.2	60.7	6	3275.7	29.6	186.9	6	6063.8	40.0	382.2	6	9681.6	50.3	646.6
7	197.3	9.0	3.9	7	1336.5	19.4	61.8	7	3305.4	29.7	188.9	7	6103.9	40.1	385.0	7	9732.1	50.5	650.4
8	206.4	9.1	4.2	8	1356.0	19.5	63.0	8	3335.3	29.9	190.9	8	6144.1	40.2	387.9	8	9782.7	50.6	654.1
9	215.6	9.3	4.5	9	1375.6	19.6	64.2	9	3365.3	30.0	192.9	9	6184.5	40.4	390.8	9	9833.4	50.7	657.9
3.0	225.0	9.4	4.9	11.0	1395.4	19.8	65.4	19.0	3395.4	30.1	195.0	27.0	6225.0	40.5	393.8	35.0	9884.3	50.9	661.7
1	234.5	9.5	5.2	1	1415.3	19.9	66.5	1	3425.6	30.3	197.0	1	6265.6	40.6	396.7	1	9935.3	51.0	665.4
2	244.1	9.6	5.5	2	1435.3	20.0	67.8	2	3456.0	30.4	199.1	2	6306.4	40.8	399.6	2	9986.4	51.1	669.2
3	253.9	9.8	5.9	3	1455.4	20.1	69.0	3	3486.5	30.5	201.2	3	6347.3	40.9	402.5	3	10037.6	51.3	673.0
4	263.8	9.9	6.2	4	1475.7	20.3	70.2	4	3517.1	30.6	203.3	4	6388.3	41.0	405.5	4	10089.0	51.4	676.9
5	273.8	10.0	6.6	5	1496.1	20.4	71.4	5	3547.9	30.8	205.4	5	6429.4	41.1	408.5	5	10140.5	51.5	680.7
6	284.0	10.2	7.0	6	1516.6	20.5	72.7	6	3578.8	30.9	207.5	6	6470.7	41.3	411.4	6	10192.1	51.6	684.5
7	294.3	10.3	7.4	7	1537.3	20.7	73.9	7	3609.8	31.0	209.6	7	6512.1	41.4	414.4	7	10243.9	51.8	688.4
8	304.7	10.4	7.8	8	1558.0	20.8	75.2	8	3641.0	31.2	211.8	8	6553.6	41.5	417.4	8	10295.8	51.9	692.2
9	315.3	10.5	8.2	9	1579.0	20.9	76.5	9	3672.3	31.3	213.9	9	6595.3	41.7	420.4	9	10347.8	52.0	696.1
4.0	325.9	10.7	8.6	12.0	1600.0	21.0	77.8	20.0	3703.7	31.4	216.0	28.0	6637.0	41.8	423.5	36.0	10400.0	52.2	700.0
1	336.7	10.8	9.1	1	1621.2	21.2	79.1	1	3735.3	31.5	218.2	1	6679.0	41.9	426.5	1	10452.3	52.3	703.9
2	347.7	10.9	9.5	2	1642.5	21.3	80.4	2	3766.9	31.7	220.4	2	6721.0	42.0	429.5	2	10504.7	52.4	707.8
3	358.7	11.1	10.0	3	1663.9	21.4	81.7	3	3798.7	31.8	222.6	3	6763.2	42.2	432.6	3	10557.3	52.5	711.7
4	369.9	11.2	10.5	4	1685.5	21.6	83.0	4	3830.7	31.9	224.8	4	6805.5	42.3	435.6	4	10609.9	52.7	715.6
5	381.3	11.3	10.9	5	1707.2	21.7	84.4	5	3862.7	32.1	227.0	5	6847.9	42.4	438.7	5	10662.7	52.8	719.6
6	392.7	11.5	11.4	6	1729.0	21.8	85.8	6	3894.9	32.2	229.2	6	6890.5	42.6	441.8	6	10715.7	52.9	723.5
7	404.3	11.6	11.9	7	1751.0	22.0	87.1	7	3927.3	32.3	231.4	7	6933.2	42.7	444.9	7	10768.7	53.1	727.5
8	416.0	11.7	12.4	8	1773.0	22.1	88.5	8	3959.7	32.5	233.7	8	6976.0	42.8	448.0	8	10821.9	53.2	731.5
9	427.8	11.8	13.0	9	1795.3	22.2	89.9	9	3992.3	32.6	235.9	9	7019.0	43.0	451.1	9	10875.3	53.3	735.4
5.0	439.8	12.0	13.5	13.0	1817.6	22.3	91.3	21.0	4025.0	32.7	238.2	29.0	7062.0	43.1	454.2	37.0	10928.7	53.5	739.4
1	451.9	12.1	14.0	1	1840.1	22.5	92.7	1	4057.8	32.8	240.5	1	7105.3	43.2	457.4	1	10982.3	53.6	743.4
2	464.1	12.2	14.6	2	1862.7	22.6	94.1	2	4090.8	33.0	242.8	2	7148.6	43.3	460.5	2	11036.0	53.7	747.4
3	476.5	12.4	15.2	3	1885.4	22.7	95.5	3	4123.9	33.1	245.0	3	7192.1	43.5	463.7	3	11089.8	53.8	751.5
4	489.0	12.5	15.8	4	1908.3	22.9	97.0	4	4157.1	33.2	247.4	4	7235.7	43.6	466.9	4	11143 8	54.0	755.5
5	501.6	12.6	16.3	5	1931.3	23.0	98.4	5	4190.5	33.4	249.7	5	7279.4	43.7	470.0	5	11197.9	54.1	759.5
6	514.4	12.8	16.9	6	1954.4	23.1	99.9	6	4224.0	33.5	252.0	6	7323.3	43.9	473.2	6	11252.1	54.2	763.6
7	527.3	12.9	17.5	7	1977.6	23.3	101.4	7	4257.6	33.6	254.3	7	7367.3	44.0	476.4	7	11306.5	54.4	767.7
8	540.3	13.0	18.2	8	2001.0	23.4	102.9	8	4291.4	33.8	256.7	8	7411.4	44.1	479.7	8	11361.0	54.5	771.7
9	553.4	13.1	18.8	9	2024.5	23.5	104.4	9	4325.3	33.9	259.0	9	7455.6	44.3	482.9	9	11415.6	54.6	775.8
6.0	566.7	13.3	19.4	14.0	2048.1	23.6	105.9	22.0	4359.3	34.0	261.4	30.0	7500.0	44.4	486.1	38.0	11470.4	54.8	779.9
1	580.1	13.4	20.1	1	2071.9	23.8	107.4	1	4393.4	34.1	263.8	1	7544.5	44.5	489.4	1	11525.3	54.9	784.0
2	593.6	13.5	20.8	2	2095.8	23.9	108.9	2	4427.7	34.3	266.2	2	7589.1	44.6	492.6	2	11580.3	55.0	788.2
3	607.3	13.7	21.4	3	2119.8	24 0	110.4	3	4462.1	34.4	268.6	3	7633.9	44.8	495.9	3	11635.4	55.1	792.3
4	621.0	13.8	22.1	4	2144.0	24.2	112.0	4	4496.6	34.5	271.0	4	7678.8	44.9	499.2	4	11690.7	55.3	796.4
5	635.0	13.9	22.8	5	2168.3	24.3	113.6	5	4531.3	34.7	273.4	5	7723.8	45.0	502.4	5	11746.1	55.4	800.6
6	649.0	14.0	23.5	6	2192.7	24.4	115.1	6	4566.0	34.8	275.9	6	7769.0	45.2	505.8	6	11801.6	55.5	804 8
7	663.2	14.2	24.2	7	2217.3	24.5	116.7	7	4601.0	34.9	278.3	7	7814.3	45.3	509.1	7	11857.3	55.7	808.9
8	677.5	14.3	25.0	8	2241.9	24.7	118.3	8	4636.0	35.0	280.8	8	7859.7	45.4	512.4	8	11913.0	55.8	813.1
9	691.9	14.4	25.7	9	2266.7	24.8	119.9	9	4671.2	35.2	283.2	9	7905.3	45.5	515.7	9	11969.0	55.9	817.3
7.0	706.5	14.6	26.5	15.0	2291.7	24.9	121.5	23.0	4706.5	35.3	285.7	31.0	7950.9	45.7	519.1	39.0	12025.0	56.0	821.5
1	721.2	14.7	27.2	1	2316.7	25.1	123.2	1	4741.9	35.4	288.2	1	7996.7	45.8	522.4	1	12081.2	56.2	825.7
2	736.0	14.8	28.0	2	2341.9	25.2	124.8	2	4777.5	35.6	290.7	2	8042.7	45.9	525.8	2	12137.5	56.3	830.0
3	751.0	15.0	28.8	3	2367.3	25.3	126.4	3	4813.2	35.7	293.2	3	8088.7	46.1	529.2	3	12193.9	56.4	834.2
4	766.0	15.1	29.6	4	2392.7	25.5	128.1	4	4849.0	35.8	295.7	4	8134.9	46.2	532.5	4	12250.5	56.6	838.5
5	781.3	15.2	30.4	5	2418.3	25.6	129.8	5	4885.0	36.0	298.3	5	8181.3	46.3	535.9	5	12307.2	56.7	842.7
6	796.6	15.3	31.2	6	2444.0	25.7	131.4	6	4921.0	36.1	300.8	6	8227.7	46.5	539.3	6	12364.0	56.8	847.0
7	812.1	15.5	32.0	7	2469.8	25.8	133.1	7	4957.3	36.2	303.4	7	8274.3	46.6	542.8	7	12421.0	57.0	851.3
8	827.7	15.6	32.9	8	2495.8	26.0	134.8	8	4993.6	36.3	305.9	8	8321.0	46.7	546.2	8	12478.0	57.1	855.6
9	843.4	15.7	33.7	9	2521.9	26.1	136.5	9	5030.1	36.5	308.5	9	8367.8	46.8	549.6	9	12535.3	57.2	859.9

TABLE No. 3.

EXCAVATION AND EMBANKMENT.

CUBIC YARDS.

Prismoids, 100 feet long. Breadth of Base, 14 feet. Slopes, 2 Horizontal to 1 Perpendicular.

Heig't	Cub. Yds.	Diff.	Cor.	Height.	Cub. Yds.	Diff.	Cor.	Height.	Cub. Yds.	Diff.	Cor.	Height.	Cub. Yds.	Diff.	Cor.	Height.	Cub. Yds.	Diff.	Cor.
0.0	0.0	0.0	0.0	8.0	888.9	17.0	39.5	16.0	2725.9	28.8	158.0	24.0	5511.1	40.7	355.6	32.0	9244.4	52.5	632.1
1	5.3	5.3	0.0	1	906.0	17.1	40.5	1	2754.9	29.0	160.0	1	5551.9	40.8	358.5	1	9297.1	52.7	636.1
2	10.7	5.4	0.0	2	923.3	17.3	41.5	2	2784.0	29.1	162.0	2	5592.9	41.0	361.5	2	9349.9	52.8	640.0
3	16.2	5.6	0.1	3	940.7	17.4	42.5	3	2813.3	29.3	164.0	3	5634.0	41.1	364.5	3	9402.9	53.0	644.0
4	21.9	5.7	0.1	4	958.2	17.6	43.6	4	2842.7	29.4	166.0	4	5675.3	41.3	367.5	4	9456.0	53.1	648.0
5	27.8	5.9	0.2	5	975.9	17.7	44.6	5	2872.2	29.6	168.1	5	5716.7	41.4	370.5	5	9509.3	53.3	652.0
6	33.8	6.0	0.2	6	993.8	17.9	45.7	6	2901.9	29.7	170.1	6	5758.2	41.6	373.6	6	9562.7	53.4	656.0
7	39.9	6.1	0.3	7	1011.8	18.0	46.7	7	2931.8	29.9	172.2	7	5799.9	41.7	376.6	7	9616.2	53.6	660.1
8	46.2	6.3	0.4	8	1029.9	18.1	47.8	8	2961.8	30.0	174.2	8	5841.8	41.9	379.7	8	9669.9	53.7	664.1
9	52.7	6.4	0.5	9	1048.2	18.3	48.9	9	2991.9	30.1	176.3	9	5883.8	42.0	382.7	9	9723.8	53.9	668.2
1.0	59.3	6.6	0.6	9.0	1066.7	18.4	50.0	17.0	3022.2	30.3	178.4	25.0	5925.9	42.1	385.8	33.0	9777.8	54.0	672.2
1	66.0	6.7	0.7	1	1085.3	18.6	51.1	1	3052.7	30.4	180.5	1	5968.2	42.3	388.9	1	9831.9	54.1	676.3
2	72.9	6.9	0.9	2	1104.0	18.7	52.2	2	3083.3	30.6	182.6	2	6010.7	42.4	392.0	2	9886.2	54.3	680.4
3	79.9	7.0	1.0	3	1122.9	18.9	53.4	3	3114.0	30.7	184.7	3	6053.3	42.6	395.1	3	9940.7	54.4	684.5
4	87.1	7.2	1.2	4	1141.9	19.0	54.5	4	3144.9	30.9	186.9	4	6096.0	42.7	398.2	4	9995.3	54.6	688.6
5	94.4	7.3	1.4	5	1161.1	19.2	55.7	5	3175.9	31.0	189.0	5	6138.9	42.9	401.4	5	10050.0	54.7	692.7
6	101.9	7.5	1.6	6	1180.4	19.3	56.9	6	3207.1	31.2	191.2	6	6181.9	43.0	404.5	6	10104.9	54.9	696.9
7	109.6	7.6	1.8	7	1199.9	19.5	58.1	7	3238.4	31.3	193.4	7	6225.1	43.2	407.7	7	10159.9	55.0	701.0
8	117.3	7.8	2.0	8	1219.6	19.6	59.3	8	3269.9	31.5	195.6	8	6268.4	43.3	410.9	8	10215.1	55.2	705.2
9	125.3	7.9	2.2	9	1239.3	19.8	60.5	9	3301.6	31.6	197.8	9	6311.9	43.5	414.1	9	10270.4	55.3	709.4
2.0	133.3	8.1	2.5	10.0	1259.3	19.9	61.7	18.0	3333.3	31.8	200.0	26.0	6355.6	43.6	417.3	34.0	10325.9	55.5	713.6
1	141.6	8.2	2.7	1	1279.3	20.1	63.0	1	3365.3	31.9	202.2	1	6399.3	43.8	420.5	1	10381.6	55.6	717.8
2	149.9	8.4	3.0	2	1299.6	20.2	64.2	2	3397.3	32.1	204.5	2	6443.3	43.9	423.7	2	10437.3	55.8	722.0
3	158.4	8.5	3.3	3	1319.9	20.4	65.5	3	3429.6	32.2	206.7	3	6487.3	44.1	427.0	3	10493.3	55.9	726.2
4	167.1	8.7	3.6	4	1340.4	20.5	66.8	4	3461.9	32.4	209.0	4	6531.6	44.2	430.2	4	10549.3	56.1	730.5
5	175.9	8.8	3.9	5	1361.1	20.7	68.1	5	3494.4	32.5	211.3	5	6575.9	44.4	433.5	5	10605.6	56.2	734.7
6	184.9	9.0	4.2	6	1381.9	20.8	69.4	6	3527.1	32.7	213.6	6	6620.4	44.5	436.8	6	10661.9	56.4	739.0
7	194.0	9.1	4.5	7	1402.9	21.0	70.7	7	3559.9	32.8	215.9	7	6665.1	44.7	440.1	7	10718.4	56.5	743.3
8	203.3	9.3	4.8	8	1424.0	21.1	72.0	8	3592.9	33.0	218.2	8	6709.9	44.8	443.4	8	10775.1	56.7	747.6
9	212.7	9.4	5.2	9	1445.3	21.3	73.3	9	3626.0	33.1	220.5	9	6754.9	45.0	446.7	9	10831.9	56.8	751.9
3.0	222.2	9.6	5.6	11.0	1466.7	21.4	74.7	19.0	3659.3	33.3	222.8	27.0	6800.0	45.1	450.0	35.0	10888.9	57.0	756.2
1	231.9	9.7	5.9	1	1488.2	21.6	76.1	1	3692.7	33.4	225.2	1	6845.3	45.3	453.3	1	10946.0	57.1	760.5
2	241.8	9.9	6.3	2	1509.9	21.7	77.4	2	3726.2	33.6	227.6	2	6890.7	45.4	456.7	2	11003.3	57.3	764.8
3	251.8	10.0	6.7	3	1531.8	21.9	78.8	3	3759.9	33.7	229.9	3	6936.2	45.6	460.1	3	11060.7	57.4	769.2
4	261.9	10.1	7.1	4	1553.8	22.0	80.2	4	3793.8	33.9	232.3	4	6981.9	45.7	463.4	4	11118.2	57.6	773.6
5	272.2	10.3	7.6	5	1575.9	22.1	81.6	5	3927.8	34.0	234.7	5	7027.8	45.9	466.8	5	11175.9	57.7	777.9
6	282.7	10.4	8.0	6	1598.2	22.3	83.1	6	3861.9	34.1	237.1	6	7073.8	46.0	470.2	6	11233.8	57.9	782.3
7	293.3	10.6	8.5	7	1620.7	22.4	84.5	7	3896.2	34.3	239.6	7	7119.9	46.1	473.6	7	11291.8	58.0	786.7
8	304.0	10.7	8.9	8	1643.3	22.6	86.0	8	3930.7	34.4	242.0	8	7166.2	46.3	477.1	8	11349.9	58.1	791.1
9	314.9	10.9	9.4	9	1666.0	22.7	87.4	9	3965.3	34.6	244.5	9	7212.7	46.4	480.5	9	11408.2	58.3	795.6
4.0	325.9	11.0	9.9	12.0	1688.9	22.9	88.9	20.0	4000.0	34.7	246.9	28.0	7259.3	46.6	484.0	36.0	11466.7	58.4	800.0
1	337.1	11.2	10.4	1	1711.9	23.0	90.4	1	4034.9	34.9	249.4	1	7306.0	46.7	487.4	1	11525.3	58.6	804.5
2	348.4	11.3	10.9	2	1735.1	23.2	91.9	2	4069.9	35.0	251.9	2	7352.9	46.9	490.9	2	11584.0	58.7	808.9
3	359.9	11.5	11.4	3	1758.4	23.3	93.4	3	4105.1	35.2	254.4	3	7399.9	47.0	494.4	3	11642.9	58.9	813.4
4	371.6	11.6	12.0	4	1781.9	23.5	94.9	4	4140.4	35.3	256.9	4	7447.1	47.2	497.9	4	11701.9	59.0	817.9
5	383.3	11.8	12.5	5	1805.6	23.6	96.5	5	4175.9	35.5	259.4	5	7494.4	47.3	501.4	5	11761.1	59.2	822.4
6	395.3	11.9	13.1	6	1829.3	23.8	98.0	6	4211.6	35.6	262.0	6	7541.9	47.5	504.9	6	11820.4	59.3	826.9
7	407.3	12.1	13.6	7	1853.3	23.9	99.6	7	4247.3	35.8	264.5	7	7589.6	47.6	508.5	7	11879.9	59.5	831.4
8	419.6	12.2	14.2	8	1877.3	24.1	101.1	8	4283.3	35.9	267.1	8	7637.3	47.8	512.0	8	11939.6	59.6	836.0
9	431.9	12.4	14.8	9	1901.6	24.2	102.7	9	4319.3	36.1	269.6	9	7685.3	47.9	515.6	9	11999.3	59.8	840.5
5.0	444.4	12.5	15.4	13.0	1925.9	24.4	104.3	21.0	4355.6	36.2	272.2	29.0	7733.3	48.1	519.1	37.0	12059.3	59.9	845.1
1	457.1	12.7	16.1	1	1950.4	24.5	105.9	1	4391.9	36.4	274.8	1	7781.6	48.2	522.7	1	12119.3	60.1	849.6
2	469.9	12.8	16.7	2	1975.1	24.7	107.6	2	4428.4	36.5	277.4	2	7829.9	48.4	526.3	2	12179.6	60.2	854.2
3	482.9	13.0	17.3	3	1999.9	24.8	109.2	3	4465.1	36.7	280.1	3	7878.4	48.5	529.9	3	12239.9	60.4	858.8
4	496.0	13.1	18.0	4	2024.9	25.0	110.8	4	4501.9	36.8	282.7	4	7927.1	48.7	533.6	4	12300.4	60.5	863.4
5	509.3	13.3	18.7	5	2050.0	25.1	112.5	5	4538.9	37.0	285.3	5	7975.9	48.8	537.2	5	12361.1	60.7	868.1
6	522.7	13.4	19.4	6	2075.3	25.3	114.2	6	4576.0	37.1	288.0	6	8024.9	49.0	540.8	6	12421.9	60.8	872.7
7	536.2	13.6	20.1	7	2100.7	25.4	115.9	7	4613.3	37.3	290.7	7	8074.0	49.1	544.5	7	12482.9	61.0	877.3
8	549.9	13.7	20.8	8	2126.2	25.6	117.6	8	4650.7	37.4	293.4	8	8123.3	49.3	548.2	8	12544.0	61.1	882.0
9	563.8	13.9	21.5	9	2151.9	25.7	119.3	9	4688.2	37.6	296.1	9	8172.7	49.4	551.9	9	12605.3	61.3	886.7
6.0	577.8	14.0	22.2	14.0	2177.8	25.9	121.0	22.0	4725.9	37.7	298.8	30.0	8222.2	49.6	555.6	38.0	12666.7	61.4	891.4
1	591.9	14.1	23.0	1	2203.8	26.0	122.7	1	4763.8	37.9	301.5	1	8271.9	49.7	559.3	1	12728.2	61.6	896.1
2	606.2	14.3	23.7	2	2229.9	26.1	124.5	2	4801.8	38.0	304.2	2	8321.8	49.9	563.0	2	12789.9	61.7	900.8
3	620.7	14.4	24.5	3	2256.2	26.3	126.2	3	4839.9	38.1	307.0	3	8371.8	50.0	566.7	3	12851.8	61.9	905.5
4	635.3	14.6	25.3	4	2282.7	26.4	128.0	4	4878.2	38.3	309.7	4	8421.9	50.1	570.5	4	12913.8	62.0	910.2
5	650.0	14.7	26.1	5	2309.3	26.6	129.8	5	4916.7	38.4	312.5	5	8472.2	50.3	574.2	5	12975.9	62.1	915.0
6	664.9	14.9	26.9	6	2336.0	26.7	131.6	6	4955.3	38.6	315.3	6	8522.7	50.4	578.0	6	13038.2	62.3	919.7
7	679.9	15.0	27.7	7	2362.9	26.9	133.4	7	4994.0	38.7	318.1	7	8573.3	50.6	581.8	7	13100.7	62.4	924.5
8	695.1	15.2	28.5	8	2389.9	27.0	135.2	8	5032.9	38.9	320.9	8	8624.0	50.7	585.6	8	13163.3	62.6	929.3
9	710.4	15.3	29.4	9	2417.1	27.2	137.0	9	5071.9	39.0	323.7	9	8674.9	50.9	589.4	9	13226.0	62.7	934.1
7.0	725.9	15.5	30.2	15.0	2444.4	27.3	138.9	23.0	5111.1	39.2	326.5	31.0	8725.9	51.0	593.2	39.0	13288.9	62.9	938.9
1	741.6	15.6	31.1	1	2471.9	27.5	140.7	1	5150.4	39.3	329.4	1	8777.1	51.2	597.0	1	13351.9	63.0	943.7
2	757.3	15.8	32.0	2	2499.6	27.6	142.6	2	5189.9	39.5	332.2	2	8828.4	51.3	600.9	2	13415.1	63.2	948.5
3	773.3	15.9	32.9	3	2527.3	27.8	144.5	3	5229.6	39.6	335.1	3	8879.9	51.5	604.7	3	13478.4	63.3	953.4
4	789.3	16.1	33.8	4	2555.3	27.9	146.4	4	5269.3	39.8	338.0	4	8931.6	51.6	608.6	4	13541.9	63.5	958.2
5	805.6	16.2	34.7	5	2583.3	28.1	148.3	5	5309.3	39.9	340.9	5	8983.3	51.8	612.5	5	13605.6	63.6	963.1
6	821.9	16.4	35.7	6	2611.6	28.2	150.2	6	5349.3	40.1	343.8	6	9035.3	51.9	616.4	6	13669.3	63.8	968.0
7	838.4	16.5	36.6	7	2639.9	28.4	152.2	7	5389.6	40.2	346.7	7	9087.3	52.1	620.3	7	13733.3	63.9	972.9
8	855.1	16.7	37.6	8	2668.4	28.5	154.1	8	5429.9	40.4	349.7	8	9139.6	52.2	624.2	8	13797.3	64.1	977.8
9	871.9	16.8	38.5	9	2697.1	28.7	156.1	9	5470.4	40.5	352.6	9	9191.9	52.4	628.2	9	13861.6	64.2	982.7

TABLE No. 4.

EXCAVATION AND EMBANKMENT.

CUBIC YARDS.

Prismoids, 100 feet long. Breadth of Base, 15 feet. Slopes, 2 Horizontal to 1 Perpendicular.

Heig't	Cub. Yds.	Diff.	Cor.	Height.	Cub. Yds.	Diff.	Cor.	Height	Cub. Yds.	Diff.	Cor.	Height.	Cub. Yds.	Diff.	Cor.	Height.	Cub. Yds.	Diff.	Cor.
0.0	0.0	0.0	0.0	8.0	918.5	17.3	39.5	16.0	2785.2	29.2	158.0	24.0	5600.0	41.0	355.6	32.0	9363.0	52.9	632.1
1	5.6	5.6	0.0	1	936.0	17.5	40.5	1	2814.5	29.3	160.0	1	5641.2	41.2	358.5	1	9416.0	53.0	636.1
2	11.4	5.8	0.0	2	953.6	17.6	41.5	2	2844.0	29.5	162.0	2	5682.5	41.3	361.5	2	9469.2	53.2	640.0
3	17.3	5.9	0.1	3	971.4	17.8	42.5	3	2873.6	29.6	164.0	3	5724.0	41.5	364.5	3	9522.5	53.3	644.0
4	23.4	6.1	0.1	4	989.3	17.9	43.6	4	2903.4	29.8	166.0	4	5765.6	41.6	367.5	4	9576.0	53.5	648.0
5	29.6	6.2	0.2	5	1007.4	18.1	44.6	5	2933.3	29.9	168.1	5	5807.4	41.8	370.5	5	9629.6	53.6	652.0
6	36.0	6.4	0.2	6	1025.6	18.2	45.7	6	2963.4	30.1	170.1	6	5849.3	41.9	373.6	6	9683.4	53.8	656.0
7	42.5	6.5	0.3	7	1044.0	18.4	46.7	7	2993.6	30.2	172.2	7	5891.4	42.1	376.6	7	9737.3	53.9	660.1
8	49.2	6.7	0.4	8	1062.5	18.5	47.8	8	3024.0	30.4	174.2	8	5933.6	42.2	379.7	8	9791.4	54.1	664.1
9	56.0	6.8	0.5	9	1081.2	18.7	48.9	9	3054.5	30.5	176.3	9	5976.0	42.4	382.7	9	9845.6	54.2	668.2
1.0	63.0	7.0	0.6	9.0	1100.0	18.8	50.0	17.0	3085.2	30.7	178.4	25.0	6018.5	42.5	385.8	33.0	9900.0	54.4	672.2
1	70.1	7.1	0.7	1	1119.0	19.0	51.1	1	3116.0	30.8	180.5	1	6061.2	42.7	388.9	1	9954.5	54.5	676.3
2	77.3	7.3	0.9	2	1138.1	19.1	52.2	2	3147.0	31.0	182.6	2	6104.0	42.8	392.0	2	10009.2	54.7	680.4
3	84.7	7.4	1.0	3	1157.3	19.3	53.4	3	3178.1	31.1	184.7	3	6147.0	43.0	395.1	3	10064.0	54.8	684.5
4	92.3	7.6	1.2	4	1176.7	19.4	54.5	4	3209.3	31.3	186.9	4	6190.1	43.1	398.2	4	10119.0	55.0	688.6
5	100.0	7.7	1.4	5	1196.3	19.6	55.7	5	3240.7	31.4	189.0	5	6233.3	43.3	401.4	5	10174.1	55.1	692.7
6	107.9	7.9	1.6	6	1216.0	19.7	56.9	6	3272.3	31.6	191.2	6	6276.7	43.4	404.5	6	10129.3	55.3	696.9
7	115.9	8.0	1.8	7	1235.9	19.9	58.1	7	3304.0	31.7	193.4	7	6320.3	43.6	407.7	7	10284.7	55.4	701.0
8	124.0	8.1	2.0	8	1255.9	20.0	59.3	8	3335.9	31.9	195.6	8	6364.0	43.7	410.9	8	10340.3	55.6	705.2
9	132.3	8.3	2.2	9	1276.0	20.1	60.5	9	3367.9	32.0	197.8	9	6407.9	43.9	414.1	9	10396.0	55.7	709.4
2.0	140.7	8.4	2.5	10.0	1296.3	20.3	61.7	18.0	3400.0	32.1	200.0	26.0	6451.9	44.0	417.3	34.0	10451.9	55.9	713.6
1	149.3	8.6	2.7	1	1316.7	20.4	63.0	1	3432.3	32.3	202.2	1	6496.0	44.1	420.5	1	10507.9	56.0	717.8
2	158.1	8.7	3.0	2	1337.3	20.6	64.2	2	3464.7	32.4	204.5	2	6540.3	44.3	423.7	2	10564.0	56.1	722.0
3	167.0	8.9	3.3	3	1358.1	20.7	65.5	3	3497.3	32.6	206.7	3	6584.7	44.4	427.0	3	10620.3	56.3	726.2
4	176.0	9.0	3.6	4	1379.0	20.9	66.8	4	3530.1	32.7	209.0	4	6629.3	44.6	430.2	4	10676.7	56.4	730.5
5	185.2	9.2	3.9	5	1400.0	21.0	68.1	5	3563.0	32.9	211.3	5	6674.1	44.7	433.5	5	10733.3	56.6	734.7
6	194.5	9.3	4.2	6	1421.2	21.2	69.4	6	3596.0	33.0	213.6	6	6719.0	44.9	436.8	6	10790.1	56.7	739.0
7	204.0	9.5	4.5	7	1442.5	21.3	70.7	7	3629.2	33.2	215.9	7	6764.0	45.0	440.1	7	10847.0	56.9	743.3
8	213.6	9.6	4.8	8	1464.0	21.5	72.0	8	3662.5	33.3	218.2	8	6809.2	45.2	443.4	8	10904.0	57.0	747.6
9	223.4	9.8	5.2	9	1485.6	21.6	73.3	9	3696.0	33.5	220.5	9	6854.5	45.3	446.7	9	10961.2	57.2	751.9
3.0	233.3	9.9	5.6	11.0	1507.4	21.8	74.7	19.0	3729.6	33.6	222.8	27.0	6900.0	45.5	450.0	35.0	11018.5	57.3	756.2
1	243.4	10.1	5.9	1	1529.3	21.9	76.1	1	3763.4	33.8	225.2	1	6945.6	45.6	453.3	1	11076.0	57.5	760.5
2	253.6	10.2	6.3	2	1551.4	22.1	77.4	2	3797.3	33.9	227.6	2	6991.4	45.8	456.7	2	11133.6	57.6	764.8
3	264.0	10.4	6.7	3	1573.6	22.2	78.8	3	3831.4	34.1	229.9	3	7037.3	45.9	460.1	3	11191.4	57.8	769.2
4	274.5	10.5	7.1	4	1596.0	22.4	80.2	4	3865.6	34.2	232.3	4	7083.4	46.1	463.4	4	11249.3	57.9	773.6
5	285.2	10.7	7.6	5	1618.5	22.5	81.6	5	3900.0	34.4	234.7	5	7129.6	46.2	466.8	5	11307.4	58.1	777.9
6	296.0	10.8	8.0	6	1641.2	22.7	83.1	6	3934.5	34.5	237.1	6	7176.0	46.4	470.2	6	11365.6	58.2	782.3
7	307.0	11.0	8.4	7	1664.0	22.8	84.5	7	3969.2	34.7	239.6	7	7222.5	46.5	473.6	7	11424.0	58.4	786.7
8	318.1	11.1	8.9	8	1687.0	23.0	86.0	8	4004.0	34.8	242.0	8	7269.2	46.7	477.1	8	11482.5	58.5	791.1
9	329.3	11.3	9.4	9	1710.1	23.1	87.4	9	4039.0	35.0	244.5	9	7316.0	46.8	480.5	9	11541.2	58.7	795.6
4.0	340.7	11.4	9.9	12.0	1733.3	23.3	88.9	20.0	4074.1	35.1	246.9	28.0	7363.0	47.0	484.0	36.0	11600.0	58.8	800.0
1	352.3	11.6	10.4	1	1756.7	23.4	90.4	1	4109.3	35.3	249.4	1	7410.1	47.1	487.4	1	11659.0	59.0	804.5
2	364.0	11.7	10.9	2	1780.3	23.6	91.9	2	4144.7	35.4	251.9	2	7457.3	47.3	490.9	2	11718.1	59.1	808.9
3	375.9	11.9	11.4	3	1804.0	23.7	93.4	3	4180.3	35.6	254.4	3	7504.7	47.4	494.4	3	11777.3	59.3	813.4
4	387.9	12.0	12.0	4	1827.9	23.9	94.9	4	4216.0	35.7	256.9	4	7552.3	47.6	497.9	4	11836.7	59.4	817.9
5	400.0	12.1	12.5	5	1851.9	24.0	96.4	5	4251.9	35.9	259.4	5	7600.0	47.7	501.4	5	11896.3	59.6	822.4
6	412.3	12.3	13.1	6	1876.0	24.1	98.0	6	4287.9	36.0	262.0	6	7647.9	47.9	504.9	6	11956.0	59.7	826.9
7	424.7	12.4	13.6	7	1900.3	24.3	99.6	7	4324.0	36.1	264.5	7	7695.9	48.0	508.4	7	12015.9	59.9	831.4
8	437.3	12.6	14.2	8	1924.7	24.4	101.1	8	4360.3	36.3	267.1	8	7744.0	48.1	512.0	8	12075.9	60.0	836.0
9	450.1	12.7	14.8	9	1949.3	24.6	102.7	9	4396.7	36.4	269.6	9	7792.3	48.3	515.6	9	12136.0	60.1	840.5
5.0	463.0	12.9	15.4	13.0	1974.1	24.7	104.3	21.0	4433.3	36.6	272.2	29.0	7840.7	48.4	519.1	37.0	12196.3	60.3	845.1
1	476.0	13.0	16.1	1	1999.0	24.9	105.9	1	4470.1	36.7	274.8	1	7889.3	48.6	522.7	1	12256.7	60.4	849.6
2	489.2	13.2	16.7	2	2024.0	25.0	107.6	2	4507.0	36.9	277.4	2	7938.1	48.7	526.3	2	12317.3	60.6	854.2
3	502.5	13.3	17.3	3	2049.2	25.2	109.2	3	4544.0	37.0	280.1	3	7987.0	48.9	529.9	3	12378.1	60.7	858.8
4	516.0	13.5	18.0	4	2074.5	25.3	110.8	4	4581.2	37.2	282.7	4	8036.0	49.0	533.6	4	12439.0	60.9	863.4
5	529.6	13.6	18.7	5	2100.0	25.5	112.5	5	4618.5	37.3	285.3	5	8085.2	49.2	537.2	5	12500.0	61.0	868.1
6	543.4	13.8	19.4	6	2125.6	25.6	114.2	6	4656.0	37.5	288.0	6	8134.5	49.3	540.8	6	12561.2	61.2	872.7
7	557.3	13.9	20.1	7	2151.4	25.8	115.9	7	4693.6	37.6	290.7	7	8184.0	49.5	544.5	7	12622.5	61.3	877.3
8	571.4	14.1	20.8	8	2177.3	25.9	117.6	8	4731.4	37.8	293.4	8	8233.6	49.6	548.2	8	12684.0	61.5	881.0
9	585.6	14.2	21.5	9	2203.4	26.1	119.3	9	4769.3	37.9	296.1	9	8283.4	49.8	551.9	9	12745.6	61.6	886.7
6.0	600.0	14.4	22.2	14.0	2229.6	26.2	121.0	22.0	4807.4	38.1	298.8	30.0	8333.3	49.9	555.6	38.0	12807.4	61.8	891.4
1	614.5	14.5	23.0	1	2256.0	26.4	122.7	1	4845.6	38.2	301.5	1	8383.4	50.1	559.3	1	12869.3	61.9	896.1
2	629.2	14.7	23.7	2	2282.5	26.5	124.5	2	4884.0	38.4	304.2	2	8433.6	50.2	563.0	2	12931.4	62.1	900.8
3	644.0	14.8	24.5	3	2309.2	26.7	126.2	3	4922.5	38.5	307.0	3	8484.0	50.4	566.7	3	12993.6	62.2	905.5
4	659.0	15.0	25.3	4	2336.0	26.8	128.0	4	4961.2	38.7	309.7	4	8534.5	50.5	570.5	4	130 6.0	62.4	910.2
5	674.1	15.1	26.1	5	2363.0	27.0	129.8	5	5000.0	38.8	312.5	5	8585.2	50.7	574.2	5	13118.5	62.5	915.0
6	689.3	15.3	26.9	6	2390.1	27.1	131.6	6	5039.0	39.0	315.3	6	8636.0	50.8	578.0	6	13181.2	62.7	919.7
7	704.7	15.4	27.7	7	2417.3	27.3	133.4	7	5078.1	39.1	318.1	7	8687.0	51.0	581.8	7	13244.0	62.8	924.5
8	720.3	15.6	28.5	8	2444.7	27.4	135.2	8	5117.3	39.3	320.0	8	8738.1	51.1	585.6	8	13307.0	63.0	929.3
9	736.0	15.7	29.4	9	2472.3	27.6	137.0	9	5156.7	39.4	323.7	9	8789.3	51.3	589.4	9	13370.1	63.1	934.1
7.0	751.9	15.9	30.2	15.0	2500.0	27.7	138.9	23.0	5196.3	39.6	326.5	31.0	8840.7	51.4	593.2	39.0	13433.3	63.3	938.9
1	767.9	16.0	31.1	1	2527.9	27.9	140.7	1	5236.0	39.7	329.4	1	8892.3	51.6	597.0	1	13496.7	63.4	943.7
2	784.0	16.1	32.0	2	2555.9	28.0	142.6	2	5275.9	39.9	332.2	2	8944.0	51.7	600.9	2	13560.3	63.6	948.5
3	800.3	16.3	32.9	3	2584.0	28.1	144.5	3	5315.9	40.0	335.1	3	8995.9	51.9	604.7	3	13624.0	63.7	953.4
4	816.7	16.4	33.8	4	2612.3	28.3	146.4	4	5356.0	40.1	338.0	4	9047.9	52.0	608.6	4	13687.9	63.9	958.2
5	833.3	16.6	34.7	5	2640.7	28.4	148.3	5	5396.3	40.3	340.9	5	9100.0	52.1	612.5	5	13751.9	64.0	963.1
6	850.1	16.7	35.7	6	2669.3	28.6	150.2	6	5436.7	40.4	343.8	6	9152.3	52.3	616.4	6	13816.0	64.1	968.0
7	867.0	16.9	36.6	7	2698.1	28.7	152.2	7	5477.3	40.6	346.7	7	9204.7	52.4	620.3	7	13880.3	64.3	972.9
8	884.0	17.0	37.6	8	2727.0	28.9	154.1	8	5518.1	40.7	349.7	8	9257.3	52.6	624.2	8	13944.7	64.4	977.8
9	901.2	17.2	38.5	9	2756.0	29.0	156.1	9	5559.0	40.9	352.6	9	9310.1	52.7	628.2	9	14009.3	64.6	982.7

TABLE No. 5.

EXCAVATION AND EMBANKMENT.

CUBIC YARDS,

Prismoids, 100 Feet Long. Breadth of Base, 9 Feet. Slopes, 1 1-2 Horizontal to 1 Perpendicular.

Height.	Cub. Yds.	Diff.	Cor.	Height.	Cub. Yds.	Diff.	Cor.	Height.	Cub. Yds.	Diff.	Cor.	Height.	Cub. Yds.	Diff.	Cor.	Height.	Cub. Yds.	Diff.	Cor.
0.0	0.0	0.0	0.0	8.0	622.2	12.2	29.6	16.0	1955.6	21.1	118.5	24.0	4000.0	29.9	266.7	32.0	6755.6	38.8	474.1
1	3.4	3.4	0.0	1	634.5	12.3	30.4	1	1976.7	21.2	120.0	1	4030.1	30.1	268.9	1	6794.5	38.9	477.0
2	6.9	3.5	0.0	2	646.9	12.4	31.1	2	1998.0	21.3	121.5	2	4060.2	30.2	271.1	2	6833.6	39.1	480.0
3	10.5	3.6	0.0	3	659.4	12.5	31.9	3	2019.4	21.4	123.0	3	4090.5	30.3	273.4	3	6872.7	39.2	483.0
4	14.2	3.7	0.1	4	672.0	12.6	32.7	4	2040.9	21.5	124.5	4	4120.9	30.4	275.6	4	6912.0	39.3	486.0
5	18.1	3.8	0.1	5	684.7	12.7	33.4	5	2062.5	21.6	126.0	5	4151.4	30.5	277.9	5	6951.4	39.4	489.0
6	22.0	3.9	0.2	6	697.6	12.8	34.2	6	2084.2	21.7	127.6	6	4182.0	30.6	280.2	6	6990.9	39.5	492.0
7	26.1	4.1	0.2	7	710.5	12.9	35.0	7	2106.1	21.8	129.1	7	4212.7	30.7	282.5	7	7030.5	39.6	495.0
8	30.2	4.2	0.3	8	723.6	13.1	35.9	8	2128.0	21.9	130.7	8	4243.6	30.8	284.7	8	7070.2	39.7	498.0
9	34.5	4.3	0.4	9	736.7	13.2	36.7	9	2150.1	22.1	132.2	9	4274.5	30.9	287.0	9	7110.1	39.8	501.1
1.0	38.9	4.4	0.5	9.0	750.0	13.3	37.5	17.0	2172.2	22.2	133.8	25.0	4305.6	31.1	289.4	33.0	7150.0	39.9	504.2
1	43.4	4.5	0.6	1	763.4	13.4	38.3	1	2194.5	22.3	135.4	1	4336.7	31.2	291.7	1	7190.1	40.1	507.2
2	48.0	4.6	0.7	2	776.9	13.5	39.2	2	2216.9	22.4	137.0	2	4368.0	31.3	294.0	2	7230.2	40.2	510.3
3	52.7	4.7	0.8	3	790.5	13.6	40.0	3	2239.4	22.5	138.6	3	4399.4	31.4	296.3	3	7270.5	40.3	513.4
4	57.6	4.8	0.9	4	804.2	13.7	40.9	4	2262.0	22.6	140.2	4	4430.9	31.5	298.7	4	7310.9	40.4	516.5
5	62.5	4.9	1.0	5	818.1	13.8	41.8	5	2284.7	22.7	141.8	5	4462.5	31.6	301.0	5	7351.4	40.5	519.6
6	67.6	5.1	1.2	6	832.0	13.9	42.7	6	2307.6	22.8	143.4	6	4494.2	31.7	303.4	6	7392.0	40.6	522.7
7	72.7	5.2	1.3	7	846.1	14.1	43.6	7	2330.5	22.9	145.0	7	4526.1	31.8	305.8	7	7432.7	40.7	525.8
8	78.0	5.3	1.5	8	860.2	14.2	44.5	8	2353.6	23.1	146.7	8	4558.0	31.9	308.2	8	7473.6	40.8	528.9
9	83.4	5.4	1.7	9	874.5	14.3	45.4	9	2376.7	23.2	148.3	9	4590.1	32.1	310.6	9	7514.5	40.9	532.0
2.0	88.9	5.5	1.9	10.0	888.9	14.4	46.3	18.0	2400.0	23.3	150.0	26.0	4622.2	32.2	313.0	34.0	7555.6	41.1	535.2
1	94.5	5.6	2.0	1	903.4	14.5	47.2	1	2423.4	23.4	151.7	1	4654.5	32.3	315.4	1	7596.7	41.2	538.3
2	100.2	5.7	2.2	2	918.0	14.6	48.2	2	2446.9	23.5	153.4	2	4686.9	32.4	317.8	2	7638.0	41.3	541.5
3	106.1	5.8	2.4	3	932.7	14.7	49.1	3	2470.5	23.6	155.0	3	4719.4	32.5	320.2	3	7679.4	41.4	544.7
4	112.0	5.9	2.7	4	947.6	14.8	50.1	4	2494.2	23.7	156.7	4	4752.0	32.6	322.7	4	7720.9	41.5	547.9
5	118.1	6.1	2.9	5	962.5	14.9	51.0	5	2518.1	23.8	158.4	5	4784.7	32.7	325.1	5	7762.5	41.6	551.0
6	124.2	6.2	3.1	6	977.6	15.1	52.0	6	2542.0	23.9	160.2	6	4817.6	32.8	327.6	6	7804.2	41.7	554.2
7	130.5	6.3	3.4	7	992.7	15.2	53.0	7	2566.1	24.1	161.9	7	4850.5	32.9	330.0	7	7846.1	41.8	557.5
8	136.9	6.4	3.6	8	1008.0	15.3	54.0	8	2590.2	24.2	163.6	8	4883.6	33.1	332.5	8	7888.0	41.9	560.7
9	143.4	6.5	3.9	9	1023.4	15.4	55.0	9	2614.5	24.3	165.4	9	4916.7	33.2	335.0	9	7930.1	42.1	563.9
3.0	150.0	6.6	4.2	11.0	1038.9	15.5	56.0	19.0	2638.9	24.4	167.1	27.0	4950.0	33.3	337.5	35.0	7972.2	42.2	567.1
1	156.7	6.7	4.4	1	1054.5	15.6	57.0	1	2663.4	24.5	168.9	1	4983.4	33.4	340.0	1	8014.5	42.3	570.4
2	163.6	6.8	4.7	2	1070.2	15.7	58.1	2	2688.0	24.6	170.7	2	5016.9	33.5	342.5	2	8056.9	42.4	573.6
3	170.5	6.9	5.0	3	1086.1	15.8	59.1	3	2712.7	24.7	172.4	3	5050.5	33.6	345.0	3	8099.4	42.5	576.9
4	177.6	7.1	5.4	4	1102.0	15.9	60.2	4	2737.6	24.8	174.2	4	5084.2	33.7	347.6	4	8142.0	42.6	580.2
5	184.7	7.2	5.7	5	1118.1	16.1	61.2	5	2762.5	24.9	176.0	5	5118.1	33.8	350.1	5	8184.7	42.7	583.4
6	192.0	7.3	6.0	6	1134.2	16.2	62.3	6	2787.6	25.1	177.9	6	5152.0	33.9	352.7	6	8227.6	42.8	586.7
7	199.4	7.4	6.3	7	1150.5	16.3	63.4	7	2812.7	25.2	179.7	7	5186.1	34.1	355.2	7	8270.5	42.9	590.0
8	206.9	7.5	6.7	8	1166.9	16.4	64.5	8	2838.0	25.3	181.5	8	5220.2	34.2	357.8	8	8313.6	43.1	593.4
9	214.5	7.6	7.0	9	1183.4	16.5	65.6	9	2863.4	25.4	183.3	9	5254.5	34.3	360.4	9	8356.7	43.2	596.7
4.0	222.2	7.7	7.4	12.0	1200.0	16.6	66.7	20.0	2888.9	25.5	185.2	28.0	5288.9	34.4	363.0	36.0	8400.0	43.3	600.0
1	230.1	7.8	7.8	1	1216.7	16.7	67.8	1	2914.5	25.6	187.0	1	5323.4	34.5	365.6	1	8443.4	43.4	603.3
2	238.0	7.9	8.1	2	1233.6	16.8	68.9	2	2940.2	25.7	188.9	2	5358.0	34.6	368.2	2	8486.9	43.5	606.7
3	246.1	8.1	8.6	3	1250.5	16.9	70.0	3	2966.1	25.8	190.8	3	5392.7	34.7	370.8	3	8530.5	43.6	610.0
4	254.2	8.2	9.0	4	1267.6	17.1	71.2	4	2992.0	25.9	192.7	4	5427.6	34.8	373.4	4	8574.2	43.7	613.4
5	262.5	8.3	9.4	5	1284.7	17.2	72.3	5	3018.1	26.1	194.6	5	5462.5	34.9	376.0	5	8618.1	43.8	616.8
6	270.9	8.4	9.8	6	1302.0	17.3	73.5	6	3044.2	26.2	196.5	6	5497.6	35.1	378.7	6	8662.0	43.9	620.2
7	279.4	8.5	10.2	7	1319.4	17.4	74.7	7	3070.5	26.3	198.4	7	5532.7	35.2	381.3	7	8706.1	44.1	623.6
8	288.0	8.6	10.7	8	1336.9	17.5	75.9	8	3096.9	26.4	200.3	8	5568.0	35.3	384.0	8	8750.2	44.2	627.0
9	296.7	8.7	11.1	9	1354.5	17.6	77.0	9	3123.4	26.5	202.2	9	5603.4	35.4	386.7	9	8794.5	44.3	630.4
5.0	305.6	8.8	11.6	13.0	1372.2	17.7	78.2	21.0	3150.0	26.6	204.2	29.0	5638.9	35.5	389.4	37.0	8838.9	44.4	633.8
1	314.5	8.9	12.0	1	1390.1	17.8	79.4	1	3176.7	26.7	206.1	1	5674.5	35.6	392.0	1	6883.4	44.5	637.2
2	323.6	9.1	12.5	2	1408.0	17.9	80.7	2	3203.6	26.8	208.1	2	5710.2	35.7	394.7	2	8928.0	44.6	640.7
3	332.7	9.2	13.0	3	1426.1	18.1	81.9	3	3230.5	26.9	210.0	3	5746.1	35.8	397.4	3	8972.7	44.7	644.1
4	342.0	9.3	13.5	4	1444.2	18.2	83.1	4	3257.6	27.1	212.0	4	5782.0	35.9	400.2	4	9017.6	44.8	647.6
5	351.4	9.4	14.0	5	1462.5	18.3	84.4	5	3284.7	27.2	214.0	5	5818.1	36.1	402.9	5	9062.5	44.9	651.0
6	360.9	9.5	14.5	6	1480.9	18.4	85.6	6	3312.0	27.3	216.0	6	5854.2	36.2	405.6	6	9107.6	45.1	654.5
7	370.5	9.6	15.0	7	1499.4	18.5	86.9	7	3339.4	27.4	218.0	7	5890.5	36.3	408.4	7	9152.7	45.2	658.0
8	380.2	9.7	15.6	8	1518.0	18.6	88.2	8	3366.9	27.5	220.0	8	5926.9	36.4	411.1	8	9198.0	45.3	661.5
9	390.1	9.8	16.1	9	1536.7	18.7	89.4	9	3394.5	27.6	222.0	9	5963.4	36.5	413.9	9	9243.4	45.4	665.0
6.0	400.0	9.9	16.7	14.0	1555.6	18.8	90.7	22.0	3422.2	27.7	224.1	30.0	6000.0	36.6	416.7	38.0	9288.9	45.5	668.5
1	410.1	10.1	17.2	1	1574.5	18.9	92.0	1	3450.1	27.8	226.1	1	6036.7	36.7	419.4	1	9334.5	45.6	672.0
2	420.2	10.2	17.8	2	1593.6	19.1	93.4	2	3478.0	27.9	228.2	2	6073.6	36.8	422.2	2	9380.2	45.7	675.6
3	430.5	10.3	18.4	3	1612.7	19.2	94.7	3	3506.1	28.1	230.2	3	6110.5	36.9	425.0	3	9426.1	45.8	679.1
4	440.9	10.4	19.0	4	1632.0	19.3	96.0	4	3534.2	28.2	232.3	4	6147.6	37.1	427.9	4	9472.0	45.9	682.7
5	451.4	10.5	19.6	5	1651.4	19.4	97.3	5	3562.5	28.3	234.4	5	6184.7	37.2	430.7	5	9518.1	46.1	686.2
6	462.0	10.6	20.2	6	1670.9	19.5	98.7	6	3590.9	28.4	236.5	6	6222.0	37.3	433.5	6	9564.2	46.2	689.8
7	472.7	10.7	20.8	7	1690.5	19.6	100.0	7	3619.4	28.5	238.6	7	6259.4	37.4	436.3	7	9610.5	46.3	693.4
8	483.6	10.8	21.4	8	1710.2	19.7	101.4	8	3648.0	28.6	240.7	8	6296.9	37.5	439.2	8	9656.9	46.4	697.0
9	494.5	10.9	22.0	9	1730.1	19.8	102.8	9	3676.7	28.7	242.8	9	6334.5	37.6	442.0	9	9703.4	46.5	700.6
7.0	505.6	11.1	22.7	15.0	1750.0	19.9	104.2	23.0	3705.6	28.8	244.9	31.0	6372.2	37.7	444.9	39.0	9750.0	46.6	704.2
1	516.7	11.2	23.3	1	1770.1	20.1	105.6	1	3734.5	28.9	247.0	1	6410.1	37.8	447.8	1	9796.7	46.7	707.8
2	528.0	11.3	24.0	2	1790.2	20.2	107.0	2	3763.6	29.1	249.2	2	6448.0	37.9	450.7	2	9843.6	46.8	711.4
3	539.4	11.4	24.7	3	1810.5	20.3	108.4	3	3792.7	29.2	251.3	3	6486.1	38.1	453.6	3	9890.5	46.9	715.0
4	550.9	11.5	25.4	4	1830.9	20.4	109.8	4	3822.0	29.3	253.5	4	6524.2	38.2	456.5	4	9937.6	47.1	718.7
5	562.5	11.6	26.0	5	1851.4	20.5	111.2	5	3851.4	29.4	255.7	5	6562.5	38.3	459.4	5	9984.7	47.2	722.3
6	574.2	11.7	26.7	6	1872.0	20.6	112.7	6	3880.9	29.5	257.9	6	6600.9	38.4	462.3	6	10032.0	47.3	726.0
7	586.1	11.8	27.4	7	1892.7	20.7	114.1	7	3910.5	29.6	260.0	7	6639.4	38.5	465.2	7	10079.4	47.4	729.7
8	598.0	11.9	28.2	8	1913.6	20.8	115.6	8	3940.2	29.7	262.2	8	6678.0	38.6	468.2	8	10126.9	47.5	733.4
9	610.1	12.1	28.9	9	1934.5	20.9	117.0	9	3970.1	29.8	264.5	9	6716.7	38.7	471.1	9	10174.5	47.6	737.0

TABLE No. 6.

EXCAVATION AND EMBANKMENT.

CUBIC YARDS.

Prismoids, 100 feet long. **Breadth of Base, 10 feet.** **Slopes, 1 1-2 Horizontal to 1 Perpendicular.**

Heig't	Cub. Yds.	Diff.	Cor.	Height.	Cub. Yds.	Diff.	Cor.	Height.	Cub. Yds.	Diff.	Cor.	Height.	Cub. Yds.	Diff.	Cor.	Height.	Cub. Yds.	Diff.	Cor.
0.0	0.0	0.0	0.0	8.0	651.9	12.5	29.6	16.0	2014.8	21.4	118.5	24.0	4088.9	30.3	266.7	32.0	6874.1	39.2	474.1
1	3.8	3.8	0.0	1	664.5	12.6	30.4	1	2036.4	21.5	120.0	1	4119.3	30.4	268.9	1	6913.4	39.3	477.0
2	7.6	3.9	0.0	2	677.3	12.8	31.1	2	2058.0	21.6	121.5	2	4149.9	30.5	271.1	2	6952.8	39.4	480.0
3	11.6	4.0	0.0	3	690.1	12.9	31.9	3	2079.8	21.8	123.0	3	4180.5	30.6	273.4	3	6992.4	39.5	483.0
4	15.7	4.1	0.1	4	703.1	13.0	32.7	4	2101.6	21.9	124.5	4	4211.3	30.8	275.6	4	7032.0	39.6	486.0
5	19.9	4.2	0.1	5	716.2	13.1	33.4	5	2123.6	22.0	126.0	5	4242.1	30.9	277.9	5	7071.8	39.8	489.0
6	24.2	4.3	0.2	6	729.4	13.2	34.2	6	2145.7	22.1	127.6	6	4273.1	31.0	280.2	6	7111.6	39.9	492.0
7	28.6	4.4	0.2	7	742.7	13.3	35.0	7	2167.9	22.2	129.1	7	4304.2	31.1	282.5	7	7151.6	40.0	495.0
8	33.2	4.5	0.3	8	756.1	13.4	35.9	8	2190.2	22.3	130.7	8	4335.4	31.2	284.7	8	7191.7	40.1	498.0
9	37.8	4.6	0.4	9	769.7	13.5	36.7	9	2212.6	22.4	132.2	9	4366.7	31.3	287.0	9	7231.9	40.2	501.1
1.0	42.6	4.8	0.5	9.0	783.3	13.6	37.5	17.0	2235.2	22.5	133.8	25.0	4398.1	31.4	289.4	33.0	7272.2	40.3	504.2
1	47.5	4.9	0.6	1	797.1	13.8	38.3	1	2257.8	22.6	135.4	1	4429.7	31.5	291.7	1	7312.6	40.4	507.2
2	52.4	5.0	0.7	2	811.0	13.9	39.2	2	2280.6	22.8	137.0	2	4461.3	31.6	294.0	2	7353.2	40.5	510.3
3	57.5	5.1	0.8	3	824.9	14.0	40.0	3	2303.5	22.9	138.6	3	4493.1	31.8	296.3	3	7393.8	40.6	513.4
4	62.7	5.2	0.9	4	839.0	14.1	40.9	4	2326.4	23.0	140.2	4	4525.0	31.9	298.7	4	7434.6	40.8	516.5
5	68.1	5.3	1.0	5	853.2	14.2	41.8	5	2349.5	23.1	141.8	5	4556.9	32.0	301.0	5	7475.5	40.9	519.6
6	73.5	5.4	1.2	6	867.6	14.3	42.7	6	2372.7	23.2	143.4	6	4589.0	32.1	303.4	6	7516.4	41.0	522.7
7	79.0	5.5	1.3	7	882.0	14.4	43.6	7	2396.1	23.3	145.0	7	4621.2	32.2	305.8	7	7557.5	41.1	525.8
8	84.7	5.6	1.5	8	896.5	14.5	44.5	8	2419.5	23.4	146.7	8	4653.6	32.3	308.2	8	7598.7	41.2	528.9
9	90.4	5.8	1.7	9	911.2	14.6	45.4	9	2443.0	23.5	148.3	9	4686.0	32.4	310.6	9	7640.1	41.3	532.0
2.0	96.3	5.9	1.9	10.0	925.9	14.8	46.3	18.0	2466.7	23.6	150.0	26.0	4718.5	32.5	313.0	34.0	7681.5	41.4	535.2
1	102.3	6.0	2.0	1	940.8	14.9	47.2	1	2490.4	23.8	151.7	1	4751.2	32.6	315.4	1	7723.0	41.5	538.3
2	108.4	6.1	2.2	2	955.8	15.0	48.2	2	2514.3	23.9	153.4	2	4783.9	32.8	317.8	2	7764.7	41.6	541.5
3	114.6	6.2	2.4	3	970.9	15.1	49.1	3	2538.3	24.0	155.0	3	4816.8	32.9	320.2	3	7806.4	41.8	544.7
4	120.9	6.3	2.7	4	986.1	15.2	50.1	4	2562.4	24.1	156.7	4	4849.8	33.0	322.7	4	7848.3	41.9	547.9
5	127.3	6.4	2.9	5	1001.4	15.3	51.0	5	2586.6	24.2	158.4	5	4882.9	33.1	325.1	5	7890.3	42.0	551.0
6	133.9	6.5	3.1	6	1016.8	15.4	52.0	6	2610.9	24.3	160.2	6	4916.1	33.2	327.6	6	7932.4	42.1	554.2
7	140.5	6.6	3.4	7	1032.4	15.5	53.0	7	2635.3	24.4	161.9	7	4949.4	33.3	330.0	7	7974.6	42.2	557.5
8	147.3	6.8	3.6	8	1048.0	15.6	54.0	8	2659.9	24.5	163.6	8	4982.8	33.4	332.5	8	8016.9	42.3	560.7
9	154.1	6.9	3.9	9	1063.8	15.8	55.0	9	2684.5	24.6	165.4	9	5016.4	33.5	335.0	9	8059.3	42.4	563.9
3.0	161.1	7.0	4.2	11.0	1079.6	15.9	56.0	19.0	2709.3	24.8	167.1	27.0	5050.0	33.6	337.5	35.0	8101.9	42.5	567.1
1	168.2	7.1	4.4	1	1095.6	16.0	57.0	1	2734.1	24.9	168.9	1	5083.8	33.8	340.0	1	8144.5	42.6	570.4
2	175.4	7.2	4.7	2	1111.7	16.1	58.1	2	2759.1	25.0	170.7	2	5117.6	33.9	342.5	2	8187.3	42.8	573.6
3	182.7	7.3	5.0	3	1127.9	16.2	59.1	3	2784.2	25.1	172.4	3	5151.6	34.0	345.0	3	8230.1	42.9	576.9
4	190.1	7.4	5.4	4	1144.2	16.3	60.2	4	2809.4	25.2	174.2	4	5185.7	34.1	347.6	4	8273.1	43.0	580.2
5	197.7	7.5	5.7	5	1160.6	16.4	61.2	5	2834.7	25.3	176.0	5	5219.9	34.2	350.1	5	8316.2	43.1	583.4
6	205.3	7.6	6.0	6	1177.2	16.5	62.3	6	2860.1	25.4	177.9	6	5254.2	34.3	352.7	6	8359.4	43.2	586.7
7	213.1	7.8	6.3	7	1193.8	16.6	63.4	7	2885.7	25.5	179.7	7	5288.6	34.4	355.2	7	8402.7	43.3	590.0
8	221.0	7.9	6.7	8	1210.6	16.8	64.5	8	2911.3	25.6	181.5	8	5323.2	34.5	357.8	8	8446.1	43.4	593.4
9	228.9	8.0	7.0	9	1227.5	16.9	65.6	9	2937.1	25.8	183.3	9	5357.8	34.6	360.4	9	8489.7	43.5	596.7
4.0	237.0	8.1	7.4	12.0	1244.4	17.0	66.7	20.0	2963.0	25.9	185.2	28.0	5392.6	34.8	363.0	36.0	8533.3	43.6	600.0
1	245.2	8.2	7.8	1	1261.5	17.1	67.8	1	2988.9	26.0	187.0	1	5427.5	34.9	365.6	1	8577.1	43.8	603.3
2	253.6	8.3	8.1	2	1278.7	17.2	68.9	2	3015.0	26.1	188.9	2	5462.4	35.0	368.2	2	8621.0	43.9	606.7
3	262.0	8.4	8.6	3	1296.1	17.3	70.0	3	3041.2	26.2	190.8	3	5497.5	35.1	370.8	3	8664.9	44.0	610.0
4	270.5	8.5	9.0	4	1313.5	17.4	71.2	4	3067.6	26.3	192.7	4	5532.7	35.2	373.4	4	8709.0	44.1	613.4
5	279.2	8.6	9.4	5	1331.0	17.5	72.3	5	3094.0	26.4	194.6	5	5568.1	35.3	376.0	5	8753.2	44.2	616.8
6	287.9	8.8	9.8	6	1348.7	17.6	73.5	6	3120.5	26.5	196.5	6	5603.5	35.4	378.7	6	8797.6	44.3	620.2
7	296.8	8.9	10.2	7	1366.4	17.8	74.7	7	3147.2	26.6	198.4	7	5639.0	35.5	381.3	7	8842.0	44.4	623.6
8	305.8	9.0	10.7	8	1384.3	17.9	75.9	8	3173.9	26.8	200.3	8	5674.7	35.6	384.0	8	8886.5	44.5	627.0
9	314.9	9.1	11.1	9	1402.3	18.0	77.0	9	3200.8	26.9	202.2	9	5710.4	35.8	386.7	9	8931.2	44.6	630.4
5.0	324.1	9.2	11.6	13.0	1420.4	18.1	78.2	21.0	3227.8	27.0	204.2	29.0	5746.3	35.9	389.4	37.0	8975.9	44.8	633.8
1	333.4	9.3	12.0	1	1438.6	18.2	79.4	1	3254.9	27.1	206.1	1	5782.3	36.0	392.0	1	9020.8	44.9	637.2
2	342.8	9.4	12.5	2	1456.9	18.3	80.7	2	3282.1	27.2	208.1	2	5818.4	36.1	394.7	2	9065.8	45.0	640.7
3	352.4	9.5	13.0	3	1475.3	18.4	81.9	3	3309.4	27.3	210.0	3	5854.6	36.2	397.4	3	9110.9	45.1	644.1
4	362.0	9.6	13.5	4	1493.9	18.5	83.1	4	3336.8	27.4	212.0	4	5890.9	36.3	400.2	4	9156.1	45.2	647.6
5	371.8	9.8	14.0	5	1512.5	18.6	84.4	5	3364.4	27.5	214.0	5	5927.3	36.4	402.9	5	9201.4	45.3	651.0
6	381.6	9.9	14.5	6	1531.3	18.8	85.6	6	3392.0	27.6	216.0	6	5963.8	36.5	405.6	6	9246.8	45.4	654.5
7	391.6	10.0	15.0	7	1550.1	18.9	86.9	7	3419.8	27.8	218.0	7	6000.5	36.6	408.4	7	9292.4	45.5	658.0
8	401.7	10.1	15.6	8	1569.1	19.0	88.2	8	3447.6	27.9	220.0	8	6037.3	36.8	411.1	8	9338.0	45.6	661.5
9	411.9	10.2	16.1	9	1588.2	19.1	89.4	9	3475.6	28.0	222.0	9	6074.1	36.9	413.9	9	9383.8	45.8	665.0
6.0	422.2	10.3	16.7	14.0	1607.4	19.2	90.7	22.0	3503.7	28.1	224.1	30.0	6111.1	37.0	416.7	38.0	9429.6	45.9	668.5
1	432.6	10.4	17.2	1	1626.7	19.3	92.0	1	3531.9	28.2	226.1	1	6148.2	37.1	419.4	1	9475.6	46.0	672.0
2	443.2	10.5	17.8	2	1646.1	19.4	93.4	2	3560.2	28.3	228.2	2	6185.4	37.2	422.2	2	9521.7	46.1	675.6
3	453.8	10.6	18.4	3	1665.7	19.5	94.7	3	3588.6	28.4	230.2	3	6222.7	37.3	425.0	3	9567.9	46.2	679.1
4	464.6	10.8	19.0	4	1685.3	19.6	96.0	4	3617.2	28.5	232.3	4	6260.1	37.4	427.9	4	9614.2	46.3	682.7
5	475.5	10.9	19.6	5	1705.1	19.8	97.3	5	3645.8	28.6	234.4	5	6297.7	37.5	430.7	5	9660.6	46.4	686.2
6	486.4	11.0	20.2	6	1725.0	19.9	98.7	6	3674.6	28.8	236.5	6	6335.3	37.6	433.5	6	9707.2	46.5	689.8
7	497.5	11.1	20.8	7	1744.9	20.0	100.0	7	3703.5	28.9	238.6	7	6373.1	37.8	436.3	7	9753.8	46.6	693.4
8	508.7	11.2	21.4	8	1765.0	20.1	101.4	8	3732.4	29.0	240.7	8	6411.0	37.9	439.2	8	9800.6	46.8	697.0
9	520.1	11.3	22.0	9	1785.2	20.2	102.8	9	3761.5	29.1	242.8	9	6448.9	38.0	442.0	9	9847.5	46.9	700.6
7.0	531.5	11.4	22.7	15.0	1805.6	20.3	104.2	23.0	3790.7	29.2	244.9	31.0	6487.0	38.1	444.9	39.0	9894.4	47.0	704.2
1	543.0	11.5	23.3	1	1826.0	20.4	105.6	1	3820.1	29.3	247.0	1	6525.2	38.2	447.8	1	9941.5	47.1	707.8
2	554.7	11.6	24.0	2	1846.5	20.5	107.0	2	3849.5	29.4	249.2	2	6563.6	38.3	450.7	2	9988.7	47.2	711.4
3	566.4	11.8	24.7	3	1867.2	20.6	108.4	3	3879.0	29.5	251.3	3	6602.0	38.4	453.6	3	10036.1	47.3	715.0
4	578.3	11.9	25.4	4	1887.9	20.8	109.8	4	3908.7	29.6	253.5	4	6640.5	38.5	456.5	4	10083.5	47.4	718.7
5	590.3	12.0	26.0	5	1908.8	20.9	111.2	5	3938.4	29.8	255.7	5	6679.2	38.6	459.4	5	10131.0	47.5	722.3
6	602.4	12.1	26.7	6	1929.8	21.0	112.7	6	3968.3	29.9	257.9	6	6717.9	38.8	462.3	6	10178.7	47.6	726.0
7	614.6	12.2	27.4	7	1950.9	21.1	114.1	7	3998.3	30.0	260.0	7	6756.8	38.9	465.2	7	10226.4	47.8	729.7
8	626.9	12.3	28.2	8	1971.1	21.2	115.6	8	4028.4	30.1	262.2	8	6795.8	39.0	468.2	8	10274.3	47.9	733.4
9	639.3	12.4	28.9	9	1993.4	21.3	117.0	9	4058.6	30.2	264.5	9	6834.9	39.1	471.1	9	10322.3	48.0	737.0

TABLE NO. 7.

EXCAVATION AND EMBANKMENT.

CUBIC YARDS.

PRISMOIDS 100 FEET LONG. **BREADTH OF BASE 12 FEET.**

Slopes 1 1-2 Horizontal to 1 Perpendicular.

Ht.	Cub. Yds.	Diff.	Cor.	Ht.	Cub. Yds.	Diff.	Cor.	Ht.	Cub. Yds.	Diff.	Cor.	Ht.	Cub. Yds.	Diff.	Cor.	Ht.	Cub. Yds.	Diff.	Cor.	Ht.	Cub. Yds.	Diff.	Cor.
0.0	0.0	0.0	0.0	8.0	711.1	13.3	29.6	16.0	2133.3	22.2	118.5	24.0	4266.7	31.0	266.7	32.0	7111.1	39.9	474.1
.1	4.5	4.5	0.0	.1	724.5	13.4	30.4	.1	2155.6	22.3	120.0	.1	4297.8	31.2	268.9	.1	7151.2	40.0	477.0
.2	9.1	4.6	0.0	.2	738.0	13.5	31.1	.2	2178.0	22.4	121.5	.2	4329.1	31.3	271.1	.2	7191.3	40.2	480.0
.3	13.8	4.7	0.0	.3	751.6	13.6	31.9	.3	2200.5	22.5	123.0	.3	4360.5	31.4	273.4	.3	7231.6	40.3	483.0
.4	18.7	4.8	0.1	.4	765.3	13.7	32.7	.4	2223.1	22.6	124.5	.4	4392.0	31.5	275.6	.4	7272.0	40.4	486.0
.5	23.6	4.9	0.1	.5	779.2	13.8	33.4	.5	2245.8	22.7	126.0	.5	4423.6	31.6	277.9	.5	7312.5	40.5	489.0
.6	28.7	5.1	0.2	.6	793.1	13.9	34.2	.6	2268.7	22.8	127.6	.6	4455.3	31.7	280.2	.6	7353.1	40.6	492.0
.7	33.9	5.2	0.2	.7	807.2	14.0	35.0	.7	2291.6	22.9	129.1	.7	4487.2	31.8	282.5	.7	7393.8	40.7	495.0
.8	39.1	5.3	0.3	.8	821.3	14.2	35.9	.8	2314.7	23.0	130.7	.8	4519.1	31.9	284.7	.8	7434.7	40.8	498.0
.9	44.5	5.4	0.4	.9	835.6	14.3	36.7	.9	2337.8	23.2	132.2	.9	4551.2	32.0	287.0	.9	7475.6	40.9	501.1
1.0	50.0	5.5	0.5	9.0	850.0	14.4	37.5	17.0	2361.1	23.3	133.8	25.0	4583.3	32.2	289.4	33.0	7516.7	41.0	504.2
.1	55.6	5.6	0.6	.1	864.5	14.5	38.3	.1	2384.5	23.4	135.4	.1	4615.6	32.3	291.7	.1	7557.8	41.1	507.2
.2	61.3	5.7	0.7	.2	879.1	14.6	39.2	.2	2408.0	23.5	137.0	.2	4648.0	32.4	294.0	.2	7599.1	41.2	510.3
.3	67.2	5.8	0.8	.3	893.8	14.7	40.0	.3	2431.6	23.6	138.6	.3	4680.5	32.5	296.3	.3	7640.5	41.3	513.4
.4	73.1	5.9	0.9	.4	908.7	14.8	40.9	.4	2455.3	23.7	140.2	.4	4713.1	32.6	298.7	.4	7682.0	41.5	516.5
.5	79.2	6.1	1.0	.5	923.6	14.9	41.8	.5	2479.2	23.8	141.8	.5	4745.8	32.7	301.0	.5	7723.6	41.6	519.6
.6	85.3	6.2	1.2	.6	938.7	15.0	42.7	.6	2503.1	23.9	143.4	.6	4778.7	32.8	303.4	.6	7765.3	41.7	522.7
.7	91.6	6.3	1.3	.7	953.8	15.1	43.6	.7	2527.2	24.0	145.0	.7	4811.6	32.9	305.8	.7	7807.2	41.8	525.8
.8	98.0	6.4	1.5	.8	969.1	15.3	44.5	.8	2551.3	24.1	146.7	.8	4844.7	33.0	308.2	.8	7849.1	41.9	528.9
.9	104.5	6.5	1.7	.9	984.5	15.4	45.4	.9	2575.6	24.3	148.3	.9	4877.8	33.2	310.6	.9	7891.2	42.0	532.0
2.0	111.1	6.6	1.9	10.0	1000.0	15.5	46.3	18.0	2600.0	24.4	150.0	26.0	4911.1	33.3	313.0	34.0	7933.3	42.2	535.2
.1	117.8	6.7	2.0	.1	1015.6	15.6	47.2	.1	2624.5	24.5	151.7	.1	4944.5	33.4	315.4	.1	7975.6	42.3	538.3
.2	124.7	6.8	2.2	.2	1031.3	15.7	48.2	.2	2649.1	24.6	153.4	.2	4978.0	33.5	317.8	.2	8018.0	42.4	541.5
.3	131.6	6.9	2.4	.3	1047.2	15.8	49.1	.3	2673.8	24.7	155.0	.3	5011.6	33.6	320.2	.3	8060.5	42.5	544.7
.4	138.7	7.0	2.7	.4	1063.1	15.9	50.1	.4	2698.7	24.8	156.7	.4	5045.3	33.7	322.7	.4	8103.1	42.6	547.9
.5	145.8	7.2	2.9	.5	1079.2	16.0	51.0	.5	2723.6	24.9	158.4	.5	5079.2	33.8	325.1	.5	8145.8	42.7	551.0
.6	153.1	7.3	3.1	.6	1095.3	16.2	52.0	.6	2748.7	25.0	160.2	.6	5113.1	33.9	327.6	.6	8188.7	42.8	554.2
.7	160.5	7.4	3.4	.7	1111.6	16.3	53.0	.7	2773.8	25.2	161.9	.7	5147.2	34.0	330.0	.7	8231.6	42.9	557.5
.8	168.0	7.5	3.6	.8	1128.0	16.4	54.0	.8	2799.1	25.3	163.6	.8	5181.3	34.2	332.5	.8	8274.7	43.0	560.7
.9	175.6	7.6	3.9	.9	1144.5	16.5	55.0	.9	2824.5	25.4	165.4	.9	5215.6	34.3	335.0	.9	8317.8	43.2	563.9
3.0	183.3	7.7	4.2	11.0	1161.1	16.6	56.0	19.0	2850.0	25.5	167.1	27.0	5250.0	34.4	337.5	35.0	8361.1	43.3	567.1
.1	191.2	7.8	4.4	.1	1177.8	16.7	57.0	.1	2875.6	25.6	168.9	.1	5284.5	34.5	340.0	.1	8404.5	43.4	570.4
.2	199.1	7.9	4.7	.2	1194.7	16.8	58.1	.2	2901.3	25.7	170.7	.2	5319.1	34.6	342.5	.2	8448.0	43.5	573.6
.3	207.2	8.0	5.0	.3	1211.6	16.9	59.1	.3	2927.2	25.8	172.4	.3	5353.8	34.7	345.0	.3	8491.6	43.6	576.9
.4	215.3	8.1	5.4	.4	1228.7	17.0	60.2	.4	2953.1	25.9	174.2	.4	5388.7	34.8	347.6	.4	8535.3	43.7	580.2
.5	223.6	8.3	5.7	.5	1245.8	17.2	61.2	.5	2979.2	26.0	176.0	.5	5423.6	34.9	350.1	.5	8579.2	43.8	583.4
.6	232.0	8.4	6.0	.6	1263.1	17.3	62.3	.6	3005.3	26.2	177.9	.6	5458.7	35.1	352.7	.6	8623.1	43.9	586.7
.7	240.5	8.5	6.3	.7	1280.5	17.4	63.4	.7	3031.6	26.3	179.7	.7	5493.8	35.2	355.2	.7	8667.2	44.0	590.0
.8	249.1	8.6	6.7	.8	1298.0	17.5	64.5	.8	3058.0	26.4	181.5	.8	5529.1	35.3	357.8	.8	8711.3	44.2	593.4
.9	257.8	8.7	7.0	.9	1315.6	17.6	65.6	.9	3084.5	26.5	183.3	.9	5564.5	35.4	360.4	.9	8755.6	44.3	596.7
4.0	266.7	8.8	7.4	12.0	1333.3	17.7	66.7	20.0	3111.1	26.6	185.2	28.0	5600.0	35.5	363.0	36.0	8800.0	44.4	600.0
.1	275.6	8.9	7.8	.1	1351.2	17.8	67.8	.1	3137.8	26.7	187.0	.1	5635.6	35.6	365.6	.1	8844.5	44.5	603.3
.2	284.7	9.0	8.1	.2	1369.1	17.9	68.9	.2	3164.7	26.8	188.9	.2	5671.3	35.7	368.2	.2	8889.1	44.6	606.7
.3	293.8	9.1	8.6	.3	1387.2	18.0	70.0	.3	3191.6	26.9	190.8	.3	5707.2	35.8	370.8	.3	8933.8	44.7	610.0
.4	303.1	9.3	9.0	.4	1405.3	18.2	71.2	.4	3218.7	27.0	192.7	.4	5743.1	35.9	373.4	.4	8978.7	44.8	613.4
.5	312.5	9.4	9.4	.5	1423.6	18.3	72.3	.5	3245.8	27.1	194.6	.5	5779.2	36.0	376.0	.5	9023.6	44.9	616.8
.6	322.0	9.5	9.8	.6	1442.0	18.4	73.5	.6	3273.1	27.3	196.5	.6	5815.3	36.2	378.7	.6	9068.7	45.0	620.2
.7	331.6	9.6	10.2	.7	1460.5	18.5	74.7	.7	3300.5	27.4	198.4	.7	5851.6	36.3	381.3	.7	9113.8	45.1	623.6
.8	341.3	9.7	10.7	.8	1479.1	18.6	75.9	.8	3328.0	27.5	200.3	.8	5888.0	36.4	384.0	.8	9159.1	45.3	627.0
.9	351.2	9.8	11.1	.9	1497.8	18.7	77.0	.9	3355.6	27.6	202.2	.9	5924.5	36.5	386.7	.9	9204.5	45.4	630.4
5.0	361.1	9.9	11.6	13.0	1516.7	18.8	78.2	21.0	3383.3	27.7	204.2	29.0	5961.1	36.6	389.4	37.0	9250.0	45.5	633.8
.1	371.2	10.0	12.0	.1	1535.6	18.9	79.4	.1	3411.2	27.8	206.1	.1	5997.8	36.7	392.0	.1	9295.6	45.6	637.2
.2	381.3	10.2	12.5	.2	1554.7	19.0	80.7	.2	3439.1	27.9	208.1	.2	6034.7	36.8	394.7	.2	9341.3	45.7	640.7
.3	391.6	10.3	13.0	.3	1573.8	19.2	81.9	.3	3467.2	28.0	210.0	.3	6071.6	36.9	397.4	.3	9387.2	45.8	644.1
.4	402.0	10.4	13.5	.4	1593.1	19.3	83.1	.4	3495.3	28.2	212.0	.4	6108.7	37.0	400.2	.4	9433.1	45.9	647.6
.5	412.5	10.5	14.0	.5	1612.5	19.4	84.4	.5	3523.6	28.3	214.0	.5	6145.8	37.2	402.9	.5	9479.2	46.0	651.0
.6	423.1	10.6	14.5	.6	1632.0	19.5	85.6	.6	3552.0	28.4	216.0	.6	6183.1	37.3	405.6	.6	9525.3	46.2	654.5
.7	433.8	10.7	15.0	.7	1651.6	19.6	86.9	.7	3580.5	28.5	218.0	.7	6220.5	37.4	408.4	.7	9571.6	46.3	658.0
.8	444.7	10.8	15.6	.8	1671.3	19.7	88.2	.8	3609.1	28.6	220.0	.8	6258.0	37.5	411.1	.8	9618.0	46.4	661.5
.9	455.6	10.9	16.1	.9	1691.2	19.8	89.4	.9	3637.8	28.7	222.0	.9	6295.6	37.6	413.9	.9	9664.5	46.5	665.0
6.0	466.7	11.0	16.7	14.0	1711.1	19.9	90.7	22.0	3666.7	28.8	224.1	30.0	6333.3	37.7	416.7	38.0	9711.1	46.6	668.5
.1	477.8	11.2	17.2	.1	1731.2	20.0	92.0	.1	3695.6	28.9	226.1	.1	6371.2	37.8	419.4	.1	9757.8	46.7	672.0
.2	489.1	11.3	17.8	.2	1751.3	20.2	93.4	.2	3724.7	29.0	228.2	.2	6409.1	37.9	422.2	.2	9804.7	46.8	675.6
.3	500.5	11.4	18.4	.3	1771.6	20.3	94.7	.3	3753.8	29.2	230.2	.3	6447.2	38.0	425.0	.3	9851.6	46.9	679.1
.4	512.0	11.5	19.0	.4	1792.0	20.4	96.0	.4	3783.1	29.3	232.3	.4	6485.3	38.2	427.9	.4	9898.7	47.0	682.7
.5	523.6	11.6	19.6	.5	1812.5	20.5	97.3	.5	3812.5	29.4	234.4	.5	6523.6	38.3	430.7	.5	9945.8	47.2	686.2
.6	535.3	11.7	20.2	.6	1833.1	20.6	98.7	.6	3842.0	29.5	236.5	.6	6562.0	38.4	433.5	.6	9993.1	47.3	689.8
.7	547.2	11.8	20.8	.7	1853.8	20.7	100.0	.7	3871.6	29.6	238.6	.7	6600.5	38.5	436.3	.7	10040.5	47.4	693.4
.8	559.1	11.9	21.4	.8	1874.7	20.8	101.4	.8	3901.3	29.7	240.7	.8	6639.1	38.6	439.2	.8	10088.0	47.5	697.0
.9	571.2	12.0	22.0	.9	1895.6	20.9	102.8	.9	3931.2	29.9	242.8	.9	6677.8	38.7	442.0	.9	10135.6	47.6	700.6
7.0	583.3	12.1	22.7	15.0	1916.7	21.0	104.2	23.0	3961.1	29.9	244.9	31.0	6716.7	38.8	444.9	39.0	10183.3	47.7	704.2
.1	595.6	12.3	23.3	.1	1937.8	21.2	105.6	.1	3991.2	30.1	247.0	.1	6755.6	38.9	447.8	.1	10231.2	47.8	707.8
.2	608.0	12.4	24.0	.2	1959.1	21.3	107.0	.2	4021.3	30.2	249.2	.2	6794.7	39.0	450.7	.2	10279.1	47.9	711.4
.3	620.5	12.5	24.7	.3	1980.5	21.4	108.4	.3	4051.6	30.3	251.3	.3	6833.8	39.2	453.6	.3	10327.2	48.0	715.0
.4	633.1	12.6	25.4	.4	2002.0	21.5	109.8	.4	4082.0	30.4	253.5	.4	6873.1	39.3	456.5	.4	10375.3	48.2	718.7
.5	645.8	12.7	26.0	.5	2023.6	21.6	111.2	.5	4112.5	30.5	255.7	.5	6912.5	39.4	459.4	.5	10423.6	48.3	722.3
.6	658.7	12.8	26.7	.6	2045.3	21.7	112.7	.6	4143.1	30.6	257.9	.6	6952.0	39.5	462.3	.6	10472.0	48.4	726.0
.7	671.6	12.9	27.4	.7	2067.2	21.8	114.1	.7	4173.8	30.7	260.0	.7	6991.6	39.6	465.2	.7	10520.5	48.5	729.7
.8	684.7	13.0	28.2	.8	2089.1	21.9	115.6	.8	4204.7	30.8	262.2	.8	7031.3	39.7	468.2	.8	10569.1	48.6	733.4
.9	697.8	13.2	29.0	.9	2111.2	22.0	117.0	.9	4235.6	30.9	264.5	.9	7071.2	39.8	471.1	.9	10617.8	48.7	737.0

TABLE NO. 8.

EXCAVATION AND EMBANKMENT.

CUBIC YARDS.

PRISMOIDS 100 FEET LONG. **BREADTH OF BASE 14 FEET.**

Slopes 1 1-2 Horizontal to 1 Perpendicular.

Ht.	Cub. Yds.	Diff.	Cor.	Ht.	Cub. Yds.	Diff.	Cor.	Ht.	Cub. Yds.	Diff.	Cor.	Ht.	Cub. Yds.	Diff.	Cor.	Ht.	Cub. Yds.	Diff.	Cor.
0.0	0.0	0.0	0.0	8.0	770.4	14.0	29.6	16.0	2251.0	22.9	118.5	24.0	4444.4	31.8	266.7	32.0	7348.1	40.7	474.1
.1	5.2	5.2	0.0	.1	784.5	14.1	30.4	.1	2274.0	23.0	120.0	.1	4476.4	31.9	268.9	.1	7388.9	40.8	477.0
.2	10.6	5.4	0.0	.2	798.7	14.2	31.1	.2	2298.0	23.1	121.5	.2	4508.4	32.0	271.1	.2	7429.9	40.9	480.0
.3	16.1	5.5	0.0	.3	813.1	14.4	31.9	.3	2321.2	23.2	123.0	.3	4540.5	32.1	273.4	.3	7470.9	41.0	483.0
.4	21.6	5.6	0.1	.4	827.6	14.5	32.7	.4	2344.6	23.4	124.5	.4	4572.7	32.2	275.6	.4	7512.0	41.1	486.0
.5	27.3	5.7	0.1	.5	842.1	14.6	33.4	.5	2368.1	23.5	126.0	.5	4605.1	32.4	277.9	.5	7553.2	41.2	489.0
.6	33.1	5.8	0.2	.6	856.8	14.7	34.2	.6	2391.6	23.6	127.6	.6	4637.6	32.5	280.2	.6	7594.6	41.4	492.0
.7	39.0	5.9	0.2	.7	871.6	14.8	35.0	.7	2415.3	23.7	129.1	.7	4670.1	32.6	282.5	.7	7636.1	41.5	495.0
.8	45.0	6.0	0.3	.8	886.5	14.9	35.9	.8	2439.1	23.8	130.7	.8	4702.8	32.7	284.7	.8	7677.6	41.6	498.0
.9	51.2	6.1	0.4	.9	901.5	15.0	36.7	.9	2463.0	23.9	132.2	.9	4735.6	32.8	287.0	.9	7719.3	41.7	501.1
1.0	57.4	6.2	0.5	9.0	916.7	15.1	37.5	17.0	2487.0	24.0	133.8	25.0	4768.5	32.9	289.4	33.0	7761.1	41.8	504.2
.1	63.8	6.4	0.6	.1	931.9	15.2	38.3	.1	2511.2	24.1	135.4	.1	4801.5	33.0	291.7	.1	7803.0	41.9	507.2
.2	70.2	6.5	0.7	.2	947.3	15.4	39.2	.2	2535.4	24.2	137.0	.2	4834.7	33.1	294.0	.2	7845.0	42.0	510.3
.3	76.8	6.6	0.8	.3	962.7	15.5	40.0	.3	2559.8	24.4	138.6	.3	4867.9	33.2	296.3	.3	7887.2	42.1	513.4
.4	83.5	6.7	0.9	.4	978.3	15.6	40.9	.4	2584.2	24.5	140.2	.4	4901.3	33.4	298.7	.4	7929.4	42.2	516.5
.5	90.3	6.8	1.0	.5	994.0	15.7	41.8	.5	2608.8	24.6	141.8	.5	4934.7	33.5	301.0	.5	7971.8	42.4	519.6
.6	97.2	6.9	1.2	.6	1009.8	15.8	42.7	.6	2633.5	24.7	143.4	.6	4968.3	33.6	303.4	.6	8014.2	42.5	522.7
.7	104.2	7.0	1.3	.7	1025.7	15.9	43.6	.7	2658.3	24.8	145.0	.7	5002.0	33.7	305.8	.7	8056.8	42.6	525.8
.8	111.3	7.1	1.5	.8	1041.7	16.0	44.5	.8	2683.2	24.9	146.7	.8	5035.8	33.8	308.2	.8	8099.5	42.7	528.9
.9	118.6	7.2	1.7	.9	1057.8	16.1	45.4	.9	2708.2	25.0	148.3	.9	5069.7	33.9	310.6	.9	8142.3	42.8	532.0
2.0	125.9	7.4	1.9	10.0	1074.1	16.2	46.3	18.0	2733.3	25.1	150.0	26.0	5103.7	34.0	313.0	34.0	8185.2	42.9	535.2
.1	133.4	7.5	2.0	.1	1090.4	16.4	47.2	.1	2758.6	25.2	151.7	.1	5137.8	34.1	315.4	.1	8228.2	43.0	538.3
.2	141.0	7.6	2.2	.2	1106.9	16.5	48.2	.2	2783.9	25.4	153.4	.2	5172.1	34.2	317.8	.2	8271.3	43.1	541.5
.3	148.6	7.7	2.4	.3	1123.5	16.6	49.1	.3	2809.4	25.5	155.0	.3	5206.4	34.4	320.2	.3	8314.6	43.2	544.7
.4	156.4	7.8	2.7	.4	1140.1	16.7	50.1	.4	2835.0	25.6	156.7	.4	5240.9	34.5	322.7	.4	8357.9	43.4	547.9
.5	164.4	7.9	2.9	.5	1156.9	16.8	51.0	.5	2860.6	25.7	158.4	.5	5275.5	34.6	325.1	.5	8401.4	43.5	551.0
.6	172.4	8.0	3.1	.6	1173.9	16.9	52.0	.6	2886.4	25.8	160.2	.6	5310.1	34.7	327.6	.6	8445.0	43.6	554.2
.7	180.5	8.1	3.4	.7	1190.9	17.0	53.0	.7	2912.4	25.9	161.9	.7	5344.9	34.8	330.0	.7	8488.6	43.7	557.5
.8	188.7	8.2	3.6	.8	1208.0	17.1	54.0	.8	2938.4	26.0	163.6	.8	5379.9	34.9	332.5	.8	8532.4	43.8	560.7
.9	197.1	8.4	3.9	.9	1225.2	17.2	55.0	.9	2964.5	26.1	165.4	.9	5414.9	35.0	335.0	.9	8576.4	43.9	563.9
3.0	205.6	8.5	4.2	11.0	1242.6	17.4	56.0	19.0	2990.7	26.2	167.1	27.0	5450.0	35.1	337.5	35.0	8620.4	44.0	567.1
.1	214.1	8.6	4.4	.1	1260.1	17.5	57.0	.1	3017.1	26.4	168.9	.1	5485.2	35.2	340.0	.1	8664.5	44.1	570.4
.2	222.8	8.7	4.7	.2	1277.6	17.6	58.1	.2	3043.6	26.5	170.7	.2	5520.6	35.4	342.5	.2	8708.7	44.2	573.6
.3	231.6	8.8	5.0	.3	1295.3	17.7	59.1	.3	3070.1	26.6	172.4	.3	5556.1	35.5	345.0	.3	8753.1	44.4	576.9
.4	240.5	8.9	5.4	.4	1313.1	17.8	60.2	.4	3096.8	26.7	174.2	.4	5591.6	35.6	347.6	.4	8797.6	44.5	580.2
.5	249.5	9.0	5.7	.5	1331.0	17.9	61.2	.5	3123.6	26.8	176.0	.5	5627.3	35.7	350.1	.5	8842.1	44.6	583.4
.6	258.7	9.1	6.0	.6	1349.0	18.0	62.3	.6	3150.5	26.9	177.9	.6	5663.1	35.8	352.7	.6	8886.8	44.7	586.7
.7	267.9	9.2	6.3	.7	1367.2	18.1	63.4	.7	3177.5	27.0	179.7	.7	5699.0	35.9	355.2	.7	8931.6	44.8	590.0
.8	277.3	9.4	6.7	.8	1385.4	18.2	64.5	.8	3204.7	27.1	181.5	.8	5735.0	36.0	357.8	.8	8976.5	44.9	593.4
.9	286.7	9.5	7.0	.9	1403.8	18.4	65.6	.9	3231.9	27.2	183.3	.9	5771.2	36.1	360.4	.9	9021.5	45.0	596.7
4.0	296.3	9.6	7.4	12.0	1422.2	18.5	66.7	20.0	3259.3	27.4	185.2	28.0	5807.4	36.2	363.0	36.0	9066.7	45.1	600.0
.1	306.0	9.7	7.8	.1	1440.8	18.6	67.8	.1	3286.7	27.5	187.0	.1	5843.8	36.4	365.6	.1	9111.9	45.2	603.3
.2	315.8	9.8	8.1	.2	1459.5	18.7	68.9	.2	3314.3	27.6	188.9	.2	5880.2	36.5	368.2	.2	9157.3	45.4	606.7
.3	325.7	9.9	8.6	.3	1478.3	18.8	70.0	.3	3342.0	27.7	190.8	.3	5916.8	36.6	370.8	.3	9202.7	45.5	610.0
.4	335.7	10.0	9.0	.4	1497.2	18.9	71.2	.4	3369.8	27.8	192.7	.4	5953.5	36.7	373.4	.4	9248.3	45.6	613.4
.5	345.8	10.1	9.4	.5	1516.2	19.0	72.3	.5	3397.7	27.9	194.6	.5	5990.3	36.8	376.0	.5	9294.0	45.7	616.8
.6	356.1	10.2	9.8	.6	1535.3	19.1	73.5	.6	3425.7	28.0	196.5	.6	6027.2	36.9	378.7	.6	9339.8	45.8	620.2
.7	366.4	10.4	10.2	.7	1554.6	19.2	74.7	.7	3453.8	28.1	198.4	.7	6064.2	37.0	381.3	.7	9385.7	45.9	623.6
.8	376.9	10.5	10.7	.8	1573.9	19.4	75.9	.8	3482.1	28.2	200.3	.8	6101.3	37.1	384.0	.8	9431.7	46.0	627.0
.9	387.5	10.6	11.1	.9	1593.4	19.5	77.0	.9	3510.4	28.4	202.2	.9	6138.6	37.2	386.7	.9	9477.8	46.1	630.4
5.0	398.1	10.7	11.6	13.0	1613.0	19.6	78.2	21.0	3538.9	28.5	204.2	29.0	6175.9	37.4	389.4	37.0	9524.1	46.2	633.8
.1	408.9	10.8	12.0	.1	1632.6	19.7	79.4	.1	3567.5	28.6	206.1	.1	6213.4	37.5	392.0	.1	9570.4	46.4	637.2
.2	419.9	10.9	12.5	.2	1652.4	19.8	80.7	.2	3596.1	28.7	208.1	.2	6251.0	37.6	394.7	.2	9616.9	46.5	640.7
.3	430.9	11.0	13.0	.3	1672.1	19.9	81.9	.3	3624.9	28.8	210.0	.3	6288.6	37.7	397.4	.3	9663.5	46.6	644.1
.4	442.0	11.1	13.5	.4	1692.4	20.0	83.1	.4	3653.9	28.9	212.0	.4	6326.4	37.8	400.2	.4	9710.1	46.7	647.6
.5	453.2	11.2	14.0	.5	1712.5	20.1	84.4	.5	3682.9	29.0	214.0	.5	6364.4	37.9	402.9	.5	9756.9	46.8	651.0
.6	464.6	11.4	14.5	.6	1732.7	20.2	85.6	.6	3712.0	29.1	216.0	.6	6402.4	38.0	405.6	.6	9803.9	46.9	654.5
.7	476.1	11.5	15.0	.7	1753.1	20.4	86.9	.7	3741.2	29.2	218.0	.7	6440.5	38.1	408.4	.7	9850.9	47.0	658.0
.8	487.6	11.6	15.6	.8	1773.6	20.5	88.2	.8	3770.6	29.4	220.0	.8	6478.7	38.2	411.1	.8	9898.0	47.1	661.5
.9	499.3	11.7	16.1	.9	1794.1	20.6	89.4	.9	3800.1	29.5	222.0	.9	6517.1	38.4	413.9	.9	9945.2	47.2	665.0
6.0	511.1	11.8	16.7	14.0	1814.8	20.7	90.7	22.0	3829.6	29.6	224.1	30.0	6555.6	38.5	416.7	38.0	9992.6	47.4	668.5
.1	523.0	11.9	17.2	.1	1835.6	20.8	92.0	.1	3859.3	29.7	226.1	.1	6594.1	38.6	419.4	.1	10040.1	47.5	672.0
.2	535.0	12.0	17.8	.2	1856.5	20.9	93.4	.2	3889.1	29.8	228.2	.2	6632.8	38.7	422.2	.2	10087.6	47.6	675.6
.3	547.2	12.1	18.4	.3	1877.5	21.0	94.7	.3	3919.0	29.9	230.2	.3	6671.6	38.8	425.0	.3	10135.3	47.7	679.1
.4	559.4	12.2	19.0	.4	1898.7	21.1	96.0	.4	3949.0	30.0	232.3	.4	6710.5	38.9	427.9	.4	10183.1	47.8	682.7
.5	571.8	12.4	19.6	.5	1919.9	21.2	97.3	.5	3979.2	30.1	234.4	.5	6749.5	39.0	430.7	.5	10231.0	47.9	686.2
.6	584.2	12.5	20.2	.6	1941.3	21.4	98.7	.6	4009.4	30.2	236.5	.6	6788.7	39.1	433.5	.6	10279.0	48.0	689.8
.7	596.8	12.6	20.8	.7	1962.7	21.5	100.0	.7	4039.8	30.4	238.6	.7	6827.9	39.2	436.3	.7	10327.2	48.1	693.4
.8	609.5	12.7	21.4	.8	1984.3	21.6	101.4	.8	4070.2	30.5	240.7	.8	6867.3	39.4	439.2	.8	10375.4	48.2	697.0
.9	622.3	12.8	22.0	.9	2006.0	21.7	102.8	.9	4100.8	30.6	242.8	.9	6906.7	39.5	442.0	.9	10423.8	48.4	700.6
7.0	635.2	12.9	22.7	15.0	2027.8	21.8	104.2	23.0	4131.5	30.7	244.9	31.0	6946.3	39.6	444.9	39.0	10472.2	48.5	704.2
.1	648.2	13.0	23.3	.1	2049.7	21.9	105.6	.1	4162.3	30.8	247.0	.1	6986.0	39.7	447.8	.1	10520.8	48.6	707.8
.2	661.3	13.1	24.0	.2	2071.7	22.0	107.0	.2	4193.2	30.9	249.2	.2	7025.8	39.8	450.7	.2	10569.5	48.7	711.4
.3	674.6	13.2	24.7	.3	2093.8	22.1	108.4	.3	4224.2	31.0	251.3	.3	7065.7	39.9	453.6	.3	10618.3	48.8	715.0
.4	687.9	13.4	25.4	.4	2116.1	22.2	109.8	.4	4255.3	31.1	253.5	.4	7105.7	40.0	456.5	.4	10667.2	48.9	718.7
.5	701.4	13.5	26.0	.5	2138.4	22.4	111.2	.5	4286.6	31.2	255.7	.5	7145.8	40.1	459.4	.5	10716.2	49.0	722.3
.6	715.0	13.6	26.7	.6	2160.9	22.5	112.7	.6	4317.9	31.4	257.9	.6	7186.1	40.2	462.3	.6	10765.3	49.1	726.0
.7	728.6	13.7	27.4	.7	2183.5	22.6	114.1	.7	4349.4	31.5	260.0	.7	7226.4	40.4	465.2	.7	10814.6	49.2	729.7
.8	742.4	13.8	28.2	.8	2206.1	22.7	115.6	.8	4381.0	31.6	262.2	.8	7266.9	40.5	468.2	.8	10863.9	49.4	733.4
.9	756.4	13.9	28.9	.9	2228.9	22.8	117.0	.9	4412.6	31.7	264.5	.9	7307.5	40.6	471.1	.9	10913.4	49.5	737.0

TABLE NO. 9.

EXCAVATION AND EMBANKMENT.

CUBIC YARDS.

PRISMOIDS 100 FEET LONG. **BREADTH OF BASE 15 FEET.**

Slopes 1 1-2 Horizontal to 1 Perpendicular.

Ht.	Cub. Yds.	Diff.	Cor.	Ht.	Cub. Yds.	Diff.	Cor.	Ht.	Cub. Yds.	Diff.	Cor.	Ht.	Cub. Yds.	Diff.	Cor.	Ht.	Cub. Yds.	Diff.	Cor.
0.0	0.0	0.0	0.0	8.0	800.0	14.4	29.6	16.0	2311.1	23.3	118.5	24.0	4533.3	32.2	266.7	32.0	7466.7	41.1	474.1
.1	5.6	5.6	0.0	.1	814.5	14.5	30.4	.1	2334.5	23.4	120.0	.1	4565.6	32.3	268.9	.1	7507.8	41.2	477.0
.2	11.3	5.7	0.0	.2	829.1	14.6	31.1	.2	2358.0	23.5	121.5	.2	4598.0	32.4	271.1	.2	7549.1	41.3	480.0
.3	17.2	5.8	0.0	.3	843.8	14.7	31.9	.3	2381.6	23.6	123.0	.3	4630.5	32.5	273.4	.3	7590.5	41.4	483.0
.4	23.1	5.9	0.1	.4	858.7	14.8	32.7	.4	2405.3	23.7	124.5	.4	4663.1	32.6	275.6	.4	7632.0	41.5	486.0
.5	29.2	6.1	0.1	.5	873.6	14.9	33.4	.5	2429.2	23.8	126.0	.5	4695.8	32.7	277.9	.5	7673.6	41.6	489.0
.6	35.3	6.2	0.2	.6	888.7	15.1	34.2	.6	2453.1	23.9	127.6	.6	4728.7	32.8	280.2	.6	7715.3	41.7	492.0
.7	41.6	6.3	0.2	.7	903.8	15.2	35.0	.7	2477.2	24.1	129.1	.7	4761.6	32.9	282.5	.7	7757.2	41.8	495.0
.8	48.0	6.4	0.3	.8	919.1	15.3	35.9	.8	2501.3	24.2	130.7	.8	4794.7	33.1	284.7	.8	7799.1	41.9	498.0
.9	54.5	6.5	0.4	.9	934.5	15.4	36.7	.9	2525.6	24.3	132.2	.9	4827.8	33.2	287.0	.9	7841.2	42.1	501.1
1.0	61.1	6.6	0.5	9.0	950.0	15.5	37.5	17.0	2550.0	24.4	133.8	25.0	4861.1	33.3	289.4	33.0	7883.3	42.2	504.2
.1	67.8	6.7	0.6	.1	965.6	15.6	38.3	.1	2574.5	24.5	135.4	.1	4894.5	33.4	291.7	.1	7925.6	42.3	507.2
.2	74.7	6.8	0.7	.2	981.3	15.7	39.2	.2	2599.1	24.6	137.0	.2	4928.0	33.5	294.0	.2	7968.0	42.4	510.3
.3	81.6	6.9	0.8	.3	997.2	15.8	40.0	.3	2623.8	24.7	138.6	.3	4961.6	33.6	296.3	.3	8010.5	42.5	513.4
.4	88.7	7.1	0.9	.4	1013.1	15.9	40.9	.4	2648.7	24.8	140.2	.4	4995.3	33.7	298.7	.4	8053.1	42.6	516.5
.5	95.8	7.2	1.0	.5	1029.2	16.1	41.8	.5	2673.6	24.9	141.8	.5	5029.2	33.8	301.0	.5	8095.8	42.7	519.6
.6	103.1	7.3	1.2	.6	1045.3	16.2	42.7	.6	2698.7	25.1	143.4	.6	5063.1	33.9	303.4	.6	8138.7	42.8	522.7
.7	110.5	7.4	1.3	.7	1061.6	16.3	43.6	.7	2723.8	25.2	145.0	.7	5097.2	34.1	305.8	.7	8181.6	42.9	525.8
.8	118.0	7.5	1.5	.8	1078.0	16.4	44.5	.8	2749.1	25.3	146.7	.8	5131.3	34.2	308.2	.8	8224.7	43.1	528.9
.9	125.6	7.6	1.7	.9	1094.5	16.5	45.4	.9	2774.5	25.4	148.3	.9	5165.6	34.3	310.6	.9	8267.8	43.2	532.0
2.0	133.3	7.7	1.9	10.0	1111.1	16.6	46.3	18.0	2800.0	25.5	150.0	26.0	5200.0	34.4	313.0	34.0	8311.1	43.3	535.2
.1	141.2	7.8	2.0	.1	1127.8	16.7	47.2	.1	2825.6	25.6	151.7	.1	5234.5	34.5	315.4	.1	8354.5	43.4	538.3
.2	149.1	7.9	2.2	.2	1144.7	16.8	48.2	.2	2851.3	25.7	153.4	.2	5269.1	34.6	317.8	.2	8398.0	43.5	541.5
.3	157.2	8.1	2.4	.3	1161.6	16.9	49.1	.3	2877.2	25.8	155.0	.3	5303.8	34.7	320.2	.3	8441.6	43.6	544.7
.4	165.3	8.2	2.7	.4	1178.7	17.1	50.1	.4	2903.1	25.9	156.7	.4	5338.6	34.8	322.7	.4	8485.3	43.7	547.9
.5	173.6	8.3	2.9	.5	1195.8	17.2	51.0	.5	2929.2	26.1	158.4	.5	5373.6	34.9	325.1	.5	8529.2	43.8	551.0
.6	182.0	8.4	3.1	.6	1213.1	17.3	52.0	.6	2955.3	26.2	160.2	.6	5408.7	35.1	327.6	.6	8573.1	43.9	554.2
.7	190.5	8.5	3.4	.7	1230.5	17.4	53.0	.7	2981.6	26.3	161.9	.7	5443.8	35.2	330.0	.7	8617.2	44.1	557.5
.8	199.1	8.6	3.6	.8	1248.0	17.5	54.0	.8	3008.0	26.4	163.6	.8	5479.1	35.3	332.5	.8	8661.3	44.2	560.7
.9	207.8	8.7	3.9	.9	1265.6	17.6	55.0	.9	3034.5	26.5	165.4	.9	5514.5	35.4	335.0	.9	8705.6	44.3	563.9
3.0	216.7	8.8	4.2	11.0	1283.3	17.7	56.0	19.0	3061.1	26.6	167.1	27.0	5550.0	35.5	337.5	35.0	8750.0	44.4	567.1
.1	225.6	8.9	4.4	.1	1301.2	17.8	57.0	.1	3087.8	26.7	168.9	.1	5585.6	35.6	340.0	.1	8794.5	44.5	570.4
.2	234.7	9.1	4.7	.2	1319.1	17.9	58.1	.2	3114.7	26.8	170.7	.2	5621.3	35.7	342.5	.2	8839.1	44.6	573.6
.3	243.8	9.2	5.0	.3	1337.2	18.1	59.1	.3	3141.6	26.9	172.4	.3	5657.2	35.8	345.0	.3	8883.8	44.7	576.9
.4	253.1	9.3	5.4	.4	1355.3	18.2	60.2	.4	3168.7	27.1	174.2	.4	5693.1	35.9	347.6	.4	8928.7	44.8	580.2
.5	262.5	9.4	5.7	.5	1373.6	18.3	61.2	.5	3195.8	27.2	176.0	.5	5729.2	36.1	350.1	.5	8973.6	44.9	583.4
.6	272.0	9.5	6.0	.6	1392.0	18.4	62.3	.6	3223.1	27.3	177.9	.6	5765.3	36.2	352.7	.6	9018.7	45.1	586.7
.7	281.6	9.6	6.3	.7	1410.5	18.5	63.4	.7	3250.5	27.4	179.7	.7	5801.6	36.3	355.2	.7	9063.8	45.2	590.0
.8	291.3	9.7	6.7	.8	1429.1	18.6	64.5	.8	3278.0	27.5	181.5	.8	5838.0	36.4	357.8	.8	9109.1	45.3	593.4
.9	301.2	9.8	7.0	.9	1447.8	18.7	65.6	.9	3305.6	27.6	183.3	.9	5874.5	36.5	360.4	.9	9154.5	45.4	596.7
4.0	311.1	9.9	7.4	12.0	1466.7	18.8	66.7	20.0	3333.3	27.7	185.2	28.0	5911.1	36.6	363.0	36.0	9200.0	45.5	600.0
.1	321.2	10.1	7.8	.1	1485.6	18.9	67.8	.1	3361.2	27.8	187.0	.1	5947.8	36.7	365.6	.1	9245.6	45.6	603.3
.2	331.3	10.2	8.1	.2	1504.7	19.1	68.9	.2	3389.1	27.9	188.9	.2	5984.7	36.8	368.2	.2	9291.3	45.7	606.7
.3	341.6	10.3	8.6	.3	1523.8	19.2	70.0	.3	3417.2	28.1	190.8	.3	6021.6	36.9	370.8	.3	9337.2	45.8	610.0
.4	352.0	10.4	9.0	.4	1543.1	19.3	71.2	.4	3445.3	28.2	192.7	.4	6058.7	37.1	373.4	.4	9383.1	45.9	613.4
.5	362.5	10.5	9.4	.5	1562.5	19.4	72.3	.5	3473.6	28.3	194.6	.5	6095.8	37.2	376.0	.5	9429.2	46.1	616.8
.6	373.1	10.6	9.8	.6	1582.0	19.5	73.5	.6	3502.0	28.4	196.5	.6	6133.1	37.3	378.7	.6	9475.3	46.2	620.2
.7	383.8	10.7	10.2	.7	1601.6	19.6	74.7	.7	3530.5	28.5	198.4	.7	6170.5	37.4	381.3	.7	9521.6	46.3	623.6
.8	394.7	10.8	10.7	.8	1621.3	19.7	75.9	.8	3559.1	28.6	200.3	.8	6208.0	37.5	384.0	.8	9568.0	46.4	627.0
.9	405.6	10.9	11.1	.9	1641.2	19.8	77.0	.9	3587.8	28.7	202.2	.9	6245.6	37.6	386.7	.9	9614.5	46.5	630.4
5.0	416.7	11.1	11.6	13.0	1661.1	19.9	78.2	21.0	3616.7	28.8	204.2	29.0	6283.3	37.7	389.4	37.0	9661.1	46.6	633.8
.1	427.8	11.2	12.0	.1	1681.2	20.1	79.4	.1	3645.6	28.9	206.1	.1	6321.2	37.8	392.0	.1	9707.8	46.7	637.2
.2	439.1	11.3	12.5	.2	1701.3	20.2	80.7	.2	3674.7	29.1	208.1	.2	6359.1	37.9	394.7	.2	9754.7	46.8	640.7
.3	450.5	11.4	13.0	.3	1721.6	20.3	81.9	.3	3703.8	29.2	210.0	.3	6397.2	38.1	397.4	.3	9801.6	46.9	644.1
.4	462.0	11.5	13.5	.4	1742.0	20.4	83.1	.4	3733.1	29.3	212.0	.4	6435.3	38.2	400.2	.4	9848.7	47.1	647.6
.5	473.6	11.6	14.0	.5	1762.5	20.5	84.4	.5	3762.5	29.4	214.0	.5	6473.6	38.3	402.9	.5	9895.8	47.2	651.0
.6	485.3	11.7	14.5	.6	1783.1	20.6	85.6	.6	3792.0	29.5	216.0	.6	6512.0	38.4	405.6	.6	9943.1	47.3	654.5
.7	497.2	11.8	15.0	.7	1803.8	20.7	86.9	.7	3821.6	29.6	218.0	.7	6550.5	38.5	408.4	.7	9990.5	47.4	658.0
.8	509.1	11.9	15.6	.8	1824.7	20.8	88.2	.8	3851.3	29.7	220.0	.8	6589.1	38.6	411.1	.8	10038.0	47.5	661.5
.9	521.2	12.1	16.1	.9	1845.6	20.9	89.4	.9	3881.2	29.9	222.0	.9	6627.8	38.7	413.9	.9	10085.6	47.6	665.0
6.0	533.3	12.2	16.7	14.0	1866.7	21.1	90.7	22.0	3911.1	29.9	224.1	30.0	6666.7	38.8	416.7	38.0	10133.3	47.7	668.5
.1	545.6	12.3	17.2	.1	1887.8	21.2	92.0	.1	3941.2	30.1	226.1	.1	6705.6	38.9	419.4	.1	10181.2	47.8	672.0
.2	558.0	12.4	17.8	.2	1909.1	21.3	93.4	.2	3971.3	30.2	228.2	.2	6744.7	39.1	422.2	.2	10229.1	47.9	675.6
.3	570.5	12.5	18.4	.3	1930.5	21.4	94.7	.3	4001.6	30.3	230.2	.3	6783.8	39.2	425.0	.3	10277.2	48.1	679.1
.4	583.1	12.6	19.0	.4	1952.0	21.5	96.0	.4	4032.0	30.4	232.3	.4	6823.1	39.3	427.9	.4	10325.3	48.2	682.7
.5	595.8	12.7	19.6	.5	1973.6	21.6	97.3	.5	4062.5	30.5	234.4	.5	6862.5	39.4	430.7	.5	10373.6	48.3	686.2
.6	608.7	12.8	20.2	.6	1995.3	21.7	98.7	.6	4093.1	30.6	236.5	.6	6902.0	39.5	433.5	.6	10422.0	48.4	689.8
.7	621.6	12.9	20.8	.7	2017.2	21.8	100.0	.7	4123.8	30.7	238.6	.7	6941.6	39.6	436.3	.7	10470.5	48.5	693.4
.8	634.7	13.1	21.4	.8	2039.1	21.9	101.4	.8	4151.7	30.8	240.7	.8	6981.3	39.7	439.2	.8	10519.1	48.6	697.0
.9	647.8	13.2	22.0	.9	2061.2	22.1	102.8	.9	4185.6	30.9	242.8	.9	7021.2	39.8	442.0	.9	10567.8	48.7	700.6
7.0	661.1	13.3	22.7	15.0	2083.3	22.2	104.2	23.0	4216.7	31.1	244.9	31.0	7061.1	39.9	444.9	39.0	10616.7	48.8	704.2
.1	674.5	13.4	23.3	.1	2105.6	22.3	105.6	.1	4247.8	31.2	247.0	.1	7101.2	40.1	447.8	.1	10665.6	48.9	707.8
.2	688.0	13.5	24.0	.2	2128.0	22.4	107.0	.2	4279.1	31.3	249.2	.2	7141.3	40.2	450.7	.2	10714.7	49.1	711.4
.3	701.6	13.6	24.7	.3	2150.5	22.5	108.4	.3	4310.5	31.4	251.3	.3	7181.6	40.3	453.6	.3	10763.8	49.2	715.0
.4	715.3	13.7	25.4	.4	2173.1	22.6	109.8	.4	4342.0	31.5	253.5	.4	7222.0	40.4	456.5	.4	10813.1	49.3	718.7
.5	729.2	13.8	26.0	.5	2195.8	22.7	111.2	.5	4373.6	31.6	255.7	.5	7262.5	40.5	459.4	.5	10862.5	49.4	722.3
.6	743.1	13.9	26.7	.6	2218.7	22.8	112.7	.6	4405.3	31.7	257.9	.6	7303.1	40.6	462.3	.6	10912.0	49.5	726.0
.7	757.2	14.1	27.4	.7	2241.6	22.9	114.1	.7	4437.2	31.8	260.0	.7	7343.8	40.7	465.2	.7	10961.6	49.6	729.7
.8	771.3	14.2	28.2	.8	2264.7	23.1	115.6	.8	4469.1	31.9	262.2	.8	7384.7	40.8	468.2	.8	11011.3	49.7	733.4
.9	785.6	14.3	28.9	.9	2287.8	23.2	117.0	.9	4501.2	32.1	264.5	.9	7425.6	40.9	471.1	.9	11061.2	49.8	737.0

TABLE NO. 9.

EXCAVATION AND EMBANKMENT.

CUBIC YARDS.

PRISMOIDS 100 FEET LONG. **BREADTH OF BASE 15 FEET.**

Slopes 1 1-2 Horizontal to 1 Perpendicular.

Ht.	Cub. Yds.	Diff.	Cor.	Ht.	Cub. Yds.	Diff.	Cor.	Ht.	Cub. Yds.	Diff.	Cor.	Ht.	Cub. Yds.	Diff.	Cor.	Ht.	Cub. Yds.	Diff.	Cor.
40.0	11111.1	49.9	740.7	48.0	15466.7	58.8	1066.7	56.0	20533.3	67.7	1451.9	64.0	26311.1	76.6	1896.3	72.0	32800.0	85.5	2400.0
.1	11161.2	50.1	744.4	.1	15525.6	58.9	1071.1	.1	20601.2	67.8	1457.0	.1	26387.8	76.7	1902.2	.1	32885.6	85.6	2406.7
.2	11211.3	50.2	748.2	.2	15584.7	59.1	1075.6	.2	20669.1	67.9	1462.2	.2	26464.7	76.8	1908.2	.2	32971.3	85.7	2413.4
.3	11261.6	50.3	751.9	.3	15643.8	59.2	1080.0	.3	20737.2	68.1	1467.5	.3	26541.6	76.9	1914.1	.3	33057.2	85.8	2420.0
.4	11312.0	50.4	755.6	.4	15703.1	59.3	1084.5	.4	20805.3	68.2	1472.7	.4	26618.7	77.1	1920.1	.4	33143.1	85.9	2426.7
.5	11362.5	50.5	759.4	.5	15762.5	59.4	1089.0	.5	20873.6	68.3	1477.9	.5	26695.8	77.2	1926.0	.5	33229.2	86.1	2433.4
.6	11413.1	50.6	763.1	.6	15822.0	59.5	1093.5	.6	20942.0	68.4	1483.1	.6	26773.1	77.3	1932.0	.6	33315.3	86.2	2440.2
.7	11463.8	50.7	766.9	.7	15881.6	59.6	1098.0	.7	21010.5	68.5	1488.4	.7	26850.5	77.4	1938.0	.7	33401.6	86.3	2446.9
.8	11514.7	50.8	770.7	.8	15941.3	59.7	1102.5	.8	21079.1	68.6	1493.6	.8	26928.0	77.5	1944.0	.8	33488.0	86.4	2453.6
.9	11565.6	50.9	774.4	.9	16001.2	59.8	1107.0	.9	21147.8	68.7	1498.9	.9	27005.6	77.6	1950.0	.9	33574.5	86.5	2460.4
41.0	11616.7	51.1	778.2	49.0	16061.1	59.9	1111.6	57.0	21216.7	68.8	1504.2	65.0	27083.3	77.7	1956.0	73.0	33661.1	86.6	2467.1
.1	11667.8	51.2	782.0	.1	16121.2	60.1	1116.1	.1	21285.6	68.9	1509.4	.1	27161.2	77.8	1962.0	.1	33747.8	86.7	2473.9
.2	11719.1	51.3	785.9	.2	16181.3	60.2	1120.7	.2	21354.7	69.1	1514.7	.2	27239.1	77.9	1968.1	.2	33834.7	86.8	2480.7
.3	11770.5	51.4	789.7	.3	16241.6	60.3	1125.2	.3	21423.8	69.2	1520.0	.3	27317.2	78.1	1974.1	.3	33921.6	86.9	2487.4
.4	11822.0	51.5	793.5	.4	16302.0	60.4	1129.8	.4	21493.1	69.3	1525.4	.4	27395.3	78.2	1980.2	.4	34008.7	87.1	2494.2
.5	11873.6	51.6	797.3	.5	16362.5	60.5	1134.4	.5	21562.5	69.4	1530.7	.5	27473.6	78.3	1986.2	.5	34095.8	87.2	2501.0
.6	11925.3	51.7	801.2	.6	16423.1	60.6	1139.0	.6	21632.0	69.5	1536.0	.6	27552.0	78.4	1992.3	.6	34183.1	87.3	2507.9
.7	11977.2	51.8	805.0	.7	16483.8	60.7	1143.6	.7	21701.6	69.6	1541.3	.7	27630.5	78.5	1998.4	.7	34270.5	87.4	2514.7
.8	12029.1	51.9	808.9	.8	16544.7	60.8	1148.2	.8	21771.3	69.7	1546.7	.8	27709.1	78.6	2004.5	.8	34358.0	87.5	2521.5
.9	12081.2	52.1	812.8	.9	16605.6	60.9	1152.8	.9	21841.2	69.8	1552.0	.9	27787.8	78.7	2010.6	.9	34445.6	87.6	2528.3
42.0	12133.3	52.2	816.7	50.0	16666.7	61.1	1157.4	58.0	21911.1	69.9	1557.4	66.0	27866.7	78.8	2016.7	74.0	34533.3	87.7	2535.2
.1	12185.6	52.3	820.6	.1	16727.8	61.2	1162.0	.1	21981.2	70.1	1562.8	.1	27945.6	78.9	2022.8	.1	34621.2	87.8	2542.0
.2	12238.0	52.4	824.5	.2	16789.1	61.3	1166.7	.2	22051.3	70.2	1568.2	.2	28024.7	79.1	2028.9	.2	34709.1	87.9	2548.9
.3	12290.5	52.5	828.4	.3	16850.5	61.4	1171.3	.3	22121.6	70.3	1573.6	.3	28103.8	79.2	2035.0	.3	34797.2	88.1	2555.8
.4	12343.1	52.6	832.3	.4	16912.0	61.5	1176.0	.4	22192.0	70.4	1579.0	.4	28183.1	79.3	2041.2	.4	34885.3	88.2	2562.7
.5	12395.8	52.7	836.2	.5	16973.6	61.6	1180.7	.5	22262.5	70.5	1584.4	.5	28262.5	79.4	2047.3	.5	34973.6	88.3	2569.6
.6	12448.7	52.8	840.2	.6	17035.3	61.7	1185.4	.6	22333.1	70.6	1589.8	.6	28342.0	79.5	2053.5	.6	35062.0	88.4	2576.5
.7	12501.6	52.9	844.1	.7	17097.2	61.8	1190.0	.7	22403.8	70.7	1595.2	.7	28421.6	79.6	2059.7	.7	35150.5	88.5	2583.4
.8	12554.7	53.1	848.1	.8	17159.1	61.9	1194.7	.8	22474.7	70.8	1600.7	.8	28501.3	79.7	2065.9	.8	35239.1	88.6	2590.3
.9	12607.8	53.2	852.0	.9	17221.2	62.1	1199.5	.9	22545.6	70.9	1606.1	.9	28581.2	79.8	2072.0	.9	35327.8	88.7	2597.2
43.0	12661.1	53.3	856.0	51.0	17283.3	62.2	1204.2	59.0	22616.7	71.1	1611.6	67.0	28661.1	79.9	2078.2	75.0	35416.7	88.8	2604.2
.1	12714.5	53.4	860.0	.1	17345.6	62.3	1208.9	.1	22687.8	71.2	1617.0	.1	28741.2	80.1	2084.4	.1	35505.6	88.9	2611.1
.2	12768.0	53.5	864.0	.2	17408.0	62.4	1213.6	.2	22759.1	71.3	1622.5	.2	28821.3	80.2	2090.7	.2	35594.7	89.1	2618.1
.3	12821.6	53.6	868.0	.3	17470.5	62.5	1218.4	.3	22830.5	71.4	1628.0	.3	28901.6	80.3	2096.9	.3	35683.8	89.2	2625.0
.4	12875.3	53.7	872.0	.4	17533.1	62.6	1223.1	.4	22902.0	71.5	1633.5	.4	28982.0	80.4	2103.1	.4	35773.1	89.3	2632.0
.5	12929.2	53.8	876.0	.5	17595.8	62.7	1227.9	.5	22973.6	71.6	1639.0	.5	29062.5	80.5	2109.4	.5	35862.5	89.4	2639.0
.6	12983.1	53.9	880.1	.6	17658.7	62.8	1232.7	.6	23045.3	71.7	1644.5	.6	29143.1	80.6	2115.6	.6	35952.0	89.5	2646.0
.7	13037.2	54.1	884.1	.7	17721.6	62.9	1237.4	.7	23117.2	71.8	1650.0	.7	29223.8	80.7	2121.9	.7	36041.6	89.6	2653.0
.8	13091.3	54.2	888.2	.8	17784.7	63.1	1242.2	.8	23189.1	71.9	1655.6	.8	29304.7	80.8	2128.2	.8	36131.3	89.7	2660.0
.9	13145.6	54.3	892.2	.9	17847.8	63.2	1247.0	.9	23261.2	72.1	1661.1	.9	29385.6	80.9	2134.5	.9	36221.2	89.8	2667.0
44.0	13200.0	54.4	896.3	52.0	17911.1	63.3	1251.9	60.0	23333.3	72.2	1666.7	68.0	29466.7	81.1	2140.7	76.0	36311.1	89.9	2674.1
.1	13254.5	54.5	900.4	.1	17974.5	63.4	1256.7	.1	23405.6	72.3	1672.2	.1	29547.8	81.2	2147.0	.1	36401.2	90.1	2681.1
.2	13309.1	54.6	904.5	.2	18038.0	63.5	1261.5	.2	23478.0	72.4	1677.8	.2	29629.1	81.3	2153.4	.2	36491.3	90.2	2688.2
.3	13363.8	54.7	908.6	.3	18101.6	63.6	1266.3	.3	23550.5	72.5	1683.4	.3	29710.5	81.4	2159.7	.3	36581.6	90.3	2695.2
.4	13418.7	54.8	912.7	.4	18165.3	63.7	1271.2	.4	23623.1	72.6	1689.0	.4	29792.0	81.5	2166.0	.4	36672.0	90.4	2702.3
.5	13473.6	54.9	916.8	.5	18229.2	63.8	1276.0	.5	23695.8	72.7	1694.6	.5	29873.6	81.6	2172.3	.5	36762.5	90.5	2709.4
.6	13528.7	55.1	920.9	.6	18293.1	63.9	1280.9	.6	23768.7	72.8	1700.2	.6	29955.3	81.7	2178.7	.6	36853.1	90.6	2716.5
.7	13583.8	55.2	925.0	.7	18357.2	64.1	1285.8	.7	23841.6	72.9	1705.8	.7	30037.2	81.8	2185.0	.7	36943.8	90.7	2723.6
.8	13639.1	55.3	929.2	.8	18421.3	64.2	1290.7	.8	23914.7	73.1	1711.4	.8	30119.1	81.9	2191.4	.8	37034.7	90.8	2730.7
.9	13694.5	55.4	933.3	.9	18485.6	64.3	1295.6	.9	23987.8	73.2	1717.0	.9	30201.2	82.1	2197.8	.9	37125.6	90.9	2737.8
45.0	13750.0	55.5	937.5	53.0	18550.0	64.4	1300.5	61.0	24061.1	73.3	1722.7	69.0	30283.3	82.2	2204.2	77.0	37216.7	91.1	2744.9
.1	13805.6	55.6	941.7	.1	18614.5	64.5	1305.4	.1	24134.5	73.4	1728.3	.1	30365.6	82.3	2210.6	.1	37307.8	91.2	2752.0
.2	13861.3	55.7	945.9	.2	18679.1	64.6	1310.3	.2	24208.0	73.5	1734.0	.2	30448.0	82.4	2217.0	.2	37399.1	91.3	2759.2
.3	13917.2	55.8	950.0	.3	18743.8	64.7	1315.2	.3	24281.6	73.6	1739.7	.3	30530.5	82.5	2223.4	.3	37490.5	91.4	2766.3
.4	13973.1	55.9	954.2	.4	18808.7	64.8	1320.2	.4	24355.3	73.7	1745.4	.4	30613.1	82.6	2229.8	.4	37582.0	91.5	2773.5
.5	14029.2	56.1	958.4	.5	18873.6	64.9	1325.1	.5	24429.2	73.8	1751.0	.5	30695.8	82.7	2236.2	.5	37673.6	91.6	2780.7
.6	14085.3	56.2	962.7	.6	18938.7	65.1	1330.1	.6	24503.1	73.9	1756.7	.6	30778.7	82.8	2242.7	.6	37765.3	91.7	2787.9
.7	14141.6	56.3	966.9	.7	19003.8	65.2	1335.0	.7	24577.2	74.1	1762.5	.7	30861.6	82.9	2249.1	.7	37857.2	91.8	2795.0
.8	14198.0	56.4	971.1	.8	19069.1	65.3	1340.0	.8	24651.3	74.2	1768.2	.8	30944.7	83.1	2255.6	.8	37949.1	91.9	2802.2
.9	14254.5	56.5	975.4	.9	19134.5	65.4	1345.0	.9	24725.6	74.3	1773.9	.9	31027.8	83.2	2262.0	.9	38041.2	92.1	2809.4
46.0	14311.1	56.6	979.6	54.0	19200.0	65.5	1350.0	62.0	24800.0	74.4	1779.6	70.0	31111.1	83.3	2268.5	78.0	38133.3	92.2	2816.7
.1	14367.8	56.7	983.9	.1	19265.6	65.6	1355.0	.1	24874.5	74.5	1785.4	.1	31194.5	83.4	2275.0	.1	38225.6	92.3	2823.9
.2	14424.7	56.8	988.2	.2	19331.3	65.7	1360.0	.2	24949.1	74.6	1791.1	.2	31278.0	83.5	2281.5	.2	38318.0	92.4	2831.1
.3	14481.6	56.9	992.4	.3	19397.2	65.8	1365.0	.3	25023.8	74.7	1796.9	.3	31361.6	83.6	2288.0	.3	38410.5	92.5	2838.4
.4	14538.7	57.1	996.7	.4	19463.1	65.9	1370.1	.4	25098.7	74.8	1802.7	.4	31445.3	83.7	2294.5	.4	38503.1	92.6	2845.6
.5	14595.8	57.2	1001.0	.5	19529.2	66.1	1375.1	.5	25173.6	74.9	1808.5	.5	31529.2	83.8	2301.0	.5	38595.8	92.7	2852.9
.6	14653.1	57.3	1005.4	.6	19595.3	66.2	1380.2	.6	25248.7	75.1	1814.2	.6	31613.1	83.9	2307.6	.6	38688.7	92.8	2860.2
.7	14710.5	57.4	1009.7	.7	19661.6	66.3	1385.2	.7	25323.8	75.2	1820.0	.7	31697.2	84.1	2314.1	.7	38781.6	92.9	2867.4
.8	14768.0	57.5	1014.0	.8	19728.0	66.4	1390.3	.8	25399.1	75.3	1825.9	.8	31781.3	84.2	2320.7	.8	38874.7	93.1	2874.7
.9	14825.6	57.6	1018.3	.9	19794.5	66.5	1395.4	.9	25474.5	75.4	1831.7	.9	31865.6	84.3	2327.2	.9	38967.8	93.2	2882.0
47.0	14883.3	57.7	1022.7	55.0	19861.1	66.6	1400.5	63.0	25550.0	75.5	1837.5	71.0	31950.0	84.4	2333.8	79.0	39061.1	93.3	2889.4
.1	14941.2	57.8	1027.0	.1	19927.8	66.7	1405.6	.1	25625.6	75.6	1843.3	.1	32034.5	84.5	2340.4	.1	39154.5	93.4	2896.7
.2	14999.1	57.9	1031.4	.2	19994.7	66.8	1410.7	.2	25701.3	75.7	1849.2	.2	32119.1	84.6	2347.0	.2	39248.0	93.5	2904.0
.3	15057.2	58.1	1035.8	.3	20061.6	66.9	1415.8	.3	25777.2	75.8	1855.0	.3	32203.8	84.7	2353.6	.3	39341.6	93.6	2911.3
.4	15115.3	58.2	1040.2	.4	20128.7	67.1	1420.9	.4	25853.1	75.9	1860.9	.4	32288.7	84.8	2360.2	.4	39435.3	93.7	2918.7
.5	15173.6	58.3	1044.6	.5	20195.8	67.2	1426.0	.5	25929.2	76.1	1866.8	.5	32373.6	84.9	2366.8	.5	39529.2	93.8	2926.0
.6	15232.0	58.4	1049.0	.6	20263.1	67.3	1431.2	.6	26005.3	76.2	1872.7	.6	32458.7	85.1	2373.4	.6	39623.1	93.9	2933.4
.7	15290.5	58.5	1053.4	.7	20330.5	67.4	1436.3	.7	26081.6	76.3	1878.6	.7	32543.8	85.2	2380.0	.7	39717.2	94.1	2940.8
.8	15349.1	58.6	1057.8	.8	20398.0	67.5	1441.5	.8	26158.0	76.4	1884.5	.8	32629.1	85.3	2386.7	.8	39811.3	94.2	2948.2
.9	15407.8	58.7	1062.2	.9	20465.6	67.6	1446.7	.9	26234.5	76.5	1890.4	.9	32714.5	85.4	2393.3	.9	39905.6	94.3	2955.6

TABLE No. 10.

EXCAVATION AND EMBANKMENT.

CUBIC YARDS.

Prismoids, 100 Feet Long. Breadth of Base, 16 Feet. Slopes, 1 1-2 Horizontal to 1 Perpendicular.

Height.	Cub. Yds.	Diff.	Cor.	Height.	Cub. Yds.	Diff.	Cor.	Height.	Cub. Yds.	Diff.	Cor.	Height.	Cub. Yds.	Diff.	Cor.	Height.	Cub. Yds.	Diff.	Cor.
0.0	0.0	0.0	0.0	8.0	829.6	14.8	29.6	16.0	2370.4	23.6	118.5	24.0	4622.2	32.5	266.7	32.0	7585.2	41.4	474.1
1	6.0	6.0	0.0	1	844.5	14.9	30.4	1	2394.1	23.8	120.0	1	4654.9	32.6	268.9	1	7626.7	41.5	477.0
2	12.1	6.1	0.0	2	859.5	15.0	31.1	2	2418.0	23.9	121.5	2	4687.6	32.8	271.1	2	7668.4	41.6	480.0
3	18.3	6.2	0.0	3	874.6	15.1	31.9	3	2442.0	24.0	123.0	3	4720.5	32.9	273.4	3	7710.1	41.8	483.0
4	24.6	6.3	0.1	4	889.8	15.2	32.7	4	2466.1	24.1	124.5	4	4753.5	33.0	275.6	4	7752.0	41.9	486.0
5	31.0	6.4	0.1	5	905.1	15.3	33.4	5	2490.3	24.2	126.0	5	4786.6	33.1	277.9	5	7791.0	42.0	489.0
6	37.6	6.5	0.2	6	920.5	15.4	34.2	6	2514.6	24.3	127.6	6	4819.8	33.2	280.2	6	7836.1	42.1	491.0
7	44.2	6.6	0.2	7	936.1	15.5	35.0	7	2539.0	24.4	129.1	7	4853.1	33.3	282.5	7	7878.3	42.2	495.0
8	51.0	6.8	0.3	8	951.7	15.6	35.9	8	2563.6	24.5	130.7	8	4886.5	33.4	284.7	8	7920.6	42.3	498.0
9	57.8	6.9	0.4	9	967.5	15.8	36.7	9	2588.2	24.6	132.2	9	4920.1	33.5	287.0	9	7963.0	42.4	501.1
1.0	64.8	7.0	0.5	9.0	983.3	15.9	37.5	17.0	2613.0	24.8	133.8	25.0	4953.7	33.6	289.4	33.0	8005.6	42.5	504.2
1	71.9	7.1	0.6	1	999.3	16.0	38.3	1	2637.8	24.9	135.4	1	4987.5	33.8	291.7	1	8048.2	42.6	507.2
2	79.1	7.2	0.7	2	1015.4	16.1	39.2	2	2662.8	25.0	137.0	2	5021.3	33.9	294.0	2	8091.0	42.8	510.3
3	86.4	7.3	0.8	3	1031.6	16.2	40.0	3	2687.9	25.1	138.6	3	5055.3	34.0	296.3	3	8133.8	42.9	513.4
4	93.9	7.4	0.9	4	1047.9	16.3	40.9	4	2713.1	25.2	140.2	4	5089.4	34.1	298.7	4	8176.8	43.0	516.5
5	101.4	7.5	1.0	5	1064.4	16.4	41.8	5	2738.4	25.3	141.8	5	5123.6	34.2	301.0	5	8219.9	43.1	519.6
6	109.0	7.6	1.2	6	1080.9	16.5	42.7	6	2763.9	25.4	143.4	6	5157.9	34.3	303.4	6	8263.1	43.2	522.7
7	116.8	7.8	1.3	7	1097.6	16.6	43.6	7	2789.4	25.5	145.0	7	5192.4	34.4	305.8	7	8306.4	43.3	525.8
8	124.7	7.9	1.5	8	1114.3	16.8	44.5	8	2815.0	25.6	146.7	8	5226.9	34.5	308.2	8	8349.9	43.4	528.9
9	132.6	8.0	1.7	9	1131.2	16.9	45.4	9	2840.8	25.8	148.3	9	5261.5	34.6	310.6	9	8393.4	43.5	532.0
2.0	140.7	8.1	1.9	10.0	1148.1	17.0	46.3	18.0	2866.7	25.9	150.0	26.0	5296.3	34.8	313.0	34.0	8437.0	43.6	535.2
1	148.9	8.2	2.0	1	1165.2	17.1	47.2	1	2892.6	26.0	151.7	1	5331.2	34.9	315.4	1	8480.8	43.8	538.3
2	157.3	8.3	2.2	2	1182.4	17.2	48.2	2	2918.7	26.1	153.4	2	5366.1	35.0	317.8	2	8524.7	43.9	541.5
3	165.7	8.4	2.4	3	1199.8	17.3	49.1	3	2944.9	26.2	155.0	3	5401.2	35.1	320.2	3	8568.6	44.0	544.7
4	174.2	8.5	2.7	4	1217.2	17.4	50.1	4	2971.3	26.3	156.7	4	5436.4	35.2	322.7	4	8612.7	44.1	547.9
5	182.9	8.6	2.9	5	1234.7	17.5	51.0	5	2997.7	26.4	158.4	5	5471.8	35.3	325.1	5	8656.9	44.2	551.0
6	191.6	8.8	3.1	6	1252.4	17.6	52.0	6	3024.2	26.5	160.2	6	5507.2	35.4	327.6	6	8701.3	44.3	554.2
7	200.5	8.9	3.4	7	1270.1	17.8	53.0	7	3050.9	26.6	161.9	7	5542.7	35.5	330.0	7	8745.7	44.4	557.5
8	209.5	9.0	3.6	8	1288.0	17.9	54.0	8	3077.6	26.8	163.6	8	5578.4	35.6	332.5	8	8790.2	44.5	560.7
9	218.6	9.1	3.9	9	1306.0	18.0	55.0	9	3104.5	26.9	165.4	9	5614.1	35.8	335.0	9	8834.9	44.6	563.9
3.0	227.8	9.2	4.2	11.0	1324.1	18.1	56.0	19.0	3131.5	27.0	167.1	27.0	5650.0	35.9	337.5	35.0	8879.6	44.8	567.1
1	237.1	9.3	4.4	1	1342.3	18.2	57.0	1	3158.6	27.1	168.9	1	5686.0	36.0	340.0	1	8924.5	44.9	570.4
2	246.5	9.4	4.7	2	1360.6	18.3	58.1	2	3185.8	27.2	170.7	2	5722.1	36.1	342.5	2	8969.5	45.0	573.6
3	256.1	9.5	5.0	3	1379.0	18.4	59.1	3	3213.1	27.3	172.4	3	5758.3	36.2	345.0	3	9014.6	45.1	576.9
4	265.7	9.6	5.4	4	1397.6	18.5	60.2	4	3240.5	27.4	174.2	4	5794.6	36.3	347.6	4	9059.8	45.2	580.2
5	275.5	9.8	5.7	5	1416.2	18.6	61.2	5	3268.1	27.5	176.0	5	5831.0	36.4	350.1	5	9105.1	45.3	583.4
6	285.3	9.9	6.0	6	1435.0	18.8	62.3	6	3295.7	27.6	177.9	6	5867.6	36.5	352.7	6	9150.5	45.4	586.7
7	295.3	10.0	6.3	7	1453.8	18.9	63.4	7	3323.5	27.8	179.7	7	5904.2	36.6	355.2	7	9196.1	45.5	590.0
8	305.4	10.1	6.7	8	1472.8	19.0	64.5	8	3351.3	27.9	181.5	8	5941.0	36.8	357.8	8	9241.7	45.6	593.4
9	315.6	10.2	7.0	9	1491.9	19.1	65.6	9	3379.3	28.0	183.3	9	5977.8	36.9	360.4	9	9287.5	45.8	596.7
4.0	325.9	10.3	7.4	12.0	1511.1	19.2	66.7	20.0	3407.4	28.1	185.2	28.0	6014.8	37.0	363.0	36.0	9333.3	45.9	600.0
1	336.4	10.4	7.8	1	1530.4	19.3	67.8	1	3435.6	28.2	187.0	1	6051.9	37.1	365.6	1	9379.3	46.0	603.3
2	346.9	10.5	8.1	2	1549.9	19.4	68.9	2	3463.9	28.3	188.9	2	6089.1	37.2	368.2	2	9425.4	46.1	606.7
3	357.5	10.6	8.6	3	1569.4	19.5	70.0	3	3492.4	28.4	190.8	3	6126.4	37.3	370.8	3	9471.6	46.2	610.0
4	368.3	10.8	9.0	4	1589.0	19.6	71.2	4	3520.9	28.5	192.7	4	6163.9	37.4	373.4	4	9517.9	46.3	613.4
5	379.2	10.9	9.4	5	1608.8	19.8	72.3	5	3549.5	28.6	194.6	5	6201.4	37.5	376.0	5	9564.4	46.4	616.8
6	390.1	11.0	9.8	6	1628.7	19.9	73.5	6	3578.3	28.8	196.5	6	6239.0	37.6	378.7	6	9610.9	46.5	620.2
7	401.2	11.1	10.3	7	1648.6	20.0	74.7	7	3607.2	28.9	198.4	7	6276.8	37.8	381.3	7	9657.5	46.6	623.6
8	412.4	11.2	10.7	8	1668.7	20.1	75.9	8	3636.1	29.0	200.3	8	6314.7	37.9	384.0	8	9704.3	46.8	627.0
9	423.8	11.3	11.1	9	1688.9	20.2	77.0	9	3665.2	29.1	202.2	9	6352.6	38.0	386.7	9	9751.2	46.9	630.4
5.0	435.2	11.4	11.6	13.0	1709.3	20.3	78.2	21.0	3694.4	29.2	204.2	29.0	6390.7	38.1	389.4	37.0	9798.1	47.0	633.8
1	446.7	11.5	12.0	1	1729.7	20.4	79.4	1	3723.8	29.3	206.1	1	6428.9	38.2	392.0	1	9845.2	47.1	637.2
2	458.4	11.6	12.5	2	1750.2	20.5	80.7	2	3753.2	29.4	208.1	2	6467.3	38.3	394.7	2	9892.4	47.2	640.7
3	470.1	11.8	13.0	3	1770.9	20.6	81.9	3	3782.7	29.5	210.0	3	6505.7	38.4	397.4	3	9939.8	47.3	644.1
4	482.0	11.9	13.5	4	1791.6	20.8	83.1	4	3812.4	29.6	212.0	4	6544.2	38.5	400.2	4	9987.2	47.4	647.6
5	494.0	12.0	14.0	5	1812.5	20.9	84.4	5	3842.1	29.8	214.0	5	6582.9	38.6	402.9	5	10034.7	47.5	651.0
6	506.1	12.1	14.5	6	1833.5	21.0	85.6	6	3872.0	29.9	216.0	6	6621.6	38.8	405.6	6	10082.4	47.6	654.5
7	518.3	12.2	15.0	7	1854.6	21.1	86.9	7	3902.0	30.0	218.0	7	6660.5	38.9	408.4	7	10130.1	47.8	658.0
8	530.6	12.3	15.6	8	1875.8	21.2	88.2	8	3932.1	30.1	220.0	8	6699.5	39.0	411.1	8	10178.0	47.9	661.5
9	543.0	12.4	16.1	9	1897.1	21.3	89.4	9	3962.3	30.2	222.0	9	6738.6	39.1	413.9	9	10226.0	48.0	665.0
6.0	555.6	12.5	16.7	14.0	1918.5	21.4	90.7	22.0	3992.6	30.3	224.1	30.0	6777.8	39.2	416.7	38.0	10274.1	48.1	668.5
1	568.2	12.6	17.2	1	1940.1	21.5	92.0	1	4023.0	30.4	226.1	1	6817.1	39.3	419.4	1	10322.3	48.2	672.0
2	581.0	12.8	17.8	2	1961.7	21.6	93.4	2	4053.6	30.5	228.2	2	6856.5	39.4	422.2	2	10370.6	48.3	675.6
3	593.8	12.9	18.4	3	1983.5	21.8	94.7	3	4084.2	30.6	230.2	3	6896.1	39.5	425.0	3	10419.0	48.4	679.1
4	606.8	13.0	19.0	4	2005.3	21.9	96.0	4	4115.0	30.8	232.3	4	6935.7	39.6	427.9	4	10467.6	48.5	682.7
5	619.9	13.1	19.6	5	2027.3	22.0	97.3	5	4145.8	30.9	234.4	5	6975.5	39.8	430.7	5	10516.2	48.6	686.2
6	633.1	13.2	20.2	6	2049.4	22.1	98.7	6	4176.8	31.0	236.5	6	7015.3	39.9	433.5	6	10565.0	48.8	689.8
7	646.4	13.3	20.8	7	2071.6	22.2	100.0	7	4207.9	31.1	238.6	7	7055.3	40.0	436.3	7	10613.8	48.9	693.4
8	659.9	13.4	21.4	8	2093.9	22.3	101.4	8	4239.1	31.2	240.7	8	7095.4	40.1	439.2	8	10662.8	49.0	697.0
9	673.4	13.5	22.0	9	2116.4	22.4	102.8	9	4270.4	31.3	242.8	9	7135.6	40.2	442.0	9	10711.9	49.1	700.6
7.0	687.0	13.6	22.7	15.0	2138.9	22.5	104.2	23.0	4301.9	31.4	244.9	31.0	7175.9	40.3	444.9	39.0	10761.1	49.2	704.2
1	700.8	13.8	23.3	1	2161.5	22.6	105.6	1	4333.4	31.5	247.0	1	7216.4	40.4	447.8	1	10810.4	49.3	707.8
2	714.7	13.9	24.0	2	2184.3	22.8	107.0	2	4365.0	31.6	249.2	2	7256.9	40.5	450.7	2	10859.9	49.4	711.4
3	728.6	14.0	24.7	3	2207.2	22.9	108.4	3	4396.8	31.8	251.3	3	7297.5	40.6	453.6	3	10909.4	49.5	715.0
4	742.7	14.1	25.4	4	2230.1	23.0	109.8	4	4428.7	31.9	253.5	4	7338.3	40.8	456.5	4	10959.0	49.6	718.7
5	756.9	14.2	26.0	5	2253.2	23.1	111.2	5	4460.6	32.0	255.7	5	7379.2	40.9	459.4	5	11008.8	49.8	722.3
6	771.3	14.3	26.7	6	2276.4	23.2	112.7	6	4492.7	32.1	257.9	6	7420.1	41.0	462.3	6	11058.7	49.9	726.0
7	785.7	14.4	27.4	7	2299.8	23.3	114.1	7	4524.9	32.2	260.0	7	7461.2	41.1	465.2	7	11108.6	50.0	729.7
8	800.2	14.5	28.2	8	2323.2	23.4	115.6	8	4557.3	32.3	262.2	8	7502.4	41.2	468.2	8	11158.7	50.1	733.4
9	814.9	14.6	28.9	9	2346.7	23.5	117.0	9	4589.7	32.4	264.5	9	7543.8	41.3	471.1	9	11208.9	50.2	737.0

TABLE No. 11.

EXCAVATION AND EMBANKMENT.

CUBIC YARDS.

Prismoids, 100 Feet Long. Breadth of Base, 18 Feet. Slopes, 1 1-2 Horizontal to 1 Perpendicular.

Height.	Cub. Yds.	Diff.	Cor.	Height.	Cub. Yds.	Diff.	Cor.	Height.	Cub. Yds.	Diff.	Cor.	Height.	Cub. Yds.	Diff.	Cor.	Height.	Cub. Yds.	Diff.	Cor.
0·0	0·0	0·0	0·0	8·0	888·9	15·5	29·6	16·0	2488·9	24·4	118·5	24·0	4800·0	33·3	266·7	32·0	7822·2	42·2	474·1
1	6·7	6·7	0·0	1	904·5	15·6	30·4	1	2513·4	24·5	120·0	1	4833·4	33·4	268·9	1	7864·5	42·3	477·0
2	13·6	6·8	0·0	2	920·2	15·7	31·1	2	2538·0	24·6	121·5	2	4866·9	33·5	271·1	2	7906·9	42·4	480·0
3	20·5	6·9	0·0	3	936·1	15·8	31·9	3	2562·7	24·7	123·0	3	4900·5	33·6	273·4	3	7949·4	42·5	483·0
4	27·6	7·1	0·1	4	952·0	15·9	32·7	4	2587·6	24·8	124·5	4	4934·2	33·7	275·6	4	7992·0	42·6	486·0
5	34·7	7·2	0·1	5	968·1	16·1	33·4	5	2612·5	24·9	126·0	5	4968·1	33·8	277·9	5	8034·7	42·7	489·0
6	42·0	7·3	0·2	6	984·2	16·2	34·2	6	2637·6	25·1	127·6	6	5002·0	33·9	280·2	6	8077·6	42·8	492·0
7	49·4	7·4	0·2	7	1000·5	16·3	35·0	7	2662·7	25·2	129·1	7	5036·1	34·1	282·5	7	8120·5	42·9	495·0
8	56·9	7·5	0·3	8	1016·9	16·4	35·9	8	2688·0	25·3	130·7	8	5070·2	34·2	284·7	8	8163·6	43·1	498·0
9	64·5	7·6	0·4	9	1033·4	16·5	36·7	9	2713·4	25·4	132·2	9	5104·5	34·3	287·0	9	8206·7	43·2	501·1
1·0	72·2	7·7	0·5	9·0	1050·0	16·6	37·5	17·0	2738·9	25·5	133·8	25·0	5138·9	34·4	289·4	33·0	8250·0	43·3	504·2
1	80·1	7·8	0·6	1	1066·7	16·7	38·3	1	2764·5	25·6	135·4	1	5173·4	34·5	291·7	1	8293·4	43·4	507·2
2	88·0	7·9	0·7	2	1083·6	16·8	39·2	2	2790·2	25·7	137·0	2	5208·0	34·6	294·0	2	8336·9	43·5	510·3
3	96·1	8·1	0·8	3	1100·5	16·9	40·0	3	2816·1	25·8	138·6	3	5242·7	34·7	296·3	3	8380·5	43·6	513·4
4	104·2	8·2	0·9	4	1117·6	17·1	40·9	4	2842·0	25·9	140·2	4	5277·6	34·8	298·7	4	8424·2	43·7	516·5
5	112·5	8·3	1·0	5	1134·7	17·2	41·8	5	2868·1	26·1	141·8	5	5312·5	34·9	301·0	5	8468·1	43·8	519·6
6	120·9	8·4	1·2	6	1152·0	17·3	42·7	6	2894·2	26·2	143·4	6	5347·6	35·1	303·4	6	8512·0	43·9	522·7
7	129·4	8·5	1·3	7	1169·4	17·4	43·6	7	2920·5	26·3	145·0	7	5382·7	35·2	305·8	7	8556·1	44·1	525·8
8	138·0	8·6	1·5	8	1186·9	17·5	44·5	8	2946·9	26·4	146·7	8	5418·0	35·3	308·2	8	8600·2	44·2	528·9
9	146·7	8·7	1·7	9	1204·5	17·6	45·4	9	2973·4	26·5	148·3	9	5453·4	35·4	310·6	9	8644·5	44·3	532·0
2·0	155·6	8·8	1·9	10·0	1222·2	17·7	46·3	18·0	3000·0	26·6	150·0	26·0	5488·9	35·5	313·0	34·0	8688·9	44·4	535·2
1	164·5	8·9	2·0	1	1240·1	17·8	47·2	1	3026·7	26·7	151·7	1	5524·5	35·6	315·4	1	8733·4	44·5	538·3
2	173·6	9·1	2·2	2	1258·0	17·9	48·2	2	3053·6	26·8	153·4	2	5560·2	35·7	317·8	2	8778·0	44·6	541·5
3	182·7	9·2	2·4	3	1276·1	18·1	49·1	3	3080·5	26·9	155·0	3	5596·1	35·8	320·2	3	8822·7	44·7	544·7
4	192·0	9·3	2·7	4	1294·2	18·2	50·1	4	3107·6	27·1	156·7	4	5632·0	35·9	322·7	4	8867·6	44·8	547·9
5	201·4	9·4	2·9	5	1312·5	18·3	51·0	5	3134·7	27·2	158·4	5	5668·1	36·1	325·1	5	8912·5	44·9	551·0
6	210·9	9·5	3·1	6	1330·9	18·4	52·0	6	3162·0	27·3	160·2	6	5704·2	36·2	327·6	6	8957·6	45·1	554·2
7	220·5	9·6	3·4	7	1349·4	18·5	53·0	7	3189·4	27·4	161·9	7	5740·5	36·3	330·0	7	9002·7	45·2	557·5
8	230·2	9·7	3·6	8	1368·0	18·6	54·0	8	3216·9	27·5	163·6	8	5776·9	36·4	332·5	8	9048·0	45·3	560·7
9	240·1	9·8	3·9	9	1386·7	18·7	55·0	9	3244·5	27·6	165·4	9	5813·4	36·5	335·0	9	9093·4	45·4	563·9
3·0	250·0	9·9	4·2	11·0	1405·6	18·8	56·0	19·0	3272·2	27·7	167·1	27·0	5850·0	36·6	337·5	35·0	9138·9	45·5	567·1
1	260·1	10·1	4·4	1	1424·5	18·9	57·0	1	3300·1	27·8	168·9	1	5886·7	36·7	340·0	1	9184·5	45·6	570·4
2	270·2	10·2	4·7	2	1443·6	19·1	58·1	2	3328·0	27·9	170·7	2	5923·6	36·8	342·5	2	9230·2	45·7	573·6
3	280·5	10·3	5·0	3	1462·7	19·2	59·1	3	3356·1	28·1	172·4	3	5960·5	36·9	345·0	3	9276·1	45·8	576·9
4	290·9	10·4	5·4	4	1482·0	19·3	60·2	4	3384·2	28·2	174·2	4	5997·6	37·1	347·6	4	9322·0	45·9	580·2
5	301·4	10·5	5·7	5	1501·4	19·4	61·2	5	3412·5	28·3	176·0	5	6034·7	37·2	350·1	5	9368·1	46·1	583·4
6	312·0	10·6	6·0	6	1520·9	19·5	62·3	6	3440·9	28·4	177·9	6	6072·0	37·3	352·7	6	9414·2	46·2	586·7
7	322·7	10·7	6·3	7	1540·5	19·6	63·4	7	3469·4	28·5	179·7	7	6109·4	37·4	355·2	7	9460·5	46·3	590·0
8	333·6	10·8	6·7	8	1560·2	19·7	64·5	8	3498·0	28·6	181·5	8	6146·9	37·5	357·8	8	9506·9	46·4	593·4
9	344·5	10·9	7·0	9	1580·1	19·8	65·6	9	3526·7	28·7	183·3	9	6184·5	37·6	360·4	9	9553·4	46·5	596·7
4·0	355·6	11·1	7·4	12·0	1600·0	19·9	66·7	20·0	3555·6	28·8	185·2	28·0	6222·2	37·7	363·0	36·0	9600·0	46·6	600·0
1	366·7	11·2	7·8	1	1620·1	20·1	67·8	1	3584·5	28·9	187·0	1	6260·1	37·8	365·6	1	9646·7	46·7	603·3
2	378·0	11·3	8·1	2	1640·2	20·2	68·9	2	3613·6	29·1	188·9	2	6298·0	37·9	368·2	2	9693·6	46·8	606·7
3	389·4	11·4	8·6	3	1660·5	20·3	70·0	3	3642·7	29·2	190·8	3	6336·1	38·1	370·8	3	9740·5	46·9	610·0
4	400·9	11·5	9·0	4	1680·9	20·4	71·2	4	3672·0	29·3	192·7	4	6374·2	38·2	373·4	4	9787·6	47·1	613·4
5	412·5	11·6	9·4	5	1701·4	20·5	72·3	5	3701·4	29·4	194·6	5	6412·5	38·3	376·0	5	9834·7	47·2	616·8
6	424·2	11·7	9·8	6	1722·0	20·6	73·5	6	3730·9	29·5	196·5	6	6450·9	38·4	378·7	6	9882·0	47·3	620·2
7	436·1	11·8	10·2	7	1742·7	20·7	74·7	7	3760·5	29·6	198·4	7	6489·4	38·5	381·3	7	9929·4	47·4	623·6
8	448·0	11·9	10·7	8	1763·6	20·8	75·9	8	3790·2	29·7	200·3	8	6528·0	38·6	384·0	8	9976·9	47·5	627·0
9	460·1	12·1	11·1	9	1784·5	20·9	77·0	9	3820·1	29·8	202·2	9	6566·7	38·7	386·7	9	10024·5	47·6	630·4
5·0	472·2	12·2	11·6	13·0	1805·6	21·1	78·2	21·0	3850·0	29·9	204·2	29·0	6605·6	38·8	389·4	37·0	10072·2	47·7	633·8
1	484·5	12·3	12·0	1	1826·7	21·2	79·4	1	3880·1	30·1	206·1	1	6644·5	38·9	392·0	1	10120·1	47·8	637·2
2	496·9	12·4	12·5	2	1848·0	21·3	80·7	2	3910·2	30·2	208·1	2	6683·6	39·1	394·7	2	10168·0	47·9	640·7
3	509·4	12·5	13·0	3	1869·4	21·4	81·9	3	3940·5	30·3	210·0	3	6722·7	39·2	397·4	3	10216·1	48·1	644·1
4	522·0	12·6	13·5	4	1890·9	21·5	83·1	4	3970·9	30·4	212·0	4	6762·0	39·3	400·2	4	10264·2	48·2	647·6
5	534·7	12·7	14·0	5	1912·5	21·6	84·4	5	4001·4	30·5	214·0	5	6801·4	39·4	402·9	5	10312·5	48·3	651·0
6	547·6	12·8	14·5	6	1934·2	21·7	85·6	6	4032·0	30·6	216·0	6	6840·9	39·5	405·6	6	10360·9	48·4	654·5
7	560·5	12·9	15·0	7	1956·1	21·8	86·9	7	4062·7	30·7	218·0	7	6880·5	39·6	408·4	7	10409·4	48·5	658·0
8	573·6	13·1	15·6	8	1978·0	21·9	88·2	8	4093·6	30·8	220·0	8	6920·2	39·7	411·1	8	10458·0	48·6	661·5
9	586·7	13·2	16·1	9	2000·1	22·1	89·4	9	4124·5	30·9	222·0	9	6960·1	39·8	413·9	9	10506·7	48·7	665·0
6·0	600·0	13·3	16·7	14·0	2022·2	22·2	90·7	22·0	4155·6	31·1	224·1	30·0	7000·0	39·9	416·7	38·0	10555·6	48·8	668·5
1	613·4	13·4	17·2	1	2044·5	22·3	92·0	1	4186·7	31·2	226·1	1	7040·1	40·1	419·4	1	10604·5	48·9	672·0
2	626·9	13·5	17·8	2	2066·9	22·4	93·4	2	4218·0	31·3	228·2	2	7080·2	40·2	422·2	2	10653·6	49·1	675·6
3	640·5	13·6	18·4	3	2089·4	22·5	94·7	3	4249·4	31·4	230·2	3	7120·5	40·3	425·0	3	10702·7	49·2	679·1
4	654·2	13·7	19·0	4	2112·0	22·6	96·0	4	4280·9	31·5	232·3	4	7160·9	40·4	427·9	4	10752·0	49·3	682·7
5	668·1	13·8	19·6	5	2134·7	22·7	97·3	5	4312·5	31·6	234·4	5	7201·4	40·5	430·7	5	10801·4	49·4	686·2
6	682·0	13·9	20·2	6	2157·6	22·8	98·7	6	4344·2	31·7	236·5	6	7242·0	40·6	433·5	6	10850·9	49·5	689·8
7	696·1	14·1	20·8	7	2180·5	22·9	100·0	7	4376·1	31·8	238·6	7	7282·7	40·7	436·3	7	10900·5	49·6	693·4
8	710·2	14·2	21·4	8	2203·6	23·1	101·4	8	4408·0	31·9	240·7	8	7323·6	40·8	439·2	8	10950·2	49·7	697·0
9	724·5	14·3	22·0	9	2226·7	23·2	102·8	9	4440·1	32·1	242·8	9	7364·5	40·9	442·0	9	11000·1	49·8	700·6
7·0	738·9	14·4	22·7	15·0	2250·0	23·3	104·2	23·0	4472·2	32·2	244·9	31·0	7405·6	41·1	444·9	39·0	11050·0	49·9	704·2
1	753·4	14·5	23·3	1	2273·4	23·4	105·6	1	4504·5	32·3	247·0	1	7446·7	41·2	447·8	1	11100·1	50·1	707·8
2	768·0	14·6	24·0	2	2296·9	23·5	107·0	2	4536·9	32·4	249·2	2	7488·0	41·3	450·7	2	11150·2	50·2	711·4
3	782·7	14·7	24·7	3	2320·5	23·6	108·4	3	4569·4	32·5	251·3	3	7529·4	41·4	453·6	3	11200·5	50·3	715·0
4	797·6	14·8	25·4	4	2344·2	23·7	109·8	4	4602·0	32·6	253·5	4	7570·9	41·5	456·5	4	11250·9	50·4	718·7
5	812·5	14·9	26·0	5	2368·1	23·8	111·2	5	4634·7	32·7	255·7	5	7612·5	41·6	459·4	5	11301·4	50·5	722·3
6	827·6	15·1	26·7	6	2392·0	23·9	112·7	6	4667·6	32·8	257·9	6	7654·2	41·7	462·3	6	11352·0	50·6	726·0
7	842·7	15·2	27·4	7	2416·1	24·1	114·1	7	4700·5	32·9	260·0	7	7696·1	41·8	465·2	7	11402·7	50·7	729·7
8	858·0	15·3	28·2	8	2440·2	24·2	115·6	8	4733·6	33·1	262·2	8	7738·0	41·9	468·2	8	11453·6	50·8	733·4
9	873·4	15·4	28·9	9	2464·5	24·3	117·0	9	4766·7	33·2	264·5	9	7780·1	42·1	471·1	9	11504·5	50·9	737·0

TABLE NO. 12.

EXCAVATION AND EMBANKMENT.

CUBIC YARDS.

PRISMOIDS 100 FEET LONG. **BREADTH OF BASE 20 FEET.**

Slopes 1 1-2 Horizontal to 1 Perpendicular.

Ht.	Cub. Yds.	Diff.	Cor.	Ht.	Cub. Yds.	Diff.	Cor.	Ht.	Cub. Yds.	Diff.	Cor.	Ht.	Cub. Yds.	Diff.	Cor.	Ht.	Cub. Yds.	Diff.	Cor.
0.0	0.0	0.0	0.0	8.0	948.1	16.2	29.6	16.0	2607.4	25.1	118.5	24.0	4977.8	34.0	266.7	32.0	8059.3	42.9	474.1
.1	7.5	7.5	0.0	.1	964.5	16.4	30.4	.1	2632.6	25.2	120.0	.1	5011.9	34.1	268.9	.1	8102.3	43.0	477.0
.2	15.0	7.6	0.0	.2	981.0	16.5	31.1	.2	2658.0	25.4	121.5	.2	5046.1	34.2	271.1	.2	8145.4	43.1	480.0
.3	22.7	7.7	0.0	.3	997.5	16.6	31.9	.3	2683.5	25.5	123.0	.3	5080.5	34.4	273.4	.3	8188.6	43.2	483.0
.4	30.5	7.8	0.1	.4	1014.2	16.7	32.7	.4	2709.0	25.6	124.5	.4	5115.0	34.5	275.6	.4	8232.0	43.4	486.0
.5	38.4	7.9	0.1	.5	1031.0	16.8	33.4	.5	2734.7	25.7	126.0	.5	5149.5	34.6	277.9	.5	8275.5	43.5	489.0
.6	46.4	8.0	0.2	.6	1047.9	16.9	34.2	.6	2760.5	25.8	127.6	.6	5184.2	34.7	280.2	.6	8319.0	43.6	492.0
.7	54.6	8.1	0.2	.7	1064.9	17.0	35.0	.7	2786.4	25.9	129.1	.7	5219.0	34.8	282.5	.7	8362.7	43.7	495.0
.8	62.8	8.2	0.3	.8	1082.1	17.1	35.9	.8	2812.4	26.0	130.7	.8	5253.9	34.9	284.7	.8	8406.5	43.8	498.0
.9	71.2	8.4	0.4	.9	1099.3	17.2	36.7	.9	2838.6	26.1	132.2	.9	5288.9	35.0	287.0	.9	8450.4	43.9	501.1
1.0	79.6	8.5	0.5	9.0	1116.7	17.4	37.5	17.0	2864.8	26.2	133.8	25.0	5324.1	35.1	289.4	33.0	8494.4	44.0	504.2
.1	88.2	8.6	0.6	.1	1134.1	17.5	38.3	.1	2891.2	26.4	135.4	.1	5359.3	35.2	291.7	.1	8538.6	44.1	507.2
.2	96.9	8.7	0.7	.2	1151.7	17.6	39.2	.2	2917.6	26.5	137.0	.2	5394.7	35.4	294.0	.2	8582.8	44.2	510.3
.3	105.7	8.8	0.8	.3	1169.4	17.7	40.0	.3	2944.2	26.6	138.6	.3	5430.1	35.5	296.3	.3	8627.2	44.4	513.4
.4	114.6	8.9	0.9	.4	1187.2	17.8	40.9	.4	2970.9	26.7	140.2	.4	5465.7	35.6	298.7	.4	8671.6	44.5	516.5
.5	123.6	9.0	1.0	.5	1205.1	17.9	41.8	.5	2997.7	26.8	141.8	.5	5501.4	35.7	301.0	.5	8716.2	44.6	519.6
.6	132.7	9.1	1.2	.6	1223.1	18.0	42.7	.6	3024.6	26.9	143.4	.6	5537.2	35.8	303.4	.6	8760.9	44.7	522.7
.7	142.0	9.2	1.3	.7	1241.2	18.1	43.6	.7	3051.6	27.0	145.0	.7	5573.1	35.9	305.8	.7	8805.7	44.8	525.8
.8	151.3	9.4	1.5	.8	1259.5	18.2	44.5	.8	3078.7	27.1	146.7	.8	5609.1	36.0	308.2	.8	8850.6	44.9	528.9
.9	160.8	9.5	1.7	.9	1277.8	18.4	45.4	.9	3106.0	27.2	148.3	.9	5645.2	36.1	310.6	.9	8895.6	45.0	532.0
2.0	170.4	9.6	1.9	10.0	1296.3	18.5	46.3	18.0	3133.3	27.4	150.0	26.0	5681.5	36.2	313.0	34.0	8940.7	45.1	535.2
.1	180.1	9.7	2.0	.1	1314.9	18.6	47.2	.1	3160.8	27.5	151.7	.1	5717.8	36.4	315.4	.1	8986.0	45.2	538.3
.2	189.9	9.8	2.2	.2	1333.6	18.7	48.2	.2	3188.4	27.6	153.4	.2	5754.3	36.5	317.8	.2	9031.3	45.4	541.5
.3	199.8	9.9	2.4	.3	1352.4	18.8	49.1	.3	3216.1	27.7	155.0	.3	5790.9	36.6	320.2	.3	9076.8	45.5	544.7
.4	209.8	10.0	2.7	.4	1371.3	18.9	50.1	.4	3243.9	27.8	156.7	.4	5827.6	36.7	322.7	.4	9122.4	45.6	547.9
.5	219.9	10.1	2.9	.5	1390.3	19.0	51.0	.5	3271.8	27.9	158.4	.5	5864.4	36.8	325.1	.5	9168.1	45.7	551.0
.6	230.1	10.2	3.1	.6	1409.4	19.1	52.0	.6	3299.8	28.0	160.2	.6	5901.3	36.9	327.6	.6	9213.9	45.8	554.2
.7	240.5	10.4	3.4	.7	1428.6	19.2	53.0	.7	3327.9	28.1	161.9	.7	5938.3	37.0	330.0	.7	9259.8	45.9	557.5
.8	251.0	10.5	3.6	.8	1448.0	19.4	54.0	.8	3356.1	28.2	163.6	.8	5975.4	37.1	332.5	.8	9305.8	46.0	560.7
.9	261.5	10.6	3.9	.9	1467.5	19.5	55.0	.9	3384.5	28.4	165.4	.9	6012.6	37.2	335.0	.9	9351.9	46.1	563.9
3.0	272.2	10.7	4.2	11.0	1487.0	19.6	56.0	19.0	3413.0	28.5	167.1	27.0	6050.0	37.4	337.5	35.0	9398.1	46.2	567.1
.1	283.0	10.8	4.4	.1	1506.7	19.7	57.0	.1	3441.5	28.6	168.9	.1	6087.5	37.5	340.0	.1	9444.5	46.4	570.4
.2	293.9	10.9	4.7	.2	1526.5	19.8	58.1	.2	3470.2	28.7	170.7	.2	6125.0	37.6	342.5	.2	9491.0	46.5	573.6
.3	304.9	11.0	5.0	.3	1546.4	19.9	59.1	.3	3499.0	28.8	172.4	.3	6162.7	37.7	345.0	.3	9537.5	46.6	576.9
.4	316.1	11.1	5.4	.4	1566.4	20.0	60.2	.4	3527.9	28.9	174.2	.4	6200.5	37.8	347.6	.4	9584.2	46.7	580.2
.5	327.3	11.2	5.7	.5	1586.6	20.1	61.2	.5	3556.9	29.0	176.0	.5	6238.4	37.9	350.1	.5	9631.0	46.8	583.4
.6	338.7	11.4	6.0	.6	1606.8	20.2	62.3	.6	3586.1	29.1	177.9	.6	6276.4	38.0	352.7	.6	9677.9	46.9	586.7
.7	350.1	11.5	6.3	.7	1627.2	20.4	63.4	.7	3615.3	29.2	179.7	.7	6314.6	38.1	355.2	.7	9724.9	47.0	590.0
.8	361.7	11.6	6.7	.8	1647.6	20.5	64.5	.8	3644.7	29.4	181.5	.8	6352.8	38.2	357.8	.8	9772.1	47.1	593.4
.9	373.4	11.7	7.0	.9	1668.2	20.6	65.6	.9	3674.1	29.5	183.3	.9	6391.2	38.4	360.4	.9	9819.3	47.2	596.7
4.0	385.2	11.8	7.4	12.0	1688.9	20.7	66.7	20.0	3703.7	29.6	185.2	28.0	6429.6	38.5	363.0	36.0	9866.7	47.4	600.0
.1	397.1	11.9	7.8	.1	1709.7	20.8	67.8	.1	3733.4	29.7	187.0	.1	6468.2	38.6	365.6	.1	9914.1	47.5	603.3
.2	409.1	12.0	8.1	.2	1730.6	20.9	68.9	.2	3763.2	29.8	188.9	.2	6506.9	38.7	368.2	.2	9961.7	47.6	606.7
.3	421.2	12.1	8.6	.3	1751.6	21.0	70.0	.3	3793.1	29.9	190.8	.3	6545.7	38.8	370.8	.3	10009.4	47.7	610.0
.4	433.5	12.2	9.0	.4	1772.7	21.1	71.2	.4	3823.1	30.0	192.7	.4	6584.6	38.9	373.4	.4	10057.2	47.8	613.4
.5	445.8	12.4	9.4	.5	1794.0	21.2	72.3	.5	3853.2	30.1	194.6	.5	6623.6	39.0	376.0	.5	10105.1	47.9	616.8
.6	458.3	12.5	9.8	.6	1815.3	21.4	73.5	.6	3883.5	30.2	196.5	.6	6662.7	39.1	378.7	.6	10153.1	48.0	620.2
.7	470.9	12.6	10.2	.7	1836.8	21.5	74.7	.7	3913.8	30.4	198.4	.7	6702.0	39.2	381.3	.7	10201.2	48.1	623.6
.8	483.6	12.7	10.7	.8	1858.4	21.6	75.9	.8	3944.3	30.5	200.3	.8	6741.3	39.4	384.0	.8	10249.5	48.2	627.0
.9	496.4	12.8	11.1	.9	1880.1	21.7	77.0	.9	3974.9	30.6	202.2	.9	6780.8	39.5	386.7	.9	10297.8	48.4	630.4
5.0	509.3	12.9	11.6	13.0	1901.9	21.8	78.2	21.0	4005.6	30.7	204.2	29.0	6820.4	39.6	389.4	37.0	10346.3	48.5	633.8
.1	522.3	13.0	12.0	.1	1923.8	21.9	79.4	.1	4036.4	30.8	206.1	.1	6860.1	39.7	392.0	.1	10394.9	48.6	637.2
.2	535.4	13.1	12.5	.2	1945.8	22.0	80.7	.2	4067.3	30.9	208.1	.2	6899.9	39.8	394.7	.2	10443.6	48.7	640.7
.3	548.6	13.2	13.0	.3	1967.9	22.1	81.9	.3	4098.3	31.0	210.0	.3	6939.8	39.9	397.4	.3	10492.4	48.8	644.1
.4	562.0	13.4	13.5	.4	1990.1	22.2	83.1	.4	4129.4	31.1	212.0	.4	6979.8	40.0	400.2	.4	10541.3	48.9	647.6
.5	575.5	13.5	14.0	.5	2012.5	22.4	84.4	.5	4160.6	31.2	214.0	.5	7019.9	40.1	402.9	.5	10590.3	49.0	651.0
.6	589.0	13.6	14.5	.6	2035.0	22.5	85.6	.6	4192.0	31.4	216.0	.6	7060.1	40.2	405.6	.6	10639.4	49.1	654.5
.7	602.7	13.7	15.0	.7	2057.6	22.6	86.9	.7	4223.5	31.5	218.0	.7	7100.5	40.4	408.4	.7	10688.6	49.2	658.0
.8	616.5	13.8	15.6	.8	2080.2	22.7	88.2	.8	4255.0	31.6	220.0	.8	7141.0	40.5	411.1	.8	10738.0	49.4	661.5
.9	630.4	13.9	16.1	.9	2103.0	22.8	89.4	.9	4286.7	31.7	222.0	.9	7181.5	40.6	413.9	.9	10787.5	49.5	665.0
6.0	644.4	14.0	16.7	14.0	2125.9	22.9	90.7	22.0	4318.5	31.8	224.1	30.0	7222.2	40.7	416.7	38.0	10837.0	49.6	668.5
.1	658.6	14.1	17.2	.1	2148.9	23.0	92.0	.1	4350.4	31.9	226.1	.1	7263.0	40.8	419.4	.1	10886.7	49.7	672.0
.2	672.8	14.2	17.8	.2	2172.1	23.1	93.4	.2	4382.4	32.0	228.2	.2	7303.9	40.9	422.2	.2	10936.5	49.8	675.6
.3	687.2	14.4	18.4	.3	2195.3	23.2	94.7	.3	4414.6	32.1	230.2	.3	7344.9	41.0	425.0	.3	10986.4	49.9	679.1
.4	701.6	14.5	19.0	.4	2218.7	23.4	96.0	.4	4446.8	32.2	232.3	.4	7386.1	41.1	427.9	.4	11036.4	50.0	682.7
.5	716.2	14.6	19.6	.5	2242.1	23.5	97.3	.5	4479.2	32.4	234.4	.5	7427.3	41.2	430.7	.5	11086.6	50.1	686.2
.6	730.9	14.7	20.2	.6	2265.7	23.6	98.7	.6	4511.6	32.5	236.5	.6	7468.7	41.4	433.5	.6	11136.8	50.2	689.8
.7	745.7	14.8	20.8	.7	2289.4	23.7	100.0	.7	4544.2	32.6	238.6	.7	7510.1	41.5	436.3	.7	11187.2	50.4	693.4
.8	760.6	14.9	21.4	.8	2313.2	23.8	101.4	.8	4576.9	32.7	240.7	.8	7551.7	41.6	439.2	.8	11237.6	50.5	697.0
.9	775.6	15.0	22.0	.9	2337.1	23.9	102.8	.9	4609.7	32.8	242.8	.9	7593.4	41.7	442.0	.9	11288.2	50.6	700.6
7.0	790.7	15.1	22.7	15.0	2361.1	24.0	104.2	23.0	4642.6	32.9	244.9	31.0	7635.2	41.8	444.9	39.0	11338.9	50.7	704.2
.1	806.0	15.2	23.3	.1	2385.2	24.1	105.6	.1	4675.6	33.0	247.0	.1	7677.1	41.9	447.8	.1	11389.7	50.8	707.8
.2	821.3	15.4	24.0	.2	2409.5	24.2	107.0	.2	4708.7	33.1	249.2	.2	7719.1	42.0	450.7	.2	11440.6	50.9	711.4
.3	836.8	15.5	24.7	.3	2433.8	24.4	108.4	.3	4742.0	33.2	251.3	.3	7761.2	42.1	453.6	.3	11491.6	51.0	715.0
.4	852.4	15.6	25.4	.4	2458.3	24.5	109.8	.4	4775.3	33.4	253.5	.4	7803.5	42.2	456.5	.4	11542.7	51.1	718.7
.5	868.1	15.7	26.0	.5	2482.9	24.6	111.2	.5	4808.8	33.5	255.7	.5	7845.8	42.4	459.4	.5	11594.0	51.2	722.3
.6	883.9	15.8	26.7	.6	2507.6	24.7	112.7	.6	4842.4	33.6	257.9	.6	7888.3	42.5	462.3	.6	11645.3	51.4	726.0
.7	899.8	15.9	27.4	.7	2532.4	24.8	114.1	.7	4876.1	33.7	260.0	.7	7930.9	42.6	465.2	.7	11696.8	51.5	729.7
.8	915.8	16.0	28.2	.8	2557.3	24.9	115.6	.8	4909.9	33.8	262.2	.8	7973.6	42.7	468.2	.8	11748.4	51.6	733.4
.9	931.9	16.1	29.0	.9	2582.3	25.0	117.0	.9	4943.8	33.9	264.5	.9	8016.4	42.8	471.1	.9	11800.1	51.7	737.0

TABLE NO. 12.

EXCAVATION AND EMBANKMENT.

CUBIC YARDS.

PRISMOIDS 100 FEET LONG. **BREADTH OF BASE 20 FEET.**

Slopes 1 1-2 Horizontal to 1 Perpendicular

Ht.	Cub. Yds.	Diff.	Cor.	Ht.	Cub. Yds.	Diff.	Cor.	Ht.	Cub. Yds.	Diff.	Cor.	Ht.	Cub. Yds.	Diff.	Cor.	Ht.	Cub. Yds.	Diff.	Cor.
40.0	11851.9	51.8	740.7	48.0	16355.6	60.7	1066.7	56.0	21570.4	69.6	1451.9	64.0	27496.3	78.5	1896.3	72.0	34133.3	87.4	2400.0
.1	11903.8	51.9	744.4	.1	16416.4	60.8	1071.1	.1	21640.1	69.7	1457.0	.1	27574.9	78.6	1902.2	.1	34220.8	87.5	2406.7
.2	11955.8	52.0	748.2	.2	16477.3	60.9	1075.6	.2	21709.9	69.8	1462.2	.2	27653.6	78.7	1908.2	.2	34308.4	87.6	2413.4
.3	12007.9	52.1	751.9	.3	16538.3	61.0	1080.0	.3	21779.8	69.9	1467.5	.3	27732.4	78.8	1914.1	.3	34396.1	87.7	2420.0
.4	12060.1	52.2	755.6	.4	16599.4	61.1	1084.5	.4	21849.8	70.0	1472.7	.4	27811.3	78.9	1920.1	.4	34483.9	87.8	2426.7
.5	12112.3	52.4	759.4	.5	16660.6	61.2	1089.0	.5	21919.9	70.1	1477.9	.5	27890.3	79.0	1926.0	.5	34571.8	87.9	2433.4
.6	12165.0	52.5	763.1	.6	16722.0	61.4	1093.5	.6	21990.1	70.2	1483.1	.6	27969.4	79.1	1932.0	.6	34659.8	88.0	2440.2
.7	12217.5	52.6	766.9	.7	16783.5	61.5	1098.0	.7	22060.5	70.4	1488.4	.7	28048.6	79.2	1938.0	.7	34747.9	88.1	2446.9
.8	12270.2	52.7	770.7	.8	16845.0	61.6	1102.5	.8	22131.0	70.5	1493.6	.8	28128.0	79.4	1944.0	.8	34836.1	88.2	2453.6
.9	12323.0	52.8	774.4	.9	16906.7	61.7	1107.0	.9	22201.5	70.6	1498.9	.9	28207.5	79.5	1950.0	.9	34924.5	88.4	2460.4
41.0	12375.9	52.9	778.2	49.0	16968.5	61.8	1111.6	57.0	22272.2	70.7	1504.2	65.0	28287.0	79.6	1956.0	73.0	35013.0	88.5	2467.1
.1	12428.9	53.0	782.0	.1	17030.4	61.9	1116.1	.1	22343.0	70.8	1509.4	.1	28366.7	79.7	1962.0	.1	35101.5	88.6	2473.9
.2	12482.1	53.1	785.9	.2	17092.4	62.0	1120.7	.2	22413.9	70.9	1514.7	.2	28446.5	79.8	1968.1	.2	35190.2	88.7	2480.7
.3	12535.3	53.2	789.7	.3	17154.6	62.1	1125.2	.3	22484.9	71.0	1520.0	.3	28526.4	79.9	1974.1	.3	35279.0	88.8	2487.4
.4	12588.7	53.4	793.5	.4	17216.8	62.2	1129.8	.4	22556.1	71.1	1525.4	.4	28606.4	80.0	1980.2	.4	35367.9	88.9	2494.2
.5	12642.1	53.5	797.3	.5	17279.2	62.4	1134.4	.5	22627.3	71.2	1530.7	.5	28686.6	80.1	1986.2	.5	35456.9	89.0	2501.0
.6	12695.7	53.6	801.2	.6	17341.6	62.5	1139.0	.6	22698.7	71.4	1536.0	.6	28766.8	80.2	1992.3	.6	35546.1	89.1	2507.9
.7	12749.4	53.7	805.0	.7	17404.2	62.6	1143.6	.7	22770.1	71.5	1541.3	.7	28847.2	80.4	1998.4	.7	35635.3	89.2	2514.7
.8	12803.2	53.8	808.9	.8	17466.9	62.7	1148.2	.8	22841.7	71.6	1546.7	.8	28927.6	80.5	2004.5	.8	35724.7	89.4	2521.5
.9	12857.1	53.9	812.8	.9	17529.7	62.8	1152.8	.9	22913.4	71.7	1552.0	.9	29008.2	80.6	2010.6	.9	35814.1	89.5	2528.3
42.0	12911.1	54.0	816.7	50.0	17592.6	62.9	1157.4	58.0	22985.2	71.8	1557.4	66.0	29088.9	80.7	2016.7	74.0	35903.7	89.6	2535.2
.1	12965.2	54.1	820.6	.1	17655.6	63.0	1162.0	.1	23057.1	71.9	1562.8	.1	29169.7	80.8	2022.8	.1	35993.4	89.7	2542.0
.2	13019.5	54.2	824.5	.2	17718.7	63.1	1166.7	.2	23129.1	72.0	1568.2	.2	29250.6	80.9	2028.9	.2	36083.2	89.8	2548.9
.3	13073.8	54.4	828.4	.3	17782.0	63.2	1171.3	.3	23201.2	72.1	1573.6	.3	29331.6	81.0	2035.0	.3	36173.1	89.9	2555.8
.4	13128.3	54.5	832.3	.4	17845.3	63.4	1176.0	.4	23273.5	72.2	1579.0	.4	29412.7	81.1	2041.2	.4	36263.1	90.0	2562.7
.5	13182.9	54.6	836.2	.5	17908.8	63.5	1180.7	.5	23345.8	72.4	1584.4	.5	29494.0	81.2	2047.3	.5	36353.2	90.1	2569.5
.6	13237.6	54.7	840.2	.6	17972.4	63.6	1185.4	.6	23418.3	72.5	1589.8	.6	29575.3	81.4	2053.5	.6	36443.5	90.2	2576.5
.7	13292.4	54.8	844.1	.7	18036.1	63.7	1190.0	.7	23490.9	72.6	1595.2	.7	29656.8	81.5	2059.7	.7	36533.8	90.4	2583.4
.8	13347.3	54.9	848.1	.8	18099.9	63.8	1194.7	.8	23563.6	72.7	1600.7	.8	29738.4	81.6	2065.9	.8	36624.3	90.5	2590.3
.9	13402.3	55.0	852.0	.9	18163.8	63.9	1199.5	.9	23636.4	72.8	1606.1	.9	29820.1	81.7	2072.0	.9	36714.9	90.6	2597.2
43.0	13457.4	55.1	856.0	51.0	18227.8	64.0	1204.2	59.0	23709.3	72.9	1611.6	67.0	29901.9	81.8	2078.2	75.0	36805.6	90.7	2604.2
.1	13512.6	55.2	860.0	.1	18291.9	64.1	1208.9	.1	23782.3	73.0	1617.0	.1	29983.8	81.9	2084.4	.1	36896.4	90.8	2611.1
.2	13568.0	55.4	864.0	.2	18356.1	64.2	1213.6	.2	23855.4	73.1	1622.5	.2	30065.8	82.0	2090.7	.2	36987.3	90.9	2618.1
.3	13623.5	55.5	868.0	.3	18420.5	64.4	1218.4	.3	23928.6	73.2	1628.0	.3	30147.9	82.1	2096.9	.3	37078.3	91.0	2625.0
.4	13679.0	55.6	872.0	.4	18485.0	64.5	1223.1	.4	24002.0	73.4	1633.5	.4	30230.1	82.2	2103.1	.4	37169.4	91.1	2632.0
.5	13734.7	55.7	876.0	.5	18549.5	64.6	1227.9	.5	24075.5	73.5	1639.0	.5	30312.5	82.4	2109.4	.5	37260.6	91.2	2639.0
.6	13790.5	55.8	880.1	.6	18614.2	64.7	1232.7	.6	24149.0	73.6	1644.5	.6	30395.0	82.5	2115.6	.6	37352.0	91.4	2646.0
.7	13846.4	55.9	884.1	.7	18679.0	64.8	1237.4	.7	24222.7	73.7	1650.0	.7	30477.5	82.6	2121.9	.7	37443.5	91.5	2653.0
.8	13902.4	56.0	888.2	.8	18743.9	64.9	1242.2	.8	24296.5	73.8	1655.6	.8	30560.2	82.7	2128.2	.8	37535.0	91.6	2660.0
.9	13958.6	56.1	892.2	.9	18808.9	65.0	1247.0	.9	24370.4	73.9	1661.1	.9	30643.0	82.8	2134.5	.9	37626.7	91.7	2667.0
44.0	14014.8	56.2	896.3	52.0	18874.1	65.1	1251.9	60.0	24444.4	74.0	1666.7	68.0	30725.9	82.9	2140.7	76.0	37718.5	91.8	2674.1
.1	14071.2	56.4	900.4	.1	18939.3	65.2	1256.7	.1	24518.6	74.1	1672.2	.1	30808.9	83.0	2147.0	.1	37810.4	91.9	2681.1
.2	14127.6	56.5	904.5	.2	19004.7	65.4	1261.5	.2	24592.8	74.2	1677.8	.2	30892.1	83.1	2153.4	.2	37902.4	92.0	2688.2
.3	14184.2	56.6	908.6	.3	19070.1	65.5	1266.3	.3	24667.2	74.4	1683.4	.3	30975.3	83.2	2159.7	.3	37994.6	92.1	2695.2
.4	14240.9	56.7	912.7	.4	19135.7	65.6	1271.2	.4	24741.6	74.5	1689.0	.4	31058.7	83.4	2166.0	.4	38086.8	92.2	2702.3
.5	14297.7	56.8	916.8	.5	19201.4	65.7	1276.0	.5	24816.2	74.6	1694.6	.5	31142.1	83.5	2172.3	.5	38179.2	92.4	2709.4
.6	14354.6	56.9	920.9	.6	19267.2	65.8	1280.9	.6	24890.9	74.7	1700.2	.6	31225.7	83.6	2178.7	.6	38271.6	92.5	2716.5
.7	14411.6	57.0	925.0	.7	19333.1	65.9	1285.8	.7	24965.7	74.8	1705.8	.7	31309.4	83.7	2185.0	.7	38364.2	92.6	2723.6
.8	14468.7	57.1	929.2	.8	19399.1	66.0	1290.7	.8	25040.6	74.9	1711.4	.8	31393.2	83.8	2191.4	.8	38456.9	92.7	2730.7
.9	14526.0	57.2	933.3	.9	19465.2	66.1	1295.6	.9	25115.6	75.0	1717.0	.9	31477.1	83.9	2197.8	.9	38549.7	92.8	2737.8
45.0	14583.3	57.4	937.5	53.0	19531.5	66.2	1300.5	61.0	25190.7	75.1	1722.7	69.0	31561.1	84.0	2204.2	77.0	38642.6	92.9	2744.9
.1	14640.8	57.5	941.7	.1	19597.8	66.4	1305.4	.1	25266.0	75.2	1728.3	.1	31645.2	84.1	2210.6	.1	38735.6	93.0	2752.0
.2	14698.4	57.6	945.9	.2	19664.3	66.5	1310.3	.2	25341.3	75.4	1734.0	.2	31729.5	84.2	2217.0	.2	38828.7	93.1	2759.2
.3	14756.1	57.7	950.0	.3	19730.9	66.6	1315.2	.3	25416.8	75.5	1739.7	.3	31813.8	84.4	2223.4	.3	38922.0	93.2	2766.3
.4	14813.9	57.8	954.2	.4	19797.6	66.7	1320.2	.4	25492.4	75.6	1745.4	.4	31898.3	84.5	2229.8	.4	39015.3	93.4	2773.5
.5	14871.8	57.9	958.4	.5	19864.4	66.8	1325.1	.5	25568.1	75.7	1751.0	.5	31982.9	84.6	2236.2	.5	39108.8	93.5	2780.7
.6	14929.8	58.0	962.7	.6	19931.3	66.9	1330.1	.6	25643.9	75.8	1756.7	.6	32067.6	84.7	2242.7	.6	39202.4	93.6	2787.9
.7	14987.9	58.1	966.9	.7	19998.3	67.0	1335.0	.7	25719.8	75.9	1762.5	.7	32152.4	84.8	2249.1	.7	39296.1	93.7	2795.0
.8	15046.1	58.2	971.1	.8	20065.4	67.1	1340.0	.8	25795.8	76.0	1768.2	.8	32237.3	84.9	2255.6	.8	39389.9	93.8	2802.2
.9	15104.5	58.4	975.4	.9	20132.6	67.2	1345.0	.9	25871.9	76.1	1773.9	.9	32322.3	85.0	2262.0	.9	39483.8	93.9	2809.4
46.0	15163.0	58.5	979.6	54.0	20200.0	67.4	1350.0	62.0	25948.1	76.2	1779.6	70.0	32407.4	85.1	2268.5	78.0	39577.8	94.0	2816.7
.1	15221.5	58.6	983.9	.1	20267.5	67.5	1355.0	.1	26024.5	76.4	1785.4	.1	32492.6	85.2	2275.0	.1	39671.9	94.1	2823.9
.2	15280.2	58.7	988.2	.2	20335.0	67.6	1360.0	.2	26101.0	76.5	1791.1	.2	32578.0	85.4	2281.5	.2	39766.1	94.2	2831.1
.3	15339.0	58.8	992.4	.3	20402.7	67.7	1365.0	.3	26177.5	76.6	1796.9	.3	32663.5	85.5	2288.0	.3	39860.5	94.4	2838.4
.4	15397.9	58.9	996.7	.4	20470.5	67.8	1370.1	.4	26254.2	76.7	1802.7	.4	32749.0	85.6	2294.5	.4	39955.0	94.5	2845.6
.5	15456.9	59.0	1001.0	.5	20538.4	67.9	1375.1	.5	26331.0	76.8	1808.5	.5	32834.7	85.7	2301.0	.5	40049.5	94.6	2852.9
.6	15516.1	59.1	1005.4	.6	20606.4	68.0	1380.2	.6	26407.9	76.9	1814.2	.6	32920.5	85.8	2307.6	.6	40144.2	94.7	2860.2
.7	15575.3	59.2	1009.7	.7	20674.6	68.1	1385.2	.7	26484.9	77.0	1820.0	.7	33006.4	85.9	2314.1	.7	40239.0	94.8	2867.4
.8	15634.7	59.4	1014.0	.8	20742.8	68.2	1390.3	.8	26562.1	77.1	1825.9	.8	33092.4	86.0	2320.7	.8	40333.9	94.9	2874.7
.9	15694.1	59.5	1018.3	.9	20811.2	68.4	1395.4	.9	26639.3	77.2	1831.7	.9	33178.6	86.1	2327.2	.9	40428.9	95.0	2882.0
47.0	15753.7	59.6	1022.7	55.0	20879.6	68.5	1400.5	63.0	26716.7	77.4	1837.5	71.0	33264.8	86.2	2333.8	79.0	40524.1	95.1	2889.4
.1	15813.4	59.7	1027.0	.1	20948.2	68.6	1405.6	.1	26794.1	77.5	1843.3	.1	33351.2	86.4	2340.4	.1	40619.3	95.2	2896.7
.2	15873.2	59.8	1031.4	.2	21016.9	68.7	1410.7	.2	26871.7	77.6	1849.2	.2	33437.6	86.5	2347.0	.2	40714.7	95.4	2904.0
.3	15933.1	59.9	1035.8	.3	21085.7	68.8	1415.8	.3	26949.4	77.7	1855.0	.3	33524.2	86.6	2353.6	.3	40810.1	95.5	2911.3
.4	15993.1	60.0	1040.2	.4	21154.6	68.9	1420.9	.4	27027.2	77.8	1860.9	.4	33610.9	86.7	2360.2	.4	40905.7	95.6	2918.7
.5	16053.2	60.1	1044.6	.5	21223.6	69.0	1426.0	.5	27105.1	77.9	1866.8	.5	33697.7	86.8	2366.8	.5	41001.4	95.7	2926.0
.6	16113.5	60.2	1049.0	.6	21292.7	69.1	1431.2	.6	27183.1	78.0	1872.7	.6	33784.6	86.9	2373.4	.6	41097.2	95.8	2933.4
.7	16173.8	60.4	1053.4	.7	21362.0	69.2	1436.3	.7	27261.2	78.1	1878.6	.7	33871.6	87.0	2380.0	.7	41193.1	95.9	2940.8
.8	16234.3	60.5	1057.8	.8	21431.3	69.4	1441.5	.8	27339.5	78.2	1884.5	.8	33958.7	87.1	2386.7	.8	41289.1	96.0	2948.2
.9	16294.9	60.6	1062.2	.9	21500.8	69.5	1446.7	.9	27417.8	78.4	1890.4	.9	34046.0	87.2	2393.3	.9	41385.2	96.1	2955.6

TABLE NO. 13.

EXCAVATION AND EMBANKMENT.

CUBIC YARDS.

PRISMOIDS 100 FEET LONG. **BREADTH OF BASE 22 FEET.**

Slopes 1 1-2 Horizontal to 1 Perpendicular.

Ht.	Cub. Yds.	Diff.	Cor.	Ht.	Cub. Yds.	Diff.	Cor.	Ht.	Cub. Yds.	Diff.	Cor.	Ht.	Cub. Yds.	Diff.	Cor.	Ht.	Cub. Yds.	Diff.	Cor.
0.0	0.0	0.0	0.0	8.0	1007.4	17.0	29.6	16.0	2725.9	25.9	118.5	24.0	5155.6	34.8	266.7	32.0	8296.3	43.6	474.1
.1	8.2	8.2	0.0	.1	1024.5	17.1	30.4	.1	2751.9	26.0	120.0	.1	5190.4	34.9	268.9	.1	8340.1	43.8	477.0
.2	16.5	8.3	0.0	.2	1041.7	17.2	31.1	.2	2778.0	26.1	121.5	.2	5225.4	35.0	271.1	.2	8383.9	43.9	480.0
.3	24.9	8.4	0.0	.3	1059.0	17.3	31.9	.3	2804.2	26.2	123.0	.3	5260.5	35.1	273.4	.3	8427.9	44.0	483.0
.4	33.5	8.5	0.1	.4	1076.4	17.4	32.7	.4	2830.5	26.3	124.5	.4	5295.7	35.2	275.6	.4	8472.0	44.1	486.0
.5	42.1	8.6	0.1	.5	1094.0	17.5	33.4	.5	2856.9	26.4	126.0	.5	5331.0	35.3	277.9	.5	8516.2	44.2	489.0
.6	50.9	8.8	0.2	.6	1111.6	17.6	34.2	.6	2883.5	26.5	127.6	.6	5366.4	35.4	280.2	.6	8560.5	44.3	492.0
.7	59.8	8.9	0.2	.7	1129.4	17.8	35.0	.7	2910.1	26.6	129.1	.7	5402.0	35.5	282.5	.7	8604.9	44.4	495.0
.8	68.7	9.0	0.3	.8	1147.3	17.9	35.9	.8	2936.9	26.8	130.7	.8	5437.6	35.6	284.7	.8	8649.5	44.5	498.0
.9	77.8	9.1	0.4	.9	1165.2	18.0	36.7	.9	2963.8	26.9	132.2	.9	5473.4	35.8	287.0	.9	8694.1	44.6	501.1
1.0	87.0	9.2	0.5	9.0	1183.3	18.1	37.5	17.0	2990.7	27.0	133.8	25.0	5509.3	35.9	289.4	33.0	8738.9	44.8	504.2
.1	96.3	9.3	0.6	.1	1201.5	18.2	38.3	.1	3017.8	27.1	135.4	.1	5545.2	36.0	291.7	.1	8783.8	44.9	507.2
.2	105.8	9.4	0.7	.2	1219.8	18.3	39.2	.2	3045.0	27.2	137.0	.2	5581.3	36.1	294.0	.2	8828.7	45.0	510.3
.3	115.3	9.5	0.8	.3	1238.3	18.4	40.0	.3	3072.3	27.3	138.6	.3	5617.5	36.2	296.3	.3	8873.8	45.1	513.4
.4	125.0	9.6	0.9	.4	1256.8	18.5	40.9	.4	3099.8	27.4	140.2	.4	5653.8	36.3	298.7	.4	8919.0	45.2	516.5
.5	134.7	9.8	1.0	.5	1275.5	18.6	41.8	.5	3127.3	27.5	141.8	.5	5690.3	36.4	301.0	.5	8964.3	45.3	519.6
.6	144.6	9.9	1.2	.6	1294.2	18.8	42.7	.6	3155.0	27.6	143.4	.6	5726.8	36.5	303.4	.6	9009.8	45.4	522.7
.7	154.6	10.0	1.3	.7	1313.1	18.9	43.6	.7	3182.7	27.8	145.0	.7	5763.5	36.6	305.8	.7	9055.3	45.5	525.8
.8	164.7	10.1	1.5	.8	1332.1	19.0	44.5	.8	3210.6	27.9	146.7	.8	5800.2	36.8	308.2	.8	9101.0	45.6	528.9
.9	174.9	10.2	1.7	.9	1351.2	19.1	45.4	.9	3238.6	28.0	148.3	.9	5837.1	36.9	310.6	.9	9146.7	45.8	532.0
2.0	185.2	10.3	1.9	10.0	1370.4	19.2	46.3	18.0	3266.7	28.1	150.0	26.0	5874.1	37.0	313.0	34.0	9192.6	45.9	535.2
.1	195.6	10.4	2.0	.1	1389.7	19.3	47.2	.1	3294.9	28.2	151.7	.1	5911.2	37.1	315.4	.1	9238.6	46.0	538.3
.2	206.1	10.5	2.2	.2	1409.1	19.4	48.2	.2	3323.2	28.3	153.4	.2	5948.4	37.2	317.8	.2	9284.7	46.1	541.5
.3	216.8	10.6	2.4	.3	1428.6	19.5	49.1	.3	3351.6	28.4	155.0	.3	5985.7	37.3	320.2	.3	9330.9	46.2	544.7
.4	227.6	10.8	2.7	.4	1448.3	19.6	50.1	.4	3380.1	28.5	156.7	.4	6023.1	37.4	322.7	.4	9377.2	46.3	547.9
.5	238.4	10.9	2.9	.5	1468.1	19.8	51.0	.5	3408.8	28.6	158.4	.5	6060.6	37.5	325.1	.5	9423.6	46.4	551.0
.6	249.4	11.0	3.1	.6	1487.9	19.9	52.0	.6	3437.6	28.8	160.2	.6	6098.3	37.6	327.6	.6	9470.1	46.5	554.2
.7	260.5	11.1	3.4	.7	1507.9	20.0	53.0	.7	3466.4	28.9	161.9	.7	6136.1	37.8	330.0	.7	9516.8	46.6	557.5
.8	271.7	11.2	3.6	.8	1528.0	20.1	54.0	.8	3495.4	29.0	163.6	.8	6173.9	37.9	332.5	.8	9563.6	46.8	560.7
.9	283.0	11.3	3.9	.9	1548.2	20.2	55.0	.9	3524.5	29.1	165.4	.9	6211.9	38.0	335.0	.9	9610.4	46.9	563.9
3.0	294.4	11.4	4.2	11.0	1568.5	20.3	56.0	19.0	3553.7	29.2	167.1	27.0	6250.0	38.1	337.5	35.0	9657.4	47.0	567.1
.1	306.0	11.5	4.4	.1	1588.9	20.4	57.0	.1	3583.0	29.3	168.9	.1	6288.2	38.2	340.0	.1	9704.5	47.1	570.4
.2	317.6	11.6	4.7	.2	1609.5	20.5	58.1	.2	3612.4	29.4	170.7	.2	6326.5	38.3	342.5	.2	9751.7	47.2	573.6
.3	329.4	11.8	5.0	.3	1630.1	20.6	59.1	.3	3642.0	29.5	172.4	.3	6364.9	38.4	345.0	.3	9799.0	47.3	576.9
.4	341.3	11.9	5.4	.4	1650.9	20.8	60.2	.4	3671.6	29.6	174.2	.4	6403.5	38.5	347.6	.4	9846.4	47.4	580.2
.5	353.2	12.0	5.7	.5	1671.8	20.9	61.2	.5	3701.4	29.8	176.0	.5	6442.1	38.6	350.1	.5	9894.0	47.5	583.4
.6	365.3	12.1	6.0	.6	1692.7	21.0	62.3	.6	3731.3	29.9	177.9	.6	6480.9	38.8	352.7	.6	9941.6	47.6	586.7
.7	377.5	12.2	6.3	.7	1713.8	21.1	63.4	.7	3761.2	30.0	179.7	.7	6519.8	38.9	355.2	.7	9989.4	47.8	590.0
.8	389.8	12.3	6.7	.8	1735.0	21.2	64.5	.8	3791.3	30.1	181.5	.8	6558.7	39.0	357.8	.8	10037.3	47.9	593.4
.9	402.3	12.4	7.0	.9	1756.3	21.3	65.6	.9	3821.5	30.2	183.3	.9	6597.8	39.1	360.4	.9	10085.2	48.0	596.7
4.0	414.8	12.5	7.4	12.0	1777.8	21.4	66.7	20.0	3851.8	30.3	185.2	28.0	6637.0	39.2	363.0	36.0	10133.3	48.1	600.0
.1	427.5	12.6	7.8	.1	1799.3	21.5	67.8	.1	3882.3	30.4	187.0	.1	6676.3	39.3	365.6	.1	10181.5	48.2	603.3
.2	440.2	12.8	8.1	.2	1821.0	21.6	68.9	.2	3912.8	30.5	188.9	.2	6715.8	39.4	368.2	.2	10229.8	48.3	606.7
.3	453.1	12.9	8.6	.3	1842.7	21.8	70.0	.3	3943.5	30.6	190.8	.3	6755.3	39.5	370.8	.3	10278.3	48.4	610.0
.4	466.1	13.0	9.0	.4	1864.6	21.9	71.2	.4	3974.2	30.8	192.7	.4	6795.0	39.6	373.4	.4	10326.8	48.5	613.4
.5	479.2	13.1	9.4	.5	1886.6	22.0	72.3	.5	4005.1	30.9	194.6	.5	6834.7	39.8	376.0	.5	10375.5	48.6	616.8
.6	492.4	13.2	9.8	.6	1908.7	22.1	73.5	.6	4036.1	31.0	196.5	.6	6874.6	39.9	378.7	.6	10424.2	48.8	620.2
.7	505.7	13.3	10.2	.7	1930.9	22.2	74.7	.7	4067.2	31.1	198.4	.7	6914.6	40.0	381.3	.7	10473.1	48.9	623.6
.8	519.1	13.4	10.7	.8	1953.2	22.3	75.9	.8	4098.4	31.2	200.3	.8	6954.7	40.1	384.0	.8	10522.1	49.0	627.0
.9	532.6	13.5	11.1	.9	1975.6	22.4	77.0	.9	4129.7	31.3	202.2	.9	6994.9	40.2	386.7	.9	10571.2	49.1	630.4
5.0	546.3	13.6	11.6	13.0	1998.1	22.5	78.2	21.0	4161.1	31.4	204.2	29.0	7035.2	40.3	389.4	37.0	10620.4	49.2	633.8
.1	560.1	13.8	12.0	.1	2020.8	22.6	79.4	.1	4192.6	31.5	206.1	.1	7075.6	40.4	392.0	.1	10669.7	49.3	637.2
.2	573.9	13.9	12.5	.2	2043.6	22.8	80.7	.2	4224.3	31.6	208.1	.2	7116.1	40.5	394.7	.2	10719.1	49.4	640.7
.3	587.9	14.0	13.0	.3	2066.4	22.9	81.9	.3	4256.1	31.8	210.0	.3	7156.8	40.6	397.4	.3	10768.6	49.5	644.1
.4	602.0	14.1	13.5	.4	2089.4	23.0	83.1	.4	4287.9	31.9	212.0	.4	7197.6	40.8	400.2	.4	10818.3	49.6	647.6
.5	616.2	14.2	14.0	.5	2112.5	23.1	84.4	.5	4319.9	32.0	214.0	.5	7238.4	40.9	402.9	.5	10868.1	49.8	651.0
.6	630.5	14.3	14.5	.6	2135.7	23.2	85.6	.6	4352.0	32.1	216.0	.6	7279.4	41.0	405.6	.6	10917.9	49.9	654.5
.7	644.9	14.4	15.0	.7	2159.0	23.3	86.9	.7	4384.2	32.2	218.0	.7	7320.5	41.1	408.4	.7	10967.9	50.0	658.0
.8	659.5	14.5	15.6	.8	2182.4	23.4	88.2	.8	4416.5	32.3	220.0	.8	7361.7	41.2	411.1	.8	11018.0	50.1	661.5
.9	674.1	14.6	16.1	.9	2206.0	23.5	89.4	.9	4448.9	32.4	222.0	.9	7403.0	41.3	413.9	.9	11068.2	50.2	665.0
6.0	688.9	14.8	16.7	14.0	2229.6	23.6	90.7	22.0	4481.5	32.5	224.1	30.0	7444.4	41.4	416.7	38.0	11118.5	50.3	668.5
.1	703.8	14.9	17.2	.1	2253.4	23.8	92.0	.1	4514.1	32.6	226.1	.1	7486.0	41.5	419.4	.1	11168.9	50.4	672.0
.2	718.7	15.0	17.8	.2	2277.3	23.9	93.4	.2	4546.9	32.8	228.2	.2	7527.6	41.6	422.2	.2	11219.5	50.5	675.6
.3	733.8	15.1	18.4	.3	2301.2	24.0	94.7	.3	4579.8	32.9	230.2	.3	7569.4	41.8	425.0	.3	11270.1	50.6	679.1
.4	749.0	15.2	19.0	.4	2325.3	24.1	96.0	.4	4612.7	33.0	232.3	.4	7611.3	41.9	427.9	.4	11320.9	50.8	682.7
.5	764.3	15.3	19.6	.5	2349.5	24.2	97.3	.5	4645.8	33.1	234.4	.5	7653.2	42.0	430.7	.5	11371.8	50.9	686.2
.6	779.8	15.4	20.2	.6	2373.8	24.3	98.7	.6	4679.0	33.2	236.5	.6	7695.3	42.1	433.5	.6	11422.7	51.0	689.8
.7	795.3	15.5	20.8	.7	2398.3	24.4	100.0	.7	4712.3	33.3	238.6	.7	7737.5	42.2	436.3	.7	11473.8	51.1	693.4
.8	811.0	15.6	21.4	.8	2422.8	24.5	101.4	.8	4745.8	33.4	240.7	.8	7779.8	42.3	439.2	.8	11525.0	51.2	697.0
.9	826.7	15.8	22.0	.9	2447.5	24.6	102.8	.9	4779.3	33.5	242.8	.9	7822.3	42.4	442.0	.9	11576.3	51.3	700.6
7.0	842.6	15.9	22.7	15.0	2472.2	24.8	104.2	23.0	4813.0	33.6	244.9	31.0	7864.8	42.5	444.9	39.0	11627.8	51.4	704.2
.1	858.6	16.0	23.3	.1	2497.1	24.9	105.6	.1	4846.7	33.8	247.0	.1	7907.5	42.6	447.8	.1	11679.3	51.5	707.8
.2	874.7	16.1	24.0	.2	2522.1	25.0	107.0	.2	4880.6	33.9	249.2	.2	7950.2	42.8	450.7	.2	11731.0	51.6	711.4
.3	890.9	16.2	24.7	.3	2547.2	25.1	108.4	.3	4914.6	34.0	251.3	.3	7993.1	42.9	453.6	.3	11782.7	51.8	715.0
.4	907.2	16.3	25.4	.4	2572.4	25.2	109.8	.4	4948.7	34.1	253.5	.4	8036.1	43.0	456.5	.4	11834.6	51.9	718.7
.5	923.6	16.4	26.0	.5	2597.7	25.3	111.2	.5	4982.9	34.2	255.7	.5	8079.2	43.1	459.4	.5	11886.6	52.0	722.3
.6	940.1	16.5	26.7	.6	2623.1	25.4	112.7	.6	5017.2	34.3	257.9	.6	8122.4	43.2	462.3	.6	11938.7	52.1	726.0
.7	956.8	16.6	27.4	.7	2648.6	25.5	114.1	.7	5051.6	34.4	260.0	.7	8165.7	43.3	465.2	.7	11990.9	52.2	729.7
.8	973.6	16.8	28.2	.8	2674.3	25.6	115.6	.8	5086.1	34.5	262.2	.8	8209.1	43.4	468.2	.8	12043.2	52.3	733.4
.9	990.4	16.9	28.9	.9	2700.1	25.8	117.0	.9	5120.8	34.6	264.5	.9	8252.6	43.5	471.1	.9	12095.6	52.4	737.0

TABLE NO. 14.

EXCAVATION AND EMBANKMENT.

CUBIC YARDS.

PRISMOIDS 100 FEET LONG. **BREADTH OF BASE 24 FEET.**

Slopes 1 1-2 Horizontal to 1 Perpendicular.

Ht.	Cub. Yds.	Diff.	Cor.	Ht.	Cub. Yds.	Diff.	Cor.	Ht.	Cub. Yds.	Diff.	Cor.	Ht.	Cub. Yds.	Diff.	Cor.	Ht.	Cub. Yds.	Diff.	Cor.
0.0	0.0	0.0	0.0	8.0	1066.7	17.7	29.6	16.0	2844.4	26.6	118.5	24.0	5333.3	35.5	266.7	32.0	8533.3	44.4	474.1
.1	8.9	8.9	0.0	.1	1084.5	17.8	30.4	.1	2871.2	26.7	120.0	.1	5368.9	35.6	268.9	.1	8577.8	44.5	477.0
.2	18.0	9.1	0.0	.2	1102.4	17.9	31.1	.2	2898.0	26.8	121.5	.2	5404.7	35.7	271.1	.2	8622.4	44.6	480.0
.3	27.2	9.2	0.0	.3	1120.5	18.1	31.9	.3	2924.9	26.9	123.0	.3	5440.5	35.8	273.4	.3	8667.2	44.7	483.0
.4	36.4	9.3	0.1	.4	1138.7	18.2	32.7	.4	2952.0	27.1	124.5	.4	5476.4	35.9	275.6	.4	8712.0	44.8	486.0
.5	45.8	9.4	0.1	.5	1156.9	18.3	33.4	.5	2979.2	27.2	126.0	.5	5512.5	36.1	277.9	.5	8756.9	44.9	489.0
.6	55.3	9.5	0.2	.6	1175.3	18.4	34.2	.6	3006.4	27.3	127.6	.6	5548.7	36.2	280.2	.6	8802.0	45.1	492.0
.7	64.9	9.6	0.2	.7	1193.8	18.5	35.0	.7	3033.8	27.4	129.1	.7	5584.9	36.3	282.5	.7	8847.2	45.2	495.0
.8	74.7	9.7	0.3	.8	1212.4	18.6	35.9	.8	3061.3	27.5	130.7	.8	5621.3	36.4	284.7	.8	8892.4	45.3	498.0
.9	84.5	9.8	0.4	.9	1231.2	18.7	36.7	.9	3088.9	27.6	132.2	.9	5657.8	36.5	287.0	.9	8937.8	45.4	501.1
1.0	94.4	9.9	0.5	9.0	1250.0	18.8	37.5	17.0	3116.7	27.7	133.8	25.0	5694.4	36.6	289.4	33.0	8983.3	45.5	504.2
.1	104.5	10.1	0.6	.1	1268.9	18.9	38.3	.1	3144.5	27.8	135.4	.1	5731.2	36.7	291.7	.1	9028.9	45.6	507.2
.2	114.7	10.2	0.7	.2	1288.0	19.1	39.2	.2	3172.4	27.9	137.0	.2	5768.0	36.8	294.0	.2	9074.7	45.7	510.3
.3	124.0	10.3	0.8	.3	1307.2	19.2	40.0	.3	3200.5	28.1	138.6	.3	5804.9	36.9	296.3	.3	9120.5	45.8	513.4
.4	135.3	10.4	0.9	.4	1326.4	19.3	40.9	.4	3228.7	28.2	140.2	.4	5842.0	37.1	298.7	.4	9166.4	45.9	516.5
.5	145.8	10.5	1.0	.5	1345.8	19.4	41.8	.5	3256.9	28.3	141.8	.5	5879.2	37.2	301.0	.5	9212.5	46.1	519.6
.6	156.4	10.6	1.2	.6	1365.3	19.5	42.7	.6	3285.3	28.4	143.4	.6	5916.4	37.3	303.4	.6	9258.7	46.2	522.7
.7	167.2	10.7	1.3	.7	1384.9	19.6	43.6	.7	3313.8	28.5	145.0	.7	5953.8	37.4	305.8	.7	9304.9	46.3	525.9
.8	178.0	10.8	1.5	.8	1404.7	19.7	44.5	.8	3342.4	28.6	146.7	.8	5991.3	37.5	308.2	.8	9351.3	46.4	529.0
.9	188.9	10.9	1.7	.9	1424.5	19.8	45.4	.9	3371.2	28.7	148.3	.9	6028.9	37.6	310.6	.9	9397.8	46.5	532.2
2.0	200.0	11.1	1.8	10.0	1444.4	19.9	46.3	18.0	3400.0	28.8	150.0	26.0	6066.7	37.7	313.0	34.0	9444.4	46.6	535.2
.1	211.2	11.2	2.0	.1	1464.5	20.1	47.2	.1	3428.9	28.9	151.7	.1	6104.5	37.8	315.4	.1	9491.2	46.7	538.3
.2	222.4	11.3	2.2	.2	1484.7	20.2	48.2	.2	3458.0	29.1	153.4	.2	6142.4	37.9	317.8	.2	9538.0	46.8	541.5
.3	233.8	11.4	2.4	.3	1504.9	20.3	49.1	.3	3487.2	29.2	155.0	.3	6180.5	38.1	320.2	.3	9584.9	46.9	544.7
.4	245.3	11.5	2.7	.4	1525.3	20.4	50.1	.4	3516.4	29.3	156.7	.4	6218.7	38.2	322.7	.4	9632.0	47.1	547.9
.5	256.9	11.6	2.9	.5	1545.8	20.5	51.0	.5	3545.8	29.4	158.4	.5	6256.9	38.3	325.1	.5	9679.2	47.2	551.0
.6	268.7	11.7	3.1	.6	1566.4	20.6	52.0	.6	3575.3	29.5	160.2	.6	6295.3	38.4	327.6	.6	9726.4	47.3	554.2
.7	280.5	11.8	3.4	.7	1587.2	20.7	53.0	.7	3604.9	29.6	161.9	.7	6333.8	38.5	330.0	.7	9773.8	47.4	557.5
.8	292.4	11.9	3.6	.8	1608.0	20.8	54.0	.8	3634.7	29.7	163.6	.8	6372.4	38.6	332.5	.8	9821.3	47.5	560.7
.9	304.5	12.1	3.9	.9	1628.9	20.9	55.0	.9	3664.5	29.8	165.4	.9	6411.2	38.7	335.0	.9	9868.9	47.6	563.9
3.0	316.7	12.2	4.2	11.0	1650.0	21.1	56.0	19.0	3694.4	29.9	167.1	27.0	6450.0	38.8	337.5	35.0	9916.7	47.7	567.1
.1	328.9	12.3	4.4	.1	1671.2	21.2	57.0	.1	3724.5	30.1	168.9	.1	6488.9	38.9	340.0	.1	9964.5	47.8	570.4
.2	341.3	12.4	4.7	.2	1692.4	21.3	58.1	.2	3754.7	30.2	170.7	.2	6528.0	39.1	342.5	.2	10012.4	47.9	573.6
.3	353.8	12.5	5.0	.3	1713.8	21.4	59.1	.3	3784.9	30.3	172.4	.3	6567.2	39.2	345.0	.3	10060.5	48.1	576.9
.4	366.4	12.6	5.4	.4	1735.3	21.5	60.2	.4	3815.3	30.4	174.2	.4	6606.4	39.3	347.6	.4	10108.7	48.2	580.2
.5	379.2	12.7	5.7	.5	1756.9	21.6	61.2	.5	3845.8	30.5	176.0	.5	6645.8	39.4	350.1	.5	10156.9	48.3	583.4
.6	392.0	12.8	6.0	.6	1778.7	21.7	62.3	.6	3876.4	30.6	177.9	.6	6685.3	39.5	352.7	.6	10205.3	48.4	586.7
.7	404.9	12.9	6.3	.7	1800.5	21.8	63.4	.7	3907.2	30.7	179.7	.7	6724.9	39.6	355.2	.7	10253.8	48.5	590.0
.8	418.0	13.1	6.7	.8	1822.4	21.9	64.5	.8	3938.0	30.8	181.5	.8	6764.7	39.7	357.8	.8	10302.4	48.6	593.4
.9	431.2	13.2	7.0	.9	1844.5	22.1	65.6	.9	3968.9	30.9	183.3	.9	6804.5	39.8	360.4	.9	10351.2	48.7	596.7
4.0	444.4	13.3	7.4	12.0	1866.7	22.2	66.7	20.0	4000.0	31.1	185.2	28.0	6844.4	39.9	363.0	36.0	10400.0	48.8	600.0
.1	457.8	13.4	7.8	.1	1888.9	22.3	67.8	.1	4031.2	31.2	187.0	.1	6884.5	40.1	365.6	.1	10448.9	48.9	603.3
.2	471.3	13.5	8.1	.2	1911.3	22.4	68.9	.2	4062.4	31.3	188.9	.2	6924.7	40.2	368.2	.2	10498.0	49.1	606.7
.3	484.9	13.6	8.6	.3	1933.8	22.5	70.0	.3	4093.8	31.4	190.8	.3	6964.9	40.3	370.8	.3	10547.2	49.2	610.0
.4	498.7	13.7	9.0	.4	1956.4	22.6	71.2	.4	4125.3	31.5	192.7	.4	7005.3	40.4	373.4	.4	10596.4	49.3	613.4
.5	512.5	13.8	9.4	.5	1979.2	22.7	72.3	.5	4156.9	31.6	194.6	.5	7045.8	40.5	376.0	.5	10645.8	49.4	616.8
.6	526.4	13.9	9.8	.6	2002.0	22.8	73.5	.6	4188.7	31.7	196.5	.6	7086.4	40.6	378.7	.6	10695.3	49.5	620.2
.7	540.5	14.1	10.2	.7	2024.9	22.9	74.7	.7	4220.5	31.8	198.4	.7	7127.2	40.7	381.3	.7	10744.9	49.6	623.6
.8	554.7	14.2	10.7	.8	2048.0	23.1	75.9	.8	4252.4	31.9	200.3	.8	7168.0	40.8	384.0	.8	10794.7	49.7	627.0
.9	568.9	14.3	11.1	.9	2071.2	23.2	77.0	.9	4284.5	32.1	202.2	.9	7208.9	40.9	386.7	.9	10844.5	49.8	630.4
5.0	583.3	14.4	11.6	13.0	2094.4	23.3	78.2	21.0	4316.7	32.2	204.2	29.0	7250.0	41.1	389.4	37.0	10894.4	49.9	633.8
.1	597.8	14.5	12.0	.1	2117.8	23.4	79.4	.1	4348.9	32.3	206.1	.1	7291.2	41.2	392.0	.1	10944.5	50.1	637.2
.2	612.4	14.6	12.5	.2	2141.3	23.5	80.7	.2	4381.3	32.4	208.1	.2	7332.4	41.3	394.7	.2	10994.7	50.2	640.7
.3	627.2	14.7	13.0	.3	2164.9	23.6	81.9	.3	4413.8	32.5	210.0	.3	7373.8	41.4	397.4	.3	11044.9	50.3	644.1
.4	642.0	14.8	13.5	.4	2188.7	23.7	83.1	.4	4446.4	32.6	212.0	.4	7415.3	41.5	400.2	.4	11095.3	50.4	647.6
.5	656.9	14.9	14.0	.5	2212.5	23.8	84.4	.5	4479.2	32.7	214.0	.5	7456.9	41.6	402.9	.5	11145.8	50.5	651.0
.6	672.0	15.1	14.5	.6	2236.4	23.9	85.6	.6	4512.0	32.8	216.0	.6	7498.7	41.7	405.6	.6	11196.4	50.6	654.5
.7	687.2	15.2	15.0	.7	2260.5	24.1	86.9	.7	4544.9	32.9	218.0	.7	7540.5	41.8	408.4	.7	11247.2	50.7	658.0
.8	702.4	15.3	15.6	.8	2284.7	24.2	88.2	.8	4578.0	33.1	220.0	.8	7582.4	41.9	411.1	.8	11298.0	50.8	661.5
.9	717.8	15.4	16.1	.9	2308.9	24.3	89.4	.9	4611.2	33.2	222.0	.9	7624.5	42.1	413.9	.9	11348.9	50.9	665.0
6.0	733.3	15.5	16.7	14.0	2333.3	24.4	90.7	22.0	4644.4	33.3	224.1	30.0	7666.7	42.2	416.7	38.0	11400.0	51.1	668.5
.1	748.9	15.6	17.2	.1	2357.8	24.5	92.0	.1	4677.8	33.4	226.1	.1	7708.9	42.3	419.4	.1	11451.2	51.2	672.0
.2	764.7	15.7	17.8	.2	2382.4	24.6	93.4	.2	4711.3	33.5	228.2	.2	7751.3	42.4	422.2	.2	11502.4	51.3	675.6
.3	780.5	15.8	18.4	.3	2407.2	24.7	94.7	.3	4744.9	33.6	230.2	.3	7793.8	42.5	425.0	.3	11553.8	51.4	679.1
.4	796.4	15.9	19.0	.4	2432.0	24.8	96.0	.4	4778.7	33.7	232.3	.4	7836.4	42.6	427.9	.4	11605.3	51.5	682.7
.5	812.5	16.1	19.6	.5	2456.9	24.9	97.3	.5	4812.5	33.8	234.4	.5	7879.2	42.7	430.7	.5	11656.9	51.6	686.2
.6	828.7	16.2	20.2	.6	2482.0	25.1	98.7	.6	4846.4	33.9	236.5	.6	7922.0	42.8	433.5	.6	11708.7	51.7	689.8
.7	844.9	16.3	20.8	.7	2507.2	25.2	100.0	.7	4880.5	34.1	238.6	.7	7964.9	42.9	436.3	.7	11760.5	51.8	693.4
.8	861.3	16.4	21.4	.8	2532.4	25.3	101.4	.8	4914.7	34.2	240.7	.8	8008.0	43.1	439.2	.8	11812.4	51.9	697.0
.9	877.8	16.5	22.0	.9	2557.8	25.4	102.8	.9	4948.9	34.3	242.8	.9	8051.2	43.2	442.0	.9	11864.5	52.1	700.6
7.0	894.4	16.6	22.7	15.0	2583.3	25.5	104.2	23.0	4983.3	34.4	244.9	31.0	8094.4	43.3	444.9	39.0	11916.7	52.2	704.2
.1	911.2	16.7	23.3	.1	2608.9	25.6	105.6	.1	5017.8	34.5	247.0	.1	8137.8	43.4	447.8	.1	11968.9	52.3	707.8
.2	928.0	16.8	24.0	.2	2634.7	25.7	107.0	.2	5052.4	34.6	249.2	.2	8181.3	43.5	450.7	.2	12021.3	52.4	711.4
.3	944.9	16.9	24.7	.3	2660.5	25.8	108.4	.3	5087.2	34.7	251.3	.3	8224.9	43.6	453.6	.3	12073.8	52.5	715.0
.4	962.0	17.1	25.4	.4	2686.4	25.9	109.8	.4	5122.0	34.8	253.5	.4	8268.7	43.7	456.5	.4	12126.4	52.6	718.7
.5	979.2	17.2	26.0	.5	2712.5	26.1	111.2	.5	5156.9	34.9	255.7	.5	8312.5	43.8	459.4	.5	12179.2	52.7	722.3
.6	996.4	17.3	26.7	.6	2738.7	26.2	112.7	.6	5192.0	35.1	257.9	.6	8356.4	43.9	462.3	.6	12232.0	52.8	726.0
.7	1013.8	17.4	27.4	.7	2764.9	26.3	114.1	.7	5227.2	35.2	260.0	.7	8400.5	44.1	465.2	.7	12284.9	52.9	729.7
.8	1031.3	17.5	28.2	.8	2791.3	26.4	115.6	.8	5262.4	35.3	262.2	.8	8444.7	44.2	468.2	.8	12338.0	53.1	733.4
.9	1048.9	17.6	28.9	.9	2817.8	26.5	117.0	.9	5297.8	35.4	264.5	.9	8488.9	44.3	471.1	.9	12391.2	53.2	737.0

TABLE NO. 15.

EXCAVATION AND EMBANKMENT.

CUBIC YARDS.

PRISMOIDS 100 FEET LONG. **BREADTH OF BASE 26 FEET.**

Slopes 1 1-2 Horizontal to 1 Perpendicular.

Ht.	Cub. Yds.	Diff.	Cor.	Ht.	Cub. Yds.	Diff.	Cor.	Ht.	Cub. Yds.	Diff.	Cor.	Ht.	Cub. Yds.	Diff.	Cor.	Ht.	Cub. Yds.	Diff.	Cor.
0.0	0.0	0.0	0.0	8.0	1125.9	18.5	29.6	16.0	2963.0	27.4	118.5	24.0	5511.1	36.2	266.7	32.0	8770.4	45.1	474.1
.1	9.7	9.7	0.0	.1	1144.5	18.6	30.4	.1	2990.4	27.5	120.0	.1	5547.3	36.4	268.9	.1	8815.6	45.2	477.0
.2	19.5	9.8	0.0	.2	1163.2	18.7	31.1	.2	3018.0	27.6	121.5	.2	5583.9	36.5	271.1	.2	8861.0	45.4	480.0
.3	29.4	9.9	0.0	.3	1182.0	18.8	31.9	.3	3045.7	27.7	123.0	.3	5620.5	36.6	273.4	.3	8906.4	45.5	483.0
.4	39.4	10.0	0.1	.4	1200.9	18.9	32.7	.4	3073.5	27.8	124.5	.4	5657.2	36.7	275.6	.4	8952.0	45.6	486.0
.5	49.5	10.1	0.1	.5	1219.9	19.0	33.4	.5	3101.4	27.9	126.0	.5	5694.0	36.8	277.9	.5	8997.7	45.7	489.0
.6	59.8	10.2	0.2	.6	1239.0	19.1	34.2	.6	3129.4	28.0	127.6	.6	5730.9	36.9	280.2	.6	9043.5	45.8	492.0
.7	70.1	10.4	0.2	.7	1258.3	19.2	35.0	.7	3157.5	28.1	129.1	.7	5767.9	37.0	282.5	.7	9089.4	45.9	495.0
.8	80.6	10.5	0.3	.8	1277.6	19.4	35.9	.8	3185.8	28.2	130.7	.8	5805.0	37.1	284.7	.8	9135.4	46.0	498.0
.9	91.2	10.6	0.4	.9	1297.1	19.5	36.7	.9	3214.1	28.4	132.2	.9	5842.3	37.2	287.0	.9	9181.5	46.1	501.1
1.0	101.9	10.7	0.5	9.0	1316.7	19.6	37.5	17.0	3242.6	28.5	133.8	25.0	5879.6	37.4	289.4	33.0	9227.8	46.2	504.2
.1	112.6	10.8	0.6	.1	1336.4	19.7	38.3	.1	3271.2	28.6	135.4	.1	5917.1	37.5	291.7	.1	9274.1	46.4	507.2
.2	123.6	10.9	0.7	.2	1356.1	19.8	39.2	.2	3299.9	28.7	137.0	.2	5954.7	37.6	294.0	.2	9320.6	46.5	510.3
.3	134.6	11.0	0.8	.3	1376.1	19.9	40.0	.3	3328.6	28.8	138.6	.3	5992.4	37.7	296.3	.3	9367.2	46.6	513.4
.4	145.7	11.1	0.9	.4	1396.1	20.0	40.9	.4	3357.6	28.9	140.2	.4	6030.1	37.8	298.7	.4	9413.9	46.7	516.5
.5	156.9	11.2	1.0	.5	1416.2	20.1	41.8	.5	3386.6	29.0	141.8	.5	6068.1	37.9	301.0	.5	9460.6	46.8	519.6
.6	168.3	11.4	1.2	.6	1436.4	20.2	42.7	.6	3415.7	29.1	143.4	.6	6106.1	38.0	303.4	.6	9507.6	46.9	522.7
.7	179.8	11.5	1.3	.7	1456.8	20.4	43.6	.7	3444.9	29.2	145.0	.7	6144.2	38.1	305.8	.7	9554.6	47.0	525.8
.8	191.3	11.6	1.5	.8	1477.3	20.5	44.5	.8	3474.3	29.4	146.7	.8	6182.4	38.2	308.2	.8	9601.7	47.1	528.9
.9	203.0	11.7	1.7	.9	1497.8	20.6	45.4	.9	3503.8	29.5	148.3	.9	6220.8	38.4	310.6	.9	9648.9	47.2	532.0
2.0	214.8	11.8	1.9	10.0	1518.5	20.7	46.3	18.0	3533.3	29.6	150.0	26.0	6259.3	38.5	313.0	34.0	9696.3	47.4	535.2
.1	226.7	11.9	2.0	.1	1539.3	20.8	47.2	.1	3563.0	29.7	151.7	.1	6297.8	38.6	315.4	.1	9743.8	47.5	538.3
.2	238.7	12.0	2.2	.2	1560.2	20.9	48.2	.2	3592.8	29.8	153.4	.2	6336.5	38.7	317.8	.2	9791.3	47.6	541.5
.3	250.9	12.1	2.4	.3	1581.2	21.0	49.1	.3	3622.7	29.9	155.0	.3	6375.3	38.8	320.2	.3	9839.0	47.7	544.7
.4	263.1	12.2	2.7	.4	1602.4	21.1	50.1	.4	3652.7	30.0	156.7	.4	6414.2	38.9	322.7	.4	9886.8	47.8	547.9
.5	275.5	12.4	2.9	.5	1623.6	21.2	51.0	.5	3682.9	30.1	158.4	.5	6453.2	39.0	325.1	.5	9934.7	47.9	551.0
.6	287.9	12.5	3.1	.6	1645.0	21.4	52.0	.6	3713.1	30.2	160.2	.6	6492.4	39.1	327.6	.6	9982.7	48.0	554.2
.7	300.5	12.6	3.4	.7	1666.4	21.5	53.0	.7	3743.5	30.4	161.9	.7	6531.6	39.2	330.0	.7	10030.9	48.1	557.5
.8	313.2	12.7	3.6	.8	1688.0	21.6	54.0	.8	3773.9	30.5	163.6	.8	6571.0	39.4	332.5	.8	10079.1	48.2	560.7
.9	326.0	12.8	3.9	.9	1709.7	21.7	55.0	.9	3804.5	30.6	165.4	.9	6610.4	39.5	335.0	.9	10127.5	48.4	563.9
3.0	338.9	12.9	4.2	11.0	1731.5	21.8	56.0	19.0	3835.2	30.7	167.1	27.0	6650.0	39.6	337.5	35.0	10175.9	48.5	567.1
.1	351.9	13.0	4.4	.1	1753.4	21.9	57.0	.1	3866.0	30.8	168.9	.1	6689.7	39.7	340.0	.1	10224.5	48.6	570.4
.2	365.0	13.1	4.7	.2	1775.4	22.0	58.1	.2	3896.9	30.9	170.7	.2	6729.5	39.8	342.5	.2	10273.2	48.7	573.6
.3	378.3	13.2	5.0	.3	1797.5	22.1	59.1	.3	3927.9	31.0	172.4	.3	6769.4	39.9	345.0	.3	10322.0	48.8	576.9
.4	391.6	13.4	5.4	.4	1819.8	22.2	60.2	.4	3959.0	31.1	174.2	.4	6809.4	40.0	347.6	.4	10370.9	48.9	580.2
.5	405.1	13.5	5.7	.5	1842.1	22.4	61.2	.5	3990.3	31.2	176.0	.5	6849.5	40.1	350.1	.5	10419.9	49.0	583.4
.6	418.7	13.6	6.0	.6	1864.6	22.5	62.3	.6	4021.6	31.4	177.9	.6	6889.8	40.2	352.7	.6	10469.0	49.1	586.7
.7	432.4	13.7	6.3	.7	1887.2	22.6	63.4	.7	4053.1	31.5	179.7	.7	6930.1	40.4	355.2	.7	10518.3	49.2	590.0
.8	446.1	13.8	6.7	.8	1909.9	22.7	64.5	.8	4084.7	31.6	181.5	.8	6970.6	40.5	357.8	.8	10567.6	49.4	593.4
.9	460.1	13.9	7.0	.9	1932.6	22.8	65.6	.9	4116.4	31.7	183.3	.9	7011.2	40.6	360.4	.9	10617.1	49.5	596.7
4.0	474.1	14.0	7.4	12.0	1955.6	22.9	66.7	20.0	4148.1	31.8	185.2	28.0	7051.9	40.7	363.0	36.0	10666.7	49.6	600.0
.1	488.2	14.1	7.8	.1	1978.6	23.0	67.8	.1	4180.1	31.9	187.0	.1	7092.6	40.8	365.6	.1	10716.4	49.7	603.3
.2	502.4	14.2	8.1	.2	2001.7	23.1	68.9	.2	4212.1	32.0	188.9	.2	7133.6	40.9	368.2	.2	10766.1	49.8	606.7
.3	516.8	14.4	8.6	.3	2024.9	23.2	70.0	.3	4244.2	32.1	190.8	.3	7174.6	41.0	370.8	.3	10816.1	49.9	610.0
.4	531.3	14.5	9.0	.4	2048.3	23.4	71.2	.4	4276.4	32.2	192.7	.4	7215.7	41.1	373.4	.4	10866.1	50.0	613.4
.5	545.8	14.6	9.4	.5	2071.8	23.5	72.3	.5	4308.8	32.4	194.6	.5	7256.9	41.2	376.0	.5	10916.2	50.1	616.8
.6	560.5	14.7	9.8	.6	2095.3	23.6	73.5	.6	4341.3	32.5	196.5	.6	7298.3	41.4	378.7	.6	10966.4	50.2	620.2
.7	575.3	14.8	10.2	.7	2119.0	23.7	74.7	.7	4373.8	32.6	198.4	.7	7339.8	41.5	381.3	.7	11016.8	50.4	623.6
.8	590.2	14.9	10.7	.8	2142.8	23.8	75.9	.8	4406.5	32.7	200.3	.8	7381.3	41.6	384.0	.8	11067.3	50.5	627.0
.9	605.2	15.0	11.1	.9	2166.7	23.9	77.0	.9	4439.3	32.8	202.2	.9	7423.0	41.7	386.7	.9	11117.8	50.6	630.4
5.0	620.4	15.1	11.6	13.0	2190.7	24.0	78.2	21.0	4472.2	32.9	204.2	29.0	7464.8	41.8	389.4	37.0	11168.5	50.7	633.8
.1	635.6	15.2	12.0	.1	2214.9	24.1	79.4	.1	4505.2	33.0	206.1	.1	7506.7	41.9	392.0	.1	11219.3	50.8	637.2
.2	651.0	15.4	12.5	.2	2239.1	24.2	80.7	.2	4538.4	33.1	208.1	.2	7548.7	42.0	394.7	.2	11270.2	50.9	640.7
.3	666.4	15.5	13.0	.3	2263.5	24.4	81.9	.3	4571.6	33.2	210.0	.3	7590.9	42.1	397.4	.3	11321.2	51.0	644.1
.4	682.0	15.6	13.5	.4	2287.9	24.5	83.1	.4	4605.0	33.4	212.0	.4	7633.1	42.2	400.2	.4	11372.4	51.1	647.6
.5	697.7	15.7	14.0	.5	2312.5	24.6	84.4	.5	4638.4	33.5	214.0	.5	7675.5	42.4	402.9	.5	11423.6	51.2	651.0
.6	713.5	15.8	14.5	.6	2337.2	24.7	85.6	.6	4672.0	33.6	216.0	.6	7717.9	42.5	405.6	.6	11475.0	51.4	654.5
.7	729.4	15.9	15.0	.7	2362.0	24.8	86.9	.7	4705.7	33.7	218.0	.7	7760.5	42.6	408.4	.7	11526.4	51.5	658.0
.8	745.4	16.0	15.6	.8	2386.9	24.9	88.2	.8	4739.5	33.8	220.0	.8	7803.2	42.7	411.1	.8	11578.0	51.6	661.5
.9	761.5	16.1	16.1	.9	2411.9	25.0	89.4	.9	4773.4	33.9	222.0	.9	7846.0	42.8	413.9	.9	11629.7	51.7	665.0
6.0	777.8	16.2	16.7	14.0	2437.0	25.1	90.7	22.0	4807.4	34.0	224.1	30.0	7888.9	42.9	416.7	38.0	11681.5	51.8	668.5
.1	794.1	16.4	17.2	.1	2462.3	25.2	92.0	.1	4841.5	34.1	226.1	.1	7931.9	43.0	419.4	.1	11733.4	51.9	672.0
.2	810.6	16.5	17.8	.2	2487.6	25.4	93.4	.2	4875.8	34.2	228.2	.2	7975.0	43.1	422.2	.2	11785.4	52.0	675.6
.3	827.2	16.6	18.4	.3	2513.1	25.5	94.7	.3	4910.1	34.4	230.2	.3	8018.3	43.2	425.0	.3	11837.5	52.1	679.1
.4	843.9	16.7	19.0	.4	2538.7	25.6	96.0	.4	4944.6	34.5	232.3	.4	8061.6	43.4	427.9	.4	11889.8	52.2	682.7
.5	860.6	16.8	19.6	.5	2564.4	25.7	97.3	.5	4979.2	34.6	234.4	.5	8105.1	43.5	430.7	.5	11942.1	52.4	686.2
.6	877.6	16.9	20.2	.6	2590.1	25.8	98.7	.6	5013.9	34.7	236.5	.6	8148.7	43.6	433.5	.6	11994.6	52.5	689.8
.7	894.6	17.0	20.8	.7	2616.1	25.9	100.0	.7	5048.6	34.8	238.6	.7	8192.4	43.7	436.3	.7	12047.2	52.6	693.4
.8	911.7	17.1	21.4	.8	2642.1	26.0	101.4	.8	5083.6	34.9	240.7	.8	8236.1	43.8	439.2	.8	12099.9	52.7	697.0
.9	928.9	17.2	22.0	.9	2668.2	26.1	102.8	.9	5118.6	35.0	242.8	.9	8280.1	43.9	442.0	.9	12152.6	52.8	700.6
7.0	946.3	17.4	22.7	15.0	2694.4	26.2	104.2	23.0	5153.7	35.1	244.9	31.0	8324.1	44.0	444.9	39.0	12205.6	52.9	704.2
.1	963.8	17.5	23.3	.1	2720.8	26.4	105.6	.1	5188.9	35.2	247.0	.1	8368.2	44.1	447.8	.1	12258.6	53.0	707.8
.2	981.3	17.6	24.0	.2	2747.3	26.5	107.0	.2	5224.3	35.4	249.2	.2	8412.4	44.2	450.7	.2	12311.7	53.1	711.4
.3	999.0	17.7	24.7	.3	2773.8	26.6	108.4	.3	5259.8	35.5	251.3	.3	8456.8	44.4	453.6	.3	12364.9	53.2	715.0
.4	1016.8	17.8	25.4	.4	2800.5	26.7	109.8	.4	5295.3	35.6	253.5	.4	8501.3	44.5	456.5	.4	12418.3	53.4	718.7
.5	1034.7	17.9	26.0	.5	2827.3	26.8	111.2	.5	5331.0	35.7	255.7	.5	8545.8	44.6	459.4	.5	12471.8	53.5	722.3
.6	1052.7	18.0	26.7	.6	2854.2	26.9	112.7	.6	5366.8	35.8	257.9	.6	8590.5	44.7	462.3	.6	12525.3	53.6	726.0
.7	1070.9	18.1	27.4	.7	2881.2	27.0	114.1	.7	5402.7	35.9	260.0	.7	8635.3	44.8	465.2	.7	12579.0	53.7	729.7
.8	1089.1	18.2	28.2	.8	2908.4	27.1	115.6	.8	5438.7	36.0	262.2	.8	8680.2	44.9	468.2	.8	12632.8	53.8	733.4
.9	1107.5	18.4	28.9	.9	2935.6	27.2	117.0	.9	5474.9	36.1	264.5	.9	8725.2	45.0	471.1	.9	12686.7	53.9	737.0

TABLE NO. 15.

EXCAVATION AND EMBANKMENT.

CUBIC YARDS.

PRISMOIDS 100 FEET LONG. **BREADTH OF BASE 26 FEET.**

Slopes 1 1-2 Horizontal to 1 Perpendicular.

Ht.	Cub. Yds.	Diff.	Cor.	Ht.	Cub. Yds.	Diff.	Cor.	Ht.	Cub. Yds.	Diff.	Cor.	Ht.	Cub. Yds.	Diff.	Cor.	Ht.	Cub. Yds.	Diff.	Cor.
40.0	12740.7	54.0	740.7	48.0	17422.2	62.9	1066.7	56.0	22814.8	71.8	1451.9	64.0	28918.5	80.7	1896.3	72.0	35733.3	89.6	2400.0
.1	12794.9	54.1	744.4	.1	17485.2	63.0	1071.1	.1	22886.7	71.9	1457.0	.1	28999.3	80.8	1902.2	.1	35823.0	89.7	2406.7
.2	12849.1	54.2	748.2	.2	17548.4	63.1	1075.6	.2	22958.7	72.0	1462.2	.2	29080.2	80.9	1908.2	.2	35912.8	89.8	2413.4
.3	12903.5	54.4	751.9	.3	17611.6	63.2	1080.0	.3	23030.9	72.1	1467.5	.3	29161.2	81.0	1914.1	.3	36002.7	89.9	2420.0
.4	12957.9	54.5	755.6	.4	17675.0	63.4	1084.5	.4	23103.1	72.2	1472.7	.4	29242.4	81.1	1920.1	.4	36092.7	90.0	2426.7
.5	13012.5	54.6	759.4	.5	17738.4	63.5	1089.0	.5	23175.5	72.4	1477.9	.5	29323.6	81.2	1926.0	.5	36182.9	90.1	2433.4
.6	13067.2	54.7	763.1	.6	17802.0	63.6	1093.5	.6	23247.9	72.5	1483.1	.6	29405.0	81.4	1932.0	.6	36273.1	90.2	2440.2
.7	13122.0	54.8	766.9	.7	17865.7	63.7	1098.0	.7	23320.5	72.6	1488.4	.7	29486.4	81.5	1938.0	.7	36363.5	90.4	2446.9
.8	13176.9	54.9	770.7	.8	17929.5	63.8	1102.5	.8	23393.2	72.7	1493.6	.8	29568.0	81.6	1944.0	.8	36453.9	90.5	2453.6
.9	13231.9	55.0	774.4	.9	17993.4	63.9	1107.0	.9	23466.0	72.8	1498.9	.9	29649.7	81.7	1950.0	.9	36544.5	90.6	2460.4
41.0	13287.0	55.1	778.2	49.0	18057.4	64.0	1111.6	57.0	23538.9	72.9	1504.2	65.0	29731.5	81.8	1956.0	73.0	36635.2	90.7	2467.1
.1	13342.3	55.2	782.0	.1	18121.5	64.1	1116.1	.1	23611.9	73.0	1509.5	.1	29813.4	81.9	1962.0	.1	36726.0	90.8	2473.9
.2	13397.6	55.4	785.9	.2	18185.8	64.2	1120.7	.2	23685.0	73.1	1514.7	.2	29895.4	82.0	1968.1	.2	36816.9	90.9	2480.7
.3	13453.1	55.5	789.7	.3	18250.1	64.4	1125.2	.3	23758.3	73.2	1520.0	.3	29977.5	82.1	1974.1	.3	36907.9	91.0	2487.4
.4	13508.7	55.6	793.5	.4	18314.6	64.5	1129.8	.4	23831.6	73.4	1525.4	.4	30059.8	82.2	1980.2	.4	36999.0	91.1	2494.2
.5	13564.4	55.7	797.3	.5	18379.2	64.6	1134.4	.5	23905.1	73.5	1530.7	.5	30142.1	82.4	1986.2	.5	37090.3	91.2	2501.0
.6	13620.1	55.8	801.2	.6	18443.9	64.7	1139.0	.6	23978.7	73.6	1536.0	.6	30224.6	82.5	1992.3	.6	37181.6	91.4	2507.9
.7	13676.1	55.9	805.0	.7	18508.6	64.8	1143.6	.7	24052.4	73.7	1541.3	.7	30307.2	82.6	1998.4	.7	37273.1	91.5	2514.7
.8	13732.1	56.0	808.9	.8	18573.6	64.9	1148.2	.8	24126.1	73.8	1546.7	.8	30389.9	82.7	2004.5	.8	37364.7	91.6	2521.5
.9	13788.2	56.1	812.8	.9	18638.6	65.0	1152.8	.9	24200.1	73.9	1552.0	.9	30472.6	82.8	2010.6	.9	37456.4	91.7	2528.3
42.0	13844.4	56.2	816.7	50.0	18703.7	65.1	1157.4	58.0	24274.1	74.0	1557.4	66.0	30555.6	82.9	2016.7	74.0	37548.1	91.8	2535.2
.1	13900.8	56.4	820.6	.1	18768.9	65.2	1162.0	.1	24348.2	74.1	1562.8	.1	30638.6	83.0	2022.8	.1	37640.1	91.9	2542.0
.2	13957.3	56.5	824.5	.2	18834.3	65.4	1166.7	.2	24422.4	74.2	1568.2	.2	30721.7	83.1	2028.9	.2	37732.1	92.0	2548.9
.3	14013.8	56.6	828.4	.3	18899.8	65.5	1171.3	.3	24496.8	74.4	1573.6	.3	30804.9	83.2	2035.0	.3	37824.2	92.1	2555.8
.4	14070.5	56.7	832.3	.4	18965.3	65.6	1176.0	.4	24571.3	74.5	1579.0	.4	30888.3	83.4	2041.2	.4	37916.4	92.2	2562.7
.5	14127.3	56.8	836.2	.5	19031.0	65.7	1180.7	.5	24645.8	74.6	1584.4	.5	30971.8	83.5	2047.3	.5	38008.8	92.4	2569.6
.6	14184.2	56.9	840.2	.6	19096.8	65.8	1185.4	.6	24720.5	74.7	1589.8	.6	31055.3	83.6	2053.5	.6	38101.3	92.5	2576.5
.7	14241.2	57.0	844.1	.7	19162.7	65.9	1190.0	.7	24795.3	74.8	1595.2	.7	31139.0	83.7	2059.7	.7	38193.8	92.6	2583.4
.8	14298.4	57.1	848.1	.8	19228.7	66.0	1194.7	.8	24870.2	74.9	1600.7	.8	31222.8	83.8	2065.9	.8	38286.5	92.7	2590.3
.9	14355.6	57.2	852.0	.9	19294.9	66.1	1199.5	.9	24945.2	75.0	1606.1	.9	31306.7	83.9	2072.0	.9	38379.3	92.8	2597.2
43.0	14413.0	57.4	856.0	51.0	19361.1	66.2	1204.2	59.0	25020.4	75.1	1611.6	67.0	31390.7	84.0	2078.2	75.0	38472.2	92.9	2604.2
.1	14470.4	57.5	860.0	.1	19427.5	66.4	1208.9	.1	25095.6	75.2	1617.0	.1	31474.9	84.1	2084.4	.1	38565.2	93.0	2611.1
.2	14528.0	57.6	864.0	.2	19493.9	66.5	1213.6	.2	25171.0	75.4	1622.5	.2	31559.1	84.2	2090.7	.2	38658.4	93.1	2618.1
.3	14585.7	57.7	868.0	.3	19560.5	66.6	1218.4	.3	25246.4	75.5	1628.0	.3	31643.5	84.4	2096.9	.3	38751.6	93.2	2625.0
.4	14643.5	57.8	872.0	.4	19627.2	66.7	1223.1	.4	25322.0	75.6	1633.5	.4	31727.9	84.5	2103.1	.4	38845.0	93.4	2632.0
.5	14701.4	57.9	876.0	.5	19694.0	66.8	1227.9	.5	25397.7	75.7	1639.0	.5	31812.5	84.6	2109.4	.5	38938.4	93.5	2639.0
.6	14759.4	58.0	880.1	.6	19760.9	66.9	1232.7	.6	25473.5	75.8	1644.5	.6	31897.2	84.7	2115.6	.6	39032.0	93.6	2646.0
.7	14817.5	58.1	884.1	.7	19827.9	67.0	1237.4	.7	25549.4	75.9	1650.0	.7	31982.0	84.8	2121.9	.7	39125.7	93.7	2653.0
.8	14875.8	58.2	888.2	.8	19895.0	67.1	1242.2	.8	25625.4	76.0	1655.6	.8	32066.9	84.9	2128.2	.8	39219.5	93.8	2660.0
.9	14934.1	58.4	892.2	.9	19962.3	67.2	1247.0	.9	25701.5	76.1	1661.1	.9	32151.9	85.0	2134.5	.9	39313.4	93.9	2667.0
44.0	14992.6	58.5	896.3	52.0	20029.6	67.4	1251.9	60.0	25777.8	76.2	1666.7	68.0	32237.0	85.1	2140.7	76.0	39407.4	94.0	2674.1
.1	15051.2	58.6	900.4	.1	20097.1	67.5	1256.7	.1	25854.1	76.4	1672.2	.1	32322.3	85.2	2147.0	.1	39501.5	94.1	2681.1
.2	15109.9	58.7	904.5	.2	20164.7	67.6	1261.5	.2	25930.6	76.5	1677.8	.2	32407.6	85.4	2153.4	.2	39595.8	94.2	2688.2
.3	15168.6	58.8	908.6	.3	20232.4	67.7	1266.3	.3	26007.2	76.6	1683.4	.3	32493.1	85.5	2159.7	.3	39690.1	94.4	2695.2
.4	15227.6	58.9	912.7	.4	20300.1	67.8	1271.2	.4	26083.9	76.7	1689.0	.4	32578.7	85.6	2166.0	.4	39784.6	94.5	2702.3
.5	15286.6	59.0	916.8	.5	20368.1	67.9	1276.0	.5	26160.6	76.8	1694.6	.5	32664.4	85.7	2172.3	.5	39879.2	94.6	2709.4
.6	15345.7	59.1	920.9	.6	20436.1	68.0	1280.9	.6	26237.6	76.9	1700.2	.6	32750.1	85.8	2178.7	.6	39973.9	94.7	2716.5
.7	15404.9	59.2	925.0	.7	20504.2	68.1	1285.8	.7	26314.6	77.0	1705.8	.7	32836.1	85.9	2185.0	.7	40068.6	94.8	2723.6
.8	15464.3	59.4	929.2	.8	20572.4	68.2	1290.7	.8	26391.7	77.1	1711.4	.8	32922.1	86.0	2191.4	.8	40163.6	94.9	2730.7
.9	15523.8	59.5	933.3	.9	20640.8	68.4	1295.6	.9	26468.9	77.2	1717.0	.9	33008.2	86.1	2197.8	.9	40258.6	95.0	2737.8
45.0	15583.3	59.6	937.5	53.0	20709.3	68.5	1300.5	61.0	26546.3	77.4	1722.7	69.0	33094.4	86.2	2204.2	77.0	40353.7	95.1	2744.9
.1	15643.0	59.7	941.7	.1	20777.8	68.6	1305.4	.1	26623.8	77.5	1728.3	.1	33180.8	86.4	2210.6	.1	40448.9	95.2	2752.0
.2	15702.8	59.8	945.9	.2	20846.5	68.7	1310.3	.2	26701.3	77.6	1734.0	.2	33267.3	86.5	2217.0	.2	40544.3	95.4	2759.2
.3	15762.7	59.9	950.0	.3	20915.3	68.8	1315.2	.3	26779.0	77.7	1739.7	.3	33353.8	86.6	2223.4	.3	40639.8	95.5	2766.3
.4	15822.7	60.0	954.2	.4	20984.2	68.9	1320.2	.4	26856.8	77.8	1745.4	.4	33440.5	86.7	2229.8	.4	40735.3	95.6	2773.5
.5	15882.9	60.1	958.4	.5	21053.2	69.0	1325.1	.5	26934.7	77.9	1751.0	.5	33527.3	86.8	2236.2	.5	40831.0	95.7	2780.7
.6	15943.1	60.2	962.7	.6	21122.4	69.1	1330.1	.6	27012.7	78.0	1756.7	.6	33614.2	86.9	2242.7	.6	40926.8	95.8	2787.9
.7	16003.5	60.4	966.9	.7	21191.6	69.2	1335.0	.7	27090.9	78.1	1762.5	.7	33701.2	87.0	2249.1	.7	41022.7	95.9	2795.0
.8	16063.9	60.5	971.1	.8	21261.0	69.4	1340.0	.8	27169.1	78.2	1768.2	.8	33788.4	87.1	2255.6	.8	41118.7	96.0	2802.2
.9	16124.5	60.6	975.4	.9	21330.4	69.5	1345.0	.9	27247.5	78.4	1773.9	.9	33875.6	87.2	2262.0	.9	41214.9	96.1	2809.4
46.0	16185.2	60.7	979.6	54.0	21400.0	69.6	1350.0	62.0	27325.9	78.5	1779.6	70.0	33963.0	87.4	2268.5	78.0	41311.1	96.2	2816.7
.1	16246.0	60.8	983.9	.1	21469.7	69.7	1355.0	.1	27404.5	78.6	1785.4	.1	34050.4	87.5	2275.0	.1	41407.5	96.4	2823.9
.2	16306.9	60.9	988.2	.2	21539.5	69.8	1360.0	.2	27483.2	78.7	1791.1	.2	34138.0	87.6	2281.5	.2	41503.9	96.5	2831.1
.3	16367.9	61.0	992.4	.3	21609.4	69.9	1365.0	.3	27562.0	78.8	1796.9	.3	34225.7	87.7	2288.0	.3	41600.5	96.6	2838.4
.4	16429.0	61.1	996.7	.4	21679.4	70.0	1370.1	.4	27640.9	78.9	1802.7	.4	34313.5	87.8	2294.5	.4	41697.2	96.7	2845.6
.5	16490.3	61.2	1001.0	.5	21749.5	70.1	1375.1	.5	27719.9	79.0	1808.5	.5	34401.4	87.9	2301.0	.5	41794.0	96.8	2852.9
.6	16551.6	61.4	1005.4	.6	21819.8	70.2	1380.2	.6	27799.0	79.1	1814.2	.6	34489.4	88.0	2307.6	.6	41890.9	96.9	2860.2
.7	16613.1	61.5	1009.7	.7	21890.1	70.4	1385.2	.7	27878.3	79.2	1820.0	.7	34577.5	88.1	2314.1	.7	41987.9	97.0	2867.4
.8	16674.7	61.6	1014.0	.8	21960.6	70.5	1390.3	.8	27957.6	79.4	1825.9	.8	34665.8	88.2	2320.7	.8	42085.0	97.1	2874.7
.9	16736.4	61.7	1018.3	.9	22031.2	70.6	1395.4	.9	28037.1	79.5	1831.7	.9	34754.1	88.4	2327.2	.9	42182.3	97.2	2882.0
47.0	16798.1	61.8	1022.7	55.0	22101.9	70.7	1400.5	63.0	28116.7	79.6	1837.5	71.0	34842.6	88.5	2333.8	79.0	42279.6	97.4	2889.4
.1	16860.1	61.9	1027.0	.1	22172.6	70.8	1405.6	.1	28196.4	79.7	1843.3	.1	34931.2	88.6	2340.4	.1	42377.1	97.5	2896.7
.2	16922.1	62.0	1031.4	.2	22243.6	70.9	1410.7	.2	28276.1	79.8	1849.2	.2	35019.9	88.7	2347.0	.2	42474.7	97.6	2904.0
.3	16984.2	62.1	1035.8	.3	22314.6	71.0	1415.8	.3	28356.1	79.9	1855.0	.3	35108.6	88.8	2353.6	.3	42572.4	97.7	2911.3
.4	17046.4	62.2	1040.2	.4	22385.7	71.1	1420.9	.4	28436.1	80.0	1860.9	.4	35197.6	88.9	2360.2	.4	42670.1	97.8	2918.7
.5	17108.8	62.4	1044.6	.5	22456.8	71.2	1426.0	.5	28516.2	80.1	1866.8	.5	35286.6	89.0	2366.8	.5	42768.1	97.9	2926.0
.6	17171.3	62.5	1049.0	.6	22528.3	71.4	1431.2	.6	28596.4	80.2	1872.7	.6	35375.7	89.1	2373.4	.6	42866.1	98.0	2933.4
.7	17233.8	62.6	1053.4	.7	22599.8	71.5	1436.3	.7	28676.8	80.4	1878.6	.7	35464.9	89.2	2380.0	.7	42964.2	98.1	2940.8
.8	17296.5	62.7	1057.8	.8	22671.3	71.6	1441.5	.8	28757.3	80.5	1884.5	.8	35554.3	89.4	2386.7	.8	43062.4	98.2	2948.2
.9	17359.3	62.8	1062.2	.9	22743.0	71.7	1446.7	.9	28837.8	80.6	1890.4	.9	35643.8	89.5	2393.3	.9	43160.8	98.4	2955.0

TABLE NO. 16.

EXCAVATION AND EMBANKMENT.

CUBIC YARDS.

PRISMOIDS 100 FEET LONG. **BREADTH OF BASE 28 FEET.**

Slopes 1 1-2 Horizontal to 1 Perpendicular.

Ht.	Cub. Yds.	Diff.	Cor.	Ht.	Cub. Yds.	Diff.	Cor.	Ht.	Cub. Yds.	Diff.	Cor.	Ht.	Cub. Yds.	Diff.	Cor.	Ht.	Cub. Yds.	Diff.	Cor.
0.0	0.0	0.0	0.0	8.0	1185.2	19.2	29.6	16.0	3080.5	28.1	118.5	24.0	5688.9	37.0	266.7	32.0	9007.4	45.9	474.1
.1	10.4	10.4	0.0	.1	1204.5	19.3	30.4	.1	3109.7	28.2	120.0	.1	5726.0	37.1	268.9	.1	9053.4	46.0	477.0
.2	21.0	10.5	0.0	.2	1223.9	19.4	31.1	.2	3138.0	28.3	121.5	.2	5763.2	37.2	271.1	.2	9099.5	46.1	480.0
.3	31.6	10.6	0.0	.3	1243.5	19.5	31.9	.3	3166.4	28.4	123.0	.3	5800.5	37.3	273.4	.3	9145.7	46.2	483.0
.4	42.4	10.8	0.1	.4	1263.1	19.6	32.7	.4	3195.0	28.5	124.5	.4	5837.9	37.4	275.6	.4	9192.0	46.3	486.0
.5	53.2	10.9	0.1	.5	1282.9	19.8	33.4	.5	3223.6	28.6	126.0	.5	5875.5	37.5	277.9	.5	9238.4	46.4	489.0
.6	64.2	11.0	0.2	.6	1302.7	19.9	34.2	.6	3252.4	28.8	127.6	.6	5913.1	37.6	280.2	.6	9285.0	46.5	492.0
.7	75.3	11.1	0.2	.7	1322.7	20.0	35.0	.7	3281.2	28.9	129.1	.7	5950.9	37.8	282.5	.7	9331.6	46.6	495.0
.8	86.5	11.2	0.3	.8	1342.8	20.1	35.9	.8	3310.2	29.0	130.7	.8	5988.7	37.9	284.7	.8	9378.4	46.8	498.0
.9	97.8	11.3	0.4	.9	1363.0	20.2	36.7	.9	3339.3	29.1	132.2	.9	6026.7	38.0	287.0	.9	9425.2	46.9	501.1
1.0	109.3	11.4	0.5	9.0	1383.3	20.3	37.5	17.0	3368.5	29.2	133.8	25.0	6064.8	38.1	289.4	33.0	9472.2	47.0	504.2
.1	120.8	11.5	0.6	.1	1403.8	20.4	38.3	.1	3397.8	29.3	135.4	.1	6103.0	38.2	291.7	.1	9519.3	47.1	507.2
.2	132.4	11.6	0.7	.2	1424.3	20.5	39.2	.2	3427.3	29.4	137.0	.2	6141.3	38.3	294.0	.2	9566.5	47.2	510.3
.3	144.2	11.8	0.8	.3	1444.9	20.6	40.0	.3	3456.8	29.5	138.6	.3	6179.8	38.4	296.3	.3	9613.8	47.3	513.4
.4	156.1	11.9	0.9	.4	1465.7	20.8	40.9	.4	3486.4	29.6	140.2	.4	6218.3	38.5	298.7	.4	9661.3	47.4	516.5
.5	168.1	12.0	1.0	.5	1486.6	20.9	41.8	.5	3516.2	29.8	141.8	.5	6256.9	38.6	301.0	.5	9708.8	47.5	519.6
.6	180.1	12.1	1.2	.6	1507.6	21.0	42.7	.6	3546.1	29.9	143.4	.6	6295.7	38.8	303.4	.6	9756.4	47.6	522.7
.7	192.4	12.2	1.3	.7	1528.6	21.1	43.6	.7	3576.1	30.0	145.0	.7	6334.6	38.9	305.8	.7	9804.2	47.8	525.8
.8	204.7	12.3	1.5	.8	1549.9	21.2	44.5	.8	3606.1	30.1	146.7	.8	6373.6	39.0	308.2	.8	9852.1	47.9	528.9
.9	217.1	12.4	1.7	.9	1571.2	21.3	45.4	.9	3636.4	30.2	148.3	.9	6412.6	39.1	310.6	.9	9900.1	48.0	532.0
2.0	229.6	12.5	1.9	10.0	1592.6	21.4	46.3	18.0	3666.7	30.3	150.0	26.0	6451.9	39.2	313.0	34.0	9948.1	48.1	535.2
.1	242.3	12.6	2.0	.1	1614.1	21.5	47.2	.1	3697.1	30.4	151.7	.1	6491.2	39.3	315.4	.1	9996.4	48.2	538.3
.2	255.0	12.8	2.2	.2	1635.8	21.6	48.2	.2	3727.6	30.5	153.4	.2	6530.6	39.4	317.8	.2	10044.7	48.3	541.5
.3	267.9	12.9	2.4	.3	1657.5	21.8	49.1	.3	3758.3	30.6	155.0	.3	6570.1	39.5	320.2	.3	10093.1	48.4	544.7
.4	280.9	13.0	2.7	.4	1679.4	21.9	50.1	.4	3789.0	30.8	156.7	.4	6609.8	39.6	322.7	.4	10141.6	48.5	547.9
.5	294.0	13.1	2.9	.5	1701.4	22.0	51.0	.5	3819.9	30.9	158.4	.5	6649.5	39.8	325.1	.5	10190.3	48.6	551.0
.6	307.2	13.2	3.1	.6	1723.5	22.1	52.0	.6	3850.9	31.0	160.2	.6	6689.4	39.9	327.6	.6	10239.0	48.8	554.2
.7	320.5	13.3	3.4	.7	1745.7	22.2	53.0	.7	3882.0	31.1	161.9	.7	6729.4	40.0	330.0	.7	10287.9	48.9	557.5
.8	333.9	13.4	3.6	.8	1768.0	22.3	54.0	.8	3913.2	31.2	163.6	.8	6769.5	40.1	332.5	.8	10336.9	49.0	560.7
.9	347.5	13.5	3.9	.9	1790.4	22.4	55.0	.9	3944.5	31.3	165.4	.9	6809.7	40.2	335.0	.9	10386.0	49.1	563.9
3.0	361.1	13.6	4.2	11.0	1813.0	22.5	56.0	19.0	3975.9	31.4	167.1	27.0	6850.0	40.3	337.5	35.0	10435.2	49.2	567.1
.1	374.9	13.8	4.4	.1	1835.6	22.6	57.0	.1	4007.5	31.5	168.9	.1	6890.4	40.4	340.0	.1	10484.5	49.3	570.4
.2	388.7	13.9	4.7	.2	1858.4	22.8	58.1	.2	4039.1	31.6	170.7	.2	6931.0	40.5	342.5	.2	10533.9	49.4	573.6
.3	402.7	14.0	5.0	.3	1881.2	22.9	59.1	.3	4070.9	31.8	172.4	.3	6971.6	40.6	345.0	.3	10583.5	49.5	576.9
.4	416.8	14.1	5.4	.4	1904.2	23.0	60.2	.4	4102.7	31.9	174.2	.4	7012.4	40.8	347.6	.4	10633.1	49.6	580.2
.5	431.0	14.2	5.7	.5	1927.3	23.1	61.2	.5	4134.7	32.0	176.0	.5	7053.2	40.9	350.1	.5	10682.9	49.8	583.4
.6	445.3	14.3	6.0	.6	1950.5	23.2	62.3	.6	4166.8	32.1	177.9	.6	7094.2	41.0	352.7	.6	10732.7	49.9	586.7
.7	459.8	14.4	6.3	.7	1973.8	23.3	63.4	.7	4199.0	32.2	179.7	.7	7135.3	41.1	355.2	.7	10782.7	50.0	590.0
.8	474.3	14.5	6.7	.8	1997.3	23.4	64.5	.8	4231.3	32.3	181.5	.8	7176.5	41.2	357.8	.8	10832.8	50.1	593.4
.9	488.9	14.6	7.0	.9	2020.8	23.5	65.6	.9	4263.8	32.4	183.3	.9	7217.8	41.3	360.4	.9	10883.0	50.2	596.7
4.0	503.7	14.8	7.4	12.0	2044.4	23.6	66.7	20.0	4296.3	32.5	185.2	28.0	7259.3	41.4	363.0	36.0	10933.3	50.3	600.0
.1	518.6	14.9	7.8	.1	2068.2	23.8	67.8	.1	4328.9	32.6	187.0	.1	7300.8	41.5	365.6	.1	10983.8	50.4	603.3
.2	533.6	15.0	8.1	.2	2092.1	23.9	68.9	.2	4361.7	32.8	188.9	.2	7342.4	41.6	368.2	.2	11034.3	50.5	606.7
.3	548.6	15.1	8.6	.3	2116.1	24.0	70.0	.3	4394.6	32.9	190.8	.3	7384.2	41.8	370.8	.3	11084.9	50.6	610.0
.4	563.9	15.2	9.0	.4	2140.1	24.1	71.2	.4	4427.6	33.0	192.7	.4	7426.1	41.9	373.4	.4	11135.7	50.8	613.4
.5	579.2	15.3	9.4	.5	2164.4	24.2	72.3	.5	4460.6	33.1	194.6	.5	7468.1	42.0	376.0	.5	11186.6	50.9	616.8
.6	594.6	15.4	9.8	.6	2188.7	24.3	73.5	.6	4493.9	33.2	196.5	.6	7510.1	42.1	378.7	.6	11237.6	51.0	620.2
.7	610.1	15.5	10.2	.7	2213.1	24.4	74.7	.7	4527.2	33.3	198.4	.7	7552.4	42.2	381.3	.7	11288.6	51.1	623.6
.8	625.8	15.6	10.7	.8	2237.6	24.5	75.9	.8	4560.6	33.4	200.3	.8	7594.7	42.3	384.0	.8	11339.9	51.2	627.0
.9	641.5	15.8	11.1	.9	2262.3	24.6	77.0	.9	4594.1	33.5	202.2	.9	7637.1	42.4	386.7	.9	11391.2	51.3	630.4
5.0	657.4	15.9	11.6	13.0	2287.0	24.8	78.2	21.0	4627.8	33.6	204.2	29.0	7679.6	42.5	389.4	37.0	11442.6	51.4	633.8
.1	673.4	16.0	12.0	.1	2311.9	24.9	79.4	.1	4661.5	33.8	206.1	.1	7722.3	42.6	392.0	.1	11494.1	51.5	637.2
.2	689.5	16.1	12.5	.2	2336.9	25.0	80.7	.2	4695.4	33.9	208.1	.2	7765.0	42.8	394.7	.2	11545.8	51.6	640.7
.3	705.7	16.2	13.0	.3	2362.0	25.1	81.9	.3	4729.4	34.0	210.0	.3	7807.9	42.9	397.4	.3	11597.5	51.8	644.1
.4	722.0	16.3	13.5	.4	2387.2	25.2	83.1	.4	4763.5	34.1	212.0	.4	7850.9	43.0	400.2	.4	11649.4	51.9	647.6
.5	738.4	16.4	14.0	.5	2412.5	25.3	84.4	.5	4797.7	34.2	214.0	.5	7894.0	43.1	402.9	.5	11701.4	52.0	651.0
.6	755.0	16.5	14.5	.6	2437.9	25.4	85.6	.6	4832.0	34.3	216.0	.6	7937.2	43.2	405.6	.6	11753.5	52.1	654.5
.7	771.6	16.6	15.0	.7	2463.5	25.5	86.9	.7	4866.4	34.4	218.0	.7	7980.5	43.3	408.4	.7	11805.7	52.2	658.0
.8	788.4	16.8	15.6	.8	2489.1	25.6	88.2	.8	4901.0	34.5	220.0	.8	8023.9	43.4	411.1	.8	11858.0	52.3	661.5
.9	805.2	16.9	16.1	.9	2514.9	25.8	89.4	.9	4935.6	34.6	222.0	.9	8067.5	43.5	413.9	.9	11910.4	52.4	665.0
6.0	822.2	17.0	16.7	14.0	2540.7	25.9	90.7	22.0	4970.4	34.8	224.1	30.0	8111.1	43.6	416.7	38.0	11963.0	52.5	668.5
.1	839.3	17.1	17.2	.1	2566.7	26.0	92.0	.1	5005.2	34.9	226.1	.1	8154.9	43.8	419.4	.1	12015.6	52.6	672.0
.2	856.5	17.2	17.8	.2	2592.8	26.1	93.4	.2	5040.2	35.0	228.2	.2	8198.7	43.9	422.2	.2	12068.4	52.8	675.6
.3	873.8	17.3	18.4	.3	2619.0	26.2	94.7	.3	5075.3	35.1	230.2	.3	8242.7	44.0	425.0	.3	12121.2	52.9	679.1
.4	891.3	17.4	19.0	.4	2645.3	26.3	96.0	.4	5110.5	35.2	232.3	.4	8286.8	44.1	427.9	.4	12174.2	53.0	682.7
.5	908.8	17.5	19.6	.5	2671.8	26.4	97.3	.5	5145.8	35.3	234.4	.5	8331.0	44.2	430.7	.5	12227.3	53.1	686.2
.6	926.4	17.6	20.2	.6	2698.3	26.5	98.7	.6	5181.3	35.4	236.5	.6	8375.3	44.3	433.5	.6	12280.5	53.2	689.8
.7	944.2	17.8	20.8	.7	2724.9	26.6	100.0	.7	5216.8	35.5	238.6	.7	8419.8	44.4	436.3	.7	12333.8	53.3	693.4
.8	962.1	17.9	21.4	.8	2751.7	26.8	101.4	.8	5252.4	35.6	240.7	.8	8464.3	44.5	439.2	.8	12387.3	53.4	697.0
.9	980.1	18.0	22.0	.9	2778.6	26.9	102.8	.9	5288.2	35.8	242.8	.9	8508.9	44.6	442.0	.9	12440.8	53.5	700.6
7.0	998.1	18.1	22.7	15.0	2805.6	27.0	104.2	23.0	5324.1	35.9	244.9	31.0	8553.7	44.8	444.9	39.0	12494.4	53.6	704.2
.1	1016.4	18.2	23.3	.1	2832.6	27.1	105.6	.1	5360.1	36.0	247.0	.1	8598.6	44.9	447.8	.1	12548.2	53.8	707.8
.2	1034.7	18.3	24.0	.2	2859.9	27.2	107.0	.2	5396.1	36.1	249.2	.2	8643.6	45.0	450.7	.2	12602.1	53.9	711.4
.3	1053.1	18.4	24.7	.3	2887.2	27.3	108.4	.3	5432.4	36.2	251.3	.3	8688.6	45.1	453.6	.3	12656.1	54.0	715.0
.4	1071.6	18.5	25.4	.4	2914.6	27.4	109.8	.4	5468.7	36.3	253.5	.4	8733.9	45.2	456.5	.4	12710.1	54.1	718.7
.5	1090.2	18.6	26.0	.5	2942.1	27.5	111.2	.5	5505.1	36.4	255.7	.5	8779.2	45.3	459.4	.5	12764.4	54.2	722.3
.6	1109.0	18.8	26.7	.6	2969.8	27.6	112.7	.6	5541.6	36.5	257.9	.6	8824.6	45.4	462.3	.6	12818.7	54.3	726.0
.7	1127.9	18.9	27.4	.7	2997.5	27.8	114.1	.7	5578.3	36.6	260.0	.7	8870.1	45.5	465.2	.7	12873.1	54.4	729.7
.8	1146.9	19.0	28.2	.8	3025.4	27.9	115.6	.8	5615.0	36.8	262.2	.8	8915.8	45.6	468.2	.8	12927.6	54.5	733.4
.9	1166.0	19.1	28.9	.9	3053.4	28.0	117.0	.9	5651.9	36.9	264.5	.9	8961.5	45.8	471.1	.9	12982.3	54.6	737.0

TABLE NO. 17.

EXCAVATION AND EMBANKMENT.

CUBIC YARDS.

PRISMOIDS 100 FEET LONG. **BREADTH OF BASE 30 FEET.**

Slopes 1 1-2 Horizontal to 1 Perpendicular.

Ht.	Cub. Yds.	Diff.	Cor.	Ht.	Cub. Yds.	Diff.	Cor.	Ht.	Cub. Yds.	Diff.	Cor.	Ht.	Cub. Yds.	Diff.	Cor.	Ht.	Cub. Yds.	Diff.	Cor.
0.0	0.0	0.0	0.0	8.0	1244.4	19.9	29.6	16.0	3200.0	28.8	118.5	24.0	5866.7	37.7	266.7	32.0	9244.4	46.6	474.1
.1	11.2	11.2	0.0	.1	1264.5	20.1	30.4	.1	3228.9	28.9	120.0	.1	5904.5	37.8	268.9	.1	9291.2	46.7	477.0
.2	22.4	11.3	0.0	.2	1284.7	20.2	31.1	.2	3258.0	29.1	121.5	.2	5942.4	37.9	271.1	.2	9338.0	46.8	480.0
.3	33.8	11.4	0.0	.3	1304.9	20.3	31.9	.3	3287.2	29.2	123.0	.3	5980.5	38.1	273.4	.3	9384.9	46.9	483.0
.4	45.3	11.5	0.1	.4	1325.3	20.4	32.7	.4	3316.4	29.3	124.5	.4	6018.7	38.2	275.6	.4	9432.0	47.1	486.0
.5	56.9	11.6	0.1	.5	1345.8	20.5	33.4	.5	3345.8	29.4	126.0	.5	6056.9	38.3	277.9	.5	9479.2	47.2	489.0
.6	68.7	11.7	0.2	.6	1366.4	20.6	34.2	.6	3375.3	29.5	127.6	.6	6095.3	38.4	280.2	.6	9526.4	47.3	492.0
.7	80.5	11.8	0.2	.7	1387.2	20.7	35.0	.7	3404.9	29.6	129.1	.7	6133.8	38.5	282.5	.7	9573.8	47.4	495.0
.8	92.4	11.9	0.3	.8	1408.0	20.8	35.9	.8	3434.7	29.7	130.7	.8	6172.4	38.6	284.7	.8	9621.3	47.5	498.0
.9	104.5	12.1	0.4	.9	1428.9	20.9	36.7	.9	3464.5	29.8	132.2	.9	6211.2	38.7	287.0	.9	9668.9	47.6	501.1
1.0	116.7	12.2	0.5	9.0	1450.0	21.1	37.5	17.0	3494.4	29.9	133.8	25.0	6250.0	38.8	289.4	33.0	9716.7	47.7	504.2
.1	128.9	12.3	0.6	.1	1471.2	21.2	38.3	.1	3524.5	30.1	135.4	.1	6288.9	38.9	291.7	.1	9764.5	47.8	507.2
.2	141.3	12.4	0.7	.2	1492.4	21.3	39.2	.2	3554.7	30.2	137.0	.2	6328.0	39.1	294.0	.2	9812.4	47.9	510.3
.3	153.8	12.5	0.8	.3	1513.8	21.4	40.0	.3	3584.9	30.3	138.6	.3	6367.2	39.2	296.3	.3	9860.5	48.1	513.4
.4	166.4	12.6	0.9	.4	1535.3	21.5	40.9	.4	3615.3	30.4	140.2	.4	6406.4	39.3	298.7	.4	9908.7	48.2	516.5
.5	179.2	12.7	1.0	.5	1556.9	21.6	41.8	.5	3645.8	30.5	141.8	.5	6445.8	39.4	301.0	.5	9956.9	48.3	519.6
.6	192.0	12.8	1.2	.6	1578.7	21.7	42.7	.6	3676.4	30.6	143.4	.6	6485.3	39.5	303.4	.6	10005.3	48.4	522.7
.7	204.9	12.9	1.3	.7	1600.5	21.8	43.6	.7	3707.2	30.7	145.0	.7	6524.9	39.6	305.8	.7	10053.8	48.5	525.8
.8	218.0	13.1	1.5	.8	1622.4	21.9	44.5	.8	3738.0	30.8	146.7	.8	6564.7	39.7	308.2	.8	10102.4	48.6	528.9
.9	231.2	13.2	1.7	.9	1644.5	22.1	45.4	.9	3768.9	30.9	148.3	.9	6604.5	39.8	310.6	.9	10151.2	48.7	532.0
2.0	244.4	13.3	1.9	10.0	1666.7	22.2	46.3	18.0	3800.0	31.1	150.0	26.0	6644.4	39.9	313.0	34.0	10200.0	48.8	535.2
.1	257.8	13.4	2.0	.1	1688.9	22.3	47.2	.1	3831.2	31.2	151.7	.1	6684.5	40.1	315.4	.1	10248.9	48.9	538.3
.2	271.3	13.5	2.2	.2	1711.3	22.4	48.2	.2	3862.4	31.3	153.4	.2	6724.7	40.2	317.8	.2	10298.0	49.1	541.5
.3	284.9	13.6	2.4	.3	1733.8	22.5	49.1	.3	3893.8	31.4	155.0	.3	6764.9	40.3	320.2	.3	10347.2	49.2	544.7
.4	298.7	13.7	2.7	.4	1756.4	22.6	50.1	.4	3925.3	31.5	156.7	.4	6805.3	40.4	322.7	.4	10396.4	49.3	547.9
.5	312.5	13.8	2.9	.5	1779.2	22.7	51.0	.5	3956.9	31.6	158.4	.5	6845.8	40.5	325.1	.5	10445.8	49.4	551.0
.6	326.4	13.9	3.1	.6	1802.0	22.8	52.0	.6	3988.7	31.7	160.2	.6	6886.4	40.6	327.6	.6	10495.3	49.5	554.2
.7	340.5	14.1	3.4	.7	1824.9	22.9	53.0	.7	4020.5	31.8	161.9	.7	6927.2	40.7	330.0	.7	10544.9	49.6	557.5
.8	354.7	14.2	3.6	.8	1848.0	23.1	54.0	.8	4052.4	31.9	163.6	.8	6968.0	40.8	332.5	.8	10594.7	49.7	560.7
.9	368.9	14.3	3.9	.9	1871.2	23.2	55.0	.9	4084.5	32.1	165.4	.9	7008.9	40.9	335.0	.9	10644.5	49.8	563.9
3.0	383.3	14.4	4.2	11.0	1894.4	23.3	56.0	19.0	4116.7	32.2	167.1	27.0	7050.0	41.1	337.5	35.0	10694.4	49.9	567.1
.1	397.8	14.5	4.4	.1	1917.8	23.4	57.0	.1	4148.9	32.3	168.9	.1	7091.2	41.2	340.0	.1	10744.5	50.1	570.4
.2	412.4	14.6	4.7	.2	1941.3	23.5	58.1	.2	4181.3	32.4	170.7	.2	7132.4	41.3	342.5	.2	10794.7	50.2	573.6
.3	427.2	14.7	5.0	.3	1964.9	23.6	59.1	.3	4213.8	32.5	172.4	.3	7173.8	41.4	345.0	.3	10844.9	50.3	576.9
.4	442.0	14.8	5.4	.4	1988.7	23.7	60.2	.4	4246.4	32.6	174.2	.4	7215.3	41.5	347.6	.4	10895.3	50.4	580.2
.5	456.9	14.9	5.7	.5	2012.5	23.8	61.2	.5	4279.2	32.7	176.0	.5	7256.9	41.6	350.1	.5	10945.8	50.5	583.4
.6	472.0	15.1	6.0	.6	2036.4	23.9	62.3	.6	4312.0	32.8	177.9	.6	7298.7	41.7	352.7	.6	10996.4	50.6	586.7
.7	487.2	15.2	6.3	.7	2060.5	24.1	63.4	.7	4344.9	32.9	179.7	.7	7340.5	41.8	355.2	.7	11047.2	50.7	590.0
.8	502.4	15.3	6.7	.8	2084.7	24.2	64.5	.8	4378.0	33.1	181.5	.8	7382.4	41.9	357.8	.8	11098.0	50.8	593.4
.9	517.8	15.4	7.0	.9	2108.9	24.3	65.6	.9	4411.2	33.2	183.3	.9	7424.5	42.1	360.4	.9	11148.9	50.9	596.7
4.0	533.3	15.5	7.4	12.0	2133.3	24.4	66.7	20.0	4444.4	33.3	185.2	28.0	7466.7	42.2	363.0	36.0	11200.0	51.1	600.0
.1	548.9	15.6	7.8	.1	2157.8	24.5	67.8	.1	4477.8	33.4	187.0	.1	7508.9	42.3	365.6	.1	11251.2	51.2	603.3
.2	564.7	15.7	8.1	.2	2182.4	24.6	68.9	.2	4511.3	33.5	188.9	.2	7551.3	42.4	368.2	.2	11302.4	51.3	606.7
.3	580.5	15.8	8.6	.3	2207.2	24.7	70.0	.3	4544.9	33.6	190.8	.3	7593.8	42.5	370.8	.3	11353.8	51.4	610.0
.4	596.4	15.9	9.0	.4	2232.0	24.8	71.2	.4	4578.7	33.7	192.7	.4	7636.4	42.6	373.4	.4	11405.3	51.5	613.4
.5	612.5	16.1	9.4	.5	2256.9	24.9	72.3	.5	4612.5	33.8	194.6	.5	7679.2	42.7	376.0	.5	11456.9	51.6	616.8
.6	628.7	16.2	9.8	.6	2282.0	25.1	73.5	.6	4646.4	33.9	196.5	.6	7722.0	42.8	378.7	.6	11508.7	51.7	620.2
.7	644.9	16.3	10.2	.7	2307.2	25.2	74.7	.7	4680.5	34.1	198.4	.7	7764.9	42.9	381.3	.7	11560.5	51.8	623.6
.8	661.3	16.4	10.7	.8	2332.4	25.3	75.9	.8	4714.7	34.2	200.3	.8	7808.0	43.1	384.0	.8	11612.4	51.9	627.0
.9	677.8	16.5	11.1	.9	2357.8	25.4	77.0	.9	4748.9	34.3	202.2	.9	7851.2	43.2	386.7	.9	11664.5	52.1	630.4
5.0	694.4	16.6	11.6	13.0	2383.3	25.5	78.2	21.0	4783.3	34.4	204.2	29.0	7894.4	43.3	389.4	37.0	11716.7	52.2	633.8
.1	711.2	16.7	12.0	.1	2408.9	25.6	79.4	.1	4817.8	34.5	206.1	.1	7937.8	43.4	392.0	.1	11768.9	52.3	637.2
.2	728.0	16.8	12.5	.2	2434.7	25.7	80.7	.2	4852.4	34.6	208.1	.2	7981.3	43.5	394.7	.2	11821.3	52.4	640.7
.3	744.9	16.9	13.0	.3	2460.5	25.8	81.9	.3	4887.2	34.7	210.0	.3	8024.9	43.6	397.4	.3	11873.8	52.5	644.1
.4	762.0	17.1	13.5	.4	2486.4	25.9	83.1	.4	4922.0	34.8	212.0	.4	8068.7	43.7	400.2	.4	11926.4	52.6	647.6
.5	779.2	17.2	14.0	.5	2512.5	26.1	84.4	.5	4956.9	34.9	214.0	.5	8112.5	43.8	402.9	.5	11979.2	52.7	651.0
.6	796.4	17.3	14.5	.6	2538.7	26.2	85.6	.6	4992.0	35.1	216.0	.6	8156.4	43.9	405.6	.6	12032.0	52.8	654.5
.7	813.8	17.4	15.0	.7	2564.9	26.3	86.9	.7	5027.2	35.2	218.0	.7	8200.5	44.1	408.4	.7	12084.9	52.9	658.0
.8	831.3	17.5	15.6	.8	2591.3	26.4	88.2	.8	5062.4	35.3	220.0	.8	8244.7	44.2	411.1	.8	12138.0	53.1	661.5
.9	848.9	17.6	16.1	.9	2617.8	26.5	89.4	.9	5097.8	35.4	222.0	.9	8288.9	44.3	413.9	.9	12191.2	53.2	665.0
6.0	866.7	17.7	16.7	14.0	2644.4	26.6	90.7	22.0	5133.3	35.5	224.1	30.0	8333.3	44.4	416.7	38.0	12244.4	53.3	668.5
.1	884.5	17.8	17.2	.1	2671.2	26.7	92.0	.1	5168.9	35.6	226.1	.1	8377.8	44.5	419.4	.1	12297.8	53.4	672.0
.2	902.4	17.9	17.8	.2	2698.0	26.8	93.4	.2	5204.7	35.7	228.2	.2	8422.4	44.6	422.2	.2	12351.3	53.5	675.6
.3	920.5	18.1	18.4	.3	2724.9	26.9	94.7	.3	5240.5	35.8	230.2	.3	8467.2	44.7	425.0	.3	12404.9	53.6	679.1
.4	938.7	18.2	19.0	.4	2752.0	27.1	96.0	.4	5276.4	35.9	232.3	.4	8512.0	44.8	427.9	.4	12458.7	53.7	682.7
.5	956.9	18.3	19.6	.5	2779.2	27.2	97.3	.5	5312.5	36.1	234.4	.5	8556.9	44.9	430.7	.5	12512.5	53.8	686.2
.6	975.3	18.4	20.2	.6	2806.4	27.3	98.7	.6	5348.7	36.2	236.5	.6	8602.0	45.1	433.5	.6	12566.4	53.9	689.8
.7	993.8	18.5	20.8	.7	2833.8	27.4	100.0	.7	5384.9	36.3	238.6	.7	8647.2	45.2	436.3	.7	12620.5	54.1	693.4
.8	1012.4	18.6	21.4	.8	2861.3	27.5	101.4	.8	5421.3	36.4	240.7	.8	8692.4	45.3	439.2	.8	12674.7	54.2	697.0
.9	1031.2	18.7	22.0	.9	2888.9	27.6	102.8	.9	5457.8	36.5	242.8	.9	8737.8	45.4	442.0	.9	12728.9	54.3	700.6
7.0	1050.0	18.8	22.7	15.0	2916.7	27.7	104.2	23.0	5494.4	36.6	244.9	31.0	8783.3	45.5	444.9	39.0	12783.3	54.4	704.2
.1	1068.9	18.9	23.3	.1	2944.5	27.8	105.6	.1	5531.2	36.7	247.0	.1	8828.9	45.6	447.8	.1	12837.8	54.5	707.8
.2	1088.0	19.1	24.0	.2	2972.4	27.9	107.0	.2	5568.0	36.8	249.2	.2	8874.7	45.7	450.7	.2	12892.4	54.6	711.4
.3	1107.2	19.2	24.7	.3	3000.5	28.1	108.4	.3	5604.9	36.9	251.3	.3	8920.5	45.8	453.6	.3	12947.2	54.7	715.0
.4	1126.4	19.3	25.4	.4	3028.7	28.2	109.8	.4	5642.0	37.1	253.5	.4	8966.4	45.9	456.5	.4	13002.0	54.8	718.7
.5	1145.8	19.4	26.0	.5	3055.9	28.3	111.2	.5	5679.2	37.2	255.7	.5	9012.5	46.1	459.4	.5	13056.9	54.9	722.3
.6	1165.3	19.5	26.7	.6	3085.3	28.4	112.7	.6	5716.4	37.3	257.9	.6	9058.7	46.2	462.3	.6	13112.0	55.1	726.0
.7	1184.9	19.6	27.4	.7	3113.8	28.5	114.1	.7	5753.8	37.4	260.0	.7	9104.9	46.3	465.2	.7	13167.2	55.2	729.7
.8	1204.7	19.7	28.2	.8	3142.4	28.6	115.6	.8	5791.3	37.5	262.2	.8	9151.3	46.4	468.2	.8	13222.4	55.3	733.4
.9	1224.5	19.8	28.9	.9	3171.2	28.7	117.0	.9	5828.9	37.6	264.5	.9	9197.8	46.5	471.1	.9	13277.8	55.4	737.0

TABLE No. 18.

EXCAVATION AND EMBANKMENT.

CUBIC YARDS

Prismoids, 100 Feet Long. **Breadth of Base, 14 Feet.** **Slopes, 1 Horizontal to 1 Perpendicular.**

Height.	Cub. Yds.	Diff.	Cor.	Height.	Cub. Yds.	Diff.	Cor.	Height.	Cub. Yds.	Diff.	Cor.	Height.	Cub. Yds.	Diff.	Cor.	Height.	Cub. Yds.	Diff.	Cor.
0.0	0.0	0.0	0.0	8.0	651.9	11.1	19.8	16.0	1777.8	17.0	79.0	24.0	3377.8	23.9	177.8	32.0	5451.9	28.9	316.0
1	5.2	5.2	0.0	1	663.0	11.1	20.2	1	1794.9	17.1	80.0	1	3400.8	23.0	179.3	1	5480.8	28.9	318.0
2	10.5	5.3	0.0	2	674.2	11.2	20.8	2	1812.0	17.1	81.0	2	3423.9	23.1	180.8	2	5509.8	29.0	320.0
3	15.9	5.4	0.0	3	685.5	11.3	21.3	3	1829.2	17.2	82.0	3	3447.0	23.1	182.2	3	5538.9	29.1	322.0
4	21.3	5.4	0.0	4	696.9	11.4	21.8	4	1846.5	17.3	83.0	4	3470.2	23.2	183.8	4	5568.0	29.1	324.0
5	26.9	5.5	0.1	5	708.3	11.4	22.3	5	1863.9	17.4	84.0	5	3493.5	23.3	185.3	5	5597.2	29.2	326.0
6	32.4	5.6	0.1	6	719.9	11.5	22.8	6	1881.3	17.4	85.0	6	3516.9	23.4	186.8	6	5626.5	29.3	328.0
7	38.1	5.7	0.2	7	731.4	11.6	23.4	7	1898.9	17.5	86.1	7	3540.3	23.4	188.3	7	5655.9	29.4	330.0
8	43.9	5.7	0.2	8	743.1	11.7	23.9	8	1916.4	17.6	87.1	8	3563.9	23.5	189.8	8	5685.3	29.4	332.0
9	49.7	5.8	0.2	9	754.9	11.7	24.4	9	1934.1	17.7	88.1	9	3587.4	23.6	191.4	9	5714.9	29.5	334.1
1.0	55.6	5.9	0.3	9.0	766.7	11.8	25.0	17.0	1951.9	17.7	89.2	25.0	3611.1	23.7	192.9	33.0	5744.4	29.6	336.1
1	61.5	6.0	0.4	1	778.6	11.9	25.6	1	1969.7	17.8	90.2	1	3634.9	23.7	194.4	1	5774.1	29.7	338.2
2	67.6	6.0	0.4	2	790.5	12.0	26.1	2	1987.6	17.9	91.3	2	3658.7	23.8	196.0	2	5803.9	29.7	340.2
3	73.7	6.1	0.5	3	802.6	12.0	26.7	3	2005.5	18.0	92.4	3	3682.6	23.9	197.6	3	5833.7	29.8	342.2
4	79.9	6.2	0.6	4	814.7	12.1	27.3	4	2023.6	18.0	93.4	4	3706.5	24.0	199.1	4	5863.6	29.9	344.3
5	86.1	6.3	0.7	5	826.9	12.2	27.9	5	2041.7	18.1	94.5	5	3730.6	24.0	200.7	5	5893.5	30.0	346.4
6	92.4	6.3	0.8	6	839.1	12.3	28.4	6	2059.9	18.2	95.6	6	3754.7	24.1	202.3	6	5923.6	30.0	348.4
7	98.9	6.4	0.9	7	851.4	12.3	29.0	7	2078.1	18.3	96.7	7	3778.9	24.2	203.8	7	5953.7	30.1	350.5
8	105.3	6.5	1.0	8	863.9	12.4	29.6	8	2096.4	18.3	97.8	8	3803.1	24.3	205.4	8	5983.9	30.2	352.6
9	111.9	6.6	1.1	9	876.3	12.5	30.3	9	2114.9	18.4	98.9	9	3827.4	24.3	207.0	9	6014.1	30.3	354.7
2.0	118.5	6.6	1.2	10.0	888.9	12.6	30.9	18.0	2133.3	18.5	100.0	26.0	3851.9	24.4	208.6	34.0	6044.4	30.3	356.8
1	125.2	6.7	1.4	1	901.5	12.6	31.5	1	2151.9	18.6	101.1	1	3876.3	24.5	210.2	1	6074.9	30.4	358.9
2	132.0	6.8	1.5	2	914.2	12.7	32.1	2	2170.5	18.6	102.2	2	3900.9	24.6	211.9	2	6105.3	30.5	361.0
3	138.9	6.9	1.6	3	927.0	12.8	32.7	3	2189.2	18.7	103.4	3	3925.5	24.6	213.5	3	6135.9	30.6	363.1
4	145.8	6.9	[illegible]	4	939.9	12.9	33.4	4	2208.0	18.8	104.5	4	3950.2	24.7	215.1	4	6166.5	30.6	365.2
5	152.8	7.0	1.9	5	952.8	12.9	34.0	5	2226.9	18.9	105.6	5	3975.0	24.8	216.7	5	6197.2	30.7	367.4
6	159.9	7.1	2.1	6	965.8	13.0	34.7	6	2245.8	18.9	106.8	6	3999.9	24.9	218.4	6	6228.0	30.8	369.5
7	167.0	7.1	2.2	7	978.9	13.1	35.3	7	2264.8	19.0	107.9	7	4024.8	24.9	220.0	7	6258.9	30.9	371.6
8	174.2	7.2	2.4	8	992.0	13.1	36.0	8	2283.9	19.1	109.1	8	4049.8	25.0	221.7	8	6289.8	30.9	373.8
9	181.5	7.3	2.6	9	1005.2	13.2	36.7	9	2303.0	19.1	110.2	9	4074.9	25.1	223.3	9	6320.8	31.0	375.9
3.0	188.9	7.4	2.8	11.0	1018.5	13.3	37.3	19.0	2322.2	19.2	111.4	27.0	4100.0	25.1	225.0	35.0	6351.9	31.1	378.1
1	196.3	7.4	3.0	1	1031.9	13.4	38.0	1	2341.5	19.3	112.6	1	4125.2	25.2	226.7	1	6383.0	31.1	380.2
2	203.9	7.5	3.2	2	1045.3	13.4	38.7	2	2360.9	19.4	113.8	2	4150.5	25.3	228.3	2	6414.2	31.2	382.4
3	211.4	7.6	3.4	3	1058.9	13.5	39.4	3	2380.3	19.4	115.0	3	4175.9	25.4	230.0	3	6445.5	31.3	384.6
4	219.1	7.7	3.6	4	1072.4	13.6	40.1	4	2399.9	19.5	116.2	4	4201.3	25.4	231.7	4	6476.9	31.4	386.8
5	226.9	7.7	3.8	5	1086.1	13.7	40.8	5	2419.4	19.6	117.4	5	4226.9	25.5	233.4	5	6508.3	31.4	389.0
6	234.7	7.8	4.0	6	1099.9	13.7	41.5	6	2439.1	19.7	118.6	6	4252.4	25.6	235.1	6	6539.9	31.5	391.2
7	242.6	7.9	4.2	7	1113.7	13.8	42.2	7	2458.9	19.7	119.8	7	4278.1	25.7	236.8	7	6571.4	31.6	393.4
8	250.5	8.0	4.5	8	1127.6	13.9	43.0	8	2478.7	19.8	121.0	8	4303.9	25.7	238.5	8	6603.1	31.7	395.6
9	258.6	8.0	4.7	9	1141.5	14.0	43.7	9	2498.6	19.9	122.2	9	4329.7	25.8	240.2	9	6634.9	31.7	397.8
4.0	266.7	8.1	4.9	12.0	1155.6	14.0	44.4	20.0	2518.5	20.0	123.5	28.0	4355.6	25.9	242.0	36.0	6666.7	31.8	400.0
1	274.9	8.2	5.2	1	1169.7	14.1	45.2	1	2538.6	20.0	124.7	1	4381.5	26.0	243.7	1	6698.6	31.9	402.2
2	283.1	8.3	5.4	2	1183.9	14.2	45.9	2	2558.7	20.1	125.9	2	4407.6	26.0	245.4	2	6730.5	32.0	404.5
3	291.4	8.3	5.7	3	1198.1	14.3	46.7	3	2578.9	20.2	127.2	3	4433.7	26.1	247.2	3	6762.6	32.0	406.7
4	299.9	8.4	6.0	4	1212.4	14.3	47.5	4	2599.1	20.3	128.4	4	4459.9	26.2	248.9	4	6794.7	32.1	408.9
5	308.3	8.5	6.2	5	1226.9	14.4	48.2	5	2619.4	20.3	129.7	5	4486.1	26.3	250.7	5	6826.9	32.2	411.2
6	316.9	8.6	6.5	6	1241.3	14.5	49.0	6	2639.9	20.4	131.0	6	4512.4	26.3	252.5	6	6859.1	32.3	413.5
7	325.5	8.6	6.8	7	1255.9	14.6	49.8	7	2660.3	20.5	132.2	7	4538.9	26.4	254.2	7	6891.4	32.3	415.7
8	334.2	8.7	7.1	8	1270.5	14.6	50.6	8	2680.9	20.6	133.5	8	4565.3	26.5	256.0	8	6923.9	32.4	418.0
9	343.0	8.8	7.4	9	1285.2	14.7	51.4	9	2701.5	20.6	134.8	9	4591.9	16.6	257.8	9	6956.3	32.5	420.2
5.0	351.9	8.9	7.7	13.0	1300.0	14.8	52.2	21.0	2722.2	20.7	136.1	29.0	4618.5	26.6	259.6	37.0	6988.9	32.6	422.5
1	360.8	8.9	8.0	1	1314.9	14.9	53.0	1	2743.0	20.8	137.4	1	4645.2	26.7	261.4	1	7021.5	32.6	424.8
2	369.8	9.0	8.3	2	1329.8	14.9	53.8	2	2763.9	20.9	138.7	2	4672.0	26.8	263.2	2	7054.2	32.7	427.1
3	378.9	9.1	8.7	3	1344.8	15.0	54.6	3	2784.8	20.9	140.0	3	4698.9	26.9	265.0	3	7087.0	32.8	429.4
4	388.0	9.1	9.0	4	1359.9	15.1	55.4	4	2805.8	21.0	141.3	4	4725.8	26.9	266.8	4	7119.9	32.9	431.7
5	397.2	9.2	9.3	5	1375.0	15.1	56.2	5	2826.9	21.1	142.7	5	4752.8	27.0	268.6	5	7152.8	32.9	434.0
6	406.5	9.3	9.7	6	1390.2	15.2	57.1	6	2848.0	21.1	144.0	6	4779.9	27.1	270.4	6	7185.8	33.0	436.3
7	415.9	9.4	10.0	7	1405.5	15.3	57.9	7	2869.2	21.2	145.3	7	4807.0	27.1	272.2	7	7218.9	33.1	438.7
8	425.3	9.4	10.4	8	1420.9	15.4	58.8	8	2890.5	21.3	146.7	8	4834.2	27.2	274.1	8	7252.0	33.1	441.0
9	434.9	9.5	10.7	9	1436.3	15.4	59.6	9	2911.9	21.4	148.0	9	4861.5	27.3	275.9	9	7285.2	33.2	443.3
6.0	444.4	9.6	11.1	14.0	1451.9	15.5	60.5	22.0	2933.3	21.4	149.4	30.0	4888.9	27.4	277.8	38.0	7318.5	33.3	445.7
1	454.1	9.7	11.5	1	1467.4	15.6	61.4	1	2954.9	21.5	150.7	1	4916.3	27.4	279.6	1	7351.9	33.4	448.0
2	463.9	9.7	11.9	2	1483.1	15.7	62.2	2	2976.4	21.6	152.1	2	4943.9	27.5	281.5	2	7385.3	33.4	450.4
3	473.7	9.8	12.2	3	1498.9	15.7	63.1	3	2998.1	21.7	153.5	3	4971.4	27.6	283.4	3	7418.9	33.5	452.7
4	483.6	9.9	12.6	4	1514.7	15.8	64.0	4	3019.9	21.7	154.9	4	4999.1	27.7	285.2	4	7452.4	33.6	455.1
5	493.5	10.0	13.0	5	1530.6	15.9	64.9	5	3041.7	21.8	156.2	5	5026.9	27.7	287.1	5	7486.1	33.7	457.5
6	503.6	10.0	13.4	6	1546.5	16.0	65.8	6	3063.6	21.9	157.6	6	5054.7	27.8	289.0	6	7519.9	33.7	459.9
7	513.7	10.1	13.9	7	1562.6	16.0	66.7	7	3085.5	22.0	159.0	7	5081.6	27.9	290.9	7	7553.7	33.8	462.2
8	523.9	10.2	14.3	8	1578.7	16.1	67.6	8	3107.6	22.0	160.4	8	5110.5	28.0	292.8	8	7587.6	33.9	464.6
9	534.1	10.3	14.7	9	1594.9	16.2	68.5	9	3129.7	22.1	161.9	9	5138.6	28.0	294.7	9	7621.5	34.0	467.0
7.0	544.4	10.3	15.1	15.0	1611.1	16.3	69.4	23.0	3151.9	22.2	163.3	31.0	5166.7	28.1	296.6	39.0	7655.6	34.0	469.4
1	554.9	10.4	15.6	1	1627.4	16.3	70.4	1	3174.1	22.3	164.7	1	5194.9	28.2	298.5	1	7689.7	34.1	471.9
2	565.3	10.5	16.0	2	1643.9	16.4	71.3	2	3196.4	22.3	166.1	2	5223.1	28.3	300.4	2	7723.9	34.2	474.3
3	575.9	10.6	16.4	3	1660.3	16.5	72.2	3	3218.9	22.4	167.6	3	5251.4	28.3	302.4	3	7758.1	34.3	476.7
4	586.5	10.6	16.9	4	1676.9	16.6	73.2	4	3241.3	22.5	169.0	4	5279.9	28.4	304.3	4	7792.4	34.3	479.1
5	597.2	10.7	17.4	5	1693.5	16.6	74.2	5	3263.9	22.6	170.4	5	5308.3	28.5	306.2	5	7826.9	34.4	481.6
6	608.0	10.8	17.8	6	1710.2	16.7	75.1	6	3286.5	22.6	171.9	6	5336.9	28.6	308.2	6	7861.3	34.5	484.0
7	618.9	10.9	18.3	7	1727.0	16.8	76.1	7	3309.2	22.7	173.4	7	5365.5	28.6	310.1	7	7895.9	34.6	486.4
8	629.8	10.9	18.8	8	1743.9	16.9	77.1	8	3332.0	22.8	174.8	8	5394.2	28.7	312.1	8	7930.5	34.6	488.9
9	640.8	11.0	19.3	9	1760.8	16.9	78.0	9	3354.9	22.9	176.3	9	5423.0	28.8	314.1	9	7965.2	34.7	491.4

TABLE No. 19.

EXCAVATION AND EMBANKMENT.

CUBIC YARDS.

Prismoids, 100 Feet Long — **Breadth of Base, 18 Feet.** — **Slopes, 1 Horizontal to 1 Perpendicular.**

Height.	Cub. Yds.	Diff.	Cor.	Height.	Cub. Yds.	Diff.	Cor.	Height.	Cub. Yds.	Diff.	Cor.	Height.	Cub. Yds.	Diff.	Cor.	Height.	Cub. Yds.	Diff.	Cor.
0.0	0.0	0.0	0.0	8.0	711.1	11.8	19.8	16.0	1896.3	17.7	79.0	24.0	3555.6	23.7	177.8	32.0	5688.9	29.6	316.0
1	6.0	6.0	0.0	1	723.0	11.9	20.2	1	1914.1	17.8	80.0	1	3579.3	23.7	179.3	1	5718.6	29.7	318.0
2	12.0	6.0	0.0	2	735.0	12.0	20.8	2	1932.0	17.9	81.0	2	3603.1	23.8	180.8	2	5748.3	29.7	320.0
3	18.1	6.1	0.0	3	747.0	12.0	21.3	3	1950.0	18.0	82.0	3	3627.0	23.9	182.2	3	5778.1	29.8	322.0
4	24.3	6.2	0.0	4	759.1	12.1	21.8	4	1968.0	18.0	83.0	4	3651.0	24.0	183.8	4	5808.0	29.9	324.0
5	30.6	6.3	0.1	5	771.3	12.2	22.3	5	1986.1	18.1	84.0	5	3675.0	24.0	185.3	5	5838.0	30.0	326.0
6	36.9	6.3	0.1	6	783.6	12.3	22.8	6	2004.3	18.2	85.0	6	3699.1	24.1	186.8	6	5868.0	30.0	328.0
7	43.3	6.4	0.2	7	795.9	12.3	23.4	7	2022.6	18.3	86.1	7	3723.3	24.2	188.3	7	5898.1	30.1	330.0
8	49.8	6.5	0.2	8	808.3	12.4	23.9	8	2040.9	18.3	87.1	8	3747.6	24.3	189.8	8	5928.3	30.2	332.0
9	56.3	6.6	0.2	9	820.8	12.5	24.4	9	2059.3	18.4	88.1	9	3771.9	24.3	191.4	9	5958.6	30.3	334.1
1.0	63.0	6.6	0.3	9.0	833.3	12.6	25.0	17.0	2077.8	18.5	89.2	25.0	3796.3	24.4	192.9	33.0	5988.9	30.3	336.1
1	69.7	6.7	0.4	1	846.0	12.6	25.6	1	2096.3	18.6	90.2	1	3820.8	24.5	194.4	1	6019.3	30.4	338.2
2	76.4	6.8	0.4	2	858.7	12.7	26.1	2	2115.0	18.6	91.3	2	3845.3	24.6	196.0	2	6049.8	30.5	340.2
3	83.3	6.9	0.5	3	871.4	12.8	26.7	3	2133.7	18.7	92.4	3	3870.0	24.6	197.6	3	6080.3	30.6	342.2
4	90.2	6.9	0.6	4	884.3	12.9	27.3	4	2152.4	18.8	93.4	4	3894.7	24.7	199.1	4	6111.0	30.6	344.3
5	97.2	7.0	0.7	5	897.2	13.0	27.9	5	2171.3	18.9	94.5	5	3919.4	24.8	200.7	5	6141.7	30.7	346.4
6	104.3	7.1	0.8	6	910.2	13.0	28.4	6	2190.2	18.9	95.6	6	3944.3	24.9	202.3	6	6172.4	30.8	348.4
7	111.4	7.1	0.9	7	923.3	13.1	29.0	7	2209.2	19.0	96.7	7	3969.2	24.9	203.8	7	6203.3	30.9	350.5
8	118.7	7.2	1.0	8	936.4	13.1	29.6	8	2228.3	19.1	97.8	8	3994.2	25.0	205.4	8	6234.2	30.9	352.6
9	126.0	7.3	1.1	9	949.7	13.2	30.3	9	2247.4	19.1	98.9	9	4019.3	25.1	207.0	9	6265.2	31.0	354.7
2.0	133.3	7.4	1.2	10.0	963.0	13.3	30.9	18.0	2266.7	19.2	100.0	26.0	4044.4	25.1	208.6	34.0	6296.3	31.1	356.8
1	140.8	7.4	1.4	1	976.3	13.4	31.5	1	2286.0	19.3	101.1	1	4069.7	25.2	210.2	1	6327.4	31.1	358.9
2	148.3	7.5	1.5	2	989.8	13.4	32.1	2	2305.3	19.4	102.2	2	4095.0	25.3	211.9	2	6358.7	31.2	361.0
3	155.9	7.6	1.6	3	1003.3	13.5	32.7	3	2324.8	19.4	103.4	3	4120.3	25.4	213.5	3	6390.0	31.3	363.1
4	163.6	7.7	1.8	4	1016.9	13.6	33.4	4	2344.3	19.5	104.5	4	4145.8	25.4	215.1	4	6421.3	31.4	365.2
5	171.3	7.7	1.9	5	1030.6	13.7	34.0	5	2363.9	19.6	105.6	5	4171.3	25.5	216.7	5	6452.8	31.4	367.4
6	179.1	7.8	2.1	6	1044.3	13.7	34.7	6	2383.6	19.7	106.8	6	4196.9	25.6	218.4	6	6484.3	31.5	369.5
7	187.0	7.9	2.2	7	1058 1	13.8	35.3	7	2403.3	19.7	107.9	7	4222.6	25.7	220.0	7	6515.9	31.6	371.6
8	195.0	8.0	2.4	8	1072.0	13.9	36.0	8	2423.1	19.8	109.1	8	4248.3	25.7	221.7	8	6547.6	31.7	373.8
9	203.0	8.0	2.6	9	1086.0	14.0	36.7	9	2443.0	19.9	110.2	9	4274.1	25.8	223.3	9	6579.3	31.7	375.9
3.0	211.1	8.1	2.8	11.0	1100.0	14.0	37.3	19.0	2463.0	20.0	111.4	27.0	4300.0	25.9	225.0	35.0	6611.1	31.8	378.1
1	219.3	8.2	3.0	1	1114.1	14.1	38.0	1	2483.0	20.0	112.6	1	4326.0	26.0	226.7	1	6643.0	31.9	380.2
2	227.6	8.3	3.2	2	1128.3	14.2	38.7	2	2503.1	20.1	113.8	2	4352.0	26.0	228.3	2	6675.0	32.0	382.4
3	235.9	8.3	3.4	3	1142.6	14.3	39.4	3	2523.3	20.2	115.0	3	4378.1	26.1	230.0	3	6707.0	32.0	384.6
4	244.3	8.4	3.6	4	1156.9	14.3	40.1	4	2543.6	20.3	116.2	4	4404.3	26.2	231.7	4	6739.1	32.1	386.8
5	252.8	8.5	3.8	5	1171.3	14.4	40 8	5	2563.9	20.3	117.4	5	4430.6	26.3	233.4	5	6771.3	32.2	389.0
6	261.3	8.6	4.0	6	1185.8	14.5	41.5	6	2584.3	20.4	118.6	6	4456.9	26.3	235.1	6	6803.6	32.3	391.2
7	270.[illegible]	8.6	4.2	7	1200.3	14.6	42.2	7	2604.8	20.5	119.8	7	4483.3	26.4	236.8	7	6835 9	32.3	393.4
8	278.7	8.7	4.5	8	1215.0	14.6	43.0	8	2625.3	20.6	121.0	8	4509.8	26.5	238.5	8	6868.3	32.4	395.6
9	287.4	8.8	4.7	9	1229.7	14.7	43.7	9	2646.0	20.6	122.2	9	4536.3	26.6	240.2	9	6900.8	32.5	397 8
4.0	296.3	8.9	4.9	12.0	1244.4	14.8	44.4	20.0	2666.7	20.7	123.5	28.0	4563.0	26.6	242.0	36.0	6933.3	32.6	400.0
1	305.2	8.9	5.2	1	1259.3	14.9	45.2	1	2687.4	20.8	124.7	1	4589.7	26.7	243.7	1	6966.0	32.6	402.2
2	31[illegible].2	9.0	5.4	2	1274.2	14.9	45.9	2	2708.3	20.9	125.9	2	4616.4	26.8	245.4	2	6998.7	32.7	404.5
3	323.3	9.1	5.7	3	1289.2	15.0	46.7	3	2729.2	20.9	127.2	3	4643.3	26.9	247.2	3	7031.4	32.8	406.7
4	332.4	9.1	6.0	4	1304.3	15.1	47.5	4	2750.2	21.0	128.4	4	4670.2	26.9	248.9	4	7064.3	32.9	408.9
5	341.7	9.2	6.2	5	1319.4	15.1	48.2	5	2771.3	21.1	129.7	5	4697.2	27.0	250.7	5	7097.2	32.9	411.2
6	351.0	9.3	6.5	6	1334.7	15.2	49.0	6	2792.4	21.1	131.0	6	4724.3	27.1	252.5	6	7130.2	33.0	413.5
7	360.3	9.4	6.8	7	1350.0	15.3	49.8	7	2813.7	21.2	132.2	7	4751.4	27.1	254.2	7	7163.3	33.1	415.7
8	369.8	9.4	7.1	8	1365.3	15.4	50.6	8	2835.0	21.3	133.5	8	4778.7	27.2	256.0	8	7196.4	33.1	418.0
9	379.3	9.5	7.4	9	1380.8	15.4	51.4	9	2856.3	21.4	134.8	9	4806.0	27.3	257.8	9	7229.7	33.2	420.2
5.0	388.9	9.6	7.7	13.0	1396.3	15.5	52.2	21.0	2877.8	21.4	136.1	29.0	4833.3	27.4	259.6	37.0	7263.0	33.3	422.5
1	398.6	9.7	8.0	1	1411.9	15.6	53.0	1	2899.3	21.5	137.4	1	4860.8	27.4	261.4	1	7296.3	33.4	424.8
2	408.3	9.7	8.3	2	1427.6	15.7	53.8	2	2920.9	21.6	138.7	2	4888.3	27.5	263.2	2	7329.8	33.4	427.1
3	418.1	9.8	8.7	3	1443.3	15.7	54.6	3	2942.6	21.7	140.0	3	4915.9	27.6	265.0	3	7363.3	33.5	429.4
4	428.0	9.9	9.0	4	1459.1	15.8	55.4	4	2964.3	21.7	141.3	4	4943.6	27.7	266.8	4	7396.9	33.6	431.7
5	438.0	10.0	9.3	5	1475.0	15.9	56.2	5	2986.1	21.8	142.7	5	4971.3	27.7	268.6	5	7430.6	33.7	434.0
6	448.0	10.0	9.7	6	1491.0	16.0	57.1	6	3008.0	21.9	144.0	6	4999.1	27.8	270.4	6	7464.3	33.7	436.3
7	458.1	10.1	10.0	7	1507.0	16.0	57.9	7	3030.0	22.0	145.3	7	5027.0	27.9	272.2	7	7498.1	33.8	438.7
8	468.3	10.2	10.4	8	1523.1	16.1	58.8	8	3052.0	22.0	146.7	8	5055.0	28.0	274.1	8	7532.0	33.9	441.0
9	478.6	10.3	10.7	9	1539.3	16.2	59.6	9	3074.1	22.1	148.0	9	5083.0	28.0	275.9	9	7566.0	34.0	443.3
6.0	488.9	10.3	11.1	14.0	1555.6	16.3	60.5	22.0	3096.3	22.2	149.4	30.0	5111.1	28.1	277.8	38.0	7600.0	34.0	445.7
1	499.3	10.4	11.5	1	1571.9	16.3	61.4	1	3118.6	22.3	150.7	1	5139.3	28.2	279.6	1	7634.1	34.1	448.0
2	509.8	10.5	11.9	2	1588.3	16.4	62.2	2	3140.9	22.3	152.1	2	5167.6	28.3	281.5	2	7668.3	34.2	450.4
3	520.3	10.6	12.2	3	1604.8	16.5	63.1	3	3163.3	22.4	153.5	3	5195.9	28.3	283.4	3	7702.6	34.3	452.7
4	531.0	10.6	12.6	4	1621.3	16.6	64.0	4	3185.8	22.5	154.9	4	5224.3	28.4	285.2	4	7736.9	34.3	455.1
5	541.7	10.7	13.0	5	1638.0	16.6	64.9	5	3208.3	22.6	156.2	5	5252.8	28.5	287.1	5	7771.3	34.4	457.5
6	552.4	10.8	13.4	6	1654.7	16.7	65.8	6	3231.0	22.6	157.6	6	5281.3	28.6	289.0	6	7805.8	34.5	459.9
7	563.3	10.9	13.9	7	1671.4	16.8	66.7	7	3253.7	22.7	159.0	7	5310.0	28.6	290.9	7	7840.3	34.6	462 2
8	574.2	10.9	14.3	8	1688.3	16.9	67.6	8	3276.4	22.8	160.4	8	5338.7	28.7	292.8	8	7875.0	34.6	464.6
9	585.2	11.0	14.7	9	1705.2	16.9	68.5	9	3299.3	22.9	161.9	9	5367.4	28.8	294.7	9	7909.7	34.7	467.0
7.0	596.3	11.1	15.1	15.0	1722.2	17.0	69.4	23.0	3322.2	22.9	163.3	31.0	5396.3	28.9	296.6	39.0	7944.4	34.8	469.4
1	607.4	11.1	15.6	1	1739.3	17.1	70.4	1	3345.2	23.0	164.7	1	5425.2	28.9	298.5	1	7979.3	34.9	471.9
2	618.7	11.2	16.0	2	1756.4	17.1	71.3	2	3368.3	23.1	166.1	2	5454.2	29.0	300.4	2	8014.2	34.9	474.3
3	630.0	11.3	16.4	3	1773.7	17.2	72.2	3	3391.4	23.1	167.6	3	5483.3	29.1	302.4	3	8049.2	35.0	476.7
4	641.3	11.4	16.9	4	1791.0	17.3	73.[illegible]	4	3414.7	23.2	169.0	4	5512.4	29.1	304.3	4	8084.3	35.1	479.1
5	652.8	11.4	17.4	5	1808.3	17.4	74.2	5	3438.0	23.3	170.4	5	5541.7	29.2	306.2	5	8119.4	35.1	481.6
6	664.3	11.5	17.8	6	1825.8	17.4	75.1	6	3461.3	23.4	171.9	6	5571.0	29.3	308.2	6	8154.7	35.2	484.0
7	675.9	11.6	18.3	7	1843.3	17.5	76.1	7	3484.8	23.4	173.4	7	5600.3	29.4	310.1	7	8190.0	35.3	486.4
8	687.6	11.7	18.8	8	1860.9	17.6	77.1	8	3508.3	23.5	174.8	8	5629.8	29.4	312.1	8	8225.3	35.4	488.9
9	699.3	11.7	19.3	9	1878.6	17.7	78.0	9	3531.9	23.6	176.3	9	5659.3	29.5	314.1	9	8260.8	35.4	491.4

TABLE No. 20.

EXCAVATION AND EMBANKMENT.

CUBIC YARDS.

Prismoids, 100 Feet Long. **Breadth of Base, 18 Feet.** **Slopes, 1 Horizontal to 1 Perpendicular.**

Height.	Cub. Yds.	Diff.	Cor.	Height.	Cub. Yds.	Diff.	Cor.	Height.	Cub. Yds.	Diff.	Cor.	Height.	Cub. Yds.	Diff.	Cor.	Height.	Cub. Yds.	Diff.	Cor.
0.0	0.0	0.0	0.0	8.0	770.4	12.6	19.8	16.0	2014.8	18.5	79.0	24.0	3733.4	24.4	177.8	32.0	5925.9	30.3	316.0
1	6.7	6.7	0.0	1	783.0	12.6	20.2	1	2033.4	18.6	80.0	1	3757.8	24.5	179.3	1	5956.3	30.4	318.0
2	13.5	6.8	0.0	2	795.7	12.7	20.8	2	2052.0	18.6	81.0	2	3782.4	24.6	180.8	2	5986.8	30.5	320.0
3	20.3	6.9	0.0	3	808.5	12.8	21.3	3	2070.7	18.7	82.0	3	3807.0	24.6	182.2	3	6017.4	30.6	322.0
4	27.3	6.9	0.0	4	821.3	12.9	21.8	4	2089.5	18.8	83.0	4	3831.7	24.7	183.8	4	6048.0	30.6	324.0
5	34.3	7.0	0.1	5	834.3	12.9	22.3	5	2108.3	18.9	84.0	5	3856.5	24.8	185.3	5	6078.7	30.7	326.0
6	41.3	7.1	0.1	6	847.3	13.0	22.8	6	2127.3	18.9	85.0	6	3881.3	24.9	186.8	6	6109.5	30.8	328.0
7	48.5	7.1	0.2	7	860.3	13.1	23.4	7	2146.3	19.0	86.1	7	3906.3	25.0	188.3	7	6140.3	30.9	330.0
8	55.7	7.2	0.2	8	873.5	13.1	23.9	8	2165.3	19.1	87.1	8	3931.3	25.0	189.8	8	6171.3	30.9	332.0
9	63.0	7.3	0.2	9	886.7	13.2	24.4	9	2184.5	19.1	88.1	9	3956.3	25.1	191.4	9	6202.3	31.0	334.1
1.0	70.4	7.4	0.3	9.0	900.0	13.3	25.0	17.0	2203.7	19.2	89.2	25.0	3981.5	25.1	192.9	33.0	6233.3	31.1	336.1
1	77.8	7.4	0.4	1	913.4	13.4	25.6	1	2223.0	19.3	90.2	1	4006.7	25.2	194.4	1	6264.5	31.1	338.2
	85.3	7.5	0.4	2	926.8	13.4	26.1	2	2242.4	19.4	91.3	2	4032.0	25.3	196.0	2	6295.7	31.2	340.2
3	92.9	7.6	0.5	3	940.3	13.5	26.7	3	2261.8	19.4	92.4	3	4057.4	25.4	197.6	3	6327.0	31.3	342.2
4	100.6	7.7	0.6	4	953.9	13.6	27.3	4	2281.3	19.5	93.4	4	4082.8	25.4	199.1	4	6358.4	31.4	344.3
5	108.3	7.7	0.7	5	967.6	13.7	27.9	5	2300.9	19.6	94.5	5	4108.3	25.5	200.7	5	6389.8	31.4	346.4
6	116.1	7.8	0.8	6	981.3	13.7	28.4	6	2320.6	19.7	95.6	6	4133.9	25.6	202.3	6	6421.3	31.5	348.4
7	124.0	7.9	0.9	7	995.1	13.8	29.0	7	2340.3	19.7	96.7	7	4159.6	25.7	203.8	7	6452.9	31.6	350.5
8	132.0	8.0	1.0	8	1009.0	13.9	29.6	8	2360.1	19.8	97.8	8	4185.3	25.7	205.4	8	6484.6	31.7	352.6
9	140.0	8.0	1.1	9	1023.0	14.0	30.3	9	2380.0	19.9	98.9	9	4211.1	25.8	207.0	9	6516.3	31.7	354.7
2.0	148.1	8.1	1.2	10.0	1037.0	14.0	30.9	18.0	2400.0	20.0	100.0	26.0	4237.0	25.9	208.6	34.0	6548.1	31.8	356.8
1	156.3	8.2	1.4	1	1051.1	14.1	31.5	1	2420.0	20.0	101.1	1	4263.0	26.0	210.2	1	6580.0	31.9	358.9
2	164.6	8.3	1.5	2	1065.3	14.2	32.1	2	2440.1	20.1	102.2	2	4289.0	26.0	211.9	2	6612.0	32.0	361.0
3	172.9	8.3	1.6	3	1079.6	14.3	32.7	3	2460.3	20.2	103.4	3	4315.1	26.1	213.5	3	6644.0	32.0	363.1
4	181.3	8.4	1.8	4	1093.9	14.3	33.4	4	2480.6	20.3	104.5	4	4341.3	26.2	215.1	4	6676.1	32.1	365.2
5	189.8	8.5	1.9	5	1108.3	14.4	34.0	5	2500.9	20.3	105.6	5	4367.6	26.3	216.7	5	6708.3	32.2	367.4
6	198.4	8.6	2.1	6	1122.8	14.5	34.7	6	2521.3	20.4	106.8	6	4393.9	26.3	218.4	6	6740.6	32.3	369.5
7	207.0	8.6	2.2	7	1137.4	14.6	35.3	7	2541.8	20.5	107.9	7	4420.3	26.4	220.0	7	6772.9	32.3	371.6
8	215.7	8.7	2.4	8	1152.0	14.6	36.0	8	2562.4	20.6	109.1	8	4446.8	26.5	221.7	8	6805.3	32.4	373.8
9	224.5	8.8	2.6	9	1166.7	14.7	36.7	9	2583.0	20.6	110.2	9	4473.4	26.6	223.3	9	6837.8	32.5	375.9
3.0	233.3	8.9	2.8	11.0	1181.5	14.8	37.3	19.0	2603.7	20.7	111.4	27.0	4500.0	26.6	225.0	35.0	6870.4	32.6	378.1
1	242.3	8.9	3.0	1	1196.3	14.9	38.0	1	2624.5	20.8	112.6	1	4526.7	26.7	226.7	1	6903.0	32.6	380.2
2	251.3	9.0	3.2	2	1211.3	14.9	38.7	2	2645.3	20.9	113.8	2	4553.5	26.8	228.3	2	6935.7	32.7	382.4
3	260.3	9.1	3.4	3	1226.3	15.0	39.4	3	2666.3	20.9	115.0	3	4580.3	26.9	230.0	3	6968.5	32.8	384.6
4	269.5	9.1	3.6	4	1241.3	15.1	40.1	4	2687.3	21.0	116.2	4	4607.3	26.9	231.7	4	7001.3	32.9	386.8
5	278.7	9.2	3.8	5	1256.5	15.1	40.8	5	2708.3	21.1	117.4	5	4634.3	27.0	233.4	5	7034.3	32.9	389.0
6	288.0	9.3	4.0	6	1271.7	15.2	41.5	6	2729.5	21.1	118.6	6	4661.3	27.1	235.1	6	7067.3	33.0	391.2
7	297.4	9.4	4.2	7	1287.0	15.3	42.2	7	2750.7	21.2	119.8	7	4688.5	27.1	236.8	7	7100.3	33.1	393.4
8	306.8	9.4	4.5	8	1302.4	15.4	43.0	8	2772.0	21.3	121.0	8	4715.7	27.2	238.5	8	7133.5	33.1	395.6
9	316.3	9.5	4.7	9	1317.8	15.4	43.7	9	2793.4	21.4	122.2	9	4743.0	27.3	240.2	9	7166.7	33.2	397.8
4.0	325.9	9.6	4.9	12.0	1333.3	15.5	44.4	20.0	2814.8	21.4	123.5	28.0	4770.4	27.4	242.0	36.0	7200.0	33.3	400.0
1	335.6	9.7	5.2	1	1348.9	15.6	45.2	1	2836.3	21.5	124.7	1	4797.8	27.4	243.7	1	7233.4	33.4	402.2
2	345.3	9.7	5.4	2	1364.6	15.7	45.9	2	2857.9	21.6	125.9	2	4825.3	27.5	245.4	2	7266.8	33.4	404.5
3	355.1	9.8	5.7	3	1380.3	15.7	46.7	3	2879.6	21.7	127.2	3	4852.9	27.6	247.2	3	7300.3	33.5	406.7
4	365.0	9.9	6.0	4	1396.1	15.8	47.5	4	2901.3	21.7	128.4	4	4880.6	27.7	248.9	4	7333.9	33.6	408.9
5	375.0	10.0	6.2	5	1412.0	15.9	48.2	5	2923.1	21.8	129.7	5	4908.3	27.7	250.7	5	7367.6	33.7	411.2
6	385.0	10.0	6.5	6	1428.0	16.0	49.0	6	2945.0	21.9	131.0	6	4936.1	27.8	252.5	6	7401.3	33.7	413.5
7	395.1	10.1	6.8	7	1444.0	16.0	49.8	7	2967.0	22.0	132.2	7	4964.0	27.9	254.2	7	7435.1	33.8	415.7
8	405.3	10.2	7.1	8	1460.1	16.1	50.6	8	2989.0	22.0	133.5	8	4992.0	28.0	256.0	8	7469.0	33.9	418.0
9	415.6	10.3	7.4	9	1476.3	16.2	51.4	9	3011.1	22.1	134.8	9	5020.0	28.0	257.8	9	7503.0	34.0	420.2
5.0	425.9	10.3	7.7	13.0	1492.6	16.3	52.2	21.0	3033.3	22.2	136.1	29.0	5048.1	28.1	259.6	37.0	7537.0	34.0	422.5
1	436.3	10.4	8.0	1	1508.9	16.3	53.0	1	3055.6	22.3	137.4	1	5076.3	28.2	261.4	1	7571.1	34.1	424.8
2	446.8	10.5	8.3	2	1525.3	16.4	53.8	2	3077.9	22.3	138.7	2	5104.6	28.3	263.2	2	7605.3	34.2	427.1
3	457.4	10.6	8.7	3	1541.8	16.5	54.6	3	3100.3	22.4	140.0	3	5132.9	28.3	265.0	3	7639.6	34.3	429.4
4	468.0	10.6	9.0	4	1558.4	16.6	55.4	4	3122.8	22.5	141.3	4	5161.3	28.4	266.8	4	7673.9	34.3	431.7
5	478.7	10.7	9.3	5	1575.0	16.6	56.2	5	3145.4	22.6	142.7	5	5189.8	28.5	268.6	5	7708.3	34.4	434.0
6	489.5	10.8	9.7	6	1591.7	16.7	57.1	6	3168.0	22.6	144.0	6	5218.4	28.6	270.4	6	7742.8	34.5	436.3
7	500.3	10.9	10.0	7	1608.5	16.8	57.9	7	3190.7	22.7	145.3	7	5247.0	28.6	272.2	7	7777.4	34.6	438.7
8	511.3	10.9	10.4	8	1625.3	16.9	58.8	8	3213.5	22.8	146.7	8	5275.7	28.7	274.1	8	7812.0	34.6	441.0
9	522.3	11.0	10.7	9	1642.3	16.9	59.6	9	3236.3	22.9	148.0	9	5304.5	28.8	275.9	9	7846.7	34.7	443.3
6.0	533.3	11.1	11.1	14.0	1659.3	17.0	60.5	22.0	3259.3	22.9	149.4	30.0	5333.3	28.9	277.8	38.0	7881.5	34.8	445.7
1	544.5	11.1	11.5	1	1676.3	17.1	61.4	1	3282.3	23.0	150.7	1	5362.3	28.9	279.6	1	7916.3	34.9	448.0
2	555.7	11.2	11.9	2	1693.5	17.1	62.2	2	3305.3	23.1	152.1	2	5391.3	29.0	281.5	2	7951.3	34.9	450.4
3	567.0	11.3	12.2	3	1710.7	17.2	63.1	3	3328.5	23.1	153.5	3	5420.3	29.1	283.4	3	7986.3	35.0	452.7
4	578.4	11.4	12.6	4	1728.0	17.3	64.0	4	3351.7	23.2	154.9	4	5449.5	29.1	285.2	4	8021.3	35.1	455.1
5	589.8	11.4	13.0	5	1745.4	17.4	64.9	5	3375.0	23.3	156.2	5	5478.7	29.2	287.1	5	8056.5	35.1	457.5
6	601.3	11.5	13.4	6	1762.8	17.4	65.8	6	3398.4	23.4	157.6	6	5508.0	29.3	289.0	6	8091.7	35.2	459.9
7	612.9	11.6	13.9	7	1780.3	17.5	66.7	7	3421.8	23.4	159.0	7	5537.4	29.4	290.9	7	8127.0	35.3	462.2
8	624.6	11.7	14.3	8	1797.9	17.6	67.6	8	3445.3	23.5	160.4	8	5566.8	29.4	292.8	8	8162.4	35.4	464.6
9	636.3	11.7	14.7	9	1815.6	17.7	68.5	9	3468.9	23.6	161.9	9	5596.3	29.5	294.7	9	8197.8	35.4	467.0
7.0	648.1	11.8	15.1	15.0	1833.3	17.7	69.4	23.0	3492.6	23.7	163.3	31.0	5625.9	29.6	296.6	39.0	8233.3	35.5	469.4
1	660.0	11.9	15.6	1	1851.1	17.8	70.4	1	3516.3	23.7	164.7	1	5655.6	29.7	298.5	1	8268.9	35.6	471.9
2	672.0	12.0	16.0	2	1869.0	17.9	71.3	2	3540.1	23.8	166.1	2	5685.3	29.7	300.4	2	8304.6	35.7	474.3
3	684.0	12.0	16.4	3	1887.0	18.0	72.2	3	3564.0	23.9	167.6	3	5715.1	29.8	302.4	3	8340.3	35.7	476.7
4	696.1	12.1	16.9	4	1905.0	18.0	73.2	4	3588.0	24.0	169.0	4	5745.0	29.9	304.3	4	8376.1	35.8	479.1
5	708.3	12.2	17.4	5	1923.1	18.1	74.2	5	3612.0	24.0	170.4	5	5775.0	30.0	306.2	5	8412.0	35.9	481.6
6	720.6	12.3	17.8	6	1941.3	18.2	75.1	6	3636.1	24.1	171.9	6	5805.0	30.0	308.2	6	8448.0	36.0	484.0
7	732.9	12.3	18.3	7	1959.6	18.3	76.1	7	3660.3	24.2	173.4	7	5835.1	30.1	310.1	7	8484.0	36.0	486.4
8	745.3	12.4	18.8	8	1977.9	18.3	77.1	8	3684.6	24.3	174.8	8	5865.3	30.2	312.1	8	8520.1	36.1	488.9
9	757.8	12.5	19.3	9	1996.3	18.4	78.0	9	3708.9	24.3	176.3	9	5895.6	30.3	314.1	9	8556.3	36.2	491.4

TABLE NO. 21.

EXCAVATION AND EMBANKMENT.

CUBIC YARDS.

PRISMOIDS 100 FEET LONG. **BREADTH OF BASE 20 FEET.**

Slopes 1 Horizontal to 1 Perpendicular.

Ht.	Cub. Yds.	Diff.	Cor.	Ht.	Cub. Yds.	Diff.	Cor.	Ht.	Cub. Yds.	Diff.	Cor.	Ht.	Cub. Yds.	Diff.	Cor.	Ht.	Cub. Yds.	Diff.	Cor.
0.0	0.0	0.0	0.0	8.0	829.6	13.3	19.8	16.0	2133.3	19.2	79.0	24.0	3911.1	25.1	177.8	32.0	6163.0	31.1	316.0
.1	7.4	7.4	0.0	.1	843.0	13.4	20.2	.1	2152.6	19.3	80.0	.1	3936.3	25.2	179.3	.1	6194.1	31.1	318.0
.2	15.0	7.5	0.0	.2	856.4	13.4	20.8	.2	2172.0	19.4	81.0	.2	3961.6	25.3	180.8	.2	6225.3	31.2	320.0
.3	22.6	7.6	0.0	.3	870.0	13.5	21.3	.3	2191.4	19.4	82.0	.3	3987.0	25.4	182.2	.3	6256.6	31.3	322.0
.4	30.2	7.7	0.0	.4	883.6	13.6	21.8	.4	2211.0	19.5	83.0	.4	4012.4	25.4	183.8	.4	6288.0	31.4	324.0
.5	38.0	7.7	0.1	.5	897.2	13.7	22.3	.5	2230.6	19.6	84.0	.5	4038.0	25.5	185.3	.5	6319.4	31.4	326.0
.6	45.8	7.8	0.1	.6	911.0	13.7	22.8	.6	2250.2	19.7	85.0	.6	4063.6	25.6	186.8	.6	6351.0	31.5	328.0
.7	53.7	7.9	0.2	.7	924.8	13.8	23.4	.7	2270.0	19.7	86.1	.7	4089.2	25.7	188.3	.7	6382.6	31.6	330.0
.8	61.6	8.0	0.2	.8	938.7	13.9	23.9	.8	2289.8	19.8	87.1	.8	4115.0	25.7	189.8	.8	6414.2	31.7	332.0
.9	69.7	8.0	0.2	.9	952.6	14.0	24.4	.9	2309.7	19.9	88.1	.9	4140.8	25.8	191.4	.9	6446.0	31.7	334.1
1.0	77.8	8.1	0.3	9.0	966.7	14.0	25.0	17.0	2329.6	20.0	89.2	25.0	4166.7	25.9	192.9	33.0	6477.8	31.8	336.1
.1	86.0	8.2	0.4	.1	980.8	14.1	25.6	.1	2349.7	20.0	90.2	.1	4192.6	26.0	194.4	.1	6509.7	31.9	338.2
.2	94.2	8.3	0.4	.2	995.0	14.2	26.1	.2	2369.8	20.1	91.3	.2	4218.7	26.0	196.0	.2	6541.6	32.0	340.2
.3	102.6	8.3	0.5	.3	1009.2	14.3	26.7	.3	2390.0	20.2	92.4	.3	4244.8	26.1	197.6	.3	6573.7	32.0	342.2
.4	111.0	8.4	0.6	.4	1023.6	14.3	27.3	.4	2410.2	20.3	93.4	.4	4271.0	26.2	199.1	.4	6605.8	32.1	344.3
.5	119.4	8.5	0.7	.5	1038.0	14.4	27.9	.5	2430.6	20.3	94.5	.5	4297.2	26.3	200.7	.5	6638.0	32.2	346.4
.6	128.0	8.6	0.8	.6	1052.4	14.5	28.4	.6	2451.0	20.4	95.6	.6	4323.6	26.3	202.3	.6	6670.2	32.3	348.4
.7	136.6	8.6	0.9	.7	1067.0	14.6	29.0	.7	2471.4	20.5	96.7	.7	4350.0	26.4	203.8	.7	6702.6	32.3	350.5
.8	145.3	8.7	1.0	.8	1081.6	14.6	29.6	.8	2492.0	20.6	97.8	.8	4376.4	26.5	205.4	.8	6735.0	32.4	352.6
.9	154.1	8.8	1.1	.9	1096.3	14.7	30.3	.9	2512.6	20.6	98.9	.9	4403.0	26.6	207.0	.9	6767.4	32.5	354.7
2.0	163.0	8.9	1.2	10.0	1111.1	14.8	30.9	18.0	2533.3	20.7	100.0	26.0	4429.6	26.6	208.6	34.0	6800.0	32.6	356.8
.1	171.9	8.9	1.4	.1	1126.0	14.9	31.5	.1	2554.1	20.8	101.1	.1	4456.3	26.7	210.2	.1	6832.6	32.6	358.9
.2	180.9	9.0	1.5	.2	1140.9	14.9	32.1	.2	2575.0	20.9	102.2	.2	4483.1	26.8	211.9	.2	6865.3	32.7	361.0
.3	190.0	9.1	1.6	.3	1155.9	15.0	32.7	.3	2595.9	20.9	103.4	.3	4510.0	26.9	213.5	.3	6898.1	32.8	363.1
.4	199.1	9.1	1.8	.4	1171.0	15.1	33.4	.4	2616.9	21.0	104.5	.4	4536.9	26.9	215.1	.4	6931.0	32.9	365.2
.5	208.3	9.2	1.9	.5	1186.1	15.1	34.0	.5	2638.0	21.1	105.6	.5	4563.9	27.0	216.7	.5	6963.9	32.9	367.4
.6	217.6	9.3	2.1	.6	1201.3	15.2	34.7	.6	2659.1	21.1	106.8	.6	4591.0	27.1	218.4	.6	6996.9	33.0	369.5
.7	227.0	9.4	2.2	.7	1216.6	15.3	35.3	.7	2680.3	21.2	107.9	.7	4618.1	27.1	220.0	.7	7030.0	33.1	371.6
.8	236.4	9.4	2.4	.8	1232.0	15.4	36.0	.8	2701.6	21.3	109.1	.8	4645.3	27.2	221.7	.8	7063.1	33.1	373.8
.9	246.0	9.5	2.6	.9	1247.4	15.4	36.7	.9	2723.0	21.4	110.2	.9	4672.6	27.3	223.3	.9	7096.3	33.2	375.9
3.0	255.6	9.6	2.8	11.0	1263.0	15.5	37.3	19.0	2744.4	21.4	111.4	27.0	4700.0	27.4	225.0	35.0	7129.6	33.3	378.1
.1	265.2	9.7	3.0	.1	1278.6	15.6	38.0	.1	2766.0	21.5	112.6	.1	4727.4	27.4	226.7	.1	7163.0	33.4	380.2
.2	275.0	9.7	3.2	.2	1294.2	15.7	38.7	.2	2787.6	21.6	113.8	.2	4755.0	27.5	228.3	.2	7196.4	33.4	382.4
.3	284.8	9.8	3.4	.3	1310.0	15.7	39.4	.3	2809.2	21.7	115.0	.3	4782.6	27.6	230.0	.3	7230.0	33.5	384.6
.4	294.7	9.9	3.6	.4	1325.8	15.8	40.1	.4	2831.0	21.7	116.2	.4	4810.2	27.7	231.7	.4	7263.6	33.6	386.8
.5	304.6	10.0	3.8	.5	1341.7	15.9	40.8	.5	2852.8	21.8	117.4	.5	4838.0	27.7	233.4	.5	7297.2	33.7	389.0
.6	314.7	10.0	4.0	.6	1357.6	16.0	41.5	.6	2874.7	21.9	118.6	.6	4865.8	27.8	235.1	.6	7331.0	33.7	391.2
.7	324.8	10.1	4.2	.7	1373.7	16.0	42.2	.7	2896.6	22.0	119.8	.7	4893.7	27.9	236.8	.7	7364.8	33.8	393.4
.8	335.0	10.2	4.5	.8	1389.8	16.1	43.0	.8	2918.7	22.0	121.0	.8	4921.6	28.0	238.5	.8	7398.7	33.9	395.6
.9	345.2	10.3	4.7	.9	1406.0	16.2	43.7	.9	2940.8	22.1	122.2	.9	4949.7	28.0	240.2	.9	7432.6	34.0	397.8
4.0	355.6	10.3	4.9	12.0	1422.2	16.3	44.4	20.0	2963.0	22.2	123.5	28.0	4977.8	28.1	242.0	36.0	7466.7	34.0	400.0
.1	366.0	10.4	5.2	.1	1438.6	16.3	45.2	.1	2985.2	22.3	124.7	.1	5006.0	28.2	243.7	.1	7500.8	34.1	402.2
.2	376.4	10.5	5.4	.2	1455.0	16.4	45.9	.2	3007.6	22.3	125.9	.2	5034.2	28.3	245.4	.2	7535.0	34.2	404.5
.3	387.0	10.6	5.7	.3	1471.4	16.5	46.7	.3	3030.0	22.4	127.2	.3	5062.6	28.3	247.2	.3	7569.2	34.3	406.7
.4	397.6	10.6	6.0	.4	1488.0	16.6	47.5	.4	3052.4	22.5	128.4	.4	5091.0	28.4	248.9	.4	7603.6	34.3	408.9
.5	408.3	10.7	6.2	.5	1504.6	16.6	48.2	.5	3075.0	22.6	129.7	.5	5119.4	28.5	250.7	.5	7638.0	34.4	411.2
.6	419.1	10.8	6.5	.6	1521.3	16.7	49.0	.6	3097.6	22.6	131.0	.6	5148.0	28.6	252.5	.6	7672.4	34.5	413.5
.7	430.0	10.9	6.8	.7	1538.1	16.8	49.8	.7	3120.3	22.7	132.2	.7	5176.6	28.6	254.2	.7	7707.0	34.6	415.7
.8	440.9	10.9	7.1	.8	1555.0	16.9	50.6	.8	3143.1	22.8	133.5	.8	5205.3	28.7	256.0	.8	7741.6	34.6	418.0
.9	451.9	11.0	7.4	.9	1571.9	16.9	51.4	.9	3166.0	22.9	134.8	.9	5234.1	28.8	257.8	.9	7776.3	34.7	420.2
5.0	463.0	11.1	7.7	13.0	1588.9	17.0	52.2	21.0	3188.9	22.9	136.1	29.0	5263.0	28.9	259.6	37.0	7811.1	34.8	422.5
.1	474.1	11.1	8.0	.1	1606.0	17.1	53.0	.1	3211.9	23.0	137.4	.1	5291.9	28.9	261.4	.1	7846.0	34.9	424.8
.2	485.3	11.2	8.3	.2	1623.1	17.1	53.8	.2	3235.0	23.1	138.7	.2	5320.9	29.0	263.2	.2	7880.9	34.9	427.1
.3	496.6	11.3	8.7	.3	1640.3	17.2	54.6	.3	3258.1	23.1	140.0	.3	5350.0	29.1	265.0	.3	7915.9	35.0	429.4
.4	508.0	11.4	9.0	.4	1657.6	17.3	55.4	.4	3281.3	23.2	141.3	.4	5379.1	29.1	266.8	.4	7951.0	35.1	431.7
.5	519.4	11.4	9.3	.5	1675.0	17.4	56.2	.5	3304.6	23.3	142.7	.5	5408.3	29.2	268.6	.5	7986.1	35.1	434.0
.6	531.0	11.5	9.7	.6	1692.4	17.4	57.1	.6	3328.0	23.4	144.0	.6	5437.6	29.3	270.4	.6	8021.3	35.2	436.3
.7	542.6	11.6	10.0	.7	1710.0	17.5	57.9	.7	3351.4	23.4	145.3	.7	5467.0	29.4	272.2	.7	8056.6	35.3	438.7
.8	554.2	11.7	10.4	.8	1727.6	17.6	58.8	.8	3375.0	23.5	146.7	.8	5496.4	29.4	274.1	.8	8092.0	35.4	441.0
.9	566.0	11.7	10.7	.9	1745.2	17.7	59.6	.9	3398.6	23.6	148.0	.9	5526.0	29.5	275.9	.9	8127.4	35.4	443.3
6.0	577.8	11.8	11.1	14.0	1763.0	17.7	60.5	22.0	3422.2	23.7	149.4	30.0	5555.6	29.6	277.8	38.0	8163.0	35.5	445.7
.1	589.7	11.9	11.5	.1	1780.8	17.8	61.4	.1	3446.0	23.7	150.7	.1	5585.2	29.7	279.6	.1	8198.6	35.6	448.0
.2	601.6	12.0	11.9	.2	1798.7	17.9	62.2	.2	3469.8	23.8	152.1	.2	5615.0	29.7	281.5	.2	8234.2	35.7	450.4
.3	613.7	12.0	12.2	.3	1816.6	18.0	63.1	.3	3493.7	23.9	153.5	.3	5644.8	29.8	283.4	.3	8270.0	35.7	452.7
.4	625.8	12.1	12.6	.4	1834.7	18.0	64.0	.4	3517.6	24.0	154.9	.4	5674.7	29.9	285.2	.4	8305.8	35.8	455.1
.5	638.0	12.2	13.0	.5	1852.8	18.1	64.9	.5	3541.7	24.0	156.2	.5	5704.6	30.0	287.1	.5	8341.7	35.9	457.5
.6	650.2	12.3	13.4	.6	1871.0	18.2	65.8	.6	3565.8	24.1	157.6	.6	5734.7	30.0	289.0	.6	8377.6	36.0	459.9
.7	662.6	12.3	13.9	.7	1889.2	18.3	66.7	.7	3590.0	24.2	159.0	.7	5764.8	30.1	290.9	.7	8413.7	36.0	462.2
.8	675.0	12.4	14.3	.8	1907.6	18.3	67.6	.8	3614.2	24.3	160.4	.8	5795.0	30.2	292.8	.8	8449.8	36.1	464.6
.9	687.4	12.5	14.7	.9	1926.0	18.4	68.5	.9	3638.6	24.3	161.9	.9	5825.2	30.3	294.7	.9	8486.0	36.2	467.0
7.0	700.0	12.6	15.1	15.0	1944.4	18.5	69.4	23.0	3663.0	24.4	163.3	31.0	5855.6	30.3	296.6	39.0	8522.2	36.3	469.4
.1	712.6	12.6	15.6	.1	1963.0	18.6	70.4	.1	3687.4	24.5	164.7	.1	5886.0	30.4	298.5	.1	8558.6	36.3	471.9
.2	725.3	12.7	16.0	.2	1981.6	18.6	71.3	.2	3712.0	24.6	166.1	.2	5916.4	30.5	300.4	.2	8595.0	36.4	474.3
.3	738.1	12.8	16.4	.3	2000.3	18.7	72.2	.3	3736.6	24.6	167.6	.3	5947.0	30.6	302.4	.3	8631.4	36.5	476.7
.4	751.0	12.9	16.9	.4	2019.1	18.8	73.2	.4	3761.3	24.7	169.0	.4	5977.6	30.6	304.3	.4	8668.0	36.6	479.1
.5	763.9	12.9	17.4	.5	2038.0	18.9	74.2	.5	3786.1	24.8	170.4	.5	6008.3	30.7	306.2	.5	8704.6	36.6	481.6
.6	776.9	13.0	17.8	.6	2056.9	18.9	75.1	.6	3811.0	24.9	171.9	.6	6039.1	30.8	308.2	.6	8741.3	36.7	484.0
.7	790.0	13.1	18.3	.7	2075.9	19.0	76.1	.7	3835.9	24.9	173.4	.7	6070.0	30.9	310.1	.7	8778.1	36.8	486.4
.8	803.1	13.1	18.8	.8	2095.0	19.1	77.1	.8	3860.9	25.0	174.8	.8	6100.9	30.9	312.1	.8	8815.0	36.9	488.9
.9	816.3	13.2	19.3	.9	2114.1	19.1	78.0	.9	3886.0	25.1	176.3	.9	6131.9	31.0	314.1	.9	8851.9	36.9	491.4

TABLE No. 22.

EXCAVATION AND EMBANKMENT.

CUBIC YARDS.

Prismoids, 100 Feet Long — Breadth of Base, 22 Feet. — Slopes, 1 Horizontal to 1 Perpendicular.

Height.	Cub. Yds.	Diff.	Cor.	Height.	Cub. Yds.	Diff.	Cor.	Height.	Cub. Yds.	Diff.	Cor.	Height.	Cub. Yds.	Diff.	Cor.	Height.	Cub. Yds.	Diff.	Cor.
0.0	0.0	0.0	0.0	8.0	888.9	14.0	19.8	16.0	2251.9	20.0	79.0	24.0	4088.9	25.9	177.8	32.0	6400.0	31.8	316.0
1	8.2	8.2	0.0	1	903.0	14.1	20.2	1	2271.9	20.0	80.0	1	4114.9	26.0	179.3	1	6431.9	31.9	318.0
2	16.4	8.3	0.0	2	917.2	14.2	20.8	2	2292.0	20.1	81.0	2	4140.9	26.0	180.8	2	6463.9	32.0	320.0
3	24.8	8.3	0.0	3	931.4	14.3	21.3	3	2312.2	20.3	82.0	3	4167.0	26.1	182.2	3	6495.9	32.0	322.0
4	33.2	8.4	0.0	4	945.8	14.3	21.8	4	2332.4	20.3	83.0	4	4193.2	26.2	183.8	4	6528.0	32.1	324.0
5	41.7	8.5	0.1	5	960.2	14.4	22.3	5	2352.8	20.3	84.0	5	4219.4	26.3	185.3	5	6560.2	32.2	326.0
6	50.2	8.6	0.1	6	974.7	14.5	22.8	6	2373.2	20.4	85.0	6	4245.8	26.3	186.8	6	6592.4	32.3	328.0
7	58.9	8.6	0.2	7	989.2	14.6	23.4	7	2393.7	20.5	86.1	7	4272.2	26.4	188.3	7	6624.8	32.3	330.0
8	67.6	8.7	0.2	8	1003.9	14.6	23.9	8	2414.2	20.6	87.1	8	4298.7	26.5	189.8	8	6657.2	32.4	332.0
9	76.3	8.8	0.2	9	1018.6	14.7	24.4	9	2434.9	20.6	88.1	9	4325.2	26.6	191.4	9	6689.7	32.5	334.1
1.0	85.2	8.9	0.3	9.0	1033.3	14.8	25.0	17.0	2455.6	20.7	89.2	25.0	4351.9	26.6	192.9	33.0	6722.2	32.6	336.1
1	94.1	8.9	0.4	1	1048.2	14.9	25.6	1	2476.3	20.8	90.2	1	4378.6	26.7	194.4	1	6754.9	32.6	338.2
2	103.1	9.0	0.4	2	1063.1	14.9	26.1	2	2497.2	20.9	91.3	2	4405.3	26.8	196.0	2	6787.6	32.7	340.2
3	112.2	9.1	0.5	3	1078.1	15.0	26.7	3	2518.1	20.9	92.4	3	4432.2	26.9	197.6	3	6820.3	32.8	342.2
4	121.3	9.1	0.6	4	1093.2	15.1	27.3	4	2539.1	21.0	93.4	4	4459.1	26.9	199.1	4	6853.2	32.9	344.3
5	130.6	9.2	0.7	5	1108.3	15.1	27.9	5	2560.2	21.1	94.5	5	4486.1	27.0	200.7	5	6886.1	32.9	346.4
6	139.9	9.3	0.8	6	1123.6	15.2	28.4	6	2581.3	21.1	95.6	6	4513.2	27.1	202.3	6	6919.1	33.0	348.4
7	149.2	9.4	0.9	7	1138.9	15.3	29.0	7	2602.6	21.2	96.7	7	4540.3	27.1	203.8	7	6952.2	33.1	350.5
8	158.7	9.4	1.0	8	1154.2	15.4	29.6	8	2623.9	21.3	97.8	8	4567.6	27.2	205.4	8	6985.3	33.1	352.6
9	168.2	9.5	1.1	9	1169.7	15.4	30.3	9	2645.2	21.4	98.9	9	4594.9	27.3	207.0	9	7018.6	33.2	354.7
2.0	177.8	9.6	1.2	10.0	1185.2	15.5	30.9	18.0	2666.7	21.4	100.0	26.0	4622.2	27.4	208.6	34.0	7051.9	33.3	356.8
1	187.4	9.7	1.4	1	1200.8	15.6	31.5	1	2688.2	21.5	101.1	1	4649.7	27.4	210.2	1	7085.2	33.4	358.9
2	197.2	9.7	1.5	2	1216.4	15.7	32.1	2	2709.8	21.6	102.2	2	4677.2	27.5	211.9	2	7118.7	33.4	361.0
3	207.0	9.8	1.6	3	1232.2	15.7	32.7	3	2731.4	21.7	103.4	3	4704.8	27.6	213.5	3	7152.2	33.5	363.1
4	216.9	9.9	1.8	4	1248.0	15.8	33.4	4	2753.2	21.7	104.5	4	4732.4	27.7	215.1	4	7185.8	33.6	365.2
5	226.9	10.0	1.9	5	1263.9	15.9	34.0	5	2775.0	21.8	105.6	5	4760.2	27.7	216.7	5	7219.4	33.7	367.4
6	236.9	10.0	2.1	6	1279.9	16.0	34.7	6	2796.9	21.9	106.8	6	4788.0	27.8	218.4	6	7253.2	33.7	369.5
7	247.0	10.1	2.2	7	1295.9	16.0	35.3	7	2818.9	22.0	107.9	7	4815.9	27.9	220.0	7	7287.0	33.8	371.6
8	257.2	10.2	2.4	8	1312.0	16.1	36.0	8	2840.9	22.0	109.1	8	4843.9	28.0	221.7	8	7320.9	33.9	373.8
9	267.4	10.3	2.6	9	1328.2	16.2	36.7	9	2863.0	22.1	110.2	9	4871.9	28.0	223.3	9	7354.9	34.0	375.9
3.0	277.8	10.3	2.8	11.0	1344.4	16.3	37.3	19.0	2885.2	22.2	111.4	27.0	4900.0	28.1	225.0	35.0	7388.9	34.0	378.1
1	288.2	10.4	3.0	1	1360.8	16.3	38.0	1	2907.4	22.3	112.6	1	4928.2	28.2	226.7	1	7423.0	34.1	380.2
2	298.7	10.5	3.2	2	1377.2	16.4	38.7	2	2929.8	22.3	113.8	2	4956.4	28.3	228.3	2	7457.2	34.2	382.4
3	309.2	10.6	3.4	3	1393.7	16.5	39.4	3	2952.2	22.4	115.0	3	4984.8	28.3	230.0	3	7491.4	34.3	384.6
4	319.9	10.6	3.6	4	1410.2	16.6	40.1	4	2974.7	22.5	116.2	4	5013.2	28.4	231.7	4	7525.8	34.3	386.8
5	330.6	10.7	3.8	5	1426.9	16.6	40.8	5	2997.2	22.6	117.4	5	5041.7	28.5	233.4	5	7560.2	34.4	389.0
6	341.3	10.8	4.0	6	1443.6	16.7	41.5	6	3019.9	22.6	118.6	6	5070.2	28.6	235.1	6	7594.7	34.5	391.2
7	352.2	10.9	4.2	7	1460.3	16.8	42.2	7	3042.6	22.7	119.8	7	5098.9	28.6	236.8	7	7629.2	34.6	393.4
8	363.1	10.9	4.5	8	1477.2	16.9	43.0	8	3065.3	22.8	121.0	8	5127.6	28.7	238.5	8	7663.9	34.6	395.6
9	374.1	11.0	4.7	9	1494.1	16.9	43.7	9	3088.2	22.9	122.2	9	5156.3	28.8	240.2	9	7698.6	34.7	397.8
4.0	385.2	11.1	4.9	12.0	1511.1	17.0	44.4	20.0	3111.1	22.9	123.5	28.0	5185.2	28.9	242.0	36.0	7733.3	34.8	400.0
1	396.3	11.1	5.2	1	1528.2	17.1	45.2	1	3134.1	23.0	124.7	1	5214.1	28.9	243.7	1	7768.2	34.9	402.2
2	407.6	11.2	5.4	2	1545.3	17.1	45.9	2	3157.2	23.1	125.9	2	5243.1	29.0	245.4	2	7803.1	34.9	404.5
3	418.9	11.3	5.7	3	1562.6	17.2	46.7	3	3180.3	23.1	127.2	3	5272.2	29.1	247.2	3	7838.1	35.0	406.7
4	430.2	11.4	6.0	4	1579.9	17.3	47.5	4	3203.6	23.2	128.4	4	5301.3	29.1	248.9	4	7873.2	35.1	408.9
5	441.7	11.4	6.2	5	1597.2	17.4	48.2	5	3226.9	23.3	129.7	5	5330.6	29.2	250.7	5	7908.3	35.1	411.2
6	453.2	11.5	6.5	6	1614.7	17.4	49.0	6	3250.2	23.4	131.0	6	5359.9	29.3	252.5	6	7943.6	35.2	413.5
7	464.8	11.6	6.8	7	1632.2	17.5	49.8	7	3273.7	23.4	132.2	7	5389.2	29.4	254.2	7	7978.9	35.3	415.7
8	476.4	11.7	7.1	8	1649.8	17.6	50.6	8	3297.2	23.5	133.5	8	5418.7	29.4	256.0	8	8014.2	35.4	418.0
9	488.2	11.7	7.4	9	1667.4	17.7	51.4	9	3320.8	23.6	134.8	9	5448.2	29.5	257.8	9	8049.7	35.4	420.2
5.0	500.0	11.8	7.7	13.0	1685.2	17.7	52.2	21.0	3344.4	23.7	136.1	29.0	5477.8	29.6	259.6	37.0	8085.2	35.5	422.5
1	511.9	11.9	8.0	1	1703.0	17.8	53.0	1	3368.2	23.7	137.4	1	5507.4	29.7	261.4	1	8120.8	35.6	424.8
2	523.9	12.0	8.3	2	1720.9	17.9	53.8	2	3392.0	23.8	138.7	2	5537.2	29.7	263.2	2	8156.4	35.7	427.1
3	535.9	12.0	8.7	3	1738.9	18.0	54.6	3	3415.9	23.9	140.0	3	5567.0	29.8	265.0	3	8192.2	35.7	429.4
4	548.0	12.1	9.0	4	1756.8	18.0	55.4	4	3439.9	24.0	141.3	4	5596.9	29.9	266.8	4	8228.0	35.8	431.7
5	560.2	12.2	9.3	5	1775.0	18.1	56.2	5	3463.9	24.0	142.7	5	5626.9	30.0	268.6	5	8263.9	35.9	434.0
6	572.4	12.3	9.7	6	1793.2	18.2	57.1	6	3488.0	24.1	144.0	6	5656.9	30.0	270.4	6	8299.9	36.0	436.3
7	584.8	12.3	10.0	7	1811.4	18.3	57.9	7	3512.2	24.2	145.3	7	5687.0	30.1	272.2	7	8335.9	36.0	438.7
8	597.2	12.4	10.4	8	1829.8	18.3	58.8	8	3536.4	24.3	146.7	8	5717.2	30.2	274.1	8	8372.0	36.1	441.0
9	609.7	12.5	10.7	9	1848.2	18.4	59.6	9	3560.8	24.3	148.0	9	5747.4	30.3	275.9	9	8408.2	36.2	443.3
6.0	622.2	12.6	11.1	14.0	1866.7	18.5	60.5	22.0	3585.2	24.4	149.4	30.0	5777.8	30.3	277.8	38.0	8444.4	36.3	445.7
1	634.9	12.6	11.5	1	1885.2	18.6	61.4	1	3609.7	24.5	150.7	1	5808.2	30.4	279.6	1	8480.8	36.3	448.0
2	647.6	12.7	11.9	2	1903.9	18.6	62.2	2	3634.2	24.6	152.1	2	5838.7	30.5	281.5	2	8517.2	36.4	450.4
3	660.3	12.8	12.2	3	1922.6	18.7	63.1	3	3658.9	24.6	153.5	3	5869.2	30.6	283.4	3	8553.7	36.5	452.7
4	673.2	12.9	12.6	4	1941.3	18.8	64.0	4	3683.6	24.7	154.9	4	5899.9	30.6	285.2	4	8590.2	36.6	455.1
5	686.1	12.9	13.0	5	1960.2	18.9	64.9	5	3708.3	24.8	156.2	5	5930.6	30.7	287.1	5	8626.9	36.6	457.5
6	699.1	13.0	13.4	6	1979.1	18.9	65.8	6	3733.2	24.9	157.6	6	5961.3	30.8	289.0	6	8663.6	36.7	459.9
7	712.2	13.1	13.9	7	1998.1	19.0	66.7	7	3758.1	24.9	159.0	7	5992.2	30.9	290.9	7	8700.3	36.8	462.2
8	725.3	13.1	14.3	8	2017.2	19.1	67.6	8	3783.1	25.0	160.4	8	6023.1	30.9	292.8	8	8737.2	36.9	464.6
9	738.6	13.2	14.7	9	2036.3	19.1	68.5	9	3808.2	25.1	161.9	9	6054.1	31.0	294.7	9	8774.1	36.9	467.0
7.0	751.9	13.3	15.1	15.0	2055.6	19.2	69.4	23.0	3833.3	25.1	163.3	31.0	6085.2	31.1	296.6	39.0	8811.1	37.0	469.4
1	765.2	13.4	15.6	1	2074.9	19.3	70.4	1	3858.6	25.2	164.7	1	6116.3	31.1	298.5	1	8848.2	37.1	471.9
2	778.7	13.4	16.0	2	2094.2	19.4	71.3	2	3883.9	25.3	166.1	2	6147.6	31.2	300.4	2	8885.3	37.1	474.3
3	792.2	13.5	16.4	3	2113.7	19.4	72.2	3	3909.2	25.4	167.6	3	6178.9	31.3	302.4	3	8922.6	37.2	476.7
4	805.8	13.6	16.9	4	2133.2	19.5	73.2	4	3934.7	25.4	169.0	4	6210.2	31.4	304.3	4	8959.9	37.3	479.1
5	819.4	13.7	17.4	5	2152.8	19.6	74.2	5	3960.2	25.5	170.4	5	6241.7	31.4	306.2	5	8997.2	37.4	481.6
6	833.2	13.7	17.8	6	2172.4	19.7	75.1	6	3985.8	25.6	171.9	6	6273.2	31.5	308.2	6	9034.7	37.4	484.0
7	847.0	13.8	18.3	7	2192.2	19.7	76.1	7	4011.4	25.7	173.4	7	6304.8	31.6	310.1	7	9072.2	37.5	486.4
8	860.9	13.9	18.8	8	2212.0	19.8	77.1	8	4037.2	25.7	174.8	8	6336.4	31.7	312.1	8	9109.8	37.6	488.9
9	874.9	14.0	19.3	9	2231.9	19.9	78.0	9	4063.0	25.8	176.3	9	6368.2	31.7	314.1	9	9147.4	37.7	491.4

TABLE No. 23.

EXCAVATION AND EMBANKMENT.

CUBIC YARDS.

Prismoids, 100 Feet Long. **Breadth of Base, 24 Feet.** **Slopes, 1 Horizontal to 1 Perpendicular.**

Height.	Cub. Yds.	Diff.	Cor.	Height.	Cub. Yds.	Diff.	Cor.	Height.	Cub. Yds.	Diff.	Cor.	Height.	Cub. Yds.	Diff.	Cor.	Height.	Cub. Yds.	Diff.	Cor.
0.0	0.0	0.0	0.0	8.0	948.1	14.8	19.8	16.0	2370.4	20.7	79.0	24.0	4266.7	26.6	177.8	32.0	6637.0	32.6	316.0
1	8.9	8.9	0.0	1	963.0	14.9	20.2	1	2391.1	20.8	80.0	1	4293.4	26.7	179.3	1	6669.7	32.6	318.0
2	17.9	9.0	0.0	2	977.9	14.9	20.8	2	2412.0	20.9	81.0	2	4320.1	26.8	180.8	2	6702.4	32.7	320.0
3	27.0	9.1	0.0	3	992.9	15.0	21.3	3	2432.9	20.9	82.0	3	4347.0	26.9	182.2	3	6735.1	32.8	322.0
4	36.1	9.1	0.0	4	1008.0	15.1	21.8	4	2453.9	21.0	83.0	4	4373.9	26.9	183.8	4	6768.0	32.9	324.0
5	45.4	9.2	0.1	5	1023.1	15.1	22.3	5	2475.0	21.1	84.0	5	4400.9	27.0	185.3	5	6800.9	32.9	326.0
6	54.7	9.3	0.1	6	1038.4	15.2	22.8	6	2496.1	21.1	85.0	6	4428.0	27.1	186.8	6	6833.9	33.0	328.0
7	64.0	9.4	0.2	7	1053.7	15.3	23.4	7	2517.4	21.2	86.1	7	4455.1	27.1	188.3	7	6867.0	33.1	330.0
8	73.5	9.4	0.2	8	1069.0	15.4	23.9	8	2538.7	21.3	87.1	8	4482.4	27.2	189.8	8	6900.1	33.1	332.0
9	83.0	9.5	0.2	9	1084.5	15.4	24.4	9	2560.0	21.4	88.1	9	4509.7	27.3	191.4	9	6933.4	33.2	334.1
1.0	92.6	9.6	0.3	9.0	1100.0	15.5	25.0	17.0	2581.5	21.4	89.2	25.0	4537.0	27.4	192.9	33.0	6966.7	33.3	336.1
1	102.3	9.7	0.4	1	1115.6	15.6	25.6	1	2603.0	21.5	90.2	1	4564.5	27.4	194.4	1	7000.0	33.4	338.2
2	112.0	9.7	0.4	2	1131.3	15.7	26.1	2	2624.6	21.6	91.3	2	4592.0	27.5	196.0	2	7033.5	33.4	340.2
3	121.8	9.8	0.5	3	1147.0	15.7	26.7	3	2646.3	21.7	92.4	3	4619.6	27.6	197.6	3	7067.0	33.5	342.2
4	131.7	9.9	0.6	4	1162.8	15.8	27.3	4	2668.0	21.7	93.4	4	4647.3	27.7	199.1	4	7100.6	33.6	344.3
5	141.7	10.0	0.7	5	1178.7	15.9	27.9	5	2689.8	21.8	94.5	5	4675.0	27.7	200.7	5	7134.3	33.7	346.4
6	151.7	10.0	0.8	6	1194.7	16.0	28.4	6	2711.7	21.9	95.6	6	4702.8	27.8	202.3	6	7168.0	33.7	348.4
7	161.8	10.1	0.9	7	1210.7	16.0	29.0	7	2733.7	22.0	96.7	7	4730.7	27.9	203.8	7	7201.8	33.8	350.5
8	172.0	10.2	1.0	8	1226.8	16.1	29.6	8	2755.7	22.0	97.8	8	4758.7	28.0	205.4	8	7235.7	33.9	352.6
9	182.3	10.3	1.1	9	1243.0	16.2	30.3	9	2777.8	22.1	98.9	9	4786.7	28.0	207.0	9	7269.7	34.0	354.7
2.0	192.6	10.3	1.2	10.0	1259.3	16.3	30.9	18.0	2800.0	22.2	100.0	26.0	4814.8	28.1	208.6	34.0	7303.7	34.0	356.8
1	203.0	10.4	1.4	1	1275.6	16.3	31.5	1	2822.3	22.3	101.1	1	4843.0	28.2	210.2	1	7337.8	34.1	358.9
2	213.5	10.5	1.5	2	1292.0	16.4	32.1	2	2844.6	22.3	102.2	2	4871.3	28.3	211.9	2	7372.0	34.2	361.0
3	224.0	10.6	1.6	3	1308.5	16.5	32.7	3	2867.0	22.4	103.4	3	4899.6	28.3	213.5	3	7406.3	34.3	363.1
4	234.7	10.6	1.8	4	1325.0	16.6	33.4	4	2889.5	22.5	104.5	4	4928.0	28.4	215.1	4	7440.6	34.3	365.2
5	245.4	10.7	1.9	5	1341.7	16.6	34.0	5	2912.0	22.6	105.6	5	4956.5	28.5	216.7	5	7475.0	34.4	367.4
6	256.1	10.8	2.1	6	1358.4	16.7	34.7	6	2934.7	22.6	106.8	6	4985.0	28.6	218.4	6	7509.5	34.5	369.5
7	267.0	10.9	2.2	7	1375.1	16.8	35.3	7	2957.4	22.7	107.9	7	5013.7	28.6	220.0	7	7544.0	34.6	371.6
8	277.9	10.9	2.4	8	1391.0	16.9	36.0	8	2980.1	22.8	109.1	8	5042.4	28.7	221.7	8	7578.7	34.6	373.8
9	288.9	11.0	2.6	9	1408.9	16.9	36.7	9	3003.0	22.9	110.2	9	5071.1	28.8	223.3	9	7613.4	34.7	375.9
3.0	300.0	11.1	2.8	11.0	1425.9	17.0	37.3	19.0	3025.9	22.9	111.4	27.0	5100.0	28.9	225.0	35.0	7648.1	34.8	378.1
1	311.1	11.1	3.0	1	1443.0	17.1	38.0	1	3048.9	23.0	112.6	1	5128.9	28.9	226.7	1	7683.0	34.9	380.2
2	322.4	11.2	3.2	2	1460.1	17.1	38.7	2	3072.0	23.1	113.8	2	5157.9	29.0	228.3	2	7717.9	34.9	382.4
3	333.7	11.3	3.4	3	1477.4	17.2	39.4	3	3095.1	23.1	115.0	3	5187.0	29.1	230.0	3	7752.9	35.0	384.6
4	345.0	11.4	3.6	4	1494.7	17.3	40.1	4	3118.4	23.2	116.2	4	5216.1	29.1	231.7	4	7788.0	35.1	386.8
5	356.5	11.4	3.8	5	1512.0	17.4	40.8	5	3141.7	23.3	117.4	5	5245.4	29.2	233.4	5	7823.1	35.1	389.0
6	368.0	11.5	4.0	6	1529.5	17.4	41.5	6	3165.0	23.4	118.6	6	5274.7	29.3	235.1	6	7858.4	35.2	391.2
7	379.6	11.6	4.2	7	1547.0	17.5	42.2	7	3188.5	23.4	119.8	7	5304.0	29.4	236.8	7	7893.7	35.3	393.4
8	391.3	11.7	4.5	8	1564.6	17.6	43.0	8	3212.0	23.5	121.0	8	5333.5	29.4	238.5	8	7929.0	35.4	395.6
9	403.0	11.7	4.7	9	1582.3	17.7	43.7	9	3235.6	23.6	122.2	9	5363.0	29.5	240.2	9	7964.5	35.4	397.8
4.0	414.8	11.8	4.9	12.0	1600.0	17.7	44.4	20.0	3259.3	23.7	123.5	28.0	5392.6	29.6	242.0	36.0	8000.0	35.5	400.0
1	426.7	11.9	5.2	1	1617.8	17.8	45.2	1	3283.0	23.7	124.7	1	5422.3	29.7	243.7	1	8035.6	35.6	402.2
2	438.7	12.0	5.4	2	1635.7	17.9	45.9	2	3306.8	23.8	125.9	2	5452.0	29.7	245.4	2	8071.3	35.7	404.5
3	450.7	12.0	5.7	3	1653.7	18.0	46.7	3	3330.7	23.9	127.2	3	5481.8	29.8	247.2	3	8107.0	35.7	406.7
4	462.8	12.1	6.0	4	1671.7	18.0	47.5	4	3354.7	24.0	128.4	4	5511.7	29.9	248.9	4	8142.8	35.8	408.9
5	475.0	12.2	6.2	5	1689.8	18.1	48.2	5	3378.7	24.0	129.7	5	5541.7	30.0	250.7	5	8178.7	35.9	411.2
6	487.3	12.3	6.5	6	1708.0	18.2	49.0	6	3402.8	24.1	131.0	6	5571.7	30.0	252.5	6	8214.7	36.0	413.5
7	499.6	12.3	6.8	7	1726.3	18.3	49.8	7	3427.0	24.2	132.2	7	5601.8	30.1	254.2	7	8250.7	36.0	415.7
8	512.0	12.4	7.1	8	1744.6	18.3	50.6	8	3451.3	24.3	133.5	8	5632.0	30.2	256.0	8	8286.8	36.1	418.0
9	524.5	12.5	7.4	9	1763.0	18.4	51.4	9	3475.6	24.3	134.8	9	5662.3	30.3	257.8	9	8323.0	36.2	420.2
5.0	537.0	12.6	7.7	13.0	1781.5	18.5	52.2	21.0	3500.0	24.4	136.1	29.0	5692.6	30.3	259.6	37.0	8359.3	36.3	422.5
1	549.7	12.6	8.0	1	1800.0	18.6	53.0	1	3524.5	24.5	137.4	1	5723.0	30.4	261.4	1	8395.6	36.3	424.8
2	562.4	12.7	8.3	2	1818.7	18.6	53.8	2	3549.0	24.6	138.7	2	5753.5	30.5	263.2	2	8432.0	36.4	427.1
3	575.1	12.8	8.7	3	1837.4	18.7	54.6	3	3573.7	24.6	140.0	3	5784.0	30.6	265.0	3	8468.5	36.5	429.4
4	588.0	12.9	9.0	4	1856.1	18.8	55.4	4	3598.4	24.7	141.3	4	5814.7	30.6	266.8	4	8505.0	36.6	431.7
5	600.9	12.9	9.3	5	1875.0	18.9	56.2	5	3623.1	24.8	142.7	5	5845.4	30.7	268.6	5	8541.7	36.6	434.0
6	613.9	13.0	9.7	6	1893.9	18.9	57.1	6	3648.0	24.9	144.0	6	5876.1	30.8	270.4	6	8578.4	36.7	436.3
7	627.0	13.1	10.0	7	1912.9	19.0	57.9	7	3672.9	24.9	145.3	7	5907.0	30.9	272.2	7	8615.1	36.8	438.7
8	640.1	13.1	10.4	8	1932.0	19.1	58.8	8	3697.9	25.0	146.7	8	5937.9	30.9	274.1	8	8652.0	36.9	441.0
9	653.4	13.2	10.7	9	1951.1	19.1	59.6	9	3723.0	25.1	148.0	9	5968.9	31.0	275.9	9	8688.9	36.9	443.3
6.0	666.7	13.3	11.1	14.0	1970.4	19.2	60.5	22.0	3748.1	25.1	149.4	30.0	6000.0	31.1	277.8	38.0	8725.9	37.0	445.7
1	680.0	13.4	11.5	1	1989.7	19.3	61.4	1	3773.4	25.2	150.7	1	6031.1	31.1	279.6	1	8763.0	37.1	448.0
2	693.5	13.4	11.9	2	2009.0	19.4	62.2	2	3798.7	25.3	152.1	2	6062.4	31.2	281.5	2	8800.1	37.1	450.4
3	707.0	13.5	12.2	3	2028.5	19.4	63.1	3	3824.0	25.4	153.5	3	6093.7	31.3	283.4	3	8837.4	37.2	452.7
4	720.6	13.6	12.6	4	2048.0	19.5	64.0	4	3849.5	25.4	154.9	4	6125.0	31.4	285.2	4	8874.7	37.3	455.1
5	734.3	13.7	13.0	5	2067.6	19.6	64.9	5	3875.0	25.5	156.2	5	6156.5	31.4	287.1	5	8912.0	37.4	457.5
6	748.0	13.7	13.4	6	2087.3	19.7	65.8	6	3900.6	25.6	157.6	6	6188.0	31.5	289.0	6	8949.5	37.4	459.9
7	761.8	13.8	13.9	7	2107.0	19.7	66.7	7	3926.3	25.7	159.0	7	6219.6	31.6	290.9	7	8987.0	37.5	462.2
8	775.7	13.9	14.3	8	2126.8	19.8	67.6	8	3952.0	25.7	160.4	8	6251.3	31.7	292.8	8	9024.6	37.6	464.6
9	789.7	14.0	14.7	9	2146.7	19.9	68.5	9	3977.8	25.8	161.9	9	6283.0	31.7	294.7	9	9062.3	37.7	467.0
7.0	803.7	14.0	15.1	15.0	2166.7	20.0	69.4	23.0	4003.7	25.9	163.3	31.0	6314.8	31.8	296.6	39.0	9100.0	37.7	469.4
1	817.8	14.1	15.6	1	2186.7	20.0	70.4	1	4029.7	26.0	164.7	1	6346.7	31.9	298.5	1	9137.8	37.8	471.9
2	832.0	14.2	16.0	2	2206.8	20.1	71.3	2	4055.7	26.0	166.1	2	6378.7	32.0	300.4	2	9175.7	37.9	474.3
3	846.3	14.3	16.4	3	2227.0	20.2	72.2	3	4081.8	26.1	167.6	3	6410.7	32.0	302.4	3	9213.7	38.0	476.7
4	860.6	14.3	16.9	4	2247.3	20.3	73.2	4	4108.0	26.2	169.0	4	6442.8	32.1	304.3	4	9251.7	38.0	479.1
5	875.0	14.4	17.4	5	2267.6	20.3	74.2	5	4134.3	26.3	170.4	5	6475.0	32.2	306.2	5	9289.8	38.1	481.6
6	889.5	14.5	17.8	6	2288.0	20.4	75.1	6	4160.6	26.3	171.9	6	6507.3	32.3	308.2	6	9328.0	38.2	484.0
7	904.0	14.6	18.3	7	2308.5	20.5	76.1	7	4187.0	26.4	173.4	7	6539.6	32.3	310.1	7	9366.3	38.3	486.4
8	918.7	14.6	18.8	8	2329.0	20.6	77.1	8	4213.5	26.5	174.8	8	6572.0	32.4	312.1	8	9404.6	38.3	488.9
9	933.4	14.7	19.3	9	2349.7	20.6	78.0	9	4240.0	26.6	176.3	9	6604.5	32.5	314.1	9	9443.0	38.4	491.4

TABLE NO. 24.

EXCAVATION AND EMBANKMENT.

CUBIC YARDS.

PRISMOIDS 100 FEET LONG. **BREADTH OF BASE 26 FEET.**

Slopes 1 Horizontal to 1 Perpendicular.

Ht.	Cub. Yds.	Diff.	Cor.	Ht.	Cub. Yds.	Diff.	Cor.	Ht.	Cub. Yds.	Diff.	Cor.	Ht.	Cub. Yds.	Diff.	Cor.	Ht.	Cub. Yds.	Diff.	Cor.
0.0	0.0	0.0	0.0	8.0	1007.4	15.5	19.8	16.0	2488.9	21.4	79.0	24.0	4444.4	27.4	177.8	32.0	6874.1	33.3	316.0
.1	9.7	9.7	0.0	.1	1023.0	15.6	20.2	.1	2510.4	21.5	80.0	.1	4471.9	27.4	179.3	.1	6907.4	33.4	318.0
.2	19.4	9.7	0.0	.2	1038.7	15.7	20.8	.2	2532.0	21.6	81.0	.2	4499.4	27.5	180.8	.2	6940.9	33.4	320.0
.3	29.2	9.8	0.0	.3	1054.4	15.7	21.3	.3	2553.7	21.7	82.0	.3	4527.0	27.6	182.2	.3	6974.4	33.5	322.0
.4	39.1	9.9	0.0	.4	1070.2	15.8	21.8	.4	2575.4	21.7	83.0	.4	4554.7	27.7	183.8	.4	7008.0	33.6	324.0
.5	49.1	10.0	0.1	.5	1086.1	15.9	22.3	.5	2597.2	21.8	84.0	.5	4582.4	27.7	185.3	.5	7041.7	33.7	326.0
.6	59.1	10.0	0.1	.6	1102.1	16.0	22.8	.6	2619.1	21.9	85.0	.6	4610.2	27.8	186.8	.6	7075.4	33.7	328.0
.7	69.2	10.1	0.2	.7	1118.1	16.0	23.4	.7	2641.1	22.0	86.1	.7	4638.1	27.9	188.3	.7	7109.2	33.8	330.0
.8	79.4	10.2	0.2	.8	1134.2	16.1	23.9	.8	2663.1	22.0	87.1	.8	4666.1	28.0	189.8	.8	7143.1	33.9	332.0
.9	89.7	10.3	0.2	.9	1150.4	16.2	24.4	.9	2685.2	22.1	88.1	.9	4694.1	28.0	191.4	.9	7177.1	34.0	334.1
1.0	100.0	10.3	0.3	9.0	1166.7	16.3	25.0	17.0	2707.4	22.2	89.2	25.0	4722.2	28.1	192.9	33.0	7211.1	34.0	336.1
.1	110.4	10.4	0.4	.1	1183.0	16.3	25.6	.1	2729.7	22.3	90.2	.1	4750.4	28.2	194.4	.1	7245.2	34.1	338.2
.2	120.9	10.5	0.4	.2	1199.4	16.4	26.1	.2	2752.0	22.3	91.3	.2	4778.7	28.3	196.0	.2	7279.4	34.2	340.2
.3	131.4	10.6	0.5	.3	1215.9	16.5	26.7	.3	2774.4	22.4	92.4	.3	4807.0	28.3	197.6	.3	7313.7	34.3	342.3
.4	142.1	10.6	0.6	.4	1232.4	16.6	27.3	.4	2796.9	22.5	93.4	.4	4835.4	28.4	199.1	.4	7348.0	34.3	344.3
.5	152.8	10.7	0.7	.5	1249.1	16.6	27.9	.5	2819.4	22.6	94.5	.5	4863.9	28.5	200.7	.5	7382.4	34.4	346.4
.6	163.6	10.8	0.8	.6	1265.8	16.7	28.4	.6	2842.1	22.6	95.6	.6	4892.4	28.6	202.3	.6	7416.9	34.5	348.4
.7	174.4	10.9	0.9	.7	1282.6	16.8	29.0	.7	2864.8	22.7	96.7	.7	4921.0	28.6	203.8	.7	7451.4	34.6	350.5
.8	185.3	10.9	1.0	.8	1299.4	16.9	29.6	.8	2887.6	22.8	97.8	.8	4949.8	28.7	205.4	.8	7486.1	34.6	352.6
.9	196.3	11.0	1.1	.9	1316.3	16.9	30.3	.9	2910.4	22.9	98.9	.9	4978.6	28.8	207.0	.9	7520.8	34.7	354.7
2.0	207.4	11.1	1.2	10.0	1333.3	17.0	30.9	18.0	2933.3	22.9	100.0	26.0	5007.4	28.9	208.6	34.0	7555.6	34.8	356.8
.1	218.6	11.1	1.4	.1	1350.4	17.1	31.5	.1	2956.3	23.0	101.1	.1	5036.3	28.9	210.2	.1	7590.4	34.9	358.9
.2	229.8	11.2	1.5	.2	1367.6	17.1	32.1	.2	2979.4	23.1	102.2	.2	5065.3	29.0	211.9	.2	7625.3	34.9	361.0
.3	241.1	11.3	1.6	.3	1384.8	17.2	32.7	.3	3002.6	23.1	103.4	.3	5094.4	29.1	213.5	.3	7660.3	35.0	363.1
.4	252.4	11.4	1.8	.4	1402.1	17.3	33.4	.4	3025.8	23.2	104.5	.4	5123.6	29.1	215.1	.4	7695.4	35.1	365.2
.5	263.9	11.4	1.9	.5	1419.4	17.4	34.0	.5	3049.1	23.3	105.6	.5	5152.8	29.2	216.7	.5	7730.6	35.1	367.4
.6	275.4	11.5	2.1	.6	1436.9	17.4	34.7	.6	3072.4	23.4	106.8	.6	5182.1	29.3	218.4	.6	7765.8	35.2	369.5
.7	287.0	11.6	2.2	.7	1454.4	17.5	35.3	.7	3095.9	23.4	107.9	.7	5211.4	29.4	220.0	.7	7801.1	35.3	371.6
.8	298.7	11.7	2.4	.8	1472.0	17.6	36.0	.8	3119.4	23.5	109.1	.8	5240.9	29.4	221.7	.8	7836.4	35.4	373.8
.9	310.4	11.7	2.6	.9	1489.7	17.7	36.7	.9	3143.0	23.6	110.2	.9	5270.4	29.5	223.3	.9	7871.9	35.4	375.9
3.0	322.2	11.8	2.8	11.0	1507.4	17.7	37.3	19.0	3166.7	23.7	111.4	27.0	5300.0	29.6	225.0	35.0	7907.4	35.5	378.1
.1	334.1	11.9	3.0	.1	1525.2	17.8	38.0	.1	3190.4	23.7	112.6	.1	5329.7	29.7	226.7	.1	7943.0	35.6	380.2
.2	346.1	12.0	3.2	.2	1543.1	17.9	38.7	.2	3214.2	23.8	113.8	.2	5359.4	29.7	228.3	.2	7978.7	35.7	382.4
.3	358.1	12.0	3.4	.3	1561.1	18.0	39.4	.3	3238.1	23.9	115.0	.3	5389.2	29.8	230.0	.3	8014.4	35.7	384.6
.4	370.2	12.1	3.6	.4	1579.1	18.0	40.1	.4	3262.1	24.0	116.2	.4	5419.1	29.9	231.7	.4	8050.2	35.8	386.8
.5	382.4	12.2	3.8	.5	1597.2	18.1	40.8	.5	3286.1	24.0	117.4	.5	5449.1	30.0	233.4	.5	8086.1	35.9	389.0
.6	394.7	12.3	4.0	.6	1615.4	18.2	41.5	.6	3310.2	24.1	118.6	.6	5479.1	30.0	235.1	.6	8122.1	36.0	391.2
.7	407.0	12.3	4.2	.7	1633.7	18.3	42.2	.7	3334.4	24.2	119.8	.7	5509.2	30.1	236.8	.7	8158.1	36.0	393.4
.8	419.4	12.4	4.5	.8	1652.0	18.3	43.0	.8	3358.7	24.3	121.0	.8	5539.4	30.2	238.5	.8	8194.2	36.1	395.6
.9	431.9	12.5	4.7	.9	1670.4	18.4	43.7	.9	3383.0	24.3	122.2	.9	5569.7	30.3	240.2	.9	8230.4	36.2	397.8
4.0	444.4	12.6	4.9	12.0	1688.9	18.5	44.4	20.0	3407.4	24.4	123.5	28.0	5600.0	30.3	242.0	36.0	8266.7	36.3	400.0
.1	457.1	12.6	5.2	.1	1707.4	18.6	45.2	.1	3431.9	24.5	124.7	.1	5630.4	30.4	243.7	.1	8303.0	36.3	402.2
.2	469.8	12.7	5.4	.2	1726.1	18.6	45.9	.2	3456.4	24.6	125.9	.2	5660.9	30.5	245.4	.2	8339.4	36.4	404.5
.3	482.6	12.8	5.7	.3	1744.8	18.7	46.7	.3	3481.1	24.6	127.2	.3	5691.4	30.6	247.2	.3	8375.9	36.5	406.7
.4	495.4	12.9	6.0	.4	1763.6	18.8	47.5	.4	3505.8	24.7	128.4	.4	5722.1	30.6	248.9	.4	8412.4	36.6	408.9
.5	508.3	12.9	6.2	.5	1782.4	18.9	48.2	.5	3530.6	24.8	129.7	.5	5752.8	30.7	250.7	.5	8449.1	36.6	411.2
.6	521.3	13.0	6.5	.6	1801.3	18.9	49.0	.6	3555.4	24.9	131.0	.6	5783.6	30.8	252.5	.6	8485.8	36.7	413.5
.7	534.4	13.1	6.8	.7	1820.3	19.0	49.8	.7	3580.3	24.9	132.2	.7	5814.4	30.9	254.2	.7	8522.6	36.8	415.7
.8	547.6	13.1	7.1	.8	1839.4	19.1	50.6	.8	3605.3	25.0	133.5	.8	5845.3	30.9	256.0	.8	8559.4	36.9	418.0
.9	560.8	13.2	7.4	.9	1858.6	19.1	51.4	.9	3630.4	25.1	134.8	.9	5876.3	31.0	257.8	.9	8596.3	36.9	420.2
5.0	574.1	13.3	7.7	13.0	1877.8	19.2	52.2	21.0	3655.6	25.1	136.1	29.0	5907.4	31.1	259.6	37.0	8633.3	37.0	422.5
.1	587.4	13.4	8.0	.1	1897.1	19.3	53.0	.1	3680.8	25.2	137.4	.1	5938.6	31.1	261.4	.1	8670.4	37.1	424.8
.2	600.9	13.4	8.3	.2	1916.4	19.4	53.8	.2	3706.1	25.3	138.7	.2	5969.8	31.2	263.2	.2	8707.6	37.1	427.1
.3	614.4	13.5	8.7	.3	1935.9	19.4	54.6	.3	3731.4	25.4	140.0	.3	6001.1	31.3	265.0	.3	8744.8	37.2	429.4
.4	628.0	13.6	9.0	.4	1955.4	19.5	55.4	.4	3756.9	25.4	141.3	.4	6032.4	31.4	266.8	.4	8782.1	37.3	431.7
.5	641.7	13.7	9.3	.5	1975.0	19.6	56.2	.5	3782.4	25.5	142.7	.5	6063.9	31.4	268.6	.5	8819.4	37.4	434.0
.6	655.4	13.7	9.7	.6	1994.7	19.7	57.1	.6	3808.0	25.6	144.0	.6	6095.4	31.5	270.4	.6	8856.9	37.4	436.3
.7	669.2	13.8	10.0	.7	2014.4	19.7	57.9	.7	3833.7	25.7	145.3	.7	6127.0	31.6	272.2	.7	8894.4	37.5	438.7
.8	683.1	13.9	10.4	.8	2034.2	19.8	58.8	.8	3859.4	25.7	146.7	.8	6158.7	31.7	274.1	.8	8932.0	37.6	441.0
.9	697.1	14.0	10.7	.9	2054.1	19.9	59.6	.9	3885.2	25.8	148.0	.9	6190.4	31.7	275.9	.9	8969.7	37.7	443.3
6.0	711.1	14.0	11.1	14.0	2074.1	20.0	60.5	22.0	3911.1	25.9	149.4	30.0	6222.2	31.8	277.8	38.0	9007.4	37.7	445.7
.1	725.2	14.1	11.5	.1	2094.1	20.0	61.4	.1	3937.1	26.0	150.7	.1	6254.1	31.9	279.6	.1	9045.2	37.8	448.0
.2	739.4	14.2	11.9	.2	2114.2	20.1	62.2	.2	3963.1	26.0	152.1	.2	6286.1	32.0	281.5	.2	9083.1	37.9	450.4
.3	753.7	14.3	12.2	.3	2134.4	20.2	63.1	.3	3989.2	26.1	153.5	.3	6318.1	32.0	283.4	.3	9121.1	38.0	452.7
.4	768.0	14.3	12.6	.4	2154.7	20.3	64.0	.4	4015.4	26.2	154.9	.4	6350.2	32.1	285.2	.4	9159.1	38.0	455.1
.5	782.4	14.4	13.0	.5	2175.0	20.3	64.9	.5	4041.7	26.3	156.2	.5	6382.4	32.2	287.1	.5	9197.2	38.1	457.5
.6	796.9	14.5	13.4	.6	2195.4	20.4	65.8	.6	4068.0	26.3	157.6	.6	6414.7	32.3	289.0	.6	9235.4	38.2	459.9
.7	811.4	14.6	13.9	.7	2215.9	20.5	66.7	.7	4094.4	26.4	159.0	.7	6447.0	32.3	290.9	.7	9273.7	38.3	462.2
.8	826.1	14.6	14.3	.8	2236.4	20.6	67.6	.8	4120.9	26.5	160.4	.8	6479.4	32.4	292.8	.8	9312.0	38.3	464.6
.9	840.8	14.7	14.7	.9	2257.1	20.6	68.5	.9	4147.4	26.6	161.9	.9	6511.9	32.5	294.7	.9	9350.4	38.4	467.0
7.0	855.6	14.8	15.1	15.0	2277.8	20.7	69.4	23.0	4174.1	26.6	163.3	31.0	6544.4	32.6	296.6	39.0	9388.9	38.5	469.4
.1	870.4	14.9	15.6	.1	2298.6	20.8	70.4	.1	4200.8	26.7	164.7	.1	6577.1	32.6	298.5	.1	9427.4	38.6	471.9
.2	885.3	14.9	16.0	.2	2319.4	20.9	71.3	.2	4227.6	26.8	166.1	.2	6609.8	32.7	300.4	.2	9466.1	38.6	474.3
.3	900.3	15.0	16.4	.3	2340.3	20.9	72.2	.3	4254.4	26.9	167.6	.3	6642.6	32.8	302.4	.3	9504.8	38.7	476.7
.4	915.4	15.1	16.9	.4	2361.3	21.0	73.2	.4	4281.3	26.9	169.0	.4	6675.4	32.9	304.3	.4	9543.6	38.8	479.1
.5	930.6	15.1	17.4	.5	2382.4	21.1	74.2	.5	4308.3	27.0	170.4	.5	6708.3	32.9	306.2	.5	9582.4	38.9	481.6
.6	945.8	15.2	17.8	.6	2403.6	21.1	75.1	.6	4335.4	27.1	171.9	.6	6741.3	33.0	308.2	.6	9621.3	38.9	484.0
.7	961.1	15.3	18.3	.7	2424.8	21.2	76.1	.7	4362.6	27.1	173.4	.7	6774.4	33.1	310.1	.7	9660.3	39.0	486.4
.8	976.4	15.4	18.8	.8	2446.1	21.3	77.1	.8	4389.8	27.2	174.8	.8	6807.6	33.1	312.1	.8	9699.4	39.1	488.9
.9	991.9	15.4	19.3	.9	2467.4	21.4	78.0	.9	4417.1	27.3	176.3	.9	6840.8	33.2	314.1	.9	9738.6	39.1	491.4

TABLE No. 25.

EXCAVATION AND EMBANKMENT.

CUBIC YARDS.

Prismoids, 100 Feet Long **Breadth of Base, 28 Feet** **Slopes, 1 Horizontal to 1 Perpendicular.**

Height.	Cub. Yds.	Diff.	Cor.	Height.	Cub. Yds.	Diff.	Cor.	Height.	Cub. Yds.	Diff.	Cor.	Height.	Cub. Yds.	Diff.	Cor.	Height.	Cub. Yds.	Diff.	Cor.
0.0	0.0	0.0	0.0	8.0	1066.7	16.3	19.8	16.0	2607.4	22.2	79.0	24.0	4622.2	28.1	177.8	32.0	7111.1	34.0	316.0
1	10.4	10.4	0.0	1	1083.0	16.3	20.2	1	2629.7	22.3	80.0	1	4650.4	28.2	179.3	1	7145.2	34.1	318.0
2	20.9	10.5	0.0	2	1099.4	16.4	20.8	2	2652.0	22.3	81.0	2	4678.7	28.3	180.8	2	7179.4	34.2	320.0
3	31.4	10.6	0.0	3	1115.9	16.5	21.3	3	2674.4	22.4	82.0	3	4707.0	28.3	182.2	3	7213.7	34.3	322.0
4	42.1	10.6	0.0	4	1132.4	16.6	21.8	4	2696.9	22.5	83.0	4	4735.4	28.4	183.8	4	7248.0	34.3	324.0
5	52.8	10.7	0.1	5	1149.1	16.6	22.3	5	2719.4	22.6	84.0	5	4763.9	28.5	185.3	5	7282.4	34.4	326.0
6	63.6	10.8	0.1	6	1165.8	16.7	22.8	6	2742.1	22.6	85.0	6	4792.4	28.6	186.8	6	7316.9	34.5	328.0
7	74.4	10.9	0.2	7	1182.6	16.8	23.4	7	2764.8	22.7	86.1	7	4821.1	28.6	188.3	7	7351.4	34.6	330.0
8	85.3	10.9	0.2	8	1199.4	16.9	23.9	8	2787.6	22.8	87.1	8	4849.8	28.7	189.8	8	7386.1	34.6	332.0
9	96.3	11.0	0.2	9	1216.3	16.9	24.4	9	2810.4	22.9	88.1	9	4878.6	28.8	191.4	9	7420.8	34.7	334.1
1.0	107.4	11.1	0.3	9.0	1233.3	17.0	25.0	17.0	2833.3	22.9	89.2	25.0	4907.4	28.9	192.9	33.0	7455.6	34.8	336.1
1	118.6	11.1	0.4	1	1250.4	17.1	25.6	1	2856.3	23.0	90.2	1	4936.3	28.9	194.4	1	7490.4	34.9	338.2
2	129.8	11.2	0.4	2	1267.6	17.1	26.1	2	2879.4	23.1	91.3	2	4965.3	29.0	196.0	2	7525.3	34.9	340.2
3	141.1	11.3	0.5	3	1284.8	17.2	26.7	3	2902.6	23.1	92.4	3	4994.4	29.1	197.6	3	7560.3	35.0	342.2
4	152.4	11.4	0.6	4	1302.1	17.3	27.3	4	2925.8	23.2	93.4	4	5023.6	29.1	199.1	4	7595.4	35.1	344.3
5	163.9	11.4	0.7	5	1319.4	17.4	27.9	5	2949.1	23.3	94.5	5	5052.8	29.2	200.7	5	7630.6	35.1	346.4
6	175.4	11.5	0.8	6	1336.9	17.4	28.4	6	2972.4	23.4	95.6	6	5082.1	29.3	202.3	6	7665.8	35.2	348.4
7	187.0	11.6	0.9	7	1354.4	17.5	29.0	7	2995.9	23.4	96.7	7	5111.4	29.4	203.8	7	7701.1	35.3	350.5
8	198.7	11.7	1.0	8	1372.0	17.6	29.6	8	3019.4	23.5	97.8	8	5140.9	29.4	205.4	8	7736.4	35.4	352.6
9	210.4	11.7	1.1	9	1389.7	17.7	30.3	9	3043.0	23.6	98.9	9	5170.4	29.5	207.0	9	7771.9	35.4	354.7
2.0	222.2	11.8	1.2	10.0	1407.4	17.7	30.9	18.0	3066.7	23.7	100.0	26.0	5200.0	29.6	208.6	34.0	7807.4	35.5	356.8
1	234.1	11.9	1.4	1	1425.2	17.8	31.5	1	3090.4	23.7	101.1	1	5229.7	29.7	210.2	1	7843.0	35.6	358.9
2	246.1	12.0	1.5	2	1443.1	17.9	32.1	2	3114.2	23.8	102.2	2	5259.4	29.7	211.9	2	7878.7	35.7	361.0
3	258.1	12.0	1.6	3	1461.1	18.0	32.7	3	3138.1	23.9	103.4	3	5289.2	29.8	213.5	3	7914.4	35.7	363.1
4	270.2	12.1	1.8	4	1479.1	18.0	33.4	4	3162.1	24.0	104.5	4	5319.1	29.9	215.1	4	7950.2	35.8	365.2
5	282.4	12.2	1.9	5	1497.2	18.1	34.0	5	3186.1	24.0	105.6	5	5349.1	30.0	216.7	5	7986.1	35.9	367.4
6	294.7	12.3	2.1	6	1515.4	18.2	34.7	6	3210.2	24.1	106.8	6	5379.1	30.0	218.4	6	8022.1	36.0	369.5
7	307.0	12.3	2.2	7	1533.7	18.3	35.3	7	3234.4	24.2	107.9	7	5409.2	30.1	220.0	7	8058.1	36.0	371.6
8	319.4	12.4	2.4	8	1552.0	18.3	36.0	8	3258.7	24.3	109.1	8	5439.4	30.2	221.7	8	8094.2	36.1	373.8
9	331.9	12.5	2.6	9	1570.4	18.4	36.7	9	3283.0	24.3	110.2	9	5469.7	30.3	223.3	9	8130.4	36.2	375.9
3.0	344.4	12.6	2.8	11.0	1588.9	18.5	37.3	19.0	3307.4	24.4	111.4	27.0	5500.0	30.3	225.0	35.0	8166.7	36.3	378.1
1	357.1	12.6	3.0	1	1607.4	18.6	38.0	1	3331.9	24.5	112.6	1	5530.4	30.4	226.7	1	8203.0	36.3	380.2
2	369.8	12.7	3.2	2	1626.1	18.6	38.7	2	3356.4	24.6	113.8	2	5560.9	30.5	228.3	2	8239.4	36.4	382.4
3	382.6	12.8	3.4	3	1644.8	18.7	39.4	3	3381.1	24.6	115.0	3	5591.4	30.6	230.0	3	8275.9	36.5	384.6
4	395.4	12.9	3.6	4	1663.6	18.8	40.1	4	3405.8	24.7	116.2	4	5622.1	30.6	231.7	4	8312.4	36.6	386.8
5	408.3	12.9	3.8	5	1682.4	18.9	40.8	5	3430.6	24.8	117.4	5	5652.8	30.7	233.4	5	8349.1	36.6	389.0
6	421.3	13.0	4.0	6	1701.3	18.9	41.5	6	3455.4	24.9	118.6	6	5683.6	30.8	235.1	6	8385.8	36.7	391.2
7	434.4	13.1	4.2	7	1720.3	19.0	42.2	7	3480.3	24.9	119.8	7	5714.4	30.9	236.8	7	8422.6	36.8	393.4
8	447.6	13.1	4.5	8	1739.4	19.1	43.0	8	3505.3	25.0	121.0	8	5745.3	30.9	238.5	8	8459.4	36.9	395.6
9	460.8	13.2	4.7	9	1758.6	19.1	43.7	9	3530.4	25.1	122.2	9	5776.3	31.0	240.2	9	8496.3	36.9	397.8
4.0	474.1	13.3	4.9	12.0	1777.8	19.2	44.4	20.0	3555.6	25.1	123.5	28.0	5807.4	31.1	242.0	36.0	8533.3	37.0	400.0
1	487.4	13.4	5.2	1	1797.1	19.3	45.2	1	3580.8	25.2	124.7	1	5838.6	31.1	243.7	1	8570.4	37.1	402.2
2	500.9	13.4	5.4	2	1816.4	19.4	45.9	2	3606.1	25.3	125.9	2	5869.8	31.2	245.4	2	8607.6	37.1	404.5
3	514.4	13.5	5.7	3	1835.9	19.4	46.7	3	3631.4	25.4	127.2	3	5901.1	31.3	247.2	3	8644.8	37.2	406.7
4	528.0	13.6	6.0	4	1855.4	19.5	47.5	4	3656.9	25.4	128.4	4	5932.4	31.4	248.9	4	8682.1	37.3	408.9
5	541.7	13.7	6.2	5	1875.0	19.6	48.2	5	3682.4	25.5	129.7	5	5963.9	31.4	250.7	5	8719.4	37.4	411.2
6	555.4	13.7	6.5	6	1894.7	19.7	49.0	6	3708.0	25.6	131.0	6	5995.4	31.5	252.5	6	8756.9	37.4	413.5
7	569.2	13.8	6.8	7	1914.4	19.7	49.8	7	3733.7	25.7	132.2	7	6027.0	31.6	254.2	7	8794.4	37.5	415.7
8	583.1	13.9	7.1	8	1934.2	19.8	50.6	8	3759.4	25.7	133.5	8	6058.7	31.7	256.0	8	8832.0	37.6	418.0
9	597.1	14.0	7.4	9	1954.1	19.9	51.4	9	3785.2	25.8	134.8	9	6090.4	31.7	257.8	9	8869.7	37.7	420.2
5.0	611.1	14.0	7.7	13.0	1974.1	20.0	52.2	21.0	3811.1	25.9	136.1	29.0	6122.2	31.8	259.6	37.0	8907.4	37.7	422.5
1	625.2	14.1	8.0	1	1994.1	20.0	53.0	1	3837.1	26.0	137.4	1	6154.1	31.9	261.4	1	8945.2	37.8	424.8
2	639.4	14.2	8.3	2	2014.2	20.1	53.8	2	3863.1	26.0	138.7	2	6186.1	32.0	263.2	2	8983.1	37.9	427.1
3	653.7	14.3	8.7	3	2034.4	20.2	54.6	3	3889.2	26.1	140.0	3	6218.1	32.0	265.0	3	9021.1	38.0	429.4
4	668.0	14.3	9.0	4	2054.7	20.3	55.4	4	3915.4	26.2	141.3	4	6250.2	32.1	266.8	4	9059.1	38.0	431.7
5	682.4	14.4	9.3	5	2075.0	20.3	56.2	5	3941.7	26.3	142.7	5	6282.4	32.2	268.6	5	9097.2	38.1	434.0
6	696.9	14.5	9.7	6	2095.4	20.4	57.1	6	3968.0	26.3	144.0	6	6314.7	32.3	270.4	6	9135.4	38.2	436.3
7	711.4	14.6	10.0	7	2115.9	20.5	57.9	7	3994.4	26.4	145.3	7	6347.0	32.3	272.2	7	9173.7	38.3	438.7
8	726.1	14.6	10.4	8	2136.4	20.6	58.8	8	4020.9	26.5	146.7	8	6379.4	32.4	274.1	8	9212.0	38.3	441.0
9	740.8	14.7	10.7	9	2157.1	20.6	59.6	9	4047.4	26.6	148.0	9	6411.9	32.5	275.9	9	9250.4	38.4	443.3
6.0	755.6	14.8	11.1	14.0	2177.8	20.7	60.5	22.0	4074.1	26.6	149.4	30.0	6444.4	32.6	277.8	38.0	9288.9	38.5	445.7
1	770.4	14.9	11.5	1	2198.6	20.8	61.4	1	4100.8	26.7	150.7	1	6477.1	32.6	279.6	1	9327.4	38.6	448.0
2	785.3	14.9	11.9	2	2219.4	20.9	62.2	2	4127.6	26.8	152.1	2	6509.8	32.7	281.5	2	9366.1	38.6	450.4
3	800.3	15.0	12.2	3	2240.3	20.9	63.1	3	4154.4	26.9	153.5	3	6542.6	32.8	283.4	3	9404.8	38.7	452.7
4	815.4	15.1	12.6	4	2261.3	21.0	64.0	4	4181.3	26.9	154.9	4	6575.4	32.9	285.2	4	9443.6	38.8	455.1
5	830.6	15.1	13.0	5	2282.4	21.1	64.9	5	4208.3	27.0	156.2	5	6608.3	32.9	287.1	5	9482.4	38.9	457.5
6	845.8	15.2	13.4	6	2303.6	21.1	65.8	6	4235.4	27.1	157.6	6	6641.3	33.0	289.0	6	9521.3	38.9	459.9
7	861.1	15.3	13.9	7	2324.8	21.2	66.7	7	4262.6	27.1	159.0	7	6674.4	33.1	290.9	7	9560.3	39.0	462.2
8	876.4	15.4	14.3	8	2346.1	21.3	67.6	8	4289.8	27.2	160.4	8	6707.6	33.1	292.8	8	9599.4	39.1	464.6
9	891.9	15.4	14.7	9	2367.4	21.4	68.5	9	4317.1	27.3	161.9	9	6740.8	33.2	294.7	9	9638.6	39.1	467.0
7.0	907.4	15.5	15.1	15.0	2388.9	21.4	69.4	23.0	4344.4	27.4	163.3	31.0	6774.1	33.3	296.6	39.0	9677.8	39.2	469.4
1	923.0	15.6	15.6	1	2410.4	21.5	70.4	1	4371.9	27.4	164.7	1	6807.4	33.4	298.5	1	9717.1	39.3	471.9
2	938.7	15.7	16.0	2	2432.0	21.6	71.3	2	4399.4	27.5	166.1	2	6840.9	33.4	300.4	2	9756.4	39.4	474.3
3	954.4	15.7	16.4	3	2453.7	21.7	72.2	3	4427.0	27.6	167.6	3	6874.4	33.5	302.4	3	9795.9	39.4	476.7
4	970.2	15.8	16.9	4	2475.4	21.7	73.2	4	4454.7	27.7	169.0	4	6908.0	33.6	304.3	4	9835.4	39.5	479.1
5	986.1	15.9	17.4	5	2497.2	21.8	74.2	5	4482.4	27.7	170.4	5	6941.7	33.7	306.2	5	9875.0	39.6	481.6
6	1002.1	16.0	17.8	6	2519.1	21.9	75.1	6	4510.2	27.8	171.9	6	6975.4	33.7	308.2	6	9914.7	39.7	484.0
7	1018.1	16.0	18.3	7	2541.1	22.0	76.1	7	4538.1	27.9	173.4	7	7009.2	33.8	310.1	7	9954.4	39.7	486.4
8	1034.2	16.1	18.8	8	2563.1	22.0	77.1	8	4566.1	28.0	174.8	8	7043.1	33.9	312.1	8	9994.2	39.8	488.9
9	1050.4	16.2	19.3	9	2585.2	22.1	78.0	9	4594.1	28.0	176.3	9	7077.1	34.0	314.1	9	10034.1	39.9	491.4

TABLE NO. 26.

EXCAVATION AND EMBANKMENT.

CUBIC YARDS.

PRISMOIDS 100 FEET LONG. **BREADTH OF BASE 30 FEET.**

Slopes 1 Horizontal to 1 Perpendicular.

Ht.	Cub. Yds.	Diff.	Cor.	Ht.	Cub. Yds.	Diff.	Cor.	Ht.	Cub. Yds.	Diff.	Cor.	Ht.	Cub. Yds.	Diff.	Cor.	Ht.	Cub. Yds.	Diff.	Cor.
0.0	0.0	0.0	0.0	8.0	1125.9	17.0	19.8	16.0	2725.9	22.9	79.0	24.0	4800.0	28.9	177.8	32.0	7348.1	34.8	316.0
.1	11.1	11.1	0.0	.1	1143.0	17.1	20.2	.1	2748.9	23.0	80.0	.1	4828.9	28.9	179.3	.1	7383.0	34.9	318.0
.2	22.4	11.2	0.0	.2	1160.1	17.1	20.8	.2	2772.0	23.1	81.0	.2	4857.9	29.0	180.8	.2	7417.9	34.9	320.0
.3	33.7	11.3	0.0	.3	1177.4	17.2	21.3	.3	2795.1	23.1	82.0	.3	4887.0	29.1	182.2	.3	7452.9	35.0	322.0
.4	45.0	11.4	0.0	.4	1194.7	17.3	21.8	.4	2818.4	23.2	83.0	.4	4916.1	29.1	183.8	.4	7488.0	35.1	324.0
.5	56.5	11.4	0.1	.5	1212.0	17.4	22.3	.5	2841.7	23.3	84.0	.5	4945.4	29.2	185.3	.5	7523.1	35.1	326.0
.6	68.0	11.5	0.1	.6	1229.5	17.4	22.8	.6	2865.0	23.4	85.0	.6	4974.7	29.3	186.8	.6	7558.4	35.2	328.0
.7	79.6	11.6	0.2	.7	1247.0	17.5	23.4	.7	2888.5	23.4	86.1	.7	5004.0	29.4	188.3	.7	7593.7	35.3	330.0
.8	91.3	11.7	0.2	.8	1264.6	17.6	23.9	.8	2912.0	23.5	87.1	.8	5033.5	29.4	189.8	.8	7629.0	35.4	332.0
.9	103.0	11.7	0.2	.9	1282.3	17.7	24.4	.9	2935.6	23.6	88.1	.9	5063.0	29.5	191.4	.9	7664.5	35.4	334.1
1.0	114.8	11.8	0.3	9.0	1300.0	17.7	25.0	17.0	2959.3	23.7	89.2	25.0	5092.6	29.6	192.9	33.0	7700.0	35.5	336.1
.1	126.7	11.9	0.4	.1	1317.8	17.8	25.6	.1	2983.0	23.7	90.2	.1	5122.3	29.7	194.4	.1	7735.6	35.6	338.2
.2	138.7	12.0	0.4	.2	1335.7	17.9	26.1	.2	3006.8	23.8	91.3	.2	5152.0	29.7	196.0	.2	7771.3	35.7	340.2
.3	150.7	12.0	0.5	.3	1353.7	18.0	26.7	.3	3030.7	23.9	92.4	.3	5181.8	29.8	197.6	.3	7807.0	35.7	342.2
.4	162.8	12.1	0.6	.4	1371.7	18.0	27.3	.4	3054.7	24.0	93.4	.4	5211.7	29.9	199.1	.4	7842.8	35.8	344.3
.5	175.0	12.2	0.7	.5	1389.8	18.1	27.9	.5	3078.7	24.0	94.5	.5	5241.7	30.0	200.7	.5	7878.7	35.9	346.4
.6	187.3	12.3	0.8	.6	1408.0	18.2	28.4	.6	3102.8	24.1	95.6	.6	5271.7	30.0	202.3	.6	7914.7	36.0	348.4
.7	199.6	12.3	0.9	.7	1426.3	18.3	29.0	.7	3127.0	24.2	96.7	.7	5301.8	30.1	203.8	.7	7950.7	36.0	350.5
.8	212.0	12.4	1.0	.8	1444.6	18.3	29.6	.8	3151.3	24.3	97.8	.8	5332.0	30.2	205.4	.8	7986.8	36.1	352.6
.9	224.5	12.5	1.1	.9	1463.0	18.4	30.3	.9	3175.6	24.3	98.9	.9	5362.3	30.3	207.0	.9	8023.0	36.2	354.7
2.0	237.0	12.6	1.2	10.0	1481.5	18.5	30.9	18.0	3200.0	24.4	100.0	26.0	5392.6	30.3	208.6	34.0	8059.3	36.3	356.8
.1	249.7	12.6	1.4	.1	1500.0	18.6	31.5	.1	3224.5	24.5	101.1	.1	5423.0	30.4	210.2	.1	8095.6	36.3	358.9
.2	262.4	12.7	1.5	.2	1518.7	18.6	32.1	.2	3249.0	24.6	102.2	.2	5453.5	30.5	211.9	.2	8132.0	36.4	361.0
.3	275.1	12.8	1.6	.3	1537.4	18.7	32.7	.3	3273.7	24.6	103.4	.3	5484.0	30.6	213.5	.3	8168.5	36.5	363.1
.4	288.0	12.9	1.8	.4	1556.1	18.8	33.4	.4	3298.4	24.7	104.5	.4	5514.7	30.6	215.1	.4	8205.0	36.6	365.2
.5	300.9	12.9	1.9	.5	1575.0	18.9	34.0	.5	3323.1	24.8	105.6	.5	5545.4	30.7	216.7	.5	8241.7	36.6	367.4
.6	313.9	13.0	2.1	.6	1593.9	18.9	34.7	.6	3348.0	24.9	106.8	.6	5576.1	30.8	218.4	.6	8278.4	36.7	369.5
.7	327.0	13.1	2.2	.7	1612.9	19.0	35.3	.7	3372.9	24.9	107.9	.7	5607.0	30.9	220.0	.7	8315.1	36.8	371.6
.8	340.1	13.1	2.4	.8	1632.0	19.1	36.0	.8	3397.9	25.0	109.1	.8	5637.9	30.9	221.7	.8	8352.0	36.9	373.8
.9	353.4	13.2	2.6	.9	1651.1	19.1	36.7	.9	3423.0	25.1	110.2	.9	5668.9	31.0	223.3	.9	8388.9	36.9	375.9
3.0	366.7	13.3	2.8	11.0	1670.4	19.2	37.3	19.0	3448.1	25.1	111.4	27.0	5700.0	31.1	225.0	35.0	8425.9	37.0	378.1
.1	380.0	13.4	3.0	.1	1689.7	19.3	38.0	.1	3473.4	25.2	112.6	.1	5731.1	31.1	226.7	.1	8463.0	37.1	380.2
.2	393.5	13.4	3.2	.2	1709.0	19.4	38.7	.2	3498.7	25.3	113.8	.2	5762.4	31.2	228.3	.2	8500.1	37.1	382.4
.3	407.0	13.5	3.4	.3	1728.5	19.4	39.4	.3	3524.0	25.4	115.0	.3	5793.7	31.3	230.0	.3	8537.4	37.2	384.6
.4	420.6	13.6	3.6	.4	1748.0	19.5	40.1	.4	3549.5	25.4	116.2	.4	5825.0	31.4	231.7	.4	8574.7	37.3	386.8
.5	434.3	13.7	3.8	.5	1767.6	19.6	40.8	.5	3575.0	25.5	117.4	.5	5856.5	31.4	233.4	.5	8612.0	37.4	389.0
.6	448.0	13.7	4.0	.6	1787.3	19.7	41.5	.6	3600.6	25.6	118.6	.6	5888.0	31.5	235.1	.6	8649.5	37.4	391.2
.7	461.8	13.8	4.2	.7	1807.0	19.7	42.2	.7	3626.3	25.7	119.8	.7	5919.6	31.6	236.8	.7	8687.0	37.5	393.4
.8	475.7	13.9	4.5	.8	1826.8	19.8	43.0	.8	3652.0	25.7	121.0	.8	5951.3	31.7	238.5	.8	8724.6	37.6	395.6
.9	489.7	14.0	4.7	.9	1846.7	19.9	43.7	.9	3677.8	25.8	122.2	.9	5983.0	31.7	240.2	.9	8762.3	37.7	397.8
4.0	503.7	14.0	4.9	12.0	1866.7	20.0	44.4	20.0	3703.7	25.9	123.5	28.0	6014.8	31.8	242.0	36.0	8800.0	37.7	400.0
.1	517.8	14.1	5.2	.1	1886.7	20.0	45.2	.1	3729.7	26.0	124.7	.1	6046.7	31.9	243.7	.1	8837.8	37.8	402.2
.2	532.0	14.2	5.4	.2	1906.8	20.1	45.9	.2	3755.7	26.0	125.9	.2	6078.7	32.0	245.4	.2	8875.7	37.9	404.5
.3	546.3	14.3	5.7	.3	1927.0	20.2	46.7	.3	3781.8	26.1	127.2	.3	6110.7	32.0	247.2	.3	8913.7	38.0	406.7
.4	560.6	14.3	6.0	.4	1947.3	20.3	47.5	.4	3808.0	26.2	128.4	.4	6142.8	32.1	248.9	.4	8951.7	38.0	408.9
.5	575.0	14.4	6.2	.5	1967.6	20.3	48.2	.5	3834.3	26.3	129.7	.5	6175.0	32.2	250.7	.5	8989.8	38.1	411.2
.6	589.5	14.5	6.5	.6	1988.0	20.4	49.0	.6	3860.6	26.3	131.0	.6	6207.3	32.3	252.5	.6	9028.0	38.2	413.4
.7	604.0	14.6	6.8	.7	2008.5	20.5	49.8	.7	3887.0	26.4	132.2	.7	6239.6	32.3	254.2	.7	9066.3	38.3	415.7
.8	618.7	14.6	7.1	.8	2029.0	20.6	50.6	.8	3913.5	26.5	133.5	.8	6272.0	32.4	256.0	.8	9104.6	38.3	418.0
.9	633.4	14.7	7.4	.9	2049.7	20.6	51.4	.9	3940.0	26.6	134.8	.9	6304.5	32.5	257.8	.9	9143.0	38.4	420.2
5.0	648.1	14.8	7.7	13.0	2070.4	20.7	52.2	21.0	3966.7	26.6	136.1	29.0	6337.0	32.6	259.6	37.0	9181.5	38.5	422.5
.1	663.0	14.9	8.0	.1	2091.1	20.8	53.0	.1	3993.4	26.7	137.4	.1	6369.7	32.6	261.4	.1	9220.0	38.6	424.8
.2	677.9	14.9	8.3	.2	2112.0	20.9	53.8	.2	4020.1	26.8	138.7	.2	6402.4	32.7	263.2	.2	9258.7	38.6	427.1
.3	692.9	15.0	8.7	.3	2132.9	20.9	54.6	.3	4047.0	26.9	140.0	.3	6435.1	32.8	265.0	.3	9297.4	38.7	429.4
.4	708.0	15.1	9.0	.4	2153.9	21.0	55.4	.4	4073.9	26.9	141.3	.4	6468.0	32.9	266.8	.4	9336.1	38.8	431.7
.5	723.1	15.1	9.3	.5	2175.0	21.1	56.2	.5	4100.9	27.0	142.7	.5	6500.9	32.9	268.6	.5	9375.0	38.9	434.0
.6	738.4	15.2	9.7	.6	2196.1	21.1	57.1	.6	4128.0	27.1	144.0	.6	6533.9	33.0	270.4	.6	9413.9	38.9	436.3
.7	753.7	15.3	10.0	.7	2217.4	21.2	57.9	.7	4155.1	27.1	145.3	.7	6567.0	33.1	272.2	.7	9452.9	39.0	438.7
.8	769.0	15.4	10.4	.8	2238.7	21.3	58.8	.8	4182.4	27.2	146.7	.8	6600.1	33.1	274.1	.8	9492.0	39.1	441.0
.9	784.5	15.4	10.7	.9	2260.0	21.4	59.6	.9	4209.7	27.3	148.0	.9	6633.4	33.2	275.9	.9	9531.1	39.1	443.3
6.0	800.0	15.5	11.1	14.0	2281.5	21.4	60.5	22.0	4237.0	27.4	149.4	30.0	6666.7	33.3	277.8	38.0	9570.4	39.2	445.7
.1	815.6	15.6	11.5	.1	2303.0	21.5	61.4	.1	4264.5	27.4	150.7	.1	6700.0	33.4	279.6	.1	9609.7	39.3	448.0
.2	831.3	15.7	11.9	.2	2324.6	21.6	62.2	.2	4292.0	27.5	152.1	.2	6733.5	33.4	281.5	.2	9649.0	39.4	450.4
.3	847.0	15.7	12.2	.3	2346.3	21.7	63.1	.3	4319.6	27.6	153.5	.3	6767.0	33.5	283.4	.3	9688.5	39.4	452.7
.4	862.8	15.8	12.6	.4	2368.0	21.7	64.0	.4	4347.3	27.7	154.9	.4	6800.6	33.6	285.2	.4	9728.0	39.5	455.1
.5	878.7	15.9	13.0	.5	2389.8	21.8	64.9	.5	4375.0	27.7	156.2	.5	6834.3	33.7	287.1	.5	9767.6	39.6	457.5
.6	894.7	16.0	13.4	.6	2411.7	21.9	65.8	.6	4402.8	27.8	157.6	.6	6868.0	33.7	289.0	.6	9807.3	39.7	459.9
.7	910.7	16.0	13.9	.7	2433.7	22.0	66.7	.7	4430.7	27.9	159.0	.7	6901.8	33.8	290.9	.7	9847.0	39.7	462.2
.8	926.8	16.1	14.3	.8	2455.7	22.0	67.6	.8	4458.7	28.0	160.4	.8	6935.7	33.9	292.8	.8	9886.8	39.8	464.6
.9	943.0	16.2	14.7	.9	2477.8	22.1	68.5	.9	4486.7	28.0	161.8	.9	6969.7	34.0	294.7	.9	9926.7	39.9	467.0
7.0	959.3	16.3	15.1	15.0	2500.0	22.2	69.4	23.0	4514.8	28.1	163.3	31.0	7003.7	34.0	296.6	39.0	9966.7	40.0	469.4
.1	975.6	16.3	15.6	.1	2522.3	22.3	70.4	.1	4543.0	28.2	164.7	.1	7037.8	34.1	298.5	.1	10006.7	40.0	471.9
.2	992.0	16.4	16.0	.2	2544.6	22.3	71.3	.2	4571.3	28.3	166.1	.2	7072.0	34.2	300.4	.2	10046.8	40.1	474.3
.3	1008.5	16.5	16.4	.3	2567.0	22.4	72.2	.3	4599.6	28.3	167.6	.3	7106.3	34.3	302.4	.3	10087.0	40.2	476.7
.4	1025.0	16.6	16.9	.4	2589.5	22.5	73.2	.4	4628.0	28.4	169.0	.4	7140.6	34.3	304.3	.4	10127.3	40.3	479.1
.5	1041.7	16.6	17.4	.5	2612.0	22.6	74.2	.5	4656.5	28.5	170.4	.5	7175.0	34.4	306.2	.5	10167.6	40.3	481.6
.6	1058.4	16.7	17.8	.6	2634.7	22.6	75.1	.6	4685.0	28.6	171.9	.6	7209.5	34.5	308.2	.6	10208.0	40.4	484.0
.7	1075.1	16.8	18.3	.7	2657.4	22.7	76.1	.7	4713.7	28.6	173.4	.7	7244.0	34.6	310.1	.7	10248.5	40.5	486.4
.8	1092.0	16.9	18.8	.8	2680.1	22.8	77.1	.8	4742.4	28.7	174.8	.8	7278.7	34.6	312.1	.8	10289.0	40.6	488.9
.9	1108.9	16.9	19.3	.9	2703.0	22.9	78.0	.9	4771.1	28.8	176.3	.9	7313.4	34.7	314.1	.9	10329.7	40.6	491.4

TABLE NO. 27.

EXCAVATION AND EMBANKMENT.

CUBIC YARDS.

PRISMOIDS 100 FEET LONG. **BREADTH OF BASE 20 FEET.**

Slopes 1-2 Horizontal to 1 Perpendicular.

Ht.	Cub. Yds.	Diff.	Cor.	Ht.	Cub. Yds.	Diff.	Cor.	Ht.	Cub. Yds.	Diff.	Cor.	Ht.	Cub. Yds.	Diff.	Cor.	Ht.	Cub. Yds.	Diff.	Cor.
0.0	0.0	0.0	0.0	8.0	711.1	10.4	9.9	16.0	1659.3	13.3	39.5	24.0	2844.4	16.3	88.9	32.0	4266.7	19.2	158.0
.1	7.4	7.4	0.0	.1	721.5	10.4	10.1	.1	1672.6	13.3	40.0	.1	2860.8	16.3	89.6	.1	4285.9	19.3	159.0
.2	14.9	7.5	0.0	.2	731.9	10.4	10.4	.2	1686.0	13.4	40.5	.2	2877.1	16.4	90.4	.2	4305.3	19.3	160.0
.3	22.4	7.5	0.0	.3	742.4	10.5	10.6	.3	1699.4	13.4	41.0	.3	2893.5	16.4	91.1	.3	4324.6	19.4	161.0
.4	29.9	7.5	0.0	.4	752.9	10.5	10.9	.4	1712.9	13.5	41.5	.4	2909.9	16.4	91.9	.4	4344.0	19.4	162.0
.5	37.5	7.6	0.0	.5	763.4	10.5	11.1	.5	1726.4	13.5	42.0	.5	2926.4	16.5	92.6	.5	4363.4	19.4	163.0
.6	45.1	7.6	0.0	.6	774.0	10.6	11.4	.6	1739.9	13.5	42.5	.6	2942.9	16.5	93.4	.6	4382.9	19.5	164.0
.7	52.8	7.6	0.1	.7	784.6	10.6	11.7	.7	1753.5	13.6	43.0	.7	2959.4	16.5	94.2	.7	4402.4	19.5	165.0
.8	60.4	7.7	0.1	.8	795.3	10.6	12.0	.8	1767.1	13.6	43.6	.8	2976.0	16.6	94.9	.8	4421.9	19.5	166.0
.9	68.2	7.7	0.1	.9	805.9	10.7	12.2	.9	1780.8	13.6	44.1	.9	2992.6	16.6	95.7	.9	4441.5	19.6	167.0
1.0	75.9	7.8	0.2	9.0	816.7	10.7	12.5	17.0	1794.4	13.7	44.6	25.0	3009.3	16.6	96.5	33.0	4461.1	19.6	168.1
.1	83.7	7.8	0.2	.1	827.4	10.8	12.8	.1	1808.2	13.7	45.1	.1	3025.9	16.7	97.2	.1	4480.8	19.6	169.1
.2	91.6	7.8	0.2	.2	838.2	10.8	13.1	.2	1821.9	13.8	45.7	.2	3042.7	16.7	98.0	.2	4500.4	19.7	170.1
.3	99.4	7.9	0.3	.3	849.1	10.8	13.3	.3	1835.7	13.8	46.2	.3	3059.4	16.8	98.8	.3	4520.2	19.7	171.1
.4	107.3	7.9	0.3	.4	859.9	10.9	13.6	.4	1849.6	13.8	46.7	.4	3076.2	16.8	99.6	.4	4539.9	19.8	172.2
.5	115.3	7.9	0.3	.5	870.8	10.9	13.9	.5	1863.4	13.9	47.3	.5	3093.1	16.8	100.3	.5	4559.7	19.8	173.2
.6	123.3	8.0	0.4	.6	881.8	10.9	14.2	.6	1877.3	13.9	47.8	.6	3109.9	16.9	101.1	.6	4579.6	19.8	174.2
.7	131.3	8.0	0.4	.7	892.8	11.0	14.5	.7	1891.3	13.9	48.3	.7	3126.8	16.9	101.9	.7	4599.4	19.9	175.3
.8	139.3	8.1	0.5	.8	903.8	11.0	14.8	.8	1905.3	14.0	48.9	.8	3143.8	16.9	102.7	.8	4619.3	19.9	176.3
.9	147.4	8.1	0.6	.9	914.8	11.1	15.1	.9	1919.3	14.0	49.4	.9	3160.8	17.0	103.5	.9	4639.3	19.9	177.4
2.0	155.6	8.1	0.6	10.0	925.9	11.1	15.4	18.0	1933.3	14.1	50.0	26.0	3177.8	17.0	104.3	34.0	4659.3	20.0	178.4
.1	163.7	8.2	0.7	.1	937.1	11.1	15.7	.1	1947.4	14.1	50.6	.1	3194.8	17.1	105.1	.1	4679.3	20.0	179.4
.2	171.9	8.2	0.7	.2	948.2	11.2	16.1	.2	1961.6	14.1	51.1	.2	3211.9	17.1	105.9	.2	4699.3	20.1	180.5
.3	180.2	8.2	0.8	.3	959.4	11.2	16.4	.3	1975.7	14.2	51.7	.3	3229.1	17.1	106.7	.3	4719.4	20.1	181.6
.4	188.4	8.3	0.9	.4	970.7	11.2	16.7	.4	1989.9	14.2	52.2	.4	3246.2	17.2	107.6	.4	4739.6	20.1	182.6
.5	196.8	8.3	1.0	.5	981.9	11.3	17.0	.5	2004.2	14.2	52.8	.5	3263.4	17.2	108.4	.5	4759.7	20.2	183.7
.6	205.1	8.4	1.0	.6	993.3	11.3	17.3	.6	2018.4	14.3	53.4	.6	3280.7	17.3	109.2	.6	4779.9	20.2	184.7
.7	213.5	8.4	1.1	.7	1004.6	11.4	17.7	.7	2032.8	14.3	54.0	.7	3297.9	17.3	110.0	.7	4800.2	20.2	185.8
.8	221.9	8.4	1.2	.8	1016.0	11.4	18.0	.8	2047.1	14.4	54.5	.8	3315.3	17.3	110.8	.8	4820.4	20.3	186.9
.9	230.4	8.5	1.3	.9	1027.4	11.4	18.3	.9	2061.5	14.4	55.1	.9	3332.6	17.4	111.7	.9	4840.8	20.3	188.0
3.0	238.9	8.5	1.4	11.0	1038.9	11.5	18.7	19.0	2075.9	14.4	55.7	27.0	3350.0	17.4	112.5	35.0	4861.1	20.4	189.0
.1	247.4	8.5	1.5	.1	1050.4	11.5	19.0	.1	2090.4	14.5	56.3	.1	3367.4	17.4	113.3	.1	4881.5	20.4	190.1
.2	256.0	8.6	1.6	.2	1061.9	11.5	19.4	.2	2104.9	14.5	56.9	.2	3384.9	17.5	114.2	.2	4901.9	20.4	191.2
.3	264.6	8.6	1.7	.3	1073.5	11.6	19.7	.3	2119.4	14.5	57.5	.3	3402.4	17.5	115.0	.3	4922.4	20.5	192.3
.4	273.3	8.6	1.8	.4	1085.1	11.6	20.1	.4	2134.0	14.6	58.1	.4	3419.9	17.5	115.9	.4	4942.9	20.5	193.4
.5	281.9	8.7	1.9	.5	1096.8	11.6	20.4	.5	2148.6	14.6	58.7	.5	3437.5	17.6	116.7	.5	4963.4	20.5	194.5
.6	290.7	8.7	2.0	.6	1108.4	11.7	20.8	.6	2163.3	14.6	59.3	.6	3455.1	17.6	117.6	.6	4984.0	20.6	195.6
.7	299.4	8.8	2.1	.7	1120.2	11.7	21.1	.7	2177.9	14.7	59.9	.7	3472.8	17.6	118.4	.7	5004.6	20.6	196.7
.8	308.2	8.8	2.2	.8	1131.9	11.8	21.5	.8	2192.7	14.7	60.5	.8	3490.4	17.7	119.3	.8	5025.3	20.6	197.8
.9	317.1	8.8	2.3	.9	1143.7	11.8	21.9	.9	2207.4	14.8	61.1	.9	3508.2	17.7	120.1	.9	5045.9	20.7	198.9
4.0	325.9	8.9	2.5	12.0	1155.6	11.8	22.2	20.0	2222.2	14.8	61.7	28.0	3525.9	17.8	121.0	36.0	5066.7	20.7	200.0
.1	334.8	8.9	2.6	.1	1167.4	11.9	22.6	.1	2237.1	14.8	62.3	.1	3543.7	17.8	121.9	.1	5087.4	20.8	201.1
.2	343.8	8.9	2.7	.2	1179.3	11.9	23.0	.2	2251.9	14.9	63.0	.2	3561.6	17.8	122.7	.2	5108.2	20.8	202.2
.3	352.8	9.0	2.9	.3	1191.3	11.9	23.3	.3	2266.8	14.9	63.6	.3	3579.4	17.9	123.6	.3	5129.1	20.8	203.3
.4	361.8	9.0	3.0	.4	1203.3	12.0	23.7	.4	2281.8	14.9	64.2	.4	3597.3	17.9	124.5	.4	5149.9	20.9	204.5
.5	370.8	9.1	3.1	.5	1215.3	12.0	24.1	.5	2296.8	15.0	64.9	.5	3615.3	17.9	125.3	.5	5170.8	20.9	205.6
.6	379.9	9.1	3.3	.6	1227.3	12.1	24.5	.6	2311.8	15.0	65.5	.6	3633.3	18.0	126.2	.6	5191.8	20.9	206.7
.7	389.1	9.1	3.4	.7	1239.4	12.1	24.9	.7	2326.8	15.1	66.1	.7	3651.3	18.0	127.1	.7	5212.8	21.0	207.9
.8	398.2	9.2	3.6	.8	1251.6	12.1	25.3	.8	2341.9	15.1	66.8	.8	3669.3	18.1	128.0	.8	5233.8	21.0	209.0
.9	407.4	9.2	3.7	.9	1263.7	12.2	25.7	.9	2357.1	15.1	67.4	.9	3687.4	18.1	128.9	.9	5254.8	21.1	210.1
5.0	416.7	9.2	3.9	13.0	1275.9	12.2	26.1	21.0	2372.2	15.2	68.1	29.0	3705.6	18.1	129.8	37.0	5275.9	21.1	211.3
.1	425.9	9.3	4.0	.1	1288.2	12.2	26.5	.1	2387.4	15.2	68.7	.1	3723.7	18.2	130.7	.1	5297.1	21.1	212.4
.2	435.3	9.3	4.2	.2	1300.4	12.3	26.9	.2	2402.7	15.2	69.4	.2	3741.9	18.2	131.6	.2	5318.2	21.2	213.6
.3	444.6	9.4	4.3	.3	1312.8	12.3	27.3	.3	2417.9	15.3	70.0	.3	3760.2	18.2	132.5	.3	5339.4	21.2	214.7
.4	454.0	9.4	4.5	.4	1325.1	12.4	27.7	.4	2433.3	15.3	70.7	.4	3778.4	18.3	133.4	.4	5360.7	21.2	215.9
.5	463.4	9.4	4.7	.5	1337.5	12.4	28.1	.5	2448.6	15.4	71.3	.5	3796.8	18.3	134.3	.5	5381.9	21.3	217.0
.6	472.9	9.5	4.8	.6	1349.9	12.4	28.5	.6	2464.0	15.4	72.0	.6	3815.1	18.4	135.2	.6	5403.3	21.3	218.2
.7	482.4	9.5	5.0	.7	1362.4	12.5	29.0	.7	2479.4	15.4	72.7	.7	3833.5	18.4	136.1	.7	5424.6	21.4	219.3
.8	491.9	9.5	5.2	.8	1374.9	12.5	29.4	.8	2494.9	15.5	73.3	.8	3851.9	18.4	137.0	.8	5446.0	21.4	220.5
.9	501.5	9.6	5.4	.9	1387.4	12.5	29.8	.9	2510.4	15.5	74.0	.9	3870.4	18.5	138.0	.9	5467.4	21.4	221.7
6.0	511.1	9.6	5.6	14.0	1400.0	12.6	30.2	22.0	2525.9	15.5	74.7	30.0	3888.9	18.5	138.9	38.0	5488.9	21.5	222.8
.1	520.8	9.6	5.7	.1	1412.6	12.6	30.7	.1	2541.5	15.6	75.4	.1	3907.4	18.5	139.8	.1	5510.4	21.5	224.0
.2	530.4	9.7	5.9	.2	1425.3	12.6	31.1	.2	2557.1	15.6	76.1	.2	3926.0	18.6	140.7	.2	5531.9	21.5	225.2
.3	540.2	9.7	6.1	.3	1437.9	12.7	31.6	.3	2572.8	15.6	76.7	.3	3944.6	18.6	141.7	.3	5553.5	21.6	226.4
.4	549.9	9.8	6.3	.4	1450.7	12.7	32.0	.4	2588.4	15.7	77.4	.4	3963.3	18.6	142.6	.4	5575.1	21.6	227.6
.5	559.7	9.8	6.5	.5	1463.4	12.8	32.4	.5	2604.2	15.7	78.1	.5	3981.9	18.7	143.6	.5	5596.8	21.6	228.7
.6	569.6	9.8	6.7	.6	1476.2	12.8	32.9	.6	2619.9	15.8	78.8	.6	4000.7	18.7	144.5	.6	5618.4	21.7	229.9
.7	579.4	9.9	6.9	.7	1489.1	12.8	33.3	.7	2635.7	15.8	79.5	.7	4019.4	18.8	145.4	.7	5640.2	21.7	231.1
.8	589.3	9.9	7.1	.8	1501.9	12.9	33.8	.8	2651.6	15.8	80.2	.8	4038.2	18.8	146.4	.8	5661.9	21.8	232.3
.9	599.3	9.9	7.3	.9	1514.8	12.9	34.3	.9	2667.4	15.9	80.9	.9	4057.1	18.8	147.3	.9	5683.7	21.8	233.5
7.0	609.3	10.0	7.6	15.0	1527.8	12.9	34.7	23.0	2683.3	15.9	81.6	31.0	4075.9	18.9	148.3	39.0	5705.6	21.8	234.7
.1	619.3	10.0	7.8	.1	1540.8	13.0	35.2	.1	2699.3	15.9	82.3	.1	4094.8	18.9	149.3	.1	5727.4	21.9	235.9
.2	629.3	10.1	8.0	.2	1553.8	13.0	35.7	.2	2715.3	16.0	83.1	.2	4113.8	18.9	150.2	.2	5749.3	21.9	237.1
.3	639.4	10.1	8.2	.3	1566.8	13.1	36.1	.3	2731.3	16.0	83.8	.3	4132.8	19.0	151.2	.3	5771.3	21.9	238.3
.4	649.6	10.1	8.5	.4	1579.9	13.1	36.6	.4	2747.3	16.1	84.5	.4	4151.8	19.0	152.2	.4	5793.3	22.0	239.6
.5	659.7	10.2	8.7	.5	1593.1	13.1	37.1	.5	2763.4	16.1	85.2	.5	4170.8	19.1	153.1	.5	5815.3	22.0	240.8
.6	669.9	10.2	8.9	.6	1606.2	13.2	37.6	.6	2779.6	16.1	86.0	.6	4189.9	19.1	154.1	.6	5837.3	22.1	242.0
.7	680.2	10.2	9.1	.7	1619.4	13.2	38.0	.7	2795.7	16.2	86.7	.7	4209.1	19.1	155.1	.7	5859.4	22.1	243.2
.8	690.4	10.3	9.4	.8	1632.7	13.2	38.5	.8	2811.9	16.2	87.4	.8	4228.2	19.2	156.1	.8	5881.6	22.1	244.5
.9	700.8	10.3	9.6	.9	1645.9	13.3	39.0	.9	2828.2	16.2	88.2	.9	4247.4	19.2	157.0	.9	5903.7	22.2	245.7

TABLE NO. 28.

EXCAVATION AND EMBANKMENT

CUBIC YARDS.

PRISMOIDS 100 FEET LONG. **BREADTH OF BASE 26 FEET.**

Slopes 1-2 Horizontal to 1 Perpendicular.

Ht.	Cub. Yds.	Diff.	Cor.	Ht.	Cub. Yds.	Diff.	Cor.	Ht.	Cub. Yds.	Diff.	Cor.	Ht.	Cub. Yds.	Diff.	Cor.	Ht.	Cub. Yds.	Diff.	Cor.
0.0	0.0	0.0	0.0	8.0	888.9	12.6	9.9	16.0	2014.8	15.5	39.5	24.0	3377.8	18.5	88.9	32.0	4977.8	21.5	158.0
.1	9.6	9.6	0.0	.1	901.5	12.6	10.1	.1	2030.4	15.6	40.0	.1	3396.3	18.5	89.6	.1	4999.3	21.5	159.0
.2	19.3	9.7	0.0	.2	914.1	12.6	10.4	.2	2046.0	15.6	40.5	.2	3414.9	18.6	90.4	.2	5020.8	21.5	160.0
.3	29.1	9.7	0.0	.3	926.8	12.7	10.6	.3	2061.6	15.6	41.0	.3	3433.5	18.6	91.1	.3	5042.4	21.6	161.0
.4	38.8	9.8	0.0	.4	939.6	12.7	10.9	.4	2077.3	15.7	41.5	.4	3452.1	18.6	91.9	.4	5064.0	21.6	162.0
.5	48.6	9.8	0.0	.5	952.3	12.8	11.1	.5	2093.1	15.7	42.0	.5	3470.8	18.7	92.6	.5	5085.6	21.6	163.0
.6	58.4	9.8	0.0	.6	965.1	12.8	11.4	.6	2108.8	15.8	42.5	.6	3489.6	18.7	93.4	.6	5107.3	21.7	164.0
.7	68.3	9.9	0.1	.7	977.9	12.8	11.7	.7	2124.6	15.8	43.0	.7	3508.3	18.8	94.2	.7	5129.1	21.7	165.0
.8	78.2	9.9	0.1	.8	990.8	12.9	12.0	.8	2140.4	15.8	43.6	.8	3527.1	18.8	94.9	.8	5150.8	21.8	166.0
.9	88.2	9.9	0.1	.9	1003.7	12.9	12.2	.9	2156.3	15.9	44.1	.9	3545.9	18.8	95.7	.9	5172.6	21.8	167.0
1.0	98.1	10.0	0.2	9.0	1016.7	12.9	12.5	17.0	2172.2	15.9	44.6	25.0	3564.8	18.9	96.5	33.0	5194.4	21.8	168.1
.1	108.2	10.0	0.2	.1	1029.6	13.0	12.8	.1	2188.2	15.9	45.1	.1	3583.7	18.9	97.2	.1	5216.3	21.9	169.1
.2	118.2	10.1	0.2	.2	1042.7	13.0	13.1	.2	2204.1	16.0	45.7	.2	3602.7	18.9	98.0	.2	5238.2	21.9	170.1
.3	128.3	10.1	0.3	.3	1055.7	13.1	13.3	.3	2220.2	16.0	46.2	.3	3621.6	19.0	98.8	.3	5260.2	21.9	171.1
.4	138.4	10.1	0.3	.4	1068.8	13.1	13.6	.4	2236.2	16.1	46.7	.4	3640.7	19.0	99.6	.4	5282.1	22.0	172.2
.5	148.6	10.2	0.3	.5	1081.9	13.1	13.9	.5	2252.3	16.1	47.3	.5	3659.7	19.1	100.3	.5	5304.2	22.0	173.2
.6	158.8	10.2	0.4	.6	1095.1	13.2	14.2	.6	2268.4	16.1	47.8	.6	3678.8	19.1	101.1	.6	5326.2	22.1	174.2
.7	169.1	10.2	0.4	.7	1108.3	13.2	14.5	.7	2284.6	16.2	48.3	.7	3697.9	19.1	101.9	.7	5348.3	22.1	175.3
.8	179.3	10.3	0.5	.8	1121.6	13.2	14.8	.8	2300.8	16.2	48.9	.8	3717.1	19.2	102.7	.8	5370.4	22.1	176.3
.9	189.6	10.3	0.6	.9	1134.8	13.3	15.1	.9	2317.1	16.2	49.4	.9	3736.3	19.2	103.5	.9	5392.6	22.2	177.4
2.0	200.0	10.4	0.6	10.0	1148.1	13.3	15.4	18.0	2333.3	16.3	50.0	26.0	3755.6	19.2	104.3	34.0	5414.8	22.2	178.4
.1	210.4	10.4	0.7	.1	1161.5	13.4	15.7	.1	2349.6	16.3	50.6	.1	3774.8	19.3	105.1	.1	5437.1	22.2	179.4
.2	220.8	10.4	0.7	.2	1174.9	13.4	16.1	.2	2366.0	16.4	51.1	.2	3794.1	19.3	105.9	.2	5459.3	22.3	180.5
.3	231.3	10.5	0.8	.3	1188.3	13.4	16.4	.3	2382.4	16.4	51.7	.3	3813.5	19.4	106.7	.3	5481.6	22.3	181.6
.4	241.8	10.5	0.9	.4	1201.8	13.5	16.7	.4	2398.8	16.4	52.2	.4	3832.9	19.4	107.6	.4	5504.0	22.4	182.6
.5	252.3	10.5	1.0	.5	1215.3	13.5	17.0	.5	2415.3	16.5	52.8	.5	3852.3	19.4	108.4	.5	5526.4	22.4	183.7
.6	262.9	10.6	1.0	.6	1228.8	13.5	17.3	.6	2431.8	16.5	53.4	.6	3871.8	19.5	109.2	.6	5548.8	22.4	184.7
.7	273.5	10.6	1.1	.7	1242.4	13.6	17.7	.7	2448.3	16.5	54.0	.7	3891.3	19.5	110.0	.7	5571.3	22.5	185.8
.8	284.1	10.6	1.2	.8	1256.0	13.6	18.0	.8	2464.9	16.6	54.5	.8	3910.8	19.5	110.8	.8	5593.8	22.5	186.9
.9	294.8	10.7	1.3	.9	1269.6	13.6	18.3	.9	2481.5	16.6	55.1	.9	3930.4	19.6	111.7	.9	5616.3	22.5	188.0
3.0	305.6	10.7	1.4	11.0	1283.3	13.7	18.7	19.0	2498.1	16.6	55.7	27.0	3950.0	19.6	112.5	35.0	5638.9	22.6	189.0
.1	316.3	10.8	1.5	.1	1297.1	13.7	19.0	.1	2514.8	16.7	56.3	.1	3969.6	19.6	113.3	.1	5661.5	22.6	190.1
.2	327.1	10.8	1.6	.2	1310.8	13.8	19.4	.2	2531.6	16.7	56.9	.2	3989.3	19.7	114.2	.2	5684.1	22.6	191.2
.3	337.9	10.8	1.7	.3	1324.6	13.8	19.7	.3	2548.3	16.8	57.5	.3	4009.1	19.7	115.0	.3	5706.8	22.7	192.3
.4	348.8	10.9	1.8	.4	1338.4	13.8	20.1	.4	2565.1	16.8	58.1	.4	4028.8	19.8	115.9	.4	5729.6	22.7	193.4
.5	359.7	10.9	1.9	.5	1352.3	13.9	20.4	.5	2581.9	16.8	58.7	.5	4048.6	19.8	116.7	.5	5752.3	22.8	194.5
.6	370.7	10.9	2.0	.6	1366.2	13.9	20.8	.6	2598.8	16.9	59.3	.6	4068.4	19.8	117.6	.6	5775.1	22.8	195.6
.7	381.6	11.0	2.1	.7	1380.2	13.9	21.1	.7	2615.7	16.9	59.9	.7	4088.3	19.9	118.4	.7	5797.9	22.8	196.7
.8	392.7	11.0	2.2	.8	1394.1	14.0	21.5	.8	2632.7	16.9	60.5	.8	4108.2	19.9	119.3	.8	5820.8	22.9	197.8
.9	403.7	11.1	2.3	.9	1408.2	14.1	21.9	.9	2649.6	17.0	61.1	.9	4128.2	19.9	120.1	.9	5843.7	22.9	198.9
4.0	414.8	11.1	2.5	12.0	1422.2	14.1	22.2	20.0	2666.7	17.0	61.7	28.0	4148.1	20.0	121.0	36.0	5866.7	22.9	200.0
.1	425.9	11.1	2.6	.1	1436.3	14.1	22.6	.1	2683.7	17.1	62.3	.1	4168.2	20.0	121.9	.1	5889.6	23.0	201.1
.2	437.1	11.2	2.7	.2	1450.4	14.1	23.0	.2	2700.8	17.1	63.0	.2	4188.2	20.1	122.7	.2	5912.7	23.0	202.2
.3	448.3	11.2	2.9	.3	1464.6	14.2	23.3	.3	2717.9	17.1	63.6	.3	4208.3	20.1	123.6	.3	5935.7	23.1	203.3
.4	459.6	11.2	3.0	.4	1478.8	14.2	23.7	.4	2735.1	17.2	64.2	.4	4228.4	20.1	124.5	.4	5958.8	23.1	204.5
.5	470.8	11.3	3.1	.5	1493.1	14.2	24.1	.5	2752.3	17.2	64.9	.5	4248.6	20.2	125.3	.5	5981.0	23.1	205.6
.6	482.1	11.3	3.3	.6	1507.3	14.3	24.5	.6	2769.6	17.2	65.5	.6	4268.8	20.2	126.2	.6	6005.1	23.2	206.7
.7	493.5	11.4	3.4	.7	1521.6	14.3	24.9	.7	2786.8	17.3	66.1	.7	4289.1	20.2	127.1	.7	6028.3	23.2	207.9
.8	504.9	11.4	3.6	.8	1536.0	14.4	25.3	.8	2804.1	17.3	66.8	.8	4309.3	20.3	128.0	.8	6051.6	23.2	209.0
.9	516.3	11.4	3.7	.9	1550.4	14.4	25.7	.9	2821.5	17.4	67.4	.9	4329.6	20.3	128.9	.9	6074.8	23.3	210.1
5.0	527.8	11.5	3.9	13.0	1564.8	14.4	26.1	21.0	2838.9	17.4	68.1	29.0	4350.0	20.4	129.8	37.0	6098.1	23.3	211.3
.1	539.3	11.5	4.0	.1	1579.3	14.5	26.5	.1	2856.3	17.4	68.7	.1	4370.4	20.4	130.7	.1	6121.5	23.4	212.4
.2	550.8	11.5	4.2	.2	1593.8	14.5	26.9	.2	2873.8	17.5	69.4	.2	4390.8	20.4	131.6	.2	6144.9	23.4	213.6
.3	562.4	11.6	4.3	.3	1608.3	14.5	27.3	.3	2891.3	17.5	70.0	.3	4411.3	20.5	132.5	.3	6168.3	23.4	214.7
.4	574.0	11.6	4.5	.4	1622.9	14.6	27.7	.4	2908.8	17.5	70.7	.4	4431.8	20.5	133.4	.4	6191.8	23.5	215.9
.5	585.6	11.6	4.7	.5	1637.5	14.6	28.1	.5	2926.4	17.6	71.3	.5	4452.3	20.5	134.3	.5	6215.3	23.5	217.0
.6	597.3	11.7	4.8	.6	1652.1	14.6	28.5	.6	2944.0	17.6	72.0	.6	4472.9	20.6	135.2	.6	6238.8	23.5	218.2
.7	609.1	11.7	5.0	.7	1666.8	14.7	29.0	.7	2961.6	17.6	72.7	.7	4493.5	20.6	136.1	.7	6262.4	23.6	219.3
.8	620.8	11.8	5.2	.8	1681.6	14.7	29.4	.8	2979.3	17.7	73.3	.8	4514.1	20.6	137.0	.8	6286.0	23.6	220.5
.9	632.6	11.8	5.4	.9	1696.3	14.8	29.8	.9	2997.1	17.7	74.0	.9	4534.8	20.7	138.0	.9	6309.6	23.6	221.7
6.0	644.4	11.8	5.6	14.0	1711.1	14.8	30.2	22.0	3014.8	17.8	74.7	30.0	4555.6	20.7	138.9	38.0	6333.3	23.7	222.8
.1	656.3	11.9	5.7	.1	1725.9	14.8	30.7	.1	3032.6	17.8	75.4	.1	4576.3	20.8	139.8	.1	6357.1	23.7	224.0
.2	668.2	11.9	5.9	.2	1740.8	14.9	31.1	.2	3050.4	17.8	76.1	.2	4597.1	20.8	140.7	.2	6380.8	23.8	225.2
.3	680.2	11.9	6.1	.3	1755.7	14.9	31.6	.3	3068.3	17.9	76.7	.3	4617.9	20.8	141.7	.3	6404.6	23.8	226.4
.4	692.1	12.0	6.3	.4	1770.7	14.9	32.0	.4	3086.2	17.9	77.4	.4	4638.8	20.9	142.6	.4	6428.4	23.8	227.6
.5	704.2	12.0	6.5	.5	1785.6	15.0	32.4	.5	3104.2	17.9	78.1	.5	4659.7	20.9	143.6	.5	6452.3	23.9	228.7
.6	716.2	12.1	6.7	.6	1800.7	15.0	32.9	.6	3122.1	18.0	78.8	.6	4680.7	20.9	144.5	.6	6476.2	23.9	229.9
.7	728.3	12.1	6.9	.7	1815.7	15.1	33.3	.7	3140.2	18.0	79.5	.7	4701.6	21.0	145.4	.7	6500.2	23.9	231.1
.8	740.4	12.1	7.1	.8	1830.8	15.1	33.8	.8	3158.2	18.1	80.2	.8	4722.7	21.0	146.4	.8	6524.1	24.0	232.3
.9	752.6	12.2	7.3	.9	1845.9	15.1	34.3	.9	3176.3	18.1	80.9	.9	4743.7	21.1	147.3	.9	6548.2	24.0	233.5
7.0	764.8	12.2	7.6	15.0	1861.1	15.2	34.7	23.0	3194.4	18.1	81.6	31.0	4764.8	21.1	148.3	39.0	6572.2	24.1	234.7
.1	777.1	12.2	7.8	.1	1876.3	15.2	35.2	.1	3212.6	18.2	82.3	.1	4785.9	21.1	149.3	.1	6596.3	24.1	235.9
.2	789.3	12.3	8.0	.2	1891.6	15.2	35.7	.2	3230.8	18.2	83.1	.2	4807.1	21.2	150.2	.2	6620.4	24.1	237.1
.3	801.6	12.3	8.2	.3	1906.8	15.3	36.1	.3	3249.1	18.2	83.8	.3	4828.3	21.2	151.2	.3	6644.6	24.2	238.3
.4	814.0	12.4	8.5	.4	1922.1	15.3	36.6	.4	3267.3	18.3	84.5	.4	4849.6	21.2	152.2	.4	6668.8	24.2	239.6
.5	826.4	12.4	8.7	.5	1937.5	15.4	37.1	.5	3285.6	18.3	85.2	.5	4870.8	21.3	153.1	.5	6693.1	24.2	240.8
.6	838.8	12.4	8.9	.6	1952.9	15.4	37.6	.6	3304.0	18.4	86.0	.6	4892.1	21.3	154.1	.6	6717.3	24.3	242.0
.7	851.3	12.5	9.1	.7	1968.3	15.4	38.0	.7	3322.4	18.4	86.7	.7	4913.5	21.4	155.1	.7	6741.6	24.3	243.2
.8	863.8	12.5	9.4	.8	1983.8	15.5	38.5	.8	3340.8	18.4	87.4	.8	4934.9	21.4	156.1	.8	6766.0	24.4	244.5
.9	876.3	12.5	9.6	.9	1999.3	15.5	39.0	.9	3359.3	18.5	88.2	.9	4956.3	21.4	157.0	.9	6790.4	24.4	245.7

TABLE NO. 29.

EXCAVATION AND EMBANKMENT.

CUBIC YARDS.

PRISMOIDS 100 FEET LONG. **BREADTH OF BASE 30 FEET.**

Slopes 1-2 Horizontal to 1 Perpendicular.

Ht.	Cub. Yds.	Dif.	Cor.	Ht.	Cub. Yds.	Dif.	Cor.	Ht.	Cub. Yds.	Dif.	Cor.	Ht.	Cub. Yds.	Dif.	Cor.	Ht.	Cub. Yds.	Dif.	Cor.
0.0	0.0	0.0	0.0	8.0	1007.4	14.1	9.9	16.0	2251.9	17.0	39.5	24.0	3733.3	20.0	88.9	32.0	5451.9	22.9	158.0
.1	11.1	11.1	0.0	.1	1021.5	14.1	10.1	.1	2268.9	17.1	40.0	.1	3753.4	20.0	89.6	.1	5474.8	23.0	159.0
.2	22.3	11.2	0.0	.2	1035.6	14.1	10.4	.2	2286.0	17.1	40.5	.2	3773.4	20.1	90.4	.2	5497.8	23.0	160.0
.3	33.5	11.2	0.0	.3	1049.8	14.2	10.6	.3	2303.1	17.1	41.0	.3	3793.5	20.1	91.1	.3	5520.9	23.1	161.0
.4	44.7	11.2	0.0	.4	1064.0	14.2	10.9	.4	2320.3	17.2	41.5	.4	3813.6	20.1	91.9	.4	5544.0	23.1	162.0
.5	56.0	11.3	0.0	.5	1078.2	14.2	11.1	.5	2337.5	17.2	42.0	.5	3833.8	20.2	92.6	.5	5567.1	23.1	163.0
.6	67.3	11.3	0.0	.6	1092.5	14.3	11.4	.6	2354.7	17.2	42.5	.6	3854.0	20.2	93.4	.6	5590.3	23.2	164.0
.7	78.7	11.4	0.1	.7	1106.8	14.3	11.7	.7	2372.0	17.3	43.0	.7	3874.2	20.2	94.2	.7	5613.5	23.2	165.0
.8	90.1	11.4	0.1	.8	1121.2	14.4	12.0	.8	2389.3	17.3	43.6	.8	3894.5	20.3	94.9	.8	5636.7	23.2	166.0
.9	101.5	11.4	0.1	.9	1135.6	14.4	12.2	.9	2406.7	17.4	44.1	.9	3914.8	20.3	95.7	.9	5660.0	23.3	167.0
1.0	113.0	11.5	0.2	9.0	1150.0	14.4	12.5	17.0	2424.1	17.4	44.6	25.0	3935.2	20.4	96.5	33.0	5683.3	23.3	168.1
.1	124.5	11.5	0.2	.1	1164.5	14.5	12.8	.1	2441.5	17.4	45.1	.1	3955.6	20.4	97.2	.1	5706.7	23.4	169.1
.2	136.0	11.5	0.2	.2	1179.0	14.5	13.1	.2	2459.0	17.5	45.7	.2	3976.0	20.4	98.0	.2	5730.1	23.4	170.1
.3	147.6	11.6	0.3	.3	1193.5	14.5	13.3	.3	2476.5	17.5	46.2	.3	3996.5	20.5	98.8	.3	5753.5	23.4	171.1
.4	159.2	11.6	0.3	.4	1208.1	14.6	13.6	.4	2494.0	17.5	46.7	.4	4017.0	20.5	99.6	.4	5777.0	23.5	172.2
.5	170.8	11.6	0.3	.5	1222.7	14.6	13.9	.5	2511.6	17.6	47.3	.5	4037.5	20.5	100.3	.5	5800.5	23.5	173.2
.6	182.5	11.7	0.4	.6	1237.3	14.6	14.2	.6	2529.2	17.6	47.8	.6	4058.1	20.6	101.1	.6	5824.0	23.5	174.2
.7	194.2	11.7	0.4	.7	1252.0	14.7	14.5	.7	2546.8	17.6	48.3	.7	4078.7	20.6	101.9	.7	5847.6	23.6	175.3
.8	206.0	11.8	0.5	.8	1266.7	14.7	14.8	.8	2564.5	17.7	48.9	.8	4099.3	20.6	102.7	.8	5871.2	23.6	176.3
.9	217.8	11.8	0.6	.9	1281.5	14.8	15.1	.9	2582.2	17.7	49.4	.9	4120.0	20.7	103.5	.9	5894.8	23.6	177.4
2.0	229.6	11.8	0.6	10.0	1296.3	14.8	15.4	18.0	2600.0	17.8	50.0	26.0	4140.7	20.7	104.3	34.0	5918.5	23.7	178.4
.1	241.5	11.9	0.7	.1	1311.1	14.8	15.7	.1	2617.8	17.8	50.6	.1	4161.5	20.8	105.1	.1	5942.2	23.7	179.4
.2	253.4	11.9	0.7	.2	1326.0	14.9	16.1	.2	2635.6	17.8	51.1	.2	4182.3	20.8	105.9	.2	5966.0	23.8	180.5
.3	265.4	11.9	0.8	.3	1340.9	14.9	16.4	.3	2653.5	17.9	51.7	.3	4203.1	20.8	106.7	.3	5989.8	23.8	181.6
.4	277.3	12.0	0.9	.4	1355.9	14.9	16.7	.4	2671.4	17.9	52.2	.4	4224.0	20.9	107.6	.4	6013.6	23.8	182.6
.5	289.4	12.0	1.0	.5	1370.8	15.0	17.0	.5	2689.4	17.9	52.8	.5	4244.9	20.9	108.4	.5	6037.5	23.9	183.7
.6	301.4	12.1	1.0	.6	1385.9	15.0	17.3	.6	2707.3	18.0	53.4	.6	4265.9	20.9	109.2	.6	6061.4	23.9	184.7
.7	313.5	12.1	1.1	.7	1400.9	15.1	17.7	.7	2725.4	18.0	54.0	.7	4286.8	21.0	110.0	.7	6085.4	23.9	185.8
.8	325.6	12.1	1.2	.8	1416.0	15.1	18.0	.8	2743.4	18.1	54.6	.8	4307.9	21.0	110.8	.8	6109.3	24.0	186.9
.9	337.8	12.2	1.3	.9	1431.1	15.1	18.3	.9	2761.5	18.1	55.1	.9	4328.9	21.1	111.7	.9	6133.4	24.0	188.0
3.0	350.0	12.2	1.4	11.0	1446.3	15.2	18.7	19.0	2779.6	18.1	55.7	27.0	4350.0	21.1	112.5	35.0	6157.4	24.1	189.0
.1	362.2	12.2	1.5	.1	1461.5	15.2	19.0	.1	2797.8	18.2	56.3	.1	4371.1	21.1	113.3	.1	6181.5	24.1	190.1
.2	374.5	12.3	1.6	.2	1476.7	15.2	19.4	.2	2816.0	18.2	56.9	.2	4392.3	21.2	114.2	.2	6205.6	24.1	191.2
.3	386.8	12.3	1.7	.3	1492.0	15.3	19.7	.3	2834.2	18.2	57.5	.3	4413.5	21.2	115.0	.3	6229.8	24.2	192.3
.4	399.2	12.4	1.8	.4	1507.3	15.3	20.1	.4	2852.5	18.3	58.1	.4	4434.7	21.2	115.9	.4	6254.0	24.2	193.4
.5	411.6	12.4	1.9	.5	1522.7	15.4	20.4	.5	2870.8	18.3	58.7	.5	4456.0	21.3	116.7	.5	6278.2	24.2	194.5
.6	424.0	12.4	2.0	.6	1538.1	15.4	20.8	.6	2889.2	18.4	59.3	.6	4477.3	21.3	117.6	.6	6302.5	24.3	195.6
.7	436.5	12.5	2.1	.7	1553.5	15.4	21.1	.7	2907.6	18.4	59.9	.7	4498.7	21.4	118.4	.7	6326.8	24.3	196.7
.8	449.0	12.5	2.2	.8	1569.0	15.5	21.5	.8	2926.0	18.4	60.5	.8	4520.1	21.4	119.3	.8	6351.2	24.4	197.8
.9	461.5	12.5	2.3	.9	1584.5	15.5	21.9	.9	2944.5	18.5	61.1	.9	4541.5	21.4	120.1	.9	6375.6	24.4	198.9
4.0	474.1	12.6	2.5	12.0	1600.0	15.5	22.2	20.0	2963.0	18.5	61.7	28.0	4563.0	21.5	121.0	36.0	6400.0	24.4	200.0
.1	486.7	12.6	2.6	.1	1615.6	15.6	22.6	.1	2981.5	18.5	62.3	.1	4584.5	21.5	121.9	.1	6424.5	24.5	201.1
.2	499.3	12.6	2.7	.2	1631.2	15.6	23.0	.2	3000.1	18.6	63.0	.2	4606.0	21.5	122.7	.2	6449.0	24.5	202.2
.3	512.0	12.7	2.9	.3	1646.8	15.6	23.3	.3	3018.7	18.6	63.6	.3	4627.6	21.6	123.6	.3	6473.5	24.5	203.3
.4	524.7	12.7	3.0	.4	1662.5	15.7	23.7	.4	3037.3	18.6	64.2	.4	4649.2	21.6	124.5	.4	6498.1	24.6	204.5
.5	537.5	12.8	3.1	.5	1678.2	15.7	24.1	.5	3056.0	18.7	64.9	.5	4670.8	21.6	125.3	.5	6522.7	24.6	205.6
.6	550.3	12.8	3.3	.6	1694.0	15.8	24.5	.6	3074.7	18.7	65.5	.6	4692.5	21.7	126.2	.6	6547.3	24.6	206.7
.7	563.1	12.8	3.4	.7	1709.8	15.8	24.9	.7	3093.5	18.8	66.1	.7	4714.2	21.7	127.1	.7	6572.0	24.7	207.9
.8	576.0	12.9	3.6	.8	1725.6	15.8	25.3	.8	3112.3	18.8	66.8	.8	4736.0	21.8	128.0	.8	6596.7	24.7	209.0
.9	588.9	12.9	3.7	.9	1741.5	15.9	25.7	.9	3131.1	18.8	67.4	.9	4757.8	21.8	128.9	.9	6621.5	24.8	210.1
5.0	601.9	12.9	3.9	13.0	1757.4	15.9	26.1	21.0	3150.0	18.9	68.1	29.0	4779.6	21.8	129.8	37.0	6646.3	24.8	211.3
.1	614.8	13.0	4.0	.1	1773.4	15.9	26.5	.1	3168.9	18.9	68.7	.1	4801.5	21.9	130.7	.1	6671.1	24.8	212.4
.2	627.9	13.0	4.2	.2	1789.3	16.0	26.9	.2	3187.9	18.9	69.4	.2	4823.4	21.9	131.6	.2	6696.0	24.9	213.6
.3	640.9	13.1	4.3	.3	1805.4	16.0	27.3	.3	3206.8	19.0	70.0	.3	4845.4	21.9	132.5	.3	6720.9	24.9	214.7
.4	654.0	13.1	4.5	.4	1821.4	16.1	27.7	.4	3225.9	19.0	70.7	.4	4867.3	22.0	133.4	.4	6745.9	24.9	215.9
.5	667.1	13.1	4.7	.5	1837.5	16.1	28.1	.5	3244.9	19.1	71.3	.5	4889.4	22.0	134.3	.5	6770.8	25.0	217.0
.6	680.3	13.2	4.8	.6	1853.6	16.1	28.5	.6	3264.0	19.1	72.0	.6	4911.4	22.1	135.2	.6	6795.9	25.0	218.2
.7	693.5	13.2	5.0	.7	1869.8	16.2	29.0	.7	3283.1	19.1	72.7	.7	4933.5	22.1	136.1	.7	6820.9	25.1	219.3
.8	706.7	13.2	5.2	.8	1886.0	16.2	29.4	.8	3302.3	19.2	73.3	.8	4955.6	22.1	137.0	.8	6846.0	25.1	220.5
.9	720.0	13.3	5.4	.9	1902.2	16.2	29.8	.9	3321.5	19.2	74.0	.9	4977.8	22.2	138.0	.9	6871.1	25.1	221.7
6.0	733.3	13.3	5.6	14.0	1918.5	16.3	30.2	22.0	3340.7	19.2	74.7	30.0	5000.0	22.2	138.9	38.0	6896.3	25.2	222.8
.1	746.7	13.4	5.7	.1	1934.8	16.3	30.7	.1	3360.0	19.3	75.4	.1	5022.2	22.2	139.8	.1	6921.5	25.2	224.0
.2	760.1	13.4	5.9	.2	1951.2	16.4	31.1	.2	3379.3	19.3	76.1	.2	5044.5	22.3	140.7	.2	6946.7	25.2	225.2
.3	773.5	13.4	6.1	.3	1967.6	16.4	31.6	.3	3398.7	19.4	76.7	.3	5066.8	22.3	141.7	.3	6972.0	25.3	226.4
.4	787.0	13.5	6.3	.4	1984.0	16.4	32.0	.4	3418.1	19.4	77.4	.4	5089.2	22.4	142.6	.4	6997.3	25.3	227.6
.5	800.5	13.5	6.5	.5	2000.5	16.5	32.4	.5	3437.5	19.4	78.1	.5	5111.6	22.4	143.6	.5	7022.7	25.4	228.7
.6	814.0	13.5	6.7	.6	2017.0	16.5	32.9	.6	3457.0	19.5	78.8	.6	5134.0	22.4	144.5	.6	7048.1	25.4	229.9
.7	827.6	13.6	6.9	.7	2033.5	16.5	33.3	.7	3476.5	19.5	79.5	.7	5156.5	22.5	145.4	.7	7073.5	25.4	231.1
.8	841.2	13.6	7.1	.8	2050.1	16.6	33.8	.8	3496.0	19.5	80.2	.8	5179.0	22.5	146.4	.8	7099.0	25.5	232.3
.9	854.8	13.6	7.3	.9	2066.7	16.6	34.3	.9	3515.6	19.6	80.9	.9	5201.5	22.5	147.3	.9	7124.5	25.5	233.5
7.0	868.5	13.7	7.6	15.0	2083.3	16.6	34.7	23.0	3535.2	19.6	81.6	31.0	5224.1	22.6	148.3	39.0	7150.0	25.5	234.7
.1	882.2	13.7	7.8	.1	2100.0	16.7	35.2	.1	3554.8	19.6	82.3	.1	5246.7	22.6	149.3	.1	7175.6	25.6	235.9
.2	896.0	13.8	8.0	.2	2116.7	16.7	35.7	.2	3574.5	19.7	83.1	.2	5269.3	22.6	150.2	.2	7201.2	25.6	237.1
.3	909.8	13.8	8.2	.3	2133.5	16.8	36.1	.3	3594.2	19.7	83.8	.3	5292.0	22.7	151.2	.3	7226.8	25.6	238.3
.4	923.6	13.8	8.5	.4	2150.3	16.8	36.6	.4	3614.0	19.8	84.5	.4	5314.7	22.7	152.2	.4	7252.5	25.7	239.6
.5	937.5	13.9	8.7	.5	2167.1	16.8	37.1	.5	3633.8	19.8	85.2	.5	5337.5	22.8	153.1	.5	7278.2	25.7	240.8
.6	951.4	13.9	8.9	.6	2184.0	16.9	37.6	.6	3653.6	19.8	86.0	.6	5360.3	22.8	154.1	.6	7304.0	25.8	242.0
.7	965.4	13.9	9.1	.7	2200.9	16.9	38.0	.7	3673.5	19.9	86.7	.7	5383.1	22.8	155.1	.7	7329.8	25.8	243.2
.8	979.3	14.0	9.4	.8	2217.9	16.9	38.5	.8	3693.4	19.9	87.4	.8	5406.0	22.9	156.1	.8	7355.6	25.8	244.5
.9	993.4	14.0	9.6	.9	2234.8	17.0	39.0	.9	3713.4	19.9	88.2	.9	5428.9	22.9	157.0	.9	7381.5	25.9	245.7

TABLE No. 30.

EXCAVATION AND EMBANKMENT.

CUBIC YARDS.

Prismoids, 100 feet long. Breadth of Base, 15 feet. Slopes, 1-4 Horizontal to 1 Perpendicular.

Heig't	Cub. Yds.	Diff.	Cor.	Height.	Cub. Yds.	Diff.	Cor.	Height.	Cub. Yds.	Diff.	Cor.	Height.	Cub. Yds.	Diff.	Cor.	Height.	Cub. Yds.	Diff.	Cor.
0.0	0.0	0.0	0.0	8.0	503.7	7.0	4.9	16.0	1125.9	8.5	19.8	24.0	1866.7	10.0	44.4	32.0	2725.9	11.5	79.0
1	5.6	5.6	0.0	1	510.8	7.0	5.1	1	1134.5	8.5	20.0	1	1876.7	10.0	44.8	1	2737.4	11.5	79.5
2	11.1	5.6	0.0	2	517.8	7.1	5.2	2	1143.0	8.5	20.2	2	1886.7	10.0	45.2	2	2748.9	11.5	80.0
3	16.8	5.6	0.0	3	524.9	7.1	5.3	3	1151.6	8.6	20.5	3	1896.8	10.0	45.6	3	2760.5	11.5	80.5
4	22.4	5.6	0.0	4	532.0	7.1	5.4	4	1160.2	8.6	20.8	4	1906.8	10.1	45.9	4	2772.0	11.5	81.0
5	28.0	5.6	0.0	5	539.1	7.1	5.6	5	1168.8	8.6	21.0	5	1916.9	10.1	46.3	5	2783.6	11.6	81.5
6	33.7	5.7	0.0	6	546.3	7.1	5.7	6	1177.4	8.6	21.3	6	1927.0	10.1	46.7	6	2795.1	11.6	82.0
7	39.3	5.7	0.0	7	553.4	7.2	5.8	7	1186.0	8.6	21.5	7	1937.1	10.1	47.1	7	2806.8	11.6	82.5
8	45.0	5.7	0.0	8	560.6	7.2	6.0	8	1194.7	8.7	21.8	8	1947.3	10.1	47.5	8	2818.4	11.6	83.0
9	50.8	5.7	0.1	9	567.8	7.2	6.1	9	1203.3	8.7	22.0	9	1957.4	10.2	47.8	9	2830.0	11.6	83.5
1.0	56.5	5.7	0.1	9.0	575.0	7.2	6.3	17.0	1212.0	8.7	22.3	25.0	1967.6	10.2	48.2	33.0	2841.7	11.7	84.0
1	62.2	5.8	0.1	1	582.2	7.2	6.4	1	1220.8	8.7	22.6	1	1977.8	10.2	48.6	1	2853.3	11.7	84.5
2	68.0	5.8	0.1	2	589.5	7.3	6.5	2	1229.5	8.7	22.8	2	1988.0	10.2	49.0	2	2865.0	11.7	85.0
3	73.8	5.8	0.1	3	596.8	7.3	6.7	3	1238.2	8.8	23.1	3	1998.2	10.2	49.4	3	2876.8	11.7	85.6
4	79.6	5.8	0.2	4	604.0	7.3	6.8	4	1247.0	8.8	23.4	4	2008.5	10.3	49.8	4	2888.5	11.7	86.1
5	85.4	5.8	0.2	5	611.3	7.3	7.0	5	1255.8	8.8	23.6	5	2018.8	10.3	50.2	5	2900.2	11.8	86.6
6	91.3	5.8	0.2	6	618.7	7.3	7.1	6	1264.6	8.8	23.9	6	2029.0	10.3	50.6	6	2912.0	11.8	87.1
7	97.1	5.9	0.2	7	626.0	7.3	7.3	7	1273.4	8.8	24.2	7	2039.3	10.3	51.0	7	2923.8	11.8	87.6
8	103.0	5.9	0.2	8	633.4	7.4	7.4	8	1282.3	8.8	24.4	8	2049.7	10.3	51.4	8	2935.6	11.8	88.2
9	108.9	5.9	0.3	9	640.8	7.4	7.6	9	1291.1	8.9	24.7	9	2060.0	10.3	51.8	9	2947.4	11.8	88.7
2.0	114.8	5.9	0.3	10.0	648.1	7.4	7.7	18.0	1300.0	8.9	25.0	26.0	2070.4	10.4	52.2	34.0	2959.3	11.8	89.2
1	120.8	5.9	0.3	1	655.6	7.4	7.9	1	1308.9	8.9	25.3	1	2080.8	10.4	52.6	1	2971.1	11.9	89.7
2	126.7	6.0	0.4	2	663.0	7.4	8.0	2	1317.8	8.9	25.6	2	2091.1	10.4	53.0	2	2983.0	11.9	90.2
3	132.7	6.0	0.4	3	670.5	7.5	8.2	3	1326.8	8.9	25.8	3	2101.6	10.4	53.4	3	2994.9	11.9	90.8
4	138.7	6.0	0.4	4	677.9	7.5	8.3	4	1335.7	9.0	26.1	4	2112.0	10.4	53.8	4	3006.8	11.9	91.3
5	144.7	6.0	0.5	5	685.4	7.5	8.5	5	1344.7	9.0	26.4	5	2122.5	10.5	54.2	5	3018.8	11.9	91.8
6	150.7	6.0	0.5	6	692.9	7.5	8.7	6	1353.7	9.0	26.7	6	2132.9	10.5	54.6	6	3030.7	12.0	92.4
7	156.8	6.0	0.6	7	700.5	7.5	8.8	7	1362.7	9.0	27.0	7	2143.4	10.5	55.0	7	3042.7	12.0	92.9
8	162.8	6.1	0.6	8	708.0	7.5	9.0	8	1371.7	9.0	27.3	8	2153.9	10.5	55.4	8	3054.7	12.0	93.4
9	168.9	6.1	0.6	9	715.6	7.6	9.2	9	1380.8	9.0	27.6	9	2164.5	10.5	55.8	9	3066.7	12.0	94.0
3.0	175.0	6.1	0.7	11.0	721.1	7.6	9.3	19.0	1389.8	9.1	27.9	27.0	2175.0	10.5	56.3	35.0	3078.7	12.0	94.5
1	181.1	6.1	0.7	1	730.8	7.6	9.5	1	1398.9	9.1	28.1	1	2185.6	10.6	56.7	1	3090.8	12.0	95.1
2	187.3	6.1	0.8	2	738.4	7.6	9.7	2	1408.0	9.1	28.4	2	2196.1	10.6	57.1	2	3102.8	12.1	95.6
3	193.4	6.2	0.8	3	746.0	7.6	9.9	3	1417.1	9.1	28.7	3	2206.8	10.6	57.5	3	3114.9	12.1	96.1
4	199.6	6.2	0.9	4	753.7	7.7	10.0	4	1426.3	9.1	29.0	4	2217.4	10.6	57.9	4	3127.0	12.1	96.7
5	205.8	6.2	0.9	5	761.3	7.7	10.2	5	1435.4	9.2	29.3	5	2228.0	10.6	58.4	5	3139.1	12.1	97.2
6	212.0	6.2	1.0	6	769.0	7.7	10.4	6	1444.6	9.2	29.6	6	2238.7	10.7	58.8	6	3151.3	12.1	97.8
7	218.2	6.2	1.1	7	776.8	7.7	10.6	7	1453.8	9.2	29.9	7	2249.3	10.7	59.2	7	3163.4	12.2	98.3
8	224.5	6.3	1.1	8	784.5	7.7	10.7	8	1463.0	9.2	30.2	8	2260.0	10.7	59.6	8	3175.6	12.2	98.9
9	230.8	6.3	1.2	9	792.2	7.8	10.9	9	1472.2	9.2	30.6	9	2270.8	10.7	60.1	9	3187.8	12.2	99.4
4.0	237.0	6.3	1.2	12.0	800.0	7.8	11.1	20.0	1481.5	9.3	30.9	28.0	2281.5	10.7	60.5	36.0	3200.0	12.2	100.0
1	243.3	6.3	1.3	1	807.8	7.8	11.3	1	1490.8	9.3	31.2	1	2292.2	10.8	60.9	1	3212.2	12.2	100.6
2	249.7	6.3	1.4	2	815.6	7.8	11.5	2	1500.0	9.3	31.5	2	2303.0	10.8	61.4	2	3224.5	12.3	101.1
3	256.0	6.3	1.4	3	823.4	7.8	11.7	3	1509.3	9.3	31.8	3	2313.8	10.8	61.8	3	3236.8	12.3	101.7
4	262.4	6.4	1.5	4	831.3	7.8	11.9	4	1518.7	9.3	32.1	4	2324.6	10.8	62.2	4	3249.0	12.3	102.2
5	268.8	6.4	1.6	5	839.1	7.9	12.1	5	1528.0	9.3	32.4	5	2335.4	10.8	62.7	5	3261.3	12.3	102.8
6	275.1	6.4	1.6	6	847.0	7.9	12.2	6	1537.4	9.4	32.7	6	2346.3	10.8	63.1	6	3273.7	12.3	103.4
7	281.6	6.4	1.7	7	854.9	7.9	12.4	7	1546.8	9.4	33.1	7	2357.1	10.9	63.6	7	3286.0	12.3	103.9
8	288.0	6.4	1.8	8	862.8	7.9	12.6	8	1556.1	9.4	33.4	8	2368.0	10.9	64.0	8	3298.4	12.4	104.5
9	294.5	6.5	1.9	9	870.8	7.9	12.8	9	1565.6	9.4	33.7	9	2378.9	10.9	64.4	9	3310.8	12.4	105.1
5.0	300.9	6.5	1.9	13.0	878.7	8.0	13.0	21.0	1575.0	9.4	34.0	29.0	2389.8	10.9	64.9	37.0	3323.1	12.4	105.6
1	307.4	6.5	2.0	1	886.7	8.0	13.2	1	1584.5	9.5	34.4	1	2400.8	10.9	65.3	1	3335.6	12.4	106.2
2	313.9	6.5	2.1	2	894.7	8.0	13.4	2	1593.9	9.5	34.7	2	2411.7	11.0	65.8	2	3348.0	12.4	106.8
3	320.5	6.5	2.2	3	902.7	8.0	13.6	3	1603.4	9.5	35.0	3	2422.7	11.0	66.2	3	3360.5	12.5	107.4
4	327.0	6.5	2.2	4	910.7	8.0	13.9	4	1612.9	9.5	35.3	4	2433.7	11.0	66.7	4	3372.9	12.5	107.9
5	333.6	6.6	2.3	5	918.8	8.0	14.1	5	1622.5	9.5	35.7	5	2444.7	11.0	67.1	5	3385.4	12.5	108.5
6	340.1	6.6	2.4	6	926.8	8.1	14.3	6	1632.0	9.5	36.0	6	2455.7	11.0	67.6	6	3397.9	12.5	109.1
7	346.8	6.6	2.5	7	934.9	8.1	14.5	7	1641.6	9.6	36.3	7	2466.8	11.0	68.1	7	3410.5	12.5	109.7
8	353.4	6.6	2.6	8	943.0	8.1	14.7	8	1651.1	9.6	36.7	8	2477.8	11.1	68.5	8	3423.0	12.5	110.2
9	360.0	6.6	2.7	9	951.1	8.1	14.9	9	1660.8	9.6	37.0	9	2488.9	11.1	69.0	9	3435.6	12.6	110.8
6.0	366.7	6.7	2.8	14.0	959.3	8.1	15.1	22.0	1670.4	9.6	37.3	30.0	2500.0	11.1	69.4	38.0	3448.1	12.6	111.4
1	373.3	6.7	2.9	1	967.4	8.2	15.3	1	1680.0	9.6	37.7	1	2511.1	11.1	69.9	1	3460.8	12.6	112.0
2	380.0	6.7	3.0	2	975.6	8.2	15.6	2	1689.7	9.7	38.0	2	2522.3	11.1	70.4	2	3473.4	12.6	112.6
3	386.8	6.7	3.1	3	983.8	8.2	15.8	3	1699.3	9.7	38.4	3	2533.4	11.2	70.8	3	3486.0	12.6	113.2
4	393.5	6.7	3.2	4	992.0	8.2	16.0	4	1709.0	9.7	38.7	4	2544.6	11.2	71.3	4	3498.7	12.7	113.8
5	400.2	6.8	3.3	5	1000.2	8.2	16.2	5	1718.8	9.7	39.1	5	2555.8	11.2	71.8	5	3511.3	12.7	114.4
6	407.0	6.8	3.4	6	1008.5	8.3	16.4	6	1728.5	9.7	39.4	6	2567.0	11.2	72.2	6	3524.0	12.7	115.0
7	413.8	6.8	3.5	7	1016.8	8.3	16.7	7	1738.2	9.8	39.8	7	2578.2	11.2	72.7	7	3536.8	12.7	115.6
8	420.6	6.8	3.6	8	1025.0	8.3	16.9	8	1748.0	9.8	40.1	8	2589.5	11.2	73.2	8	3549.5	12.7	116.2
9	427.4	6.8	3.7	9	1033.3	8.3	17.1	9	1757.8	9.8	40.5	9	2600.8	11.3	73.7	9	3562.2	12.8	116.8
7.0	434.3	6.8	3.8	15.0	1041.7	8.3	17.4	23.0	1767.6	9.8	40.8	31.0	2612.0	11.3	74.2	39.0	3575.0	12.8	117.4
1	441.1	6.9	3.9	1	1050.0	8.3	17.6	1	1777.4	9.8	41.2	1	2623.3	11.3	74.6	1	3587.8	12.8	118.0
2	448.0	6.9	4.0	2	1058.4	8.4	17.8	2	1787.3	9.8	41.5	2	2634.7	11.3	75.1	2	3600.6	12.8	118.6
3	454.9	6.9	4.1	3	1066.8	8.4	18.1	3	1797.1	9.9	41.9	3	2646.0	11.3	75.6	3	3613.4	12.8	119.2
4	461.8	6.9	4.2	4	1075.1	8.4	18.3	4	1807.0	9.9	42.2	4	2657.4	11.4	76.1	4	3626.3	12.8	119.8
5	468.8	6.9	4.3	5	1083.6	8.4	18.5	5	1816.9	9.9	42.6	5	2668.8	11.4	76.6	5	3639.1	12.9	120.4
6	475.7	7.0	4.5	6	1092.0	8.4	18.8	6	1826.8	9.9	43.0	6	2680.1	11.4	77.0	6	3652.0	12.9	121.0
7	482.7	7.0	4.6	7	1100.5	8.5	19.0	7	1836.8	9.9	43.3	7	2691.6	11.4	77.5	7	3664.9	12.9	121.6
8	489.7	7.0	4.7	8	1108.9	8.5	19.3	8	1846.7	10.0	43.7	8	2703.0	11.4	78.0	8	3677.8	12.9	122.2
9	496.7	7.0	4.8	9	1117.4	8.5	19.5	9	1856.7	10.0	44.1	9	2714.5	11.5	78.5	9	3690.8	12.9	122.8

TABLE No. 31.

EXCAVATION AND EMBANKMENT.

CUBIC YARDS.

Prismoids, 100 feet long. **Breadth of Base, 18 feet.** **Slopes, 1-4 Horizontal to 1 Perpendicular.**

Heig't	Cub. Yds.	Diff.	Cor.	Height.	Cub. Yds.	Diff.	Cor.	Height.	Cub. Yds.	Diff.	Cor.	Height	Cub. Yds.	Diff.	Cor.	Height.	Cub. Yds.	Diff.	Cor.
0.0	0.0	0.0	0.0	8.0	592.6	8.1	4.9	16.0	1303.7	9.6	19.8	24.0	2133.3	11.1	44.4	32.0	3081.5	12.6	79.0
1	6.7	6.7	0.0	1	600.8	8.2	5.1	1	1313.3	9.6	20.0	1	2144.5	11.1	44.8	1	3094.1	12.6	79.5
2	13.4	6.7	0.0	2	608.9	8.2	5.2	2	1323.0	9.7	20.2	2	2155.6	11.1	45.2	2	3106.7	12.6	80.0
3	20.1	6.7	0.0	3	617.1	8.2	5.3	3	1332.7	9.7	20.5	3	2166.8	11.2	45.6	3	3119.3	12.6	80.5
4	26.8	6.7	0.0	4	625.3	8.2	5.4	4	1342.4	9.7	20.8	4	2177.9	11.2	45.9	4	3132.0	12.7	81.0
5	33.6	6.8	0.0	5	633.6	8.2	5.6	5	1352.1	9.7	21.0	5	2189.1	11.2	46.3	5	3144.7	12.7	81.5
6	40.3	6.8	0.0	6	641.8	8.3	5.7	6	1361.8	9.7	21.3	6	2200.3	11.2	46.7	6	3157.4	12.7	82.0
7	47.1	6.8	0.0	7	650.1	8.3	5.8	7	1371.6	9.8	21.5	7	2211.6	11.2	47.1	7	3170.1	12.7	82.5
8	53.9	6.8	0.0	8	658.4	8.3	6.0	8	1381.3	9.8	21.8	8	2222.8	11.3	47.5	8	3182.8	12.7	83.0
9	60.8	6.8	0.1	9	666.7	8.3	6.1	9	1391.1	9.8	22.0	9	2234.1	11.3	47.8	9	3195.6	12.8	83.5
1.0	67.6	6.8	0.1	9.0	675.0	8.3	6.3	17.0	1400.9	9.8	22.3	25.0	2245.4	11.3	48.2	33.0	3208.3	12.8	84.0
1	74.5	6.9	0.1	1	683.3	8.3	6.4	1	1410.8	9.8	22.6	1	2256.7	11.3	48.6	1	3221.1	12.8	84.5
2	81.3	6.9	0.1	2	691.7	8.4	6.5	2	1420.6	9.8	22.8	2	2268.0	11.3	49.0	2	3233.9	12.8	85.0
3	88.2	6.9	0.1	3	700.1	8.4	6.7	3	1430.5	9.9	23.1	3	2279.3	11.3	49.4	3	3246.8	12.8	85.6
4	95.1	6.9	0.2	4	708.5	8.4	6.8	4	1440.3	9.9	23.4	4	2290.7	11.4	49.8	4	3259.6	12.8	86.1
5	102.1	6.9	0.2	5	716.9	8.4	7.0	5	1450.2	9.9	23.6	5	2302.1	11.4	50.2	5	3272.5	12.9	86.6
6	109.0	7.0	0.2	6	725.3	8.4	7.1	6	1460.1	9.9	23.9	6	2313.5	11.4	50.6	6	3285.3	12.9	87.1
7	116.0	7.0	0.2	7	733.8	8.5	7.3	7	1470.1	9.9	24.2	7	2324.9	11.4	51.0	7	3298.2	12.9	87.6
8	123.0	7.0	0.2	8	742.3	8.5	7.4	8	1480.0	10.0	24.4	8	2336.3	11.4	51.4	8	3311.1	12.9	88.2
9	130.0	7.0	0.3	9	750.8	8.5	7.6	9	1490.0	10.0	24.7	9	2347.8	11.5	51.8	9	3324.1	12.9	88.7
2.0	137.0	7.0	0.3	10.0	759.3	8.5	7.7	18.0	1500.0	10.0	25.0	26.0	2359.3	11.5	52.2	34.0	3337.0	13.0	89.2
1	144.1	7.0	0.3	1	767.8	8.5	7.9	1	1510.0	10.0	25.3	1	2370.8	11.5	52.6	1	3350.0	13.0	89.7
2	151.1	7.1	0.4	2	776.3	8.5	8.0	2	1520.0	10.0	25.6	2	2382.3	11.5	53.0	2	3363.0	13.0	90.2
3	158.2	7.1	0.4	3	784.9	8.6	8.2	3	1530.1	10.0	25.8	3	2393.8	11.5	53.4	3	3376.0	13.0	90.8
4	165.3	7.1	0.4	4	793.5	8.6	8.3	4	1540.1	10.1	26.1	4	2405.3	11.5	53.8	4	3389.0	13.0	91.3
5	172.5	7.1	0.5	5	802.1	8.6	8.5	5	1550.2	10.1	26.4	5	2416.9	11.6	54.2	5	3402.1	13.0	91.8
6	179.6	7.1	0.5	6	810.7	8.6	8.7	6	1560.3	10.1	26.7	6	2428.5	11.6	54.6	6	3415.1	13.1	92.4
7	186.8	7.2	0.6	7	819.3	8.6	8.8	7	1570.5	10.1	27.0	7	2440.1	11.6	55.0	7	3428.2	13.1	92.9
8	193.9	7.2	0.6	8	828.0	8.7	9.0	8	1580.6	10.1	27.3	8	2451.7	11.6	55.4	8	3441.3	13.1	93.4
9	201.1	7.2	0.6	9	836.7	8.7	9.2	9	1590.8	10.2	27.6	9	2463.3	11.6	55.8	9	3454.5	13.1	94.0
3.0	208.3	7.2	0.7	11.0	845.4	8.7	9.3	19.0	1600.9	10.2	27.9	27.0	2475.0	11.7	56.3	35.0	3467.6	13.1	94.5
1	215.6	7.2	0.7	1	854.1	8.7	9.5	1	1611.1	10.2	28.1	1	2486.7	11.7	56.7	1	3480.8	13.2	95.1
2	222.8	7.3	0.8	2	862.8	8.7	9.7	2	1621.3	10.2	28.4	2	2498.4	11.7	57.1	2	3493.9	13.2	95.6
3	230.1	7.3	0.8	3	871.6	8.8	9.9	3	1631.6	10.2	28.7	3	2510.1	11.7	57.5	3	3507.1	13.2	96.1
4	237.4	7.3	0.9	4	880.3	8.8	10.0	4	1641.8	10.3	29.0	4	2521.8	11.7	57.9	4	3520.3	13.2	96.7
5	244.7	7.3	0.9	5	889.1	8.8	10.2	5	1652.1	10.3	29.3	5	2533.6	11.8	58.4	5	3533.6	13.2	97.2
6	252.0	7.3	1.0	6	897.9	8.8	10.4	6	1662.4	10.3	29.6	6	2545.3	11.8	58.8	6	3546.8	13.3	97.8
7	259.3	7.3	1.1	7	906.8	8.8	10.6	7	1672.7	10.3	29.9	7	2557.1	11.8	59.2	7	3560.1	13.3	98.3
8	266.7	7.4	1.1	8	915.6	8.8	10.7	8	1683.0	10.3	30.2	8	2568.9	11.8	59.6	8	3573.4	13.3	98.9
9	274.1	7.4	1.2	9	924.5	8.9	10.9	9	1693.3	10.3	30.6	9	2580.8	11.8	60.1	9	3586.7	13.3	99.4
4.0	281.5	7.4	1.2	12.0	933.3	8.9	11.1	20.0	1703.7	10.4	30.9	28.0	2592.6	11.8	60.5	36.0	3600.0	13.3	100.0
1	288.9	7.4	1.3	1	942.2	8.9	11.3	1	1714.1	10.4	31.2	1	2604.5	11.9	60.9	1	3613.3	13.3	100.6
2	296.3	7.4	1.4	2	951.1	8.9	11.5	2	1724.5	10.4	31.5	2	2616.3	11.9	61.4	2	3626.7	13.4	101.1
3	303.8	7.5	1.4	3	960.1	8.9	11.7	3	1734.9	10.4	31.8	3	2628.2	11.9	61.8	3	3640.1	13.4	101.7
4	311.3	7.5	1.5	4	969.0	9.0	11.9	4	1745.3	10.4	32.1	4	2640.1	11.9	62.3	4	3653.5	13.4	102.2
5	318.8	7.5	1.6	5	978.0	9.0	12.1	5	1755.8	10.5	32.4	5	2652.1	11.9	62.7	5	3666.9	13.4	102.8
6	326.3	7.5	1.6	6	987.0	9.0	12.2	6	1766.3	10.5	32.7	6	2664.0	12.0	63.1	6	3680.3	13.4	103.4
7	333.8	7.5	1.7	7	996.0	9.0	12.4	7	1776.8	10.5	33.1	7	2676.0	12.0	63.6	7	3693.8	13.5	103.9
8	341.3	7.5	1.8	8	1005.0	9.0	12.6	8	1787.3	10.5	33.4	8	2688.0	12.0	64.0	8	3707.3	13.5	104.5
9	348.9	7.6	1.9	9	1014.1	9.0	12.8	9	1797.8	10.5	33.7	9	2700.0	12.0	64.4	9	3720.8	13.5	105.1
5.0	356.5	7.6	1.9	13.0	1023.1	9.1	13.0	21.0	1808.3	10.5	34.0	29.0	2712.0	12.0	64.9	37.0	3734.3	13.5	105.6
1	364.1	7.6	2.0	1	1032.2	9.1	13.2	1	1818.9	10.6	34.4	1	2724.1	12.0	65.3	1	3747.8	13.5	106.2
2	371.7	7.6	2.1	2	1041.3	9.1	13.4	2	1829.5	10.6	34.7	2	2736.1	12.1	65.8	2	3761.3	13.5	106.8
3	379.3	7.6	2.2	3	1050.5	9.1	13.6	3	1840.1	10.6	35.0	3	2748.2	12.1	66.2	3	3774.9	13.6	107.4
4	387.0	7.7	2.2	4	1059.6	9.1	13.9	4	1850.7	10.6	35.3	4	2760.3	12.1	66.7	4	3788.5	13.6	107.9
5	394.7	7.7	2.3	5	1068.8	9.2	14.1	5	1861.3	10.6	35.7	5	2772.5	12.1	67.1	5	3802.1	13.6	108.5
6	402.4	7.7	2.4	6	1077.9	9.2	14.3	6	1872.0	10.7	36.0	6	2784.6	12.1	67.6	6	3815.7	13.6	109.1
7	410.1	7.7	2.5	7	1087.1	9.2	14.5	7	1882.7	10.7	36.3	7	2796.8	12.2	68.1	7	3829.3	13.6	109.7
8	417.8	7.7	2.6	8	1096.3	9.2	14.7	8	1893.4	10.7	36.7	8	2808.9	12.2	68.5	8	3843.0	13.7	110.2
9	425.6	7.8	2.7	9	1105.6	9.2	14.9	9	1904.1	10.7	37.0	9	2821.1	12.2	69.0	9	3856.7	13.7	110.8
6.0	433.3	7.8	2.8	14.0	1114.8	9.3	15.1	22.0	1914.8	10.7	37.3	30.0	2833.3	12.2	69.4	38.0	3870.4	13.7	111.4
1	441.1	7.8	2.9	1	1124.1	9.3	15.3	1	1925.6	10.8	37.7	1	2845.6	12.2	69.9	1	3884.1	13.7	112.0
2	448.9	7.8	3.0	2	1133.4	9.3	15.6	2	1936.3	10.8	38.0	2	2857.8	12.3	70.4	2	3897.8	13.7	112.6
3	456.8	7.8	3.1	3	1142.7	9.3	15.8	3	1947.1	10.8	38.4	3	2870.1	12.3	70.8	3	3911.6	13.8	113.2
4	464.6	7.8	3.2	4	1152.0	9.3	16.0	4	1957.9	10.8	38.7	4	2882.4	12.3	71.3	4	3925.3	13.8	113.8
5	472.5	7.9	3.3	5	1161.3	9.3	16.2	5	1968.8	10.8	39.1	5	2894.7	12.3	71.8	5	3939.1	13.8	114.4
6	480.3	7.9	3.4	6	1170.7	9.4	16.4	6	1979.6	10.8	39.4	6	2907.0	12.3	72.2	6	3952.9	13.8	115.0
7	488.2	7.9	3.5	7	1180.1	9.4	16.7	7	1990.5	10.9	39.8	7	2919.3	12.3	72.7	7	3966.8	13.8	115.6
8	496.1	7.9	3.6	8	1189.5	9.4	16.9	8	2001.3	10.9	40.1	8	2931.7	12.4	73.2	8	3980.6	13.8	116.2
9	504.1	7.9	3.7	9	1198.9	9.4	17.1	9	2012.2	10.9	40.5	9	2944.1	12.4	73.7	9	3994.5	13.9	116.8
7.0	512.0	8.0	3.8	15.0	1208.3	9.4	17.4	23.0	2023.1	10.9	40.8	31.0	2956.5	12.4	74.2	39.0	4008.3	13.9	117.4
1	520.0	8.0	3.9	1	1217.8	9.5	17.6	1	2034.1	10.9	41.2	1	2968.9	12.4	74.6	1	4022.2	13.9	118.0
2	528.0	8.0	4.0	2	1227.3	9.5	17.8	2	2045.0	11.0	41.5	2	2981.3	12.4	75.1	2	4036.1	13.9	118.6
3	536.0	8.0	4.1	3	1236.8	9.5	18.1	3	2056.0	11.0	41.9	3	2993.8	12.5	75.6	3	4050.1	13.9	119.2
4	544.0	8.0	4.2	4	1246.3	9.5	18.3	4	2067.0	11.0	42.2	4	3006.3	12.5	76.1	4	4064.0	14.0	119.8
5	552.1	8.0	4.3	5	1255.8	9.5	18.5	5	2078.0	11.0	42.6	5	3018.8	12.5	76.6	5	4078.0	14.0	120.4
6	560.1	8.1	4.5	6	1265.3	9.5	18.8	6	2089.0	11.0	43.0	6	3031.3	12.5	77.0	6	4092.0	14.0	121.0
7	568.2	8.1	4.6	7	1274.9	9.6	19.0	7	2100.1	11.0	43.3	7	3043.8	12.5	77.5	7	4106.0	14.0	121.6
8	576.3	8.1	4.7	8	1284.5	9.6	19.3	8	2111.1	11.1	43.7	8	3056.3	12.5	78.0	8	4120.0	14.0	122.2
9	584.5	8.1	4.8	9	1294.1	9.6	19.5	9	2122.2	11.1	44.1	9	3068.9	12.5	78.5	9	4134.1	14.0	122.8

TABLE No. 32.

EXCAVATION AND EMBANKMENT.

CUBIC YARDS.

Prismoids, 100 feet long. **Breadth of Base, 20 feet.** **Slopes, 1-4 Horizontal to 1 Perpendicular.**

Height.	Cub. Yds.	Diff.	Cor.	Height.	Cub. Yds.	Diff.	Cor.	Height.	Cub. Yds.	Diff.	Cor.	Height.	Cub. Yds.	Diff.	Cor.	Height.	Cub. Yds.	Diff.	Cor.
0.0	0.0	0.0	0.0	8.0	651.9	8.9	4.9	16.0	1422.2	10.4	19.8	24.0	2311.1	11.8	44.4	32.0	3318.5	13.3	79.0
1	7.4	7.4	0.0	1	660.8	8.9	5.1	1	1432.6	10.4	20.0	1	2323.0	11.9	44.8	1	3331.9	13.3	79.5
2	14.9	7.4	0.0	2	669.7	8.9	5.2	2	1443.0	10.4	20.3	2	2334.9	11.9	45.2	2	3345.2	13.4	80.0
3	22.3	7.5	0.0	3	678.6	8.9	5.3	3	1453.4	10.4	20.5	3	2346.8	11.9	45.6	3	3358.6	13.4	80.5
4	29.8	7.5	0.0	4	687.6	9.0	5.4	4	1463.9	10.4	20.8	4	2358.7	11.9	45.9	4	3372.0	13.4	81.0
5	37.3	7.5	0.0	5	696.5	9.0	5.6	5	1474.3	10.5	21.0	5	2370.6	11.9	46.3	5	3385.4	13.4	81.5
6	44.8	7.5	0.0	6	705.5	9.0	5.7	6	1484.8	10.5	21.3	6	2382.6	12.0	46.7	6	3398.9	13.4	82.0
7	52.3	7.5	0.0	7	714.5	9.0	5.8	7	1495.3	10.5	21.5	7	2394.5	12.0	47.1	7	3412.3	13.5	82.5
8	59.9	7.5	0.0	8	723.6	9.0	6.0	8	1505.8	10.5	21.8	8	2406.5	12.0	47.5	8	3425.8	13.5	83.0
9	67.4	7.6	0.1	9	732.6	9.0	6.1	9	1516.3	10.5	22.0	9	2418.5	12.0	47.8	9	3439.3	13.5	83.5
1.0	75.0	7.6	0.1	9.0	741.7	9.1	6.3	17.0	1526.9	10.5	22.3	25.0	2430.6	12.0	48.2	33.0	3452.8	13.5	84.0
1	82.6	7.6	0.1	1	750.8	9.1	6.4	1	1537.4	10.6	22.6	1	2442.6	12.0	48.6	1	3466.3	13.5	84.5
2	90.2	7.6	0.1	2	759.9	9.1	6.5	2	1548.0	10.6	22.8	2	2454.7	12.1	49.0	2	3479.9	13.5	85.0
3	97.9	7.6	0.1	3	769.0	9.1	6.7	3	1558.6	10.6	23.1	3	2466.8	12.1	49.4	3	3491.4	13.6	85.6
4	105.5	7.7	0.2	4	778.1	9.1	6.8	4	1569.2	10.6	23.4	4	2478.9	12.1	49.8	4	3507.0	13.6	86.1
5	113.2	7.7	0.2	5	787.3	9.2	7.0	5	1579.9	10.6	23.6	5	2491.0	12.1	50.2	5	3520.6	13.6	86.6
6	120.9	7.7	0.2	6	796.4	9.2	7.1	6	1590.5	10.7	23.9	6	2503.1	12.1	50.6	6	3534.2	13.6	87.1
7	128.6	7.7	0.2	7	805.6	9.2	7.3	7	1601.2	10.7	24.2	7	2515.3	12.2	51.0	7	3547.9	13.6	87.6
8	136.3	7.7	0.2	8	814.9	9.2	7.4	8	1611.9	10.7	24.4	8	2527.4	12.2	51.4	8	3561.5	13.7	88.2
9	144.1	7.8	0.3	9	824.1	9.2	7.6	9	1622.6	10.7	24.7	9	2539.6	12.2	51.8	9	3575.2	13.7	88.7
2.0	151.9	7.8	0.3	10.0	833.3	9.3	7.7	18.0	1633.3	10.7	25.0	26.0	2551.9	12.2	52.2	34.0	3588.9	13.7	89.2
1	159.6	7.8	0.3	1	842.6	9.3	7.9	1	1644.1	10.8	25.3	1	2564.1	12.2	52.6	1	3602.6	13.7	89.7
2	167.4	7.8	0.4	2	851.9	9.3	8.0	2	1654.9	10.8	25.6	2	2576.3	12.3	53.0	2	3616.3	13.7	90.2
3	175.3	7.8	0.4	3	861.2	9.3	8.2	3	1665.6	10.8	25.8	3	2588.6	12.3	53.4	3	3630.1	13.8	90.8
4	183.1	7.8	0.4	4	870.5	9.3	8.3	4	1676.4	10.8	26.1	4	2600.9	12.3	53.8	4	3643.9	13.8	91.3
5	191.0	7.9	0.5	5	879.9	9.3	8.5	5	1687.3	10.8	26.4	5	2613.2	12.3	54.2	5	3657.6	13.8	91.8
6	198.9	7.9	0.5	6	889.2	9.4	8.7	6	1698.1	10.8	26.7	6	2625.5	12.3	54.6	6	3671.4	15.8	92.4
7	206.8	7.9	0.6	7	898.6	9.4	8.8	7	1709.0	10.9	27.0	7	2637.9	12.3	55.0	7	3685.3	13.8	92.9
8	214.7	7.9	0.6	8	908.0	9.4	9.0	8	1719.9	10.9	27.3	8	2650.2	12.4	55.4	8	3699.1	13.8	93.4
9	222.6	7.9	0.6	9	917.4	9.4	9.2	9	1730.8	10.9	27.6	9	2662.6	12.4	55.8	9	3713.0	13.9	94.0
3.0	230.6	8.0	0.7	11.0	926.9	9.4	9.3	19.0	1741.7	10.9	27.9	27.0	2675.0	12.4	56.3	35.0	3726.9	13.9	94.5
1	238.5	8.0	0.7	1	936.3	9.5	9.5	1	1752.6	10.9	28.1	1	2687.4	12.4	56.7	1	3740.8	13.9	95.1
2	246.5	8.0	0.8	2	945.8	9.5	9.7	2	1763.6	11.0	28.4	2	2699.9	12.4	57.1	2	3754.7	13.9	95.6
3	254.5	8.0	0.8	3	955.3	9.5	9.9	3	1774.5	11.0	28.7	3	2712.3	12.5	57.5	3	3768.6	13.9	96.1
4	262.6	8.0	0.9	4	964.8	9.5	10.0	4	1785.5	11.0	29.0	4	2724.8	12.5	57.9	4	3782.6	14.0	96.7
5	270.6	8.0	0.9	5	974.3	9.5	10.2	5	1796.5	11.0	29.3	5	2737.3	12.5	58.4	5	3796.5	14.0	97.2
6	278.7	8.1	1.0	6	983.9	9.5	10.4	6	1807.6	11.0	29.6	6	2749.8	12.5	58.8	6	3810.5	14.0	97.8
7	286.8	8.1	1.1	7	993.4	9.6	10.6	7	1818.6	11.0	29.9	7	2762.3	12.5	59.2	7	3824.5	14.0	98.3
8	291.9	8.1	1.1	8	1003.0	9.6	10.7	8	1829.7	11.1	30.2	8	2774.9	12.5	59.6	8	3838.6	14.0	98.9
9	303.0	8.1	1.2	9	1012.6	9.6	10.9	9	1840.8	11.1	30.6	9	2787.4	12.6	60.1	9	3852.6	14.0	99.4
4.0	311.1	8.1	1.2	12.0	1022.2	9.6	11.1	20.0	1851.9	11.1	30.9	28.0	2800.0	12.6	60.5	36.0	3866.7	14.1	100.0
1	319.3	8.1	1.3	1	1031.9	9.6	11.3	1	1863.0	11.1	31.2	1	2812.6	12.6	60.9	1	3880.8	14.1	100.6
2	327.4	8.2	1.4	2	1041.5	9.7	11.5	2	1874.1	11.1	31.5	2	2825.2	12.6	61.4	2	3894.9	14.1	101.1
3	335.6	8.2	1.4	3	1051.2	9.7	11.7	3	1885.3	11.2	31.8	3	2837.9	12.6	61.8	3	3909.0	14.1	101.7
4	343.9	8.2	1.5	4	1060.9	9.7	11.9	4	1896.4	11.2	32.1	4	2850.5	12.7	62.2	4	3923.1	14.1	102.2
5	352.1	8.2	1.6	5	1070.6	9.7	12.1	5	1907.6	11.2	32.4	5	2863.2	12.7	62.7	5	3937.3	14.2	102.8
6	360.3	8.3	1.6	6	1080.3	9.7	12.2	6	1918.9	11.2	32.7	6	2875.9	12.7	63.1	6	3951.4	14.2	103.4
7	368.6	8.3	1.7	7	1090.1	9.8	12.4	7	1930.1	11.2	33.1	7	2888.6	12.7	63.6	7	3965.6	14.2	103.9
8	376.9	8.3	1.8	8	1099.9	9.8	12.6	8	1941.3	11.3	33.4	8	2901.3	12.7	64.0	8	3979.9	14.2	104.5
9	385.2	8.3	1.9	9	1109.6	9.8	12.8	9	1952.6	11.3	33.7	9	2914.1	12.8	64.4	9	3994.1	14.2	105.1
5.0	393.5	8.3	1.9	13.0	1119.4	9.8	13.0	21.0	1963.9	11.3	34.0	29.0	2926.9	12.8	64.9	37.0	4008.3	14.3	105.6
1	401.9	8.3	2.0	1	1129.3	9.8	13.2	1	1975.2	11.3	34.4	1	2939.6	12.8	65.3	1	4022.6	14.3	106.2
2	410.2	8.4	2.1	2	1139.1	9.8	13.4	2	1986.5	11.3	34.7	2	2952.4	12.8	65.8	2	4036.9	14.3	106.8
3	418.6	8.4	2.2	3	1149.0	9.9	13.6	3	1997.9	11.3	35.0	3	2965.3	12.8	66.2	3	4051.2	14.3	107.4
4	427.0	8.4	2.2	4	1158.9	9.9	13.9	4	2009.2	11.4	35.3	4	2978.1	12.8	66.7	4	4065.5	14.3	107.9
5	435.4	8.4	2.3	5	1168.8	9.9	14.1	5	2020.6	11.4	35.7	5	2991.0	12.9	67.1	5	4079.9	14.3	108.5
6	443.9	8.4	2.4	6	1178.7	9.9	14.3	6	2032.0	11.4	36.0	6	3003.9	12.9	67.6	6	4094.2	14.4	109.1
7	452.3	8.5	2.5	7	1188.6	9.9	14.5	7	2043.4	11.4	36.3	7	3016.8	12.9	68.1	7	4108.6	14.4	109.7
8	460.8	8.5	2.6	8	1198.6	10.0	14.7	8	2054.9	11.4	36.7	8	3029.7	12.9	68.5	8	4123.0	14.4	110.2
9	469.3	8.5	2.7	9	1208.5	10.0	14.9	9	2066.3	11.5	37.0	9	3042.6	12.9	69.0	9	4137.4	14.4	110.8
6.0	477.8	8.5	2.8	14.0	1218.5	10.0	15.1	22.0	2077.8	11.5	37.3	30.0	3055.6	13.0	69.4	38.0	4151.9	14.4	111.4
1	486.3	8.5	2.9	1	1228.5	10.0	15.3	1	2089.3	11.5	37.7	1	3068.5	13.0	69.9	1	4166.3	14.5	112.0
2	494.9	8.5	3.0	2	1238.6	10.0	15.6	2	2100.8	11.5	38.0	2	3081.5	13.0	70.4	2	4180.8	14.5	112.6
3	503.4	8.6	3.1	3	1248.6	10.0	15.8	3	2112.3	11.5	38.4	3	3094.5	13.0	70.8	3	4195.3	14.5	113.2
4	512.0	8.6	3.2	4	1258.7	10.1	16.0	4	2123.9	11.5	38.7	4	3107.6	13.0	71.3	4	4209.8	14.5	113.8
5	520.6	8.6	3.3	5	1268.8	10.1	16.2	5	2135.4	11.6	39.1	5	3120.6	13.0	71.8	5	4224.3	14.5	114.4
6	529.2	8.6	3.4	6	1278.9	10.1	16.4	6	2147.0	11.6	39.4	6	3133.7	13.1	72.2	6	4238.9	14.5	115.0
7	537.9	8.6	3.5	7	1289.0	10.1	16.7	7	2158.6	11.6	39.8	7	3146.8	13.1	72.7	7	4253.4	14.6	115.6
8	546.5	8.7	3.6	8	1299.1	10.1	16.9	8	2170.2	11.6	40.1	8	3159.9	13.1	73.2	8	4268.0	14.6	116.2
9	555.2	8.7	3.7	9	1309.3	10.2	17.1	9	2181.9	11.6	40.5	9	3173.0	13.1	73.7	9	4282.6	14.6	116.8
7.0	563.9	8.7	3.8	15.0	1319.4	10.2	17.4	23.0	2193.5	11.7	40.8	31.0	3186.1	13.1	74.2	39.0	4297.2	14.6	117.4
1	572.6	8.7	3.9	1	1329.6	10.2	17.6	1	2205.2	11.7	41.2	1	3199.3	13.2	74.6	1	4311.9	14.6	118.0
2	581.3	8.7	4.0	2	1339.9	10.2	17.8	2	2216.9	11.7	41.5	2	3212.4	13.2	75.1	2	4326.5	14.7	118.6
3	590.1	8.8	4.1	3	1350.1	10.2	18.1	3	2228.6	11.7	41.9	3	3225.6	13.2	75.6	3	4341.2	14.7	119.2
4	598.9	8.8	4.2	4	1360.3	10.3	18.3	4	2240.3	11.7	42.2	4	3238.9	13.2	76.1	4	4355.9	14.7	119.8
5	607.6	8.8	4.3	5	1370.6	10.3	18.5	5	2252.1	11.8	42.6	5	3252.1	13.2	76.6	5	4370.6	14.7	120.4
6	616.4	8.8	4.5	6	1380.9	10.3	18.8	6	2263.9	11.8	43.0	6	3265.3	13.3	77.0	6	4385.3	14.7	121.0
7	625.3	8.8	4.6	7	1391.2	10.3	19.0	7	2275.6	11.8	43.3	7	3278.6	13.3	77.5	7	4400.1	14.8	121.6
8	634.1	8.8	4.7	8	1401.5	10.3	19.3	8	2287.4	11.8	43.7	8	3291.9	13.3	78.0	8	4414.9	14.8	122.2
9	643.0	8.9	4.8	9	1411.9	10.3	19.5	9	2299.3	11.8	44.1	9	3305.2	13.3	78.5	9	4429.6	14.8	122.8

TABLE NO. 33.

EXCAVATION AND EMBANKMENT.

CUBIC YARDS.

PRISMOIDS 100 FEET LONG. **BREADTH OF BASE 10 FEET.**

Slopes 1 Horizontal to 1 Perpendicular.

Ht.	Cubic Yards. Base.	Cubic Yards. Slope.	Ht.	Cubic Yards. Base.	Cubic Yards. Slope.	Ht.	Cubic Yards. Base.	Cubic Yards. Slope.	Ht.	Cubic Yards. Base.	Cubic Yards. Slope.	Ht.	Cubic Yards. Base.	Cubic Yards. Slope.
0.0	0.0	0.0	8.0	296.3	237.0	16.0	592.6	948.1	24.0	888.9	2133.3	32.0	1185.2	3792.6
.1	3.7	0.0	.1	300.0	243.0	.1	596.3	960.0	.1	892.6	2151.1	.1	1188.9	3816.3
.2	7.4	0.1	.2	303.7	249.0	.2	600.0	972.0	.2	896.3	2169.0	.2	1192.6	3840.1
.3	11.1	0.3	.3	307.4	255.1	.3	603.7	984.0	.3	900.0	2187.0	.3	1196.3	3864.0
.4	14.8	0.6	.4	311.1	261.3	.4	607.4	996.1	.4	903.7	2205.0	.4	1200.0	3888.0
.5	18.5	0.9	.5	314.8	267.6	.5	611.1	1008.3	.5	907.4	2223.1	.5	1203.7	3912.0
.6	22.2	1.3	.6	318.5	273.9	.6	614.8	1020.6	.6	911.1	2241.3	.6	1207.4	3936.1
.7	25.9	1.8	.7	322.2	280.3	.7	618.5	1032.9	.7	914.8	2259.6	.7	1211.1	3960.3
.8	29.6	2.4	.8	325.9	286.8	.8	622.2	1045.3	.8	918.5	2277.9	.8	1214.8	3984.6
.9	33.3	3.0	.9	329.6	293.4	.9	625.9	1057.8	.9	922.2	2296.3	.9	1218.5	4008.9
1.0	37.0	3.7	9.0	333.3	300.0	17.0	629.6	1070.4	25.0	925.9	2314.8	33.0	1222.2	4033.3
.1	40.7	4.5	.1	337.0	306.7	.1	633.3	1083.0	.1	929.6	2333.4	.1	1225.9	4057.8
.2	44.4	5.3	.2	340.7	313.5	.2	637.0	1095.7	.2	933.3	2352.0	.2	1229.6	4082.4
.3	48.1	6.3	.3	344.4	320.3	.3	640.7	1108.5	.3	937.0	2370.7	.3	1233.3	4107.0
.4	51.9	7.3	.4	348.1	327.3	.4	644.4	1121.3	.4	940.7	2389.5	.4	1237.0	4131.7
.5	55.6	8.3	.5	351.9	334.3	.5	648.1	1134.3	.5	944.4	2408.3	.5	1240.7	4156.5
.6	59.3	9.5	.6	355.6	341.3	.6	651.9	1147.3	.6	948.1	2427.3	.6	1244.4	4181.3
.7	63.0	10.7	.7	359.3	348.5	.7	655.6	1160.3	.7	951.9	2446.3	.7	1248.1	4206.3
.8	66.7	12.0	.8	363.0	355.7	.8	659.3	1173.5	.8	955.6	2465.3	.8	1251.9	4231.3
.9	70.4	13.4	.9	366.7	363.0	.9	663.0	1186.7	.9	959.3	2484.5	.9	1255.6	4256.3
2.0	74.1	14.8	10.0	370.4	370.4	18.0	666.7	1200.0	26.0	963.0	2503.7	34.0	1259.3	4281.5
.1	77.8	16.3	.1	374.1	377.8	.1	670.4	1213.4	.1	966.7	2523.0	.1	1263.0	4306.7
.2	81.5	17.9	.2	377.8	385.3	.2	674.1	1226.8	.2	970.4	2542.4	.2	1266.7	4332.0
.3	85.2	19.6	.3	381.5	392.9	.3	677.8	1240.3	.3	974.1	2561.8	.3	1270.4	4357.4
.4	88.9	21.3	.4	385.2	400.6	.4	681.5	1253.9	.4	977.8	2581.3	.4	1274.1	4382.8
.5	92.6	23.1	.5	388.9	408.3	.5	685.2	1267.6	.5	981.5	2600.9	.5	1277.8	4408.3
.6	96.3	25.0	.6	392.6	416.1	.6	688.9	1281.3	.6	985.2	2620.6	.6	1281.5	4433.9
.7	100.0	27.0	.7	396.3	424.0	.7	692.6	1295.1	.7	988.9	2640.3	.7	1285.2	4459.6
.8	103.7	29.0	.8	400.0	432.0	.8	696.3	1309.0	.8	992.6	2660.1	.8	1288.9	4485.3
.9	107.4	31.1	.9	403.7	440.0	.9	700.0	1323.0	.9	996.3	2680.0	.9	1292.6	4511.1
3.0	111.1	33.3	11.0	407.4	448.1	19.0	703.7	1337.0	27.0	1000.0	2700.0	35.0	1296.3	4537.0
.1	114.8	35.6	.1	411.1	456.3	.1	707.4	1351.1	.1	1003.7	2720.0	.1	1300.0	4563.0
.2	118.5	37.9	.2	414.8	464.6	.2	711.1	1365.3	.2	1007.4	2740.1	.2	1303.7	4589.0
.3	122.2	40.3	.3	418.5	472.9	.3	714.8	1379.6	.3	1011.1	2760.3	.3	1307.4	4615.1
.4	125.9	42.8	.4	422.2	481.3	.4	718.5	1393.9	.4	1014.8	2780.6	.4	1311.1	4641.3
.5	129.6	45.4	.5	425.9	489.8	.5	722.2	1408.3	.5	1018.5	2800.9	.5	1314.8	4667.6
.6	133.3	48.0	.6	429.6	498.4	.6	725.9	1422.8	.6	1022.2	2821.3	.6	1318.5	4693.9
.7	137.0	50.7	.7	433.3	507.0	.7	729.6	1437.4	.7	1025.9	2841.8	.7	1322.2	4720.3
.8	140.7	53.5	.8	437.0	515.7	.8	733.3	1452.0	.8	1029.6	2862.4	.8	1325.9	4746.8
.9	144.4	56.3	.9	440.7	524.5	.9	737.0	1466.7	.9	1033.3	2883.0	.9	1329.6	4773.4
4.0	148.1	59.3	12.0	444.4	533.3	20.0	740.7	1481.5	28.0	1037.0	2903.7	36.0	1333.3	4800.0
.1	151.9	62.3	.1	448.1	542.3	.1	744.4	1496.3	.1	1040.7	2924.5	.1	1337.0	4826.7
.2	155.6	65.3	.2	451.9	551.3	.2	748.1	1511.3	.2	1044.4	2945.3	.2	1340.7	4853.5
.3	159.3	68.5	.3	455.6	560.3	.3	751.9	1526.3	.3	1048.1	2966.3	.3	1344.4	4880.3
.4	163.0	71.7	.4	459.3	569.5	.4	755.6	1541.3	.4	1051.9	2987.3	.4	1348.1	4907.3
.5	166.7	75.0	.5	463.0	578.7	.5	759.3	1556.5	.5	1055.6	3008.3	.5	1351.9	4934.3
.6	170.4	78.4	.6	466.7	588.0	.6	763.0	1571.7	.6	1059.3	3029.5	.6	1355.6	4961.3
.7	174.1	81.9	.7	470.4	597.4	.7	766.7	1587.0	.7	1063.0	3050.7	.7	1359.3	4988.5
.8	177.8	85.3	.8	474.1	606.8	.8	770.4	1602.4	.8	1066.7	3072.0	.8	1363.0	5015.7
.9	181.5	88.9	.9	477.8	616.3	.9	774.1	1617.8	.9	1070.4	3093.4	.9	1366.7	5043.0
5.0	185.2	92.6	13.0	481.5	625.9	21.0	777.8	1633.3	29.0	1074.1	3114.8	37.0	1370.4	5070.4
.1	188.9	96.3	.1	485.2	635.6	.1	781.5	1648.9	.1	1077.8	3136.3	.1	1374.1	5097.8
.2	192.6	100.1	.2	488.9	645.3	.2	785.2	1664.6	.2	1081.5	3157.9	.2	1377.8	5125.3
.3	196.3	104.0	.3	492.6	655.1	.3	788.9	1680.3	.3	1085.2	3179.6	.3	1381.5	5152.9
.4	200.0	108.0	.4	496.3	665.0	.4	792.6	1696.1	.4	1088.9	3201.3	.4	1385.2	5180.6
.5	203.7	112.0	.5	500.0	675.0	.5	796.3	1712.0	.5	1092.6	3223.1	.5	1388.9	5208.3
.6	207.4	116.1	.6	503.7	685.0	.6	800.0	1728.0	.6	1096.3	3245.0	.6	1392.6	5236.1
.7	211.1	120.3	.7	507.4	695.1	.7	803.7	1744.0	.7	1100.0	3267.0	.7	1396.3	5264.0
.8	214.8	124.6	.8	511.1	705.3	.8	807.4	1760.1	.8	1103.7	3289.0	.8	1400.0	5292.0
.9	218.5	128.9	.9	514.8	715.6	.9	811.1	1776.3	.9	1107.4	3311.1	.9	1403.7	5320.0
6.0	222.2	133.3	14.0	518.5	725.9	22.0	814.8	1792.6	30.0	1111.1	3333.3	38.0	1407.4	5348.1
.1	225.9	137.8	.1	522.2	736.3	.1	818.5	1808.9	.1	1114.8	3355.6	.1	1411.1	5376.3
.2	229.6	142.4	.2	525.9	746.8	.2	822.2	1825.3	.2	1118.5	3377.9	.2	1414.8	5404.6
.3	233.3	147.0	.3	529.6	757.4	.3	825.9	1841.8	.3	1122.2	3400.3	.3	1418.5	5432.9
.4	237.0	151.7	.4	533.3	768.0	.4	829.6	1858.4	.4	1125.9	3422.8	.4	1422.2	5461.3
.5	240.7	156.5	.5	537.0	778.7	.5	833.3	1875.0	.5	1129.6	3445.4	.5	1425.9	5489.8
.6	244.4	161.3	.6	540.7	789.5	.6	837.0	1891.7	.6	1133.3	3468.0	.6	1429.6	5518.4
.7	248.1	166.3	.7	544.4	800.3	.7	840.7	1908.5	.7	1137.0	3490.7	.7	1433.3	5547.0
.8	251.9	171.3	.8	548.1	811.3	.8	844.4	1925.3	.8	1140.7	3513.5	.8	1437.0	5575.7
.9	255.6	176.3	.9	551.9	822.3	.9	848.1	1942.3	.9	1144.4	3536.3	.9	1440.7	5604.5
7.0	259.3	181.5	15.0	555.6	833.3	23.0	851.9	1959.3	31.0	1148.1	3559.3	39.0	1444.4	5633.3
.1	263.0	186.7	.1	559.3	844.5	.1	855.6	1976.3	.1	1151.9	3582.3	.1	1448.1	5662.3
.2	266.7	192.0	.2	563.0	855.7	.2	859.3	1993.5	.2	1155.6	3605.3	.2	1451.9	5691.3
.3	270.4	197.4	.3	566.7	867.0	.3	863.0	2010.7	.3	1159.3	3628.5	.3	1455.6	5720.3
.4	274.1	202.8	.4	570.4	878.4	.4	866.7	2028.0	.4	1163.0	3651.7	.4	1459.3	5749.5
.5	277.8	208.3	.5	574.1	889.8	.5	870.4	2045.4	.5	1166.7	3675.0	.5	1463.0	5778.7
.6	281.5	213.9	.6	577.8	901.3	.6	874.1	2062.8	.6	1170.4	3698.4	.6	1466.7	5808.0
.7	285.2	219.6	.7	581.5	912.9	.7	877.8	2080.3	.7	1174.1	3721.8	.7	1470.4	5837.4
.8	288.9	225.3	.8	585.2	924.6	.8	881.5	2097.9	.8	1177.8	3745.3	.8	1474.1	5866.8
.9	292.6	231.1	.9	588.9	936.3	.9	885.2	2115.6	.9	1181.5	3768.9	.9	1477.8	5896.3

TABLE NO. 33.

EXCAVATION AND EMBANKMENT.

CUBIC YARDS.

PRISMOIDS 100 FEET LONG. **BREADTH OF BASE 10 FEET.**

Slopes 1 Horizontal to 1 Perpendicular.

Ht.	Cubic Yards. Base.	Cubic Yards. Slope.	Ht.	Cubic Yards. Base.	Cubic Yards. Slope.	Ht.	Cubic Yards. Base.	Cubic Yards. Slope.	Ht.	Cubic Yards. Base.	Cubic Yards. Slope.	Ht.	Cubic Yards. Base.	Cubic Yards. Slope.
40.0	1481.5	5925.9	48.0	1777.8	8533.3	56.0	2074.1	11614.8	64.0	2370.4	15170.4	72.0	2666.7	19200.0
.1	1485.2	5955.6	.1	1781.5	8568.9	.1	2077.8	11656.3	.1	2374.1	15217.8	.1	2670.4	19253.4
.2	1488.9	5985.3	.2	1785.2	8604.6	.2	2081.5	11697.9	.2	2377.8	15265.3	.2	2674.1	19306.8
.3	1492.6	6015.1	.3	1788.9	8640.3	.3	2085.2	11739.6	.3	2381.5	15312.9	.3	2677.8	19360.3
.4	1496.3	6045.0	.4	1792.6	8676.1	.4	2088.9	11781.3	.4	2385.2	15360.6	.4	2681.5	19413.9
.5	1500.0	6075.0	.5	1796.3	8712.0	.5	2092.6	11823.1	.5	2388.9	15408.3	.5	2685.2	19467.6
.6	1503.7	6105.0	.6	1800.0	8748.0	.6	2096.3	11865.0	.6	2392.6	15456.1	.6	2688.9	19521.3
.7	1507.4	6135.1	.7	1803.7	8784.0	.7	2100.0	11907.0	.7	2396.3	15504.0	.7	2692.6	19575.1
.8	1511.1	6165.3	.8	1807.4	8820.1	.8	2103.7	11949.0	.8	2400.0	15552.0	.8	2696.3	19629.0
.9	1514.8	6195.6	.9	1811.1	8856.3	.9	2107.4	11991.1	.9	2403.7	15600.0	.9	2700.0	19683.0
41.0	1518.5	6225.9	49.0	1814.8	8892.6	57.0	2111.1	12033.3	65.0	2407.4	15648.1	73.0	2703.7	19737.0
.1	1522.2	6256.3	.1	1818.5	8928.9	.1	2114.8	12075.6	.1	2411.1	15696.3	.1	2707.4	19791.1
.2	1525.9	6286.8	.2	1822.2	8965.3	.2	2118.5	12117.9	.2	2414.8	15744.6	.2	2711.1	19845.3
.3	1529.6	6317.4	.3	1825.9	9001.8	.3	2122.2	12160.3	.3	2418.5	15792.9	.3	2714.8	19899.6
.4	1533.3	6348.0	.4	1829.6	9038.4	.4	2125.9	12202.8	.4	2422.2	15841.3	.4	2718.5	19953.9
.5	1537.0	6378.7	.5	1833.3	9075.0	.5	2129.6	12245.4	.5	2425.9	15889.8	.5	2722.2	20008.3
.6	1540.7	6409.5	.6	1837.0	9111.7	.6	2133.3	12288.0	.6	2429.6	15938.4	.6	2725.9	20062.8
.7	1544.4	6440.3	.7	1840.7	9148.5	.7	2137.0	12330.7	.7	2433.3	15987.0	.7	2729.6	20117.4
.8	1548.1	6471.3	.8	1844.4	9185.3	.8	2140.7	12373.5	.8	2437.0	16035.7	.8	2733.3	20172.0
.9	1551.9	6502.3	.9	1848.1	9222.3	.9	2144.4	12416.3	.9	2440.7	16084.5	.9	2737.0	20226.7
42.0	1555.6	6533.3	50.0	1851.9	9259.3	58.0	2148.1	12459.3	66.0	2444.4	16133.3	74.0	2740.7	20281.5
.1	1559.3	6564.5	.1	1855.6	9296.3	.1	2151.9	12502.3	.1	2448.1	16182.3	.1	2744.4	20336.3
.2	1563.0	6595.7	.2	1859.3	9333.5	.2	2155.6	12545.3	.2	2451.9	16231.3	.2	2748.1	20391.3
.3	1566.7	6627.0	.3	1863.0	9370.7	.3	2159.3	12588.5	.3	2455.6	16280.3	.3	2751.9	20446.3
.4	1570.4	6658.4	.4	1866.7	9408.0	.4	2163.0	12631.7	.4	2459.3	16329.5	.4	2755.6	20501.3
.5	1574.1	6689.8	.5	1870.4	9445.4	.5	2166.7	12675.0	.5	2463.0	16378.7	.5	2759.3	20556.5
.6	1577.8	6721.3	.6	1874.1	9482.8	.6	2170.4	12718.4	.6	2466.7	16428.0	.6	2763.0	20611.7
.7	1581.5	6752.9	.7	1877.8	9520.3	.7	2174.1	12761.8	.7	2470.4	16477.4	.7	2766.7	20667.0
.8	1585.2	6784.6	.8	1881.5	9557.9	.8	2177.8	12805.3	.8	2474.1	16526.8	.8	2770.4	20722.4
.9	1588.9	6816.3	.9	1885.2	9595.6	.9	2181.5	12848.9	.9	2477.8	16576.3	.9	2774.1	20777.8
43.0	1592.6	6848.1	51.0	1888.9	9633.3	59.0	2185.2	12892.6	67.0	2481.5	16625.9	75.0	2777.8	20833.3
.1	1596.3	6880.0	.1	1892.6	9671.1	.1	2188.9	12936.3	.1	2485.2	16675.6	.1	2781.5	20888.9
.2	1600.0	6912.0	.2	1896.3	9709.0	.2	2192.6	12980.1	.2	2488.9	16725.3	.2	2785.2	20944.6
.3	1603.7	6944.0	.3	1900.0	9747.0	.3	2196.3	13024.0	.3	2492.6	16775.1	.3	2788.9	21000.3
.4	1607.4	6976.1	.4	1903.7	9785.0	.4	2200.0	13068.0	.4	2496.3	16825.0	.4	2792.6	21056.1
.5	1611.1	7008.3	.5	1907.4	9823.1	.5	2203.7	13112.0	.5	2500.0	16875.0	.5	2796.3	21112.0
.6	1614.8	7040.6	.6	1911.1	9861.3	.6	2207.4	13156.1	.6	2503.7	16925.0	.6	2800.0	21168.0
.7	1618.5	7072.9	.7	1914.8	9899.6	.7	2211.1	13200.3	.7	2507.4	16975.1	.7	2803.7	21224.0
.8	1622.2	7105.3	.8	1918.5	9937.9	.8	2214.8	13244.6	.8	2511.1	17025.3	.8	2807.4	21280.1
.9	1625.9	7137.8	.9	1922.2	9976.3	.9	2218.5	13288.9	.9	2514.8	17075.6	.9	2811.1	21336.3
44.0	1629.6	7170.4	52.0	1925.9	10014.8	60.0	2222.2	13333.3	68.0	2518.5	17125.9	76.0	2814.8	21392.6
.1	1633.3	7203.0	.1	1929.6	10053.4	.1	2225.9	13377.8	.1	2522.2	17176.3	.1	2818.5	21448.9
.2	1637.0	7235.7	.2	1933.3	10092.0	.2	2229.6	13422.4	.2	2525.9	17226.8	.2	2822.2	21505.3
.3	1640.7	7268.5	.3	1937.0	10130.7	.3	2233.3	13467.0	.3	2529.6	17277.4	.3	2825.9	21561.8
.4	1644.4	7301.3	.4	1940.7	10169.5	.4	2237.0	13511.7	.4	2533.3	17328.0	.4	2829.6	21618.4
.5	1648.1	7334.3	.5	1944.4	10208.3	.5	2240.7	13556.5	.5	2537.0	17378.7	.5	2833.3	21675.0
.6	1651.9	7367.3	.6	1948.1	10247.3	.6	2244.4	13601.3	.6	2540.7	17429.5	.6	2837.0	21731.7
.7	1655.6	7400.3	.7	1951.9	10286.3	.7	2248.1	13646.3	.7	2544.4	17480.3	.7	2840.7	21788.5
.8	1659.3	7433.5	.8	1955.6	10325.3	.8	2251.9	13691.3	.8	2548.1	17531.3	.8	2844.4	21845.3
.9	1663.0	7466.7	.9	1959.3	10364.5	.9	2255.6	13736.3	.9	2551.9	17582.3	.9	2848.1	21902.3
45.0	1666.7	7500.0	53.0	1963.0	10403.7	61.0	2259.3	13781.5	69.0	2555.6	17633.3	77.0	2851.9	21959.3
.1	1670.4	7533.4	.1	1966.7	10443.0	.1	2263.0	13826.7	.1	2559.3	17684.5	.1	2855.6	22016.3
.2	1674.1	7566.8	.2	1970.4	10482.4	.2	2266.7	13872.0	.2	2563.0	17735.7	.2	2859.3	22073.5
.3	1677.8	7600.3	.3	1974.1	10521.8	.3	2270.4	13917.4	.3	2566.7	17787.0	.3	2863.0	22130.7
.4	1681.5	7633.9	.4	1977.8	10561.3	.4	2274.1	13962.8	.4	2570.4	17838.4	.4	2866.7	22188.0
.5	1685.2	7667.6	.5	1981.5	10600.9	.5	2277.8	14008.3	.5	2574.1	17889.8	.5	2870.4	22245.4
.6	1688.9	7701.3	.6	1985.2	10640.6	.6	2281.5	14053.9	.6	2577.8	17941.3	.6	2874.1	22302.8
.7	1692.6	7735.1	.7	1988.9	10680.3	.7	2285.2	14099.6	.7	2581.5	17992.9	.7	2877.8	22360.3
.8	1696.3	7769.0	.8	1992.6	10720.1	.8	2288.9	14145.3	.8	2585.2	18044.6	.8	2881.5	22417.9
.9	1700.0	7803.0	.9	1996.3	10760.0	.9	2292.6	14191.1	.9	2588.9	18096.3	.9	2885.2	22475.6
46.0	1703.7	7837.0	54.0	2000.0	10800.0	62.0	2296.3	14237.0	70.0	2592.6	18148.1	78.0	2888.9	22533.3
.1	1707.4	7871.1	.1	2003.7	10840.0	.1	2300.0	14283.0	.1	2596.3	18200.0	.1	2892.6	22591.1
.2	1711.1	7905.3	.2	2007.4	10880.1	.2	2303.7	14329.0	.2	2600.0	18252.0	.2	2896.3	22649.0
.3	1714.8	7939.6	.3	2011.1	10920.3	.3	2307.4	14375.1	.3	2603.7	18304.0	.3	2900.0	22707.0
.4	1718.5	7973.9	.4	2014.8	10960.6	.4	2311.1	14421.3	.4	2607.4	18356.1	.4	2903.7	22765.0
.5	1722.2	8008.3	.5	2018.5	11000.9	.5	2314.8	14467.6	.5	2611.1	18408.3	.5	2907.4	22823.1
.6	1725.9	8042.8	.6	2022.2	11041.3	.6	2318.5	14513.9	.6	2614.8	18460.6	.6	2911.1	22881.3
.7	1729.6	8077.4	.7	2025.9	11081.8	.7	2322.2	14560.3	.7	2618.5	18512.9	.7	2914.8	22939.6
.8	1733.3	8112.0	.8	2029.6	11122.4	.8	2325.9	14606.8	.8	2622.2	18565.3	.8	2918.5	22997.9
.9	1737.0	8146.7	.9	2033.3	11163.0	.9	2329.6	14653.4	.9	2625.9	18617.8	.9	2922.2	23056.3
47.0	1740.7	8181.5	55.0	2037.0	11203.7	63.0	2333.3	14700.0	71.0	2629.6	18670.4	79.0	2925.9	23114.8
.1	1744.4	8216.3	.1	2040.7	11244.5	.1	2337.0	14746.7	.1	2633.3	18723.0	.1	2929.6	23173.4
.2	1748.1	8251.3	.2	2044.4	11285.3	.2	2340.7	14793.5	.2	2637.0	18775.7	.2	2933.3	23232.0
.3	1751.9	8286.3	.3	2048.1	11326.3	.3	2344.4	14840.3	.3	2640.7	18828.5	.3	2937.0	23290.7
.4	1755.6	8321.3	.4	2051.9	11367.3	.4	2348.1	14887.3	.4	2644.4	18881.3	.4	2940.7	23349.5
.5	1759.3	8356.5	.5	2055.6	11408.3	.5	2351.9	14934.3	.5	2648.1	18934.3	.5	2944.4	23408.3
.6	1763.0	8391.7	.6	2059.3	11449.5	.6	2355.6	14981.3	.6	2651.9	18987.3	.6	2948.1	23467.3
.7	1766.7	8427.0	.7	2063.0	11490.7	.7	2359.3	15028.5	.7	2655.6	19040.3	.7	2951.9	23526.3
.8	1770.4	8462.4	.8	2066.7	11532.0	.8	2363.0	15075.7	.8	2659.3	19093.5	.8	2955.6	23585.3
.9	1774.1	8497.8	.9	2070.4	11573.4	.9	2366.7	15123.0	.9	2663.0	19146.7	.9	2959.3	23644.5

TABLE No. 1.

EXCAVATION AND EMBANKMENT.

AREAS.

Breadth of Base, 15 feet. **Slopes, 3 and 2 Horizontal to 1 Perpendicular.**

Height.	Area in Feet.	Difference.	Height.	Area in Feet.	Difference.	Height.	Area in Feet.	Difference.	Height.	Area in Feet.	Difference.	Height.	Area in Feet.	Difference.
0.0	0.0	0.0	8.0	280.0	5.5	16.0	880.0	9.5	24.0	1800.0	13.5	32.0	3040.0	17.5
1	1.5	1.5	1	285.5	5.5	1	889.5	9.5	1	1813.5	13.5	1	3057.5	17.5
2	3.1	1.6	2	291.1	5.6	2	899.1	9.6	2	1827.1	13.6	2	3075.1	17.6
3	4.7	1.6	3	296.7	5.6	3	908.7	9.6	3	1840.7	13.6	3	3092.7	17.6
4	6.4	1.7	4	302.4	5.7	4	918.4	9.7	4	1854.4	13.7	4	3110.4	17.7
5	8.1	1.7	5	308.1	5.7	5	928.1	9.7	5	1868.1	13.7	5	3128.1	17.7
6	9.9	1.8	6	313.9	5.8	6	937.9	9.8	6	1881.9	13.8	6	3145.9	17.8
7	11.7	1.8	7	319.7	5.8	7	947.7	9.8	7	1895.7	13.8	7	3163.7	17.8
8	13.6	1.9	8	325.6	5.9	8	957.6	9.9	8	1909.6	13.9	8	3181.6	17.9
9	15.5	1.9	9	331.5	5.9	9	967.5	9.9	9	1923.5	13.9	9	3199.5	17.9
1.0	17.5	2.0	9.0	337.5	6.0	17.0	977.5	10.0	25.0	1937.5	14.0	33.0	3217.5	18.0
1	19.5	2.0	1	343.5	6.0	1	987.5	10.0	1	1951.5	14.0	1	3235.5	18.0
2	21.6	2.1	2	349.6	6.1	2	997.6	10.1	2	1965.6	14.1	2	3253.6	18.1
3	23.7	2.1	3	355.7	6.1	3	1007.7	10.1	3	1979.7	14.1	3	3271.7	18.1
4	25.9	2.2	4	361.9	6.2	4	1017.9	10.2	4	1993.9	14.2	4	3289.9	18.2
5	28.1	2.2	5	368.1	6.2	5	1028.1	10.2	5	2008.1	14.2	5	3308.1	18.2
6	30.4	2.3	6	374.4	6.3	6	1038.4	10.3	6	2022.4	14.3	6	3326.4	18.3
7	32.7	2.3	7	380.7	6.3	7	1048.7	10.3	7	2036.7	14.3	7	3344.7	18.3
8	35.1	2.4	8	387.1	6.4	8	1059.1	10.4	8	2051.1	14.4	8	3363.1	18.4
9	37.5	2.4	9	393.5	6.4	9	1069.5	10.4	9	2065.5	14.4	9	3381.5	18.4
2.0	40.0	2.5	10.0	400.0	6.5	18.0	1080.0	10.5	26.0	2080.0	14.5	34.0	3400.0	18.5
1	42.5	2.5	1	406.5	6.5	1	1090.5	10.5	1	2094.5	14.5	1	3418.5	18.5
2	45.1	2.6	2	413.1	6.6	2	1101.1	10.6	2	2109.1	14.6	2	3437.1	18.6
3	47.7	2.6	3	419.7	6.6	3	1111.7	10.6	3	2123.7	14.6	3	3455.7	18.6
4	50.4	2.7	4	426.4	6.7	4	1122.4	10.7	4	2138.4	14.7	4	3474.4	18.7
5	53.1	2.7	5	433.1	6.7	5	1133.1	10.7	5	2153.1	14.7	5	3493.1	18.7
6	55.9	2.8	6	439.9	6.8	6	1143.9	10.8	6	2167.9	14.8	6	3511.9	18.8
7	58.7	2.8	7	446.7	6.8	7	1154.7	10.8	7	2182.7	14.8	7	3530.7	18.8
8	61.6	2.9	8	453.6	6.9	8	1165.6	10.9	8	2197.6	14.9	8	3549.6	18.9
9	64.5	2.9	9	460.5	6.9	9	1176.5	10.9	9	2212.5	14.9	9	3568.5	18.9
3.0	67.5	3.0	11.0	467.5	7.0	19.0	1187.5	11.0	27.0	2227.5	15.0	35.0	3587.5	19.0
1	70.5	3.0	1	474.5	7.0	1	1198.5	11.0	1	2242.5	15.0	1	3606.5	19.0
2	73.6	3.1	2	481.6	7.1	2	1209.6	11.1	2	2257.6	15.1	2	3625.6	19.1
3	76.7	3.1	3	488.7	7.1	3	1220.7	11.1	3	2272.7	15.1	3	3644.7	19.1
4	79.9	3.2	4	495.9	7.2	4	1231.9	11.2	4	2287.9	15.2	4	3663.9	19.2
5	83.1	3.2	5	503.1	7.2	5	1243.1	11.2	5	2303.1	15.2	5	3683.1	19.2
6	86.4	3.3	6	510.4	7.3	6	1254.4	11.3	6	2318.4	15.3	6	3702.4	19.3
7	89.7	3.3	7	517.7	7.3	7	1265.7	11.3	7	2333.7	15.3	7	3721.7	19.3
8	93.1	3.4	8	525.1	7.4	8	1277.1	11.4	8	2349.1	15.4	8	3741.1	19.4
9	96.5	3.4	9	532.5	7.4	9	1288.5	11.4	9	2364.5	15.4	9	3760.5	19.4
4.0	100.0	3.5	12.0	540.0	7.5	20.0	1300.0	11.5	28.0	2380.0	15.5	36.0	3780.0	19.5
1	103.5	3.5	1	547.5	7.5	1	1311.5	11.5	1	2395.5	15.5	1	3799.5	19.5
2	107.1	3.6	2	555.1	7.6	2	1323.1	11.6	2	2411.1	15.6	2	3819.1	19.6
3	110.7	3.6	3	562.7	7.6	3	1334.7	11.6	3	2426.7	15.6	3	3838.7	19.6
4	114.4	3.7	4	570.4	7.7	4	1346.4	11.7	4	2442.4	15.7	4	3858.4	19.7
5	118.1	3.7	5	578.1	7.7	5	1358.1	11.7	5	2458.1	15.7	5	3878.1	19.7
6	121.9	3.8	6	585.9	7.8	6	1369.9	11.8	6	2473.9	15.8	6	3897.9	19.8
7	125.7	3.8	7	593.7	7.8	7	1381.7	11.8	7	2489.7	15.8	7	3917.7	19.8
8	129.6	3.9	8	601.6	7.9	8	1393.6	11.9	8	2505.6	15.9	8	3937.6	19.9
9	133.5	3.9	9	609.5	7.9	9	1405.5	11.9	9	2521.5	15.9	9	3957.5	19.9
5.0	137.5	4.0	13.0	617.5	8.0	21.0	1417.5	12.0	29.0	2537.5	16.0	37.0	3977.5	20.0
1	141.5	4.0	1	625.5	8.0	1	1429.5	12.0	1	2553.5	16.0	1	3997.5	20.0
2	145.6	4.1	2	633.6	8.1	2	1441.6	12.1	2	2569.6	16.1	2	4017.6	20.1
3	149.7	4.1	3	641.7	8.1	3	1453.7	12.1	3	2585.7	16.1	3	4037.7	20.1
4	153.9	4.2	4	649.9	8.2	4	1465.9	12.2	4	2601.9	16.2	4	4057.9	20.2
5	158.1	4.2	5	658.1	8.2	5	1478.1	12.2	5	2618.1	16.2	5	4078.1	20.2
6	162.4	4.3	6	666.4	8.3	6	1490.4	12.3	6	2634.4	16.3	6	4098.4	20.3
7	166.7	4.3	7	674.7	8.3	7	1502.7	12.3	7	2650.7	16.3	7	4118.7	20.3
8	171.1	4.4	8	683.1	8.4	8	1515.1	12.4	8	2667.1	16.4	8	4139.1	20.4
9	175.5	4.4	9	691.5	8.4	9	1527.5	12.4	9	2683.5	16.4	9	4159.5	20.4
6.0	180.0	4.5	14.0	700.0	8.5	22.0	1540.0	12.5	30.0	2700.0	16.5	38.0	4180.0	20.5
1	184.5	4.5	1	708.5	8.5	1	1552.5	12.5	1	2716.5	16.5	1	4200.5	20.5
2	189.1	4.6	2	717.1	8.6	2	1565.1	12.6	2	2733.1	16.6	2	4221.1	20.6
3	193.7	4.6	3	725.7	8.6	3	1577.7	12.6	3	2749.7	16.6	3	4241.7	20.6
4	198.4	4.7	4	734.4	8.7	4	1590.4	12.7	4	2766.4	16.7	4	4262.4	20.7
5	203.1	4.7	5	743.1	8.7	5	1603.1	12.7	5	2783.1	16.7	5	4283.1	20.7
6	207.9	4.8	6	751.9	8.8	6	1615.9	12.8	6	2799.9	16.8	6	4303.9	20.8
7	212.7	4.8	7	760.7	8.8	7	1628.7	12.8	7	2816.7	16.8	7	4324.7	20.8
8	217.6	4.9	8	769.6	8.9	8	1641.6	12.9	8	2833.6	16.9	8	4345.6	20.9
9	222.5	4.9	9	778.5	8.9	9	1654.5	12.9	9	2850.5	16.9	9	4366.5	20.9
7.0	227.5	5.0	15.0	787.5	9.0	23.0	1667.5	13.0	31.0	2867.5	17.0	39.0	4387.5	21.0
1	232.5	5.0	1	796.5	9.0	1	1680.5	13.0	1	2884.5	17.0	1	4408.5	21.0
2	237.6	5.1	2	805.6	9.1	2	1693.6	13.1	2	2901.6	17.1	2	4429.6	21.1
3	242.7	5.1	3	814.7	9.1	3	1706.7	13.1	3	2918.7	17.1	3	4450.7	21.1
4	247.9	5.2	4	823.9	9.2	4	1719.9	13.2	4	2935.9	17.2	4	4471.9	21.2
5	253.1	5.2	5	833.1	9.2	5	1733.1	13.2	5	2953.1	17.2	5	4493.1	21.2
6	258.4	5.3	6	842.4	9.3	6	1746.4	13.3	6	2970.4	17.3	6	4514.4	21.3
7	263.7	5.3	7	851.7	9.3	7	1759.7	13.3	7	2987.7	17.3	7	4535.7	21.3
8	269.1	5.4	8	861.1	9.4	8	1773.1	13.4	8	3005.1	17.4	8	4557.1	21.4
9	274.5	5.4	9	870.5	9.4	9	1786.5	13.4	9	3022.5	17.4	9	4578.5	21.4

TABLE No. 2.

EXCAVATION AND EMBANKMENT.

AREAS.

Broadth of Base, 15 feet. **Slopes, 2 and 1 1-2 Horizontal to 1 Perpendicular.**

Height.	Area in Feet.	Difference.	Height.	Area in Feet.	Difference.	Height.	Area in Feet.	Difference.	Height.	Area in Feet.	Difference.	Height.	Area in Feet.	Differcare.
0.0	0.0	0.0	8.0	232.0	4.3	16.0	688.0	7.1	24.0	1368.0	9.9	32.0	2272.0	12.7
1	1.5	1.5	1	236.3	4.3	1	695.1	7.1	1	1377.9	9.9	1	2284.7	12.7
2	3.1	1.6	2	240.7	4.4	2	702.3	7.2	2	1387.9	10.0	2	2297.5	12.8
3	4.7	1.6	3	245.1	4.4	3	709.5	7.2	3	1397.9	10.0	3	2310.3	12.8
4	6.3	1.6	4	249.5	4.4	4	716.7	7.2	4	1407.9	10.0	4	2323.1	12.8
5	7.9	1.7	5	253.9	4.5	5	723.9	7.3	5	1417.9	10.1	5	2335.9	12.9
6	9.6	1.7	6	258.4	4.5	6	731.2	7.3	6	1428.0	10.1	6	2348.8	12.9
7	11.4	1.7	7	263.0	4.5	7	738.6	7.3	7	1438.2	10.1	7	2361.8	12.9
8	13.1	1.8	8	267.5	4.6	8	745.9	7.4	8	1448.3	10.2	8	2374.7	13.0
9	14.9	1.8	9	272.1	4.6	9	753.3	7.4	9	1458.5	10.2	9	2387.7	13.0
1.0	16.8	1.8	9.0	276.8	4.6	17.0	760.8	7.4	25.0	1468.8	10.2	33.0	2400.8	13.0
1	18.6	1.9	1	281.4	4.7	1	768.2	7.5	1	1479.0	10.3	1	2413.8	13.1
2	20.5	1.9	2	286.1	4.7	2	775.7	7.5	2	1489.3	10.3	2	2426.9	13.1
3	22.5	1.9	3	290.9	4.7	3	783.3	7.5	3	1499.7	10.3	3	2440.1	13.1
4	24.4	2.0	4	295.6	4.8	4	790.8	7.6	4	1510.0	10.4	4	2453.2	13.2
5	26.4	2.0	5	300.4	4.8	5	798.4	7.6	5	1520.4	10.4	5	2466.4	13.2
6	28.5	2.0	6	305.3	4.8	6	806.1	7.6	6	1530.9	10.4	6	2479.7	13.2
7	30.6	2.1	7	310.2	4.9	7	813.8	7.7	7	1541.4	10.5	7	2493.0	13.3
8	32.7	2.1	8	315.1	4.9	8	821.5	7.7	8	1551.9	10.5	8	2506.3	13.3
9	34.8	2.1	9	320.0	4.9	9	829.2	7.7	9	1562.4	10.5	9	2519.6	13.3
2.0	37.0	2.2	10.0	325.0	5.0	18.0	837.0	7.8	26.0	1573.0	10.6	34.0	2533.0	13.4
1	39.2	2.2	1	330.0	5.0	1	844.8	7.8	1	1583.6	10.6	1	2546.4	13.4
2	41.5	2.3	2	335.1	5.1	2	852.7	7.9	2	1594.3	10.7	2	2559.9	13.5
3	43.8	2.3	3	340.2	5.1	3	860.6	7.9	3	1605.0	10.7	3	2573.4	13.5
4	46.1	2.3	4	345.3	5.1	4	868.5	7.9	4	1615.7	10.7	4	2586.9	13.5
5	48.4	2.4	5	350.4	5.2	5	876.4	8.0	5	1626.4	10.8	5	2600.4	13.6
6	50.8	2.4	6	355.6	5.2	6	884.4	8.0	6	1637.2	10.8	6	2614.0	13.6
7	53.3	2.4	7	360.9	5.2	7	892.5	8.0	7	1648.1	10.8	7	2627.7	13.6
8	55.7	2.5	8	366.1	5.3	8	900.5	8.1	8	1658.9	10.9	8	2641.3	13.7
9	58.2	2.5	9	371.4	5.3	9	908.6	8.1	9	1669.8	10.9	9	2655.0	13.7
3.0	60.8	2.5	11.0	376.8	5.3	19.0	916.8	8.1	27.0	1680.8	10.9	35.0	2668.8	13.7
1	63.3	2.6	1	382.1	5.4	1	924.9	8.2	1	1691.7	11.0	1	2682.5	13.8
2	65.9	2.6	2	387.5	5.4	2	933.1	8.2	2	1702.7	11.0	2	2696.3	13.8
3	68.6	2.6	3	393.0	5.4	3	941.4	8.2	3	1713.8	11.0	3	2710.2	13.8
4	71.2	2.7	4	398.4	5.5	4	949.6	8.3	4	1724.8	11.1	4	2724.0	13.9
5	73.9	2.7	5	403.9	5.5	5	957.9	8.3	5	1735.9	11.1	5	2737.9	13.9
6	76.7	2.7	6	409.5	5.5	6	966.3	8.3	6	1747.1	11.1	6	2751.9	13.9
7	79.5	2.8	7	415.1	5.6	7	974.7	8.4	7	1758.3	11.2	7	2765.9	14.0
8	82.3	2.8	8	420.7	5.6	8	983.1	8.4	8	1769.5	11.2	8	2779.9	14.0
9	85.1	2.8	9	426.3	5.6	9	991.5	8.4	9	1780.7	11.2	9	2793.9	14.0
4.0	88.0	2.9	12.0	432.0	5.7	20.0	1000.0	8.5	28.0	1792.0	11.3	36.0	2808.0	14.1
1	90.9	2.9	1	437.7	5.7	1	1008.5	8.5	1	1803.3	11.3	1	2822.1	14.1
2	93.9	3.0	2	443.5	5.8	2	1017.1	8.6	2	1814.7	11.4	2	2836.3	14.2
3	96.9	3.0	3	449.3	5.8	3	1025.7	8.6	3	1826.1	11.4	3	2850.5	14.2
4	99.9	3.0	4	455.1	5.8	4	1034.3	8.6	4	1837.5	11.4	4	2864.7	14.2
5	102.9	3.1	5	460.9	5.9	5	1042.9	8.7	5	1848.9	11.5	5	2878.9	14.3
6	106.0	3.1	6	466.8	5.9	6	1051.6	8.7	6	1860.4	11.5	6	2893.2	14.3
7	109.2	3.1	7	472.8	5.9	7	1060.4	8.7	7	1872.0	11.5	7	2907.6	14.3
8	112.3	3.2	8	478.7	6.0	8	1069.1	8.8	8	1883.5	11.6	8	2921.9	14.4
9	115.5	3.2	9	484.7	6.0	9	1077.9	8.8	9	1895.1	11.6	9	2936.3	14.4
5.0	118.8	3.2	13.0	490.8	6.0	21.0	1086.8	8.8	29.0	1906.8	11.6	37.0	2950.8	14.4
1	122.0	3.3	1	496.8	6.1	1	1095.6	8.9	1	1918.4	11.7	1	2965.2	14.5
2	125.3	3.3	2	502.9	6.1	2	1104.5	8.9	2	1930.1	11.7	2	2979.7	14.5
3	128.7	3.3	3	509.1	6.1	3	1113.5	8.9	3	1941.9	11.7	3	2994.3	14.5
4	132.0	3.4	4	515.2	6.2	4	1122.4	9.0	4	1953.6	11.8	4	3008.8	14.6
5	135.4	3.4	5	521.4	6.2	5	1131.4	9.0	5	1965.4	11.8	5	3023.4	14.6
6	138.9	3.4	6	527.7	6.2	6	1140.5	9.0	6	1977.3	11.8	6	3038.1	14.6
7	142.4	3.5	7	534.0	6.3	7	1149.6	9.1	7	1989.2	11.9	7	3052.8	14.7
8	145.9	3.5	8	540.3	6.3	8	1158.7	9.1	8	2001.1	11.9	8	3067.5	14.7
9	149.4	3.5	9	546.6	6.3	9	1167.8	9.1	9	2013.0	11.9	9	3082.2	14.7
6.0	153.0	3.6	14.0	553.0	6.4	22.0	1177.0	9.2	30.0	2025.0	12.0	38.0	3097.0	14.8
1	156.6	3.6	1	559.4	6.4	1	1186.2	9.2	1	2037.0	12.0	1	3111.8	14.8
2	160.3	3.7	2	565.9	6.5	2	1195.5	9.3	2	2049.1	12.1	2	3126.7	14.9
3	164.0	3.7	3	572.4	6.5	3	1204.8	9.3	3	2061.2	12.1	3	3141.6	14.9
4	167.7	3.7	4	578.9	6.5	4	1214.1	9.3	4	2073.3	12.1	4	3156.5	14.9
5	171.4	3.8	5	585.4	6.6	5	1223.4	9.4	5	2085.4	12.2	5	3171.4	15.0
6	175.2	3.8	6	592.0	6.6	6	1232.8	9.4	6	2097.6	12.2	6	3186.4	15.0
7	179.1	3.8	7	598.7	6.6	7	1242.3	9.4	7	2109.9	12.2	7	3201.5	15.0
8	181.9	3.9	8	605.3	6.7	8	1251.7	9.5	8	2122.1	12.3	8	3216.5	15.1
9	186.8	3.9	9	612.0	6.7	9	1261.2	9.5	9	2134.4	12.3	9	3231.6	15.1
7.0	190.8	3.9	15.0	618.8	6.7	23.0	1270.8	9.5	31.0	2146.8	12.3	39.0	3246.8	15.1
1	194.7	4.0	1	625.5	6.8	1	1280.3	9.6	1	2159.1	12.4	1	3261.9	15.2
2	198.7	4.0	2	632.3	6.8	2	1289.9	9.6	2	2171.5	12.4	2	3277.1	15.2
3	202.8	4.0	3	639.2	6.8	3	1299.6	9.6	3	2184.0	12.4	3	3292.4	15.2
4	206.8	4.1	4	646.0	6.9	4	1309.2	9.7	4	2196.4	12.5	4	3307.6	15.3
5	210.9	4.1	5	652.9	6.9	5	1318.9	9.7	5	2208.9	12.5	5	3322.9	15.3
6	215.1	4.1	6	659.9	6.9	6	1328.7	9.7	6	2221.5	12.5	6	3338.3	15.3
7	219.3	4.2	7	666.9	7.0	7	1338.5	9.8	7	2234.1	12.6	7	3353.7	15.4
8	223.5	4.2	8	673.9	7.0	8	1348.3	9.8	8	2246.7	12.6	8	3369.1	15.4
9	227.7	4.2	9	680.9	7.0	9	1358.1	9.8	9	2259.3	12.6	9	3384.5	15.4

TABLE No 3.

EXCAVATION AND EMBANKMENT.

AREAS.

Breadth of Base, 14 feet. **Slopes, 2 Horizontal to 1 Perpendicular.**

Height.	Area in Feet.	Difference.	Height.	Area in Feet.	Difference.	Height.	Area in Feet.	Difference.	Height.	Area in Feet.	Difference.	Height.	Area in Feet.	Difference.
0.0	0.0	0.0	8.0	240.0	4.6	16.0	736.0	7.8	24.0	1488.0	11.0	32.0	2496.0	14.2
1	1.4	1.4	1	244.6	4.6	1	743.8	7.8	1	1499.0	11.0	1	2510.2	14.2
2	2.9	1.5	2	249.3	4.7	2	751.7	7.9	2	1510.1	11.1	2	2524.5	14.3
3	4.4	1.5	3	254.0	4.7	3	759.6	7.9	3	1521.2	11.1	3	2538.8	14.3
4	5.9	1.5	4	258.7	4.7	4	767.5	7.9	4	1532.3	11.1	4	2553.1	14.3
5	7.5	1.6	5	263.5	4.8	5	775.5	8.0	5	1543.5	11.2	5	2567.5	14.4
6	9.1	1.6	6	268.3	4.8	6	783.5	8.0	6	1554.7	11.2	6	2581.9	14.4
7	10.8	1.7	7	273.2	4.9	7	791.6	8.1	7	1566.0	11.3	7	2596.4	14.5
8	12.5	1.7	8	278.1	4.9	8	799.7	8.1	8	1577.3	11.3	8	2610.9	14.5
9	14.2	1.7	9	283.0	4.9	9	807.8	8.1	9	1588.6	11.3	9	2625.4	14.5
1.0	16.0	1.8	9.0	288.0	5.0	17.0	816.0	8.2	25.0	1600.0	11.4	33.0	2640.0	14.6
1	17.8	1.8	1	293.0	5.0	1	824.2	8.2	1	1611.4	11.4	1	2654.6	14.6
2	19.7	1.9	2	298.1	5.1	2	832.5	8.3	2	1622.9	11.5	2	2669.3	14.7
3	21.6	1.9	3	303.2	5.1	3	840.8	8.3	3	1634.4	11.5	3	2684.0	14.7
4	23.5	1.9	4	308.3	5.1	4	849.1	8.3	4	1645.9	11.5	4	2698.7	14.7
5	25.5	2.0	5	313.5	5.2	5	857.5	8.4	5	1657.5	11.6	5	2713.5	14.8
6	27.5	2.0	6	318.7	5.2	6	865.9	8.4	6	1669.1	11.6	6	2728.3	14.8
7	29.6	2.1	7	324.0	5.3	7	874.4	8.5	7	1680.8	11.7	7	2743.2	14.9
8	31.7	2.1	8	329.3	5.3	8	882.9	8.5	8	1692.5	11.7	8	2758.1	14.9
9	33.8	2.1	9	334.6	5.3	9	891.4	8.5	9	1704.2	11.7	9	2773.0	14.9
2.0	36.0	2.2	10.0	340.0	5.4	18.0	900.0	8.6	26.0	1716.0	11.8	34.0	2788.0	15.0
1	38.2	2.2	1	345.4	5.4	1	908.6	8.6	1	1727.8	11.8	1	2803.0	15.0
2	40.5	2.3	2	350.9	5.5	2	917.3	8.7	2	1739.7	11.9	2	2818.1	15.1
3	42.8	2.3	3	356.4	5.5	3	926.0	8.7	3	1751.6	11.9	3	2833.2	15.1
4	45.1	2.3	4	361.9	5.5	4	934.7	8.7	4	1763.5	11.9	4	2848.3	15.1
5	47.5	2.4	5	367.5	5.6	5	943.5	8.8	5	1775.5	12.0	5	2863.5	15.2
6	49.9	2.4	6	373.1	5.6	6	952.3	8.8	6	1787.5	12.0	6	2878.7	15.2
7	52.4	2.5	7	378.8	5.7	7	961.2	8.9	7	1799.6	12.1	7	2894.0	15.3
8	54.9	2.5	8	384.5	5.7	8	970.1	8.9	8	1811.7	12.1	8	2909.3	15.3
9	57.4	2.5	9	390.2	5.7	9	979.0	8.9	9	1823.8	12.1	9	2924.6	15.3
3.0	60.0	2.6	11.0	396.0	5.8	19.0	988.0	9.0	27.0	1836.0	12.2	35.0	2940.0	15.4
1	62.6	2.6	1	401.8	5.8	1	997.0	9.0	1	1848.2	12.2	1	2955.4	15.4
2	65.3	2.7	2	407.7	5.9	2	1006.1	9.1	2	1860.5	12.3	2	2970.9	15.5
3	68.0	2.7	3	413.6	5.9	3	1015.2	9.1	3	1872.8	12.3	3	2986.4	15.5
4	70.7	2.7	4	419.5	5.9	4	1024.3	9.1	4	1885.1	12.3	4	3001.9	15.5
5	73.5	2.8	5	425.5	6.0	5	1033.5	9.2	5	1897.5	12.4	5	3017.5	15.6
6	76.3	2.8	6	431.5	6.0	6	1042.7	9.2	6	1909.9	12.4	6	3033.1	15.6
7	79.2	2.9	7	437.6	6.1	7	1052.0	9.3	7	1922.4	12.5	7	3048.8	15.7
8	82.1	2.9	8	443.7	6.1	8	1061.3	9.3	8	1934.9	12.5	8	3064.5	15.7
9	85.0	2.9	9	449.8	6.1	9	1070.6	9.3	9	1947.4	12.5	9	3080.2	15.7
4.0	88.0	3.0	12.0	456.0	6.2	20.0	1080.0	9.4	28.0	1960.0	12.6	36.0	3096.0	15.8
1	91.0	3.0	1	462.2	6.2	1	1089.4	9.4	1	1972.6	12.6	1	3111.8	15.8
2	94.1	3.1	2	468.5	6.3	2	1098.9	9.5	2	1985.3	12.7	2	3127.7	15.9
3	97.2	3.1	3	474.8	6.3	3	1108.4	9.5	3	1998.0	12.7	3	3143.6	15.9
4	100.3	3.1	4	481.1	6.3	4	1117.9	9.5	4	2010.7	12.7	4	3159.5	15.9
5	103.5	3.2	5	487.5	6.4	5	1127.5	9.6	5	2023.5	12.8	5	3175.5	16.0
6	106.7	3.2	6	493.9	6.4	6	1137.1	9.6	6	2036.3	12.8	6	3191.5	16.0
7	110.0	3.3	7	500.4	6.5	7	1146.8	9.7	7	2049.2	12.9	7	3207.6	16.1
8	113.3	3.3	8	506.9	6.5	8	1156.5	9.7	8	2062.1	12.9	8	3223.7	16.1
9	116.6	3.3	9	513.4	6.5	9	1166.2	9.7	9	2075.0	12.9	9	3239.8	16.1
5.0	120.0	3.4	13.0	520.0	6.6	21.0	1176.0	9.8	29.0	2088.0	13.0	37.0	3256.0	16.2
1	123.4	3.4	1	526.6	6.6	1	1185.8	9.8	1	2101.0	13.0	1	3272.2	16.2
2	126.9	3.5	2	533.3	6.7	2	1195.7	9.9	2	2114.1	13.1	2	3288.5	16.3
3	130.4	3.5	3	540.0	6.7	3	1205.6	9.9	3	2127.2	13.1	3	3304.8	16.3
4	133.9	3.5	4	546.7	6.7	4	1215.5	9.9	4	2140.3	13.1	4	3321.1	16.3
5	137.5	3.6	5	553.5	6.8	5	1225.5	10.0	5	2153.5	13.2	5	3337.5	16.4
6	141.1	3.6	6	560.3	6.8	6	1235.5	10.0	6	2166.7	13.2	6	3353.9	16.4
7	144.8	3.7	7	567.2	6.9	7	1245.6	10.1	7	2180.0	13.3	7	3370.4	16.5
8	148.5	3.7	8	574.1	6.9	8	1255.7	10.1	8	2193.3	13.3	8	3386.9	16.5
9	152.2	3.7	9	581.0	6.9	9	1265.8	10.1	9	2206.6	13.3	9	3403.4	16.5
6.0	156.0	3.8	14.0	588.0	7.0	22.0	1276.0	10.2	30.0	2220.0	13.4	38.0	3420.0	16.6
1	159.8	3.8	1	595.0	7.0	1	1286.2	10.2	1	2233.4	13.4	1	3436.6	16.6
2	163.7	3.9	2	602.1	7.1	2	1296.5	10.3	2	2246.9	13.5	2	3453.3	16.7
3	167.6	3.9	3	609.2	7.1	3	1306.8	10.3	3	2260.4	13.5	3	3470.0	16.7
4	171.5	3.9	4	616.3	7.1	4	1317.1	10.3	4	2273.9	13.5	4	3486.7	16.7
5	175.5	4.0	5	623.5	7.2	5	1327.5	10.4	5	2287.5	13.6	5	3503.5	16.8
6	179.5	4.0	6	630.7	7.2	6	1337.9	10.4	6	2301.1	13.6	6	3520.3	16.8
7	183.6	4.1	7	638.0	7.3	7	1348.4	10.5	7	2314.8	13.7	7	3537.2	16.9
8	187.7	4.1	8	645.3	7.3	8	1358.9	10.5	8	2328.5	13.7	8	3554.1	16.9
9	191.8	4.1	9	652.6	7.3	9	1369.4	10.5	9	2342.2	13.7	9	3571.0	16.9
7.0	196.0	4.2	15.0	660.0	7.4	23.0	1380.0	10.6	31.0	2356.0	13.8	39.0	3588.0	17.0
1	200.2	4.2	1	667.4	7.4	1	1390.6	10.6	1	2369.8	13.8	1	3605.0	17.0
2	204.5	4.3	2	674.9	7.5	2	1401.3	10.7	2	2383.7	13.9	2	3622.1	17.1
3	208.8	4.3	3	682.4	7.5	3	1412.0	10.7	3	2397.6	13.9	3	3639.2	17.1
4	213.1	4.3	4	689.9	7.5	4	1422.7	10.7	4	2411.5	13.9	4	3656.3	17.1
5	217.5	4.4	5	697.5	7.6	5	1433.5	10.8	5	2425.5	14.0	5	3673.5	17.2
6	221.9	4.4	6	705.1	7.6	6	1444.3	10.8	6	2439.5	14.0	6	3690.7	17.2
7	226.4	4.5	7	712.8	7.7	7	1455.2	10.9	7	2453.6	14.1	7	3708.0	17.3
8	230.9	4.5	8	720.5	7.7	8	1466.1	10.9	8	2467.7	14.1	8	3725.3	17.3
9	235.4	4.5	9	728.2	7.7	9	1477.0	10.9	9	2481.8	14.1	9	3742.6	17.3

TABLE No. 4.

EXCAVATION AND EMBANKMENT.

AREAS.

Breadth of Base, 15 feet. **Slopes, 2 Horizontal to 1 Perpendicul**

Height.	Area in Feet.	Difference.	Height.	Area in Feet.	Difference.	Height.	Area in Feet.	Difference.	Height.	Area in Feet.	Difference.	Height.	Area in Feet.	Difference.
0.0	0.0	0.0	8.0	248.0	4.7	16.0	752.0	7.9	24.0	1512.0	11.1	32.0	2528.0	14.3
1	1.5	1.5	1	252.7	4.7	1	759.9	7.9	1	1523.1	11.1	1	2542.3	14.3
2	3.1	1.6	2	257.5	4.8	2	767.9	8.0	2	1534.3	11.2	2	2556.7	14.4
3	4.7	1.6	3	262.3	4.8	3	775.9	8.0	3	1545.5	11.2	3	2571.1	14.4
4	6.3	1.6	4	267.1	4.8	4	783.9	8.0	4	1556.7	11.2	4	2585.5	14.4
5	8.0	1.7	5	272.0	4.9	5	792.0	8.1	5	1568.0	11.3	5	2600.0	14.5
6	9.7	1.7	6	276.9	4.9	6	800.1	8.1	6	1579.3	11.3	6	2614.5	14.5
7	11.5	1.8	7	281.9	5.0	7	808.3	8.2	7	1590.7	11.4	7	2629.1	14.6
8	13.3	1.8	8	286.9	5.0	8	816.5	8.2	8	1602.1	11.4	8	2643.7	14.6
9	15.1	1.9	9	291.9	5.0	9	824.7	8.2	9	1613.5	11.4	9	2658.3	14.6
1.0	17.0	1.9	9.0	297.0	5.1	17.0	833.0	8.3	25.0	1625.0	11.5	33.0	2673.0	14.7
1	18.9	1.9	1	302.1	5.1	1	841.3	8.3	1	1636.5	11.5	1	2687.7	14.7
2	20.9	2.0	2	307.3	5.2	2	849.7	8.4	2	1648.1	11.6	2	2702.5	14.8
3	22.9	2.0	3	312.5	5.2	3	858.1	8.4	3	1659.7	11.6	3	2717.3	14.8
4	24.9	2.0	4	317.7	5.2	4	866.5	8.4	4	1671.3	11.6	4	2732.1	14.8
5	27.0	2.1	5	323.0	5.3	5	875.0	8.5	5	1683.0	11.7	5	2747.0	14.9
6	29.1	2.1	6	328.3	5.3	6	883.5	8.5	6	1694.7	11.7	6	2761.9	14.9
7	31.3	2.2	7	333.7	5.4	7	892.1	8.6	7	1706.5	11.8	7	2776.9	15.0
8	33.5	2.2	8	339.1	5.4	8	900.7	8.6	8	1718.3	11.8	8	2791.9	15.0
9	35.7	2.2	9	344.5	5.4	9	909.3	8.6	9	1730.1	11.8	9	2806.9	15.0
2.0	38.0	2.3	10.0	350.0	5.5	18.0	918.0	8.7	26.0	1742.0	11.9	34.0	2822.0	15.1
1	40.3	2.3	1	355.5	5.5	1	926.7	8.7	1	1753.9	11.9	1	2837.1	15.1
2	42.7	2.4	2	361.1	5.6	2	935.5	8.8	2	1765.9	12.0	2	2852.3	15.2
3	45.1	2.4	3	366.7	5.6	3	944.3	8.8	3	1777.9	12.0	3	2867.5	15.2
4	47.5	2.4	4	372.3	5.6	4	953.1	8.8	4	1789.9	12.0	4	2882.7	15.2
5	50.0	2.5	5	378.0	5.7	5	962.0	8.9	5	1802.0	12.1	5	2898.0	15.3
6	52.5	2.5	6	383.7	5.7	6	970.9	8.9	6	1814.1	12.1	6	2913.3	15.3
7	55.1	2.6	7	389.5	5.8	7	979.9	9.0	7	1826.3	12.2	7	2928.7	15.4
8	57.7	2.6	8	395.3	5.8	8	988.9	9.0	8	1838.5	12.2	8	2944.1	15.4
9	60.3	2.6	9	401.1	5.8	9	997.9	9.0	9	1850.7	12.2	9	2959.5	15.4
3.0	63.0	2.7	11.0	407.0	5.9	19.0	1007.0	9.1	27.0	1863.0	12.3	35.0	2975.0	15.5
1	65.7	2.7	1	412.9	5.9	1	1016.1	9.1	1	1875.3	12.3	1	2990.5	15.5
2	68.5	2.8	2	418.9	6.0	2	1025.3	9.2	2	1887.7	12.4	2	3006.1	15.6
3	71.3	2.8	3	424.9	6.0	3	1034.5	9.2	3	1900.1	12.4	3	3021.7	15.6
4	74.1	2.8	4	430.9	6.0	4	1043.7	9.2	4	1912.5	12.4	4	3037.3	15.6
5	77.0	2.9	5	437.0	6.1	5	1053.0	9.3	5	1925.0	12.5	5	3053.0	15.7
6	79.9	2.9	6	443.1	6.1	6	1062.3	9.3	6	1937.5	12.5	6	3068.7	15.7
7	82.9	3.0	7	449.3	6.2	7	1071.7	9.4	7	1950.1	12.6	7	3084.5	15.8
8	85.9	3.0	8	455.5	6.2	8	1081.1	9.4	8	1962.7	12.6	8	3100.3	15.8
9	88.9	3.0	9	461.7	6.2	9	1090.5	9.4	9	1975.3	12.6	9	3116.1	15.8
4.0	92.0	3.1	12.0	468.0	6.3	20.0	1100.0	9.5	28.0	1988.0	12.7	36.0	3132.0	15.9
1	95.1	3.1	1	474.3	6.3	1	1109.5	9.5	1	2000.7	12.7	1	3147.9	15.9
2	98.3	3.2	2	480.7	6.4	2	1119.1	9.6	2	2013.5	12.8	2	3163.9	16.0
3	101.5	3.2	3	487.1	6.4	3	1128.7	9.6	3	2026.3	12.8	3	3179.9	16.0
4	104.7	3.2	4	493.5	6.4	4	1138.3	9.6	4	2039.1	12.8	4	3195.9	16.0
5	108.0	3.3	5	500.0	6.5	5	1148.0	9.7	5	2052.0	12.9	5	3212.0	16.1
6	111.3	3.3	6	506.5	6.5	6	1157.7	9.7	6	2064.9	12.9	6	3228.1	16.1
7	114.7	3.4	7	513.1	6.6	7	1167.5	9.8	7	2077.9	13.0	7	3244.3	16.2
8	118.1	3.4	8	519.7	6.6	8	1177.3	9.8	8	2090.9	13.0	8	3260.5	16.2
9	121.5	3.4	9	526.3	6.6	9	1187.1	9.8	9	2103.9	13.0	9	3276.7	16.2
5.0	125.0	3.5	13.0	533.0	6.7	21.0	1197.0	9.9	29.0	2117.0	13.1	37.0	3293.0	16.3
1	128.5	3.5	1	539.7	6.7	1	1206.9	9.9	1	2130.1	13.1	1	3309.3	16.3
2	132.1	3.6	2	546.5	6.8	2	1216.9	10.0	2	2143.3	13.2	2	3325.7	16.4
3	135.7	3.6	3	553.3	6.8	3	1226.9	10.0	3	2156.5	13.2	3	3342.1	16.4
4	139.3	3.6	4	560.1	6.8	4	1236.9	10.0	4	2169.7	13.2	4	3358.5	16.4
5	143.0	3.7	5	567.0	6.9	5	1247.0	10.1	5	2183.0	13.3	5	3375.0	16.5
6	146.7	3.7	6	573.9	6.9	6	1257.1	10.1	6	2196.3	13.3	6	3391.5	16.5
7	150.5	3.8	7	580.9	7.0	7	1267.3	10.2	7	2209.7	13.4	7	3408.1	16.6
8	154.3	3.8	8	587.9	7.0	8	1277.5	10.2	8	2223.1	13.4	8	3424.7	16.6
9	158.1	3.8	9	594.9	7.0	9	1287.7	10.2	9	2236.5	13.4	9	3441.3	16.6
6.0	162.0	3.9	14.0	602.0	7.1	22.0	1298.0	10.3	30.0	2250.0	13.5	38.0	3458.0	16.7
1	165.9	3.9	1	609.1	7.1	1	1308.3	10.3	1	2263.5	13.5	1	3474.7	16.7
2	169.9	4.0	2	616.3	7.2	2	1318.7	10.4	2	2277.1	13.6	2	3491.5	16.8
3	173.9	4.0	3	623.5	7.2	3	1329.1	10.4	3	2290.7	13.6	3	3508.3	16.8
4	177.9	4.0	4	630.7	7.2	4	1339.5	10.4	4	2304.3	13.6	4	3525.1	16.8
5	182.0	4.1	5	638.0	7.3	5	1350.0	10.5	5	2318.0	13.7	5	3542.0	16.9
6	186.1	4.1	6	645.3	7.3	6	1360.5	10.5	6	2331.7	13.7	6	3558.9	16.9
7	190.3	4.2	7	652.7	7.4	7	1371.1	10.6	7	2345.5	13.8	7	3575.9	17.0
8	194.5	4.2	8	660.1	7.4	8	1381.7	10.6	8	2359.3	13.8	8	3592.9	17.0
9	198.7	4.2	9	667.5	7.4	9	1392.3	10.6	9	2373.1	13.8	9	3609.9	17.0
7.0	203.0	4.3	15.0	675.0	7.5	23.0	1403.0	10.7	31.0	2387.0	13.9	39.0	3627.0	17.1
1	207.3	4.3	1	682.5	7.5	1	1413.7	10.7	1	2400.9	13.9	1	3644.1	17.1
2	211.7	4.4	2	690.1	7.6	2	1424.5	10.8	2	2414.9	14.0	2	3661.3	17.2
3	216.1	4.4	3	697.7	7.6	3	1435.3	10.8	3	2428.9	14.0	3	3678.5	17.2
4	220.5	4.4	4	705.3	7.6	4	1446.1	10.8	4	2442.9	14.0	4	3695.7	17.2
5	225.0	4.5	5	713.0	7.7	5	1457.0	10.9	5	2457.0	14.1	5	3713.0	17.3
6	229.5	4.5	6	720.7	7.7	6	1467.9	10.9	6	2471.1	14.1	6	3730.3	17.3
7	234.1	4.6	7	728.5	7.8	7	1478.9	11.0	7	2485.3	14.2	7	3747.7	17.4
8	238.7	4.6	8	736.3	7.8	8	1489.9	11.0	8	2499.5	14.2	8	3765.1	17.4
9	243.3	4.6	9	744.1	7.8	9	1500.9	11.0	9	2513.7	14.2	9	3782.5	17.4

TABLE No. 5.

EXCAVATION AND EMBANKMENT.

AREAS.

Breadth of Base, 9 feet. **Slopes, 1 1-2 Horizontal to 1 Perpendicular.**

Height.	Area in Feet.	Difference.	Height.	Area in Feet.	Difference.	Height.	Area in Feet.	Difference.	Height.	Area in Feet.	Difference.	Height.	Area in Feet.	Difference.
0.0	0.0	0.0	8.0	168.0	3.3	16.0	528.0	5.7	24.0	1080.0	8.1	32.0	1824.0	10.5
1	0.9	0.9	1	171.3	3.3	1	533.7	5.7	1	1088.1	8.1	1	1834.5	10.5
2	1.9	0.9	2	174.7	3.3	2	539.5	5.7	2	1096.3	8.1	2	1845.1	10.5
3	2.8	1.0	3	178.0	3.4	3	545.2	5.8	3	1104.4	8.2	3	1855.6	10.6
4	3.8	1.0	4	181.4	3.4	4	551.0	5.8	4	1112.6	8.2	4	1866.2	10.6
5	4.9	1.0	5	184.9	3.4	5	556.9	5.8	5	1120.9	8.2	5	1876.9	10.6
6	5.9	1.1	6	188.3	3.5	6	562.7	5.9	6	1129.1	8.3	6	1887.5	10.7
7	7.0	1.1	7	191.8	3.5	7	568.6	5.9	7	1137.4	8.3	7	1898.2	10.7
8	8.2	1.1	8	195.4	3.5	8	574.6	5.9	8	1145.8	8.3	8	1909.0	10.7
9	9.3	1.2	9	198.9	3.6	9	580.5	6.0	9	1154.1	8.4	9	1919.7	10.8
1.0	10.5	1.2	9.0	202.5	3.6	17.0	586.5	6.0	25.0	1162.5	8.4	33.0	1930.5	10.8
1	11.7	1.2	1	206.1	3.6	1	592.5	6.0	1	1170.9	8.4	1	1941.3	10.8
2	13.0	1.2	2	209.8	3.6	2	598.6	6.0	2	1179.4	8.4	2	1952.2	10.8
3	14.2	1.3	3	213.4	3.7	3	604.6	6.1	3	1187.8	8.5	3	1963.0	10.9
4	15.5	1.3	4	217.1	3.7	4	610.7	6.1	4	1196.3	8.5	4	1973.9	10.9
5	16.9	1.3	5	220.9	3.7	5	616.9	6.1	5	1204.9	8.5	5	1984.9	10.9
6	18.2	1.4	6	224.6	3.8	6	623.0	6.2	6	1213.4	8.6	6	1995.8	11.0
7	19.6	1.4	7	228.4	3.8	7	629.2	6.2	7	1222.0	8.6	7	2006.8	11.0
8	21.1	1.4	8	232.3	3.8	8	635.5	6.2	8	1230.7	8.6	8	2017.9	11.0
9	22.5	1.5	9	236.1	3.9	9	641.7	6.3	9	1239.3	8.7	9	2028.9	11.1
2.0	24.0	1.5	10.0	240.0	3.9	18.0	648.0	6.3	26.0	1248.0	8.7	34.0	2040.0	11.1
1	25.5	1.5	1	243.9	3.9	1	654.3	6.3	1	1256.7	8.7	1	2051.1	11.1
2	27.1	1.5	2	247.9	3.9	2	660.7	6.3	2	1265.5	8.7	2	2062.3	11.1
3	28.6	1.6	3	251.8	4.0	3	667.0	6.4	3	1274.2	8.8	3	2073.4	11.2
4	30.2	1.6	4	255.8	4.0	4	673.4	6.4	4	1283.0	8.8	4	2084.6	11.2
5	31.9	1.6	5	259.9	4.0	5	679.9	6.4	5	1291.9	8.8	5	2095.9	11.2
6	33.5	1.7	6	263.9	4.1	6	686.3	6.5	6	1300.7	8.9	6	2107.1	11.3
7	35.2	1.7	7	268.0	4.1	7	692.8	6.5	7	1309.6	8.9	7	2118.4	11.3
8	37.0	1.7	8	272.2	4.1	8	699.4	6.5	8	1318.6	8.9	8	2129.8	11.3
9	38.7	1.8	9	276.3	4.2	9	705.9	6.6	9	1327.5	9.0	9	2141.1	11.4
3.0	40.5	1.8	11.0	280.5	4.2	19.0	712.5	6.6	27.0	1336.5	9.0	35.0	2152.5	11.4
1	42.3	1.8	1	284.7	4.2	1	719.1	6.6	1	1345.5	9.0	1	2163.9	11.4
2	44.2	1.8	2	289.0	4.2	2	725.8	6.6	2	1354.6	9.0	2	2175.4	11.4
3	46.0	1.9	3	293.2	4.3	3	732.4	6.7	3	1363.6	9.1	3	2186.8	11.5
4	47.9	1.9	4	297.5	4.3	4	739.1	6.7	4	1372.7	9.1	4	2198.3	11.5
5	49.9	1.9	5	301.9	4.3	5	745.9	6.7	5	1381.9	9.1	5	2209.9	11.5
6	51.8	2.0	6	306.2	4.4	6	752.6	6.8	6	1391.0	9.2	6	2221.4	11.6
7	53.8	2.0	7	310.6	4.4	7	759.4	6.8	7	1400.2	2.2	7	2233.0	11.6
8	55.9	2.0	8	315.1	4.4	8	766.3	6.8	8	1409.5	9.2	8	2244.7	11.6
9	57.9	2.1	9	319.5	4.5	9	773.1	6.9	9	1418.7	9.3	9	2256.3	11.7
4.0	60.0	2.1	12.0	324.0	4.5	20.0	780.0	6.9	28.0	1428.0	9.3	36.0	2268.0	11.7
1	62.1	2.1	1	328.5	4.5	1	786.9	6.9	1	1437.3	9.3	1	2279.7	11.7
2	64.3	2.1	2	333.1	4.5	2	793.9	6.9	2	1446.7	9.3	2	2291.5	11.7
3	66.4	2.2	3	337.6	4.6	3	800.8	7.0	3	1456.0	9.4	3	2303.2	11.8
4	68.6	2.2	4	342.2	4.6	4	807.8	7.0	4	1465.4	9.4	4	2315.0	11.8
5	70.9	2.2	5	346.9	4.6	5	814.9	7.0	5	1474.9	9.4	5	2326.9	11.8
6	73.1	2.3	6	351.5	4.7	6	821.9	7.1	6	1484.3	9.5	6	2338.7	11.9
7	75.4	2.3	7	356.2	4.7	7	829.0	7.1	7	1493.8	9.5	7	2350.6	11.9
8	77.8	2.3	8	361.0	4.7	8	836.2	7.1	8	1503.4	9.5	8	2362.6	11.9
9	80.1	2.4	9	365.7	4.8	9	843.3	7.2	9	1512.9	9.6	9	2374.5	12.0
5.0	82.5	2.4	13.0	370.5	4.8	21.0	850.5	7.2	29.0	1522.5	9.6	37.0	2386.5	12.0
1	84.9	2.4	1	375.3	4.8	1	857.7	7.2	1	1532.1	9.6	1	2398.5	12.0
2	87.4	2.4	2	380.2	4.8	2	865.0	7.2	2	1541.8	9.6	2	2410.6	12.0
3	89.8	2.5	3	385.0	4.9	3	872.2	7.3	3	1551.4	9.7	3	2422.6	12.1
4	92.3	2.5	4	389.9	4.9	4	879.5	7.3	4	1561.1	9.7	4	2434.7	12.1
5	94.9	2.5	5	394.9	4.9	5	886.9	7.3	5	1570.9	9.7	5	2446.9	12.1
6	97.4	2.6	6	399.8	5.0	6	894.2	7.4	6	1580.6	9.8	6	2459.0	12.2
7	100.0	2.6	7	404.8	5.0	7	901.6	7.4	7	1590.4	9.8	7	2471.2	12.2
8	102.7	2.6	8	409.9	5.0	8	909.1	7.4	8	1600.3	9.8	8	2483.5	12.2
9	105.3	2.7	9	414.9	5.1	9	916.5	7.5	9	1610.1	9.9	9	2495.7	12.3
6.0	108.0	2.7	14.0	420.0	5.1	22.0	924.0	7.5	30.0	1620.0	9.9	38.0	2508.0	12.3
1	110.7	2.7	1	425.1	5.1	1	931.5	7.5	1	1629.9	9.9	1	2520.3	12.3
2	113.5	2.7	2	430.3	5.1	2	939.1	7.5	2	1639.9	9.9		2532.7	12.3
3	116.2	2.8	3	435.4	5.2	3	946.6	7.6	3	1649.8	10.0	3	2545.0	12.4
4	119.0	2.8	4	440.6	5.2	4	954.2	7.6	4	1659.8	10.0	4	2557.4	12.4
5	121.9	2.8	5	445.9	5.2	5	961.9	7.6	5	1669.9	10.0	5	2569.9	12.4
6	124.7	2.9	6	451.1	5.3	6	969.5	7.7	6	1679.9	10.1	6	2582.3	12.5
7	127.6	2.9	7	456.4	5.3	7	977.2	7.7	7	1690.0	10.1	7	2594.8	12.5
8	130.6	2.9	8	461.8	5.3	8	985.0	7.7	8	1700.2	10.1	8	2607.4	12.5
9	133.5	3.0	9	467.1	5.4	9	992.7	7.8	9	1710.3	10.2	9	2619.9	12.6
7.0	136.5	3.0	15.0	472.5	5.4	23.0	1000.5	7.8	31.0	1720.5	10.2	39.0	2632.5	12.6
1	139.5	3.0	1	477.9	5.4	1	1008.3	7.8	1	1730.7	10.2	1	2645.1	12.6
2	142.6	3.0	2	483.4	5.4	2	1016.2	7.8	2	1741.0	10.2	2	2657.8	12.6
3	145.6	3.1	3	488.8	5.5	3	1024.0	7.9	3	1751.2	10.3	3	2670.4	12.7
4	148.7	3.1	4	494.3	5.5	4	1031.9	7.9	4	1761.5	10.3	4	2683.1	12.7
5	151.9	3.1	5	499.9	5.5	5	1039.9	7.9	5	1771.9	10.3	5	2695.9	12.7
6	155.0	3.2	6	505.4	5.6	6	1047.8	8.0	6	1782.2	10.4	6	2708.6	12.8
7	158.2	3.2	7	511.0	5.6	7	1055.8	8.0	7	1792.6	10.4	7	2721.4	12.8
8	161.5	3.2	8	516.7	5.6	8	1063.9	8.0	8	1803.1	10.4	8	2734.3	12.8
9	164.7	3.3	9	522.3	5.7	9	1071.9	8.1	9	1813.5	10.5	9	2747.1	12.9

TABLE No. 6.

EXCAVATION AND EMBANKMENT.

AREAS.

Breadth of Base, 10 feet. **Slopes, 1 1-2 Horizontal to 1 Perpendicular.**

Height.	Area in Feet.	Difference.	Height.	Area in Feet.	Difference.	Height.	Area in Feet.	Difference.	Height.	Area in Feet.	Difference.	Height.	Area in Feet.	Difference.
0.0	0.0	0.0	8.0	176.0	3.4	16.0	544.0	5.8	24.0	1104.0	8.2	32.0	1856.0	10.6
1	1.0	1.0	1	179.4	3.4	1	549.8	5.8	1	1112.2	8.2	1	1866.6	10.6
2	2.1	1.0	2	182.9	3.4	2	555.7	5.8	2	1120.5	8.2	2	1877.3	10.6
3	3.1	1.1	3	186.3	3.5	3	561.5	5.9	3	1128.7	8.3	3	1887.9	10.7
4	4.2	1.1	4	189.8	3.5	4	567.4	5.9	4	1137.0	8.3	4	1898.6	10.7
5	5.4	1.1	5	193.4	3.5	5	573.4	5.9	5	1145.4	8.3	5	1909.4	10.7
6	6.5	1.2	6	196.9	3.6	6	579.3	6.0	6	1153.7	8.4	6	1920.1	10.8
7	7.7	1.2	7	200.5	3.6	7	585.3	6.0	7	1162.1	8.4	7	1930.9	10.8
8	9.0	1.2	8	204.2	3.6	8	591.4	6.0	8	1170.6	8.4	8	1941.8	10.8
9	10.2	1.3	9	207.8	3.7	9	597.4	6.1	9	1179.0	8.5	9	1952.6	10.9
1.0	11.5	1.3	9.0	211.5	3.7	17.0	603.5	6.1	25.0	1187.5	8.5	33.0	1963.5	10.9
1	12.8	1.3	1	215.2	3.7	1	609.6	6.1	1	1196.0	8.5	1	1974.4	10.9
2	14.2	1.3	2	219.0	3.7	2	615.8	6.1	2	1204.6	8.5	2	1985.4	10.9
3	15.5	1.4	3	222.7	3.8	3	621.9	6.2	3	1213.1	8.6	3	1996.3	11.0
4	16.9	1.4	4	226.5	3.8	4	628.1	6.2	4	1221.7	8.6	4	2007.3	11.0
5	18.4	1.4	5	230.4	3.8	5	634.4	6.2	5	1230.4	8.6	5	2018.4	11.0
6	19.8	1.5	6	234.2	3.9	6	640.6	6.3	6	1239.0	8.7	6	2029.4	11.1
7	21.3	1.5	7	238.1	3.9	7	646.9	6.3	7	1247.7	8.7	7	2040.5	11.1
8	22.9	1.5	8	242.1	3.9	8	653.3	6.3	8	1256.5	8.7	8	2051.7	11.1
9	24.4	1.6	9	246.0	4.0	9	659.6	6.4	9	1265.2	8.8	9	2062.8	11.2
2.0	26.0	1.6	10.0	250.0	4.0	18.0	666.0	6.4	26.0	1274.0	8.8	34.0	2074.0	11.2
1	27.6	1.6	1	254.0	4.0	1	672.4	6.4	1	1282.8	8.8	1	2085.2	11.2
2	29.3	1.6	2	258.1	4.0	2	678.9	6.4	2	1291.7	8.8	2	2096.5	11.2
3	30.9	1.7	3	262.1	4.1	3	685.3	6.5	3	1300.5	8.9	3	2107.7	11.3
4	32.6	1.7	4	266.2	4.1	4	691.8	6.5	4	1309.4	8.9	4	2119.0	11.3
5	34.4	1.7	5	270.4	4.1	5	698.4	6.5	5	1318.4	8.9	5	2130.4	11.3
6	36.1	1.8	6	274.5	4.2	6	704.9	6.6	6	1327.3	9.0	6	2141.7	11.4
7	37.9	1.8	7	278.7	4.2	7	711.5	6.6	7	1336.3	9.0	7	2153.1	11.4
8	39.8	1.8	8	283.0	4.2	8	718.2	6.6	8	1345.4	9.0	8	2164.6	11.4
9	41.6	1.9	9	287.2	4.3	9	724.8	6.7	9	1354.4	9.1	9	2176.0	11.5
3.0	43.5	1.9	11.0	291.5	4.3	19.0	731.5	6.7	27.0	1363.5	9.1	35.0	2187.5	11.5
1	45.4	1.9	1	295.8	4.3	1	738.2	6.7	1	1372.6	9.1	1	2199.0	11.5
2	47.4	1.9	2	300.2	4.3	2	745.0	6.7	2	1381.8	9.1	2	2210.6	11.5
3	49.3	2.0	3	304.5	4.4	3	751.7	6.8	3	1390.9	9.2	3	2222.1	11.6
4	51.3	2.0	4	308.9	4.4	4	758.5	6.8	4	1400.1	9.2	4	2233.7	11.6
5	53.4	2.0	5	313.4	4.4	5	765.4	6.8	5	1409.4	9.2	5	2245.4	11.6
6	55.4	2.1	6	317.8	4.5	6	772.2	6.9	6	1418.6	9.3	6	2257.0	11.7
7	57.5	2.1	7	322.3	4.5	7	779.1	6.9	7	1427.9	9.3	7	2268.7	11.7
8	59.7	2.1	8	326.9	4.5	8	786.1	6.9	8	1437.3	9.3	8	2280.5	11.7
9	61.8	2.2	9	331.4	4.6	9	793.0	7.0	9	1446.6	9.4	9	2292.2	11.8
4.0	64.0	2.2	12.0	336.0	4.6	20.0	800.0	7.0	28.0	1456.0	9.4	36.0	2304.0	11.8
1	66.2	2.2	1	340.6	4.6	1	807.0	7.0	1	1465.4	9.4	1	2315.8	11.8
2	68.5	2.2	2	345.3	4.6	2	814.1	7.0	2	1474.9	9.4	2	2327.7	11.8
3	70.7	2.3	3	349.9	4.7	3	811.1	7.1	3	1484.3	9.5	3	2339.5	11.9
4	73.0	2.3	4	354.6	4.7	4	828.2	7.1	4	1493.8	9.5	4	2351.4	11.9
5	75.4	2.3	5	359.4	4.7	5	835.4	7.1	5	1503.4	9.5	5	2363.4	11.9
6	77.7	2.4	6	364.1	4.8	6	842.5	7.2	6	1512.9	9.6	6	2375.3	12.0
7	80.1	2.4	7	368.9	4.8	7	849.7	7.2	7	1522.5	9.6	7	2387.3	12.0
8	82.6	2.4	8	373.8	4.8	8	857.0	7.2	8	1532.2	9.6	8	2399.4	12.0
9	85.0	2.5	9	378.6	4.9	9	864.2	7.3	9	1541.8	9.7	9	2411.4	12.1
5.0	87.5	2.5	13.0	383.5	4.9	21.0	871.5	7.3	29.0	1551.5	9.7	37.0	2423.5	12.1
1	90.0	2.5	1	388.4	4.9	1	878.8	7.3	1	1561.2	9.7	1	2435.6	12.1
2	92.6	2.5	2	393.4	4.9	2	886.2	7.3	2	1571.0	9.7	2	2447.8	12.1
3	95.1	2.6	3	398.3	5.0	3	893.5	7.4	3	1580.7	9.8	3	2459.9	12.2
4	97.7	2.6	4	403.3	5.0	4	900.9	7.4	4	1590.5	9.8	4	2472.1	12.2
5	100.4	2.6	5	408.4	5.0	5	908.4	7.4	5	1600.4	9.8	5	2484.4	12.2
6	103.0	2.7	6	413.4	5.1	6	915.8	7.5	6	1610.2	9.9	6	2496.6	12.3
7	105.7	2.7	7	418.5	5.1	7	923.3	7.5	7	1620.1	9.9	7	2508.9	12.3
8	108.5	2.7	8	423.7	5.1	8	930.9	7.5	8	1630.1	9.9	8	2521.3	12.3
9	111.2	2.8	9	428.8	5.2	9	938.4	7.6	9	1640.0	10.0	9	2533.6	12.4
6.0	114.0	2.8	14.0	434.0	5.2	22.0	946.0	7.6	30.0	1650.0	10.0	38.0	2546.0	12.4
1	116.8	2.8	1	439.2	5.2	1	953.6	7.6	1	1660.0	10.0	1	2558.4	12.4
2	119.7	2.8	2	444.5	5.2	2	961.3	7.6	2	1670.1	10.0	2	2570.9	12.4
3	122.5	2.9	3	449.7	5.3	3	968.9	7.7	3	1680.1	10.1	3	2583.3	12.5
4	125.4	2.9	4	455.0	5.3	4	976.6	7.7	4	1690.2	10.1	4	2595.8	12.5
5	128.4	2.9	5	460.4	5.3	5	984.4	7.7	5	1700.4	10.1	5	2608.4	12.5
6	131.3	3.0	6	465.7	5.4	6	992.1	7.8	6	1710.5	10.2	6	2620.9	12.6
7	134.2	3.0	7	471.1	5.4	7	999.9	7.8	7	1720.7	10.2	7	2633.5	12.6
8	137.4	3.0	8	476.6	5.4	8	1007.8	7.8	8	1731.0	10.2	8	2646.2	12.6
9	140.4	3.1	9	482.0	5.5	9	1015.6	7.9	9	1741.2	10.3	9	2658.8	12.7
7.0	143.5	3.1	15.0	487.5	5.5	23.0	1023.5	7.9	31.0	1751.5	10.3	39.0	2671.5	12.7
1	146.6	3.1	1	493.0	5.5	1	1031.4	7.9	1	1761.8	10.3	1	2684.2	12.7
2	149.8	3.1	2	498.6	5.5	2	1039.4	7.9	2	1772.2	10.3	2	2697.0	12.7
3	152.9	3.2	3	504.1	5.6	3	1047.3	8.0	3	1782.5	10.4	3	2709.7	12.8
4	156.1	3.2	4	509.7	5.6	4	1055.3	8.0	4	1792.9	10.4	4	2722.5	12.8
5	159.4	3.2	5	515.4	5.6	5	1063.4	8.0	5	1803.4	10.4	5	2735.4	12.8
6	162.6	3.3	6	521.0	5.7	6	1071.4	8.1	6	1813.8	10.5	6	2748.2	12.9
7	165.9	3.3	7	526.7	5.7	7	1079.5	8.1	7	1824.3	10.5	7	2761.1	12.9
8	169.3	3.3	8	532.5	5.7	8	1087.7	8.1	8	1834.9	10.5	8	2774.1	12.9
9	172.6	3.4	9	538.2	5.8	9	1095.8	8.2	9	1845.4	10.6	9	2787.0	13.0

TABLE NO. 7.

EXCAVATION AND EMBANKMENT.

AREAS.

Breadth of Base, Twelve Feet. Slope 1 1-2 Horizontal to 1 Perpendicular.

Height.	Area in Feet.	Diff.	Height.	Area in Feet.	Diff.	Height.	Area in Feet.	Diff.	Height.	Area in Feet.	Diff.	Height.	Area in Feet.	Diff.
0.0	0.0	0.0	8.0	192.0	3.6	16.0	576.0	6.0	24.0	1152.0	8.4	32.0	1920.0	10.8
.1	1.2	1.2	.1	195.6	3.6	.1	582.0	6.0	.1	1160.4	8.4	.1	1930.8	10.8
.2	2.5	1.2	.2	199.3	3.6	.2	588.1	6.0	.2	1168.9	8.4	.2	1941.7	10.8
.3	3.7	1.3	.3	202.9	3.7	.3	594.1	6.1	.3	1177.3	8.5	.3	1952.5	10.9
.4	5.0	1.3	.4	206.6	3.7	.4	600.2	6.1	.4	1185.8	8.5	.4	1963.4	10.9
.5	6.4	1.3	.5	210.4	3.7	.5	606.4	6.1	.5	1194.4	8.5	.5	1974.4	10.9
.6	7.7	1.3	.6	214.1	3.8	.6	612.5	6.2	.6	1202.9	8.6	.6	1985.3	11.0
.7	9.1	1.4	.7	217.9	3.8	.7	618.7	6.2	.7	1211.5	8.6	.7	1996.3	11.0
.8	10.6	1.4	.8	221.8	3.8	.8	625.0	6.2	.8	1220.2	8.6	.8	2007.4	11.0
.9	12.0	1.5	.9	225.6	3.9	.9	631.2	6.3	.9	1228.8	8.7	.9	2018.4	11.0
1.0	13.5	1.5	9.0	229.5	3.9	17.0	637.5	6.3	25.0	1237.5	8.7	33.0	2029.5	11.1
.1	15.0	1.5	.1	233.4	3.9	.1	643.8	6.3	.1	1246.2	8.7	.1	2040.6	11.1
.2	16.6	1.5	.2	237.4	3.9	.2	650.2	6.3	.2	1255.0	8.7	.2	2051.8	11.1
.3	18.1	1.6	.3	241.3	4.0	.3	656.5	6.4	.3	1263.7	8.8	.3	2062.9	11.2
.4	19.7	1.6	.4	245.3	4.0	.4	662.9	6.4	.4	1272.5	8.8	.4	2074.1	11.2
.5	21.4	1.6	.5	249.4	4.0	.5	669.4	6.4	.5	1281.4	8.8	.5	2085.4	11.2
.6	23.0	1.7	.6	253.4	4.1	.6	675.8	6.5	.6	1290.2	8.9	.6	2096.6	11.3
.7	24.7	1.7	.7	257.5	4.1	.7	682.3	6.5	.7	1299.1	8.9	.7	2107.9	11.3
.8	26.5	1.7	.8	261.7	4.1	.8	688.9	6.5	.8	1308.0	8.9	.8	2119.3	11.3
.9	28.2	1.8	.9	265.8	4.2	.9	695.4	6.6	.9	1317.0	9.0	.9	2130.6	11.3
2.0	30.0	1.8	10.0	270.0	4.2	18.0	702.0	6.6	26.0	1326.0	9.0	34.0	2142.0	11.4
.1	31.8	1.8	.1	274.2	4.2	.1	708.6	6.6	.1	1335.0	9.0	.1	2153.4	11.4
.2	33.7	1.8	.2	278.5	4.2	.2	715.3	6.6	.2	1344.1	9.0	.2	2164.9	11.4
.3	35.5	1.9	.3	282.7	4.3	.3	721.9	6.7	.3	1353.1	9.1	.3	2176.3	11.5
.4	37.4	1.9	.4	287.0	4.3	.4	728.6	6.7	.4	1362.2	9.1	.4	2187.8	11.5
.5	39.4	1.9	.5	291.4	4.3	.5	735.4	6.7	.5	1371.4	9.1	.5	2199.4	11.5
.6	41.3	2.0	.6	295.7	4.4	.6	742.1	6.7	.6	1380.5	9.2	.6	2210.9	11.6
.7	43.3	2.0	.7	300.1	4.4	.7	748.9	6.8	.7	1389.7	9.2	.7	2222.5	11.6
.8	45.4	2.0	.8	304.6	4.4	.8	755.8	6.8	.8	1399.0	9.2	.8	2234.2	11.6
.9	47.4	2.1	.9	309.0	4.5	.9	762.6	6.9	.9	1408.2	9.3	.9	2245.8	11.6
3.0	49.5	2.1	11.0	313.5	4.5	19.0	769.5	6.9	27.0	1417.5	9.3	35.0	2257.5	11.7
.1	51.6	2.1	.1	318.0	4.5	.1	776.4	6.9	.1	1426.8	9.3	.1	2269.2	11.7
.2	53.8	2.1	.2	322.6	4.5	.2	783.4	6.9	.2	1436.2	9.3	.2	2281.0	11.7
.3	55.9	2.2	.3	327.1	4.6	.3	790.3	7.0	.3	1445.5	9.4	.3	2292.7	11.8
.4	58.1	2.2	.4	331.7	4.6	.4	797.3	7.0	.4	1454.9	9.4	.4	2304.5	11.8
.5	60.4	2.2	.5	336.4	4.6	.5	804.4	7.0	.5	1464.4	9.4	.5	2316.4	11.8
.6	62.6	2.2	.6	341.0	4.7	.6	811.4	7.1	.6	1473.8	9.5	.6	2328.2	11.9
.7	64.9	2.3	.7	345.7	4.7	.7	818.5	7.1	.7	1483.3	9.5	.7	2340.1	11.9
.8	67.3	2.3	.8	350.5	4.7	.8	825.7	7.1	.8	1492.9	9.5	.8	2352.1	11.9
.9	69.6	2.4	.9	355.2	4.8	.9	832.8	7.2	.9	1502.4	9.6	.9	2364.0	11.9
4.0	72.0	2.4	12.0	360.0	4.8	20.0	840.0	7.2	28.0	1512.0	9.6	36.0	2376.0	12.0
.1	74.4	2.4	.1	364.8	4.8	.1	847.2	7.2	.1	1521.6	9.6	.1	2388.0	12.0
.2	76.8	2.4	.2	369.7	4.8	.2	854.5	7.2	.2	1531.3	9.6	.2	2400.1	12.0
.3	79.3	2.5	.3	374.5	4.9	.3	861.7	7.3	.3	1540.9	9.7	.3	2412.1	12.1
.4	81.8	2.5	.4	379.4	4.9	.4	869.0	7.3	.4	1550.6	9.7	.4	2424.2	12.1
.5	84.4	2.5	.5	384.4	4.9	.5	876.4	7.3	.5	1560.4	9.7	.5	2436.4	12.1
.6	86.9	2.6	.6	389.3	5.0	.6	883.7	7.4	.6	1570.1	9.8	.6	2448.5	12.2
.7	89.5	2.6	.7	394.3	5.0	.7	891.1	7.4	.7	1579.9	9.8	.7	2460.7	12.2
.8	92.2	2.6	.8	399.4	5.0	.8	898.6	7.4	.8	1589.8	9.8	.8	2473.0	12.2
.9	94.8	2.7	.9	404.4	5.1	.9	906.0	7.5	.9	1599.6	9.9	.9	2485.2	12.2
5.0	97.5	2.7	13.0	409.5	5.1	21.0	913.5	7.5	29.0	1609.5	9.9	37.0	2497.5	12.3
.1	100.2	2.7	.1	414.6	5.1	.1	921.0	7.5	.1	1619.4	9.9	.1	2509.8	12.3
.2	103.0	2.7	.2	419.8	5.1	.2	928.6	7.5	.2	1629.4	9.9	.2	2522.2	12.3
.3	105.7	2.8	.3	424.9	5.2	.3	936.1	7.6	.3	1639.3	10.0	.3	2534.5	12.4
.4	108.5	2.8	.4	430.1	5.2	.4	943.7	7.6	.4	1649.3	10.0	.4	2546.9	12.4
.5	111.4	2.8	.5	435.4	5.2	.5	951.4	7.6	.5	1659.4	10.0	.5	2559.4	12.4
.6	114.2	2.9	.6	440.6	5.3	.6	959.0	7.7	.6	1669.4	10.1	.6	2571.8	12.5
.7	117.1	2.9	.7	445.9	5.3	.7	966.7	7.7	.7	1679.5	10.1	.7	2584.3	12.5
.8	120.1	2.9	.8	451.3	5.3	.8	974.5	7.7	.8	1689.7	10.1	.8	2596.9	12.5
.9	123.0	3.0	.9	456.6	5.4	.9	982.2	7.8	.9	1699.8	10.1	.9	2609.4	12.5
6.0	126.0	3.0	14.0	462.0	5.4	22.0	990.0	7.8	30.0	1710.0	10.2	38.0	2622.0	12.6
.1	129.0	3.0	.1	467.4	5.4	.1	997.8	7.8	.1	1720.2	10.2	.1	2634.6	12.6
.2	132.1	3.0	.2	472.9	5.4	.2	1005.7	7.9	.2	1730.5	10.2	.2	2647.3	12.6
.3	135.1	3.1	.3	478.3	5.5	.3	1013.5	7.9	.3	1740.7	10.3	.3	2659.9	12.7
.4	138.2	3.1	.4	483.8	5.5	.4	1021.4	7.9	.4	1751.0	10.3	.4	2672.6	12.7
.5	141.4	3.1	.5	489.4	5.5	.5	1029.4	7.9	.5	1761.4	10.3	.5	2685.4	12.7
.6	144.5	3.2	.6	494.9	5.6	.6	1037.3	8.0	.6	1771.7	10.4	.6	2698.1	12.8
.7	147.7	3.2	.7	500.5	5.6	.7	1045.3	8.0	.7	1782.1	10.4	.7	2710.9	12.8
.8	151.0	3.2	.8	506.2	5.6	.8	1053.4	8.0	.8	1792.6	10.4	.8	2723.8	12.8
.9	154.2	3.3	.9	511.8	5.7	.9	1061.4	8.1	.9	1803.0	10.4	.9	2736.6	12.8
7.0	157.5	3.3	15.0	517.5	5.7	23.0	1069.5	8.1	31.0	1813.5	10.5	39.0	2749.5	12.9
.1	160.8	3.3	.1	523.2	5.7	.1	1077.6	8.1	.1	1824.0	10.5	.1	2762.4	12.9
.2	164.2	3.3	.2	529.0	5.7	.2	1085.8	8.1	.2	1834.6	10.5	.2	2775.4	12.9
.3	167.5	3.4	.3	534.7	5.8	.3	1093.9	8.2	.3	1845.1	10.6	.3	2788.3	13.0
.4	170.9	3.4	.4	540.5	5.8	.4	1102.1	8.2	.4	1855.7	10.6	.4	2801.3	13.0
.5	174.4	3.4	.5	546.4	5.8	.5	1110.4	8.2	.5	1866.4	10.6	.5	2814.4	13.0
.6	177.8	3.5	.6	552.2	5.9	.6	1118.6	8.3	.6	1877.0	10.7	.6	2827.4	13.1
.7	181.3	3.5	.7	558.1	5.9	.7	1126.9	8.3	.7	1887.7	10.7	.7	2840.5	13.1
.8	184.8	3.5	.8	564.1	5.9	.8	1135.3	8.3	.8	1898.5	10.7	.8	2853.7	13.1
.9	188.4	3.6	.9	570.0	6.0	.9	1143.6	8.4	.9	1909.2	10.7	.9	2866.8	13.2

TABLE NO. 8.

EXCAVATION AND EMBANKMENT.

AREAS.

Breadth of Base, Fourteen Feet. Slope 1 1-2 Horizontal to 1 Perpendicular.

Height.	Area in Feet.	Diff.	Height.	Area in Feet.	Diff.	Height.	Area in Feet.	Diff.	Height.	Area in Feet.	Diff.	Height.	Area in Feet.	Diff.
0.0	0.0	0.0	8.0	208.0	3.8	16.0	608.0	6.2	24.0	1200.0	8.6	32.0	1984.0	11.0
.1	1.4	1.4	.1	211.8	3.8	.1	614.2	6.2	.1	1208.6	8.6	.1	1995.0	11.0
.2	2.9	1.4	.2	215.7	3.8	.2	620.5	6.2	.2	1217.3	8.6	.2	2006.1	11.0
.3	4.3	1.5	.3	219.5	3.9	.3	626.7	6.3	.3	1225.9	8.7	.3	2017.1	11.1
.4	5.8	1.5	.4	223.4	3.9	.4	633.0	6.3	.4	1234.6	8.7	.4	2028.2	11.1
.5	7.4	1.5	.5	227.4	3.9	.5	639.4	6.3	.5	1243.4	8.7	.5	2039.4	11.1
.6	8.9	1.6	.6	231.3	4.0	.6	645.7	6.4	.6	1252.1	8.8	.6	2050.5	11.2
.7	10.5	1.6	.7	235.3	4.0	.7	652.1	6.4	.7	1260.9	8.8	.7	2061.7	11.2
.8	12.2	1.6	.8	239.4	4.0	.8	658.6	6.4	.8	1269.8	8.8	.8	2073.0	11.2
.9	13.8	1.7	.9	243.4	4.1	.9	665.0	6.5	.9	1278.6	8.9	.9	2084.2	11.3
1.0	15.5	1.7	9.0	247.5	4.1	17.0	671.5	6.5	25.0	1287.5	8.9	33.0	2095.5	11.3
.1	17.2	1.7	.1	251.6	4.1	.1	678.0	6.5	.1	1296.4	8.9	.1	2106.8	11.3
.2	19.0	1.7	.2	255.8	4.1	.2	684.6	6.5	.2	1305.4	8.9	.2	2118.2	11.3
.3	20.7	1.8	.3	259.9	4.2	.3	691.1	6.6	.3	1314.3	9.0	.3	2129.5	11.4
.4	22.5	1.8	.4	264.1	4.2	.4	697.7	6.6	.4	1323.3	9.0	.4	2140.9	11.4
.5	24.4	1.8	.5	268.4	4.2	.5	704.4	6.6	.5	1332.4	9.0	.5	2152.4	11.4
.6	26.2	1.9	.6	272.6	4.3	.6	711.0	6.7	.6	1341.4	9.1	.6	2163.8	11.5
.7	28.1	1.9	.7	276.9	4.3	.7	717.7	6.7	.7	1350.5	9.1	.7	2175.3	11.5
.8	30.1	1.9	.8	281.3	4.3	.8	724.5	6.7	.8	1359.7	9.1	.8	2186.9	11.5
.9	32.0	2.0	.9	285.6	4.4	.9	731.2	6.8	.9	1368.8	9.2	.9	2198.4	11.6
2.0	34.0	2.0	10.0	290.0	4.4	18.0	738.0	6.8	26.0	1378.0	9.2	34.0	2210.0	11.6
.1	36.0	2.0	.1	294.4	4.4	.1	744.8	6.8	.1	1387.2	9.2	.1	2221.6	11.6
.2	38.1	2.0	.2	298.9	4.4	.2	751.7	6.8	.2	1396.5	9.2	.2	2233.3	11.6
.3	40.1	2.1	.3	303.3	4.5	.3	758.5	6.9	.3	1405.7	9.3	.3	2244.9	11.7
.4	42.2	2.1	.4	307.8	4.5	.4	765.4	6.9	.4	1415.0	9.3	.4	2256.6	11.7
.5	44.4	2.1	.5	312.4	4.5	.5	772.4	6.9	.5	1424.4	9.3	.5	2268.4	11.7
.6	46.5	2.2	.6	316.9	4.6	.6	779.3	7.0	.6	1433.7	9.4	.6	2280.1	11.8
.7	48.7	2.2	.7	321.5	4.6	.7	786.3	7.0	.7	1443.1	9.4	.7	2291.9	11.8
.8	51.0	2.2	.8	326.2	4.6	.8	793.4	7.0	.8	1452.6	9.4	.8	2303.8	11.8
.9	53.2	2.3	.9	330.8	4.7	.9	800.4	7.1	.9	1462.0	9.5	.9	2315.6	11.9
3.0	55.5	2.3	11.0	335.5	4.7	19.0	807.5	7.1	27.0	1471.5	9.5	35.0	2327.5	11.9
.1	57.8	2.3	.1	340.2	4.7	.1	814.6	7.1	.1	1481.0	9.5	.1	2339.4	11.9
.2	60.2	2.3	.2	345.0	4.7	.2	821.8	7.1	.2	1490.6	9.5	.2	2351.4	11.9
.3	62.5	2.4	.3	349.7	4.8	.3	828.9	7.2	.3	1500.1	9.6	.3	2363.3	12.0
.4	64.9	2.4	.4	354.5	4.8	.4	836.1	7.2	.4	1509.7	9.6	.4	2375.3	12.0
.5	67.4	2.4	.5	359.4	4.8	.5	843.4	7.2	.5	1519.4	9.6	.5	2387.4	12.0
.6	69.8	2.5	.6	364.2	4.9	.6	850.6	7.3	.6	1529.0	9.7	.6	2399.4	12.1
.7	72.3	2.5	.7	369.1	4.9	.7	857.9	7.3	.7	1538.7	9.7	.7	2411.5	12.1
.8	74.9	2.5	.8	374.1	4.9	.8	865.3	7.3	.8	1548.5	9.7	.8	2423.7	12.1
.9	77.4	2.6	.9	379.0	5.0	.9	872.6	7.4	.9	1558.2	9.8	.9	2435.8	12.2
4.0	80.0	2.6	12.0	384.0	5.0	20.0	880.0	7.4	28.0	1568.0	9.8	36.0	2448.0	12.2
.1	82.6	2.6	.1	389.0	5.0	.1	887.4	7.4	.1	1577.8	9.8	.1	2460.2	12.2
.2	85.3	2.6	.2	394.1	5.0	.2	894.9	7.4	.2	1587.7	9.8	.2	2472.5	12.2
.3	87.9	2.7	.3	399.1	5.1	.3	902.3	7.5	.3	1597.5	9.9	.3	2484.7	12.3
.4	90.6	2.7	.4	404.2	5.1	.4	909.8	7.5	.4	1607.4	9.9	.4	2497.0	12.3
.5	93.4	2.7	.5	409.4	5.1	.5	917.4	7.5	.5	1617.4	9.9	.5	2509.4	12.3
.6	96.1	2.8	.6	414.5	5.2	.6	924.9	7.6	.6	1627.3	10.0	.6	2521.7	12.4
.7	98.9	2.8	.7	419.7	5.2	.7	932.5	7.6	.7	1637.3	10.0	.7	2534.1	12.4
.8	101.8	2.8	.8	425.0	5.2	.8	940.2	7.6	.8	1647.4	10.0	.8	2546.6	12.4
.9	104.6	2.9	.9	430.2	5.3	.9	947.8	7.7	.9	1657.4	10.1	.9	2559.0	12.5
5.0	107.5	2.9	13.0	435.5	5.3	21.0	955.5	7.7	29.0	1667.5	10.1	37.0	2571.5	12.5
.1	110.4	2.9	.1	440.8	5.3	.1	963.2	7.7	.1	1677.6	10.1	.1	2584.0	12.5
.2	113.4	2.9	.2	446.2	5.3	.2	971.0	7.7	.2	1687.8	10.1	.2	2596.6	12.5
.3	116.3	3.0	.3	451.5	5.4	.3	978.7	7.8	.3	1697.9	10.2	.3	2609.1	12.6
.4	119.3	3.0	.4	456.9	5.4	.4	986.5	7.8	.4	1708.1	10.2	.4	2621.7	12.6
.5	122.4	3.0	.5	462.4	5.4	.5	994.4	7.8	.5	1718.4	10.2	.5	2634.4	12.6
.6	125.4	3.1	.6	467.8	5.5	.6	1002.2	7.9	.6	1728.6	10.3	.6	2647.0	12.7
.7	128.5	3.1	.7	473.3	5.5	.7	1010.1	7.9	.7	1738.9	10.3	.7	2659.7	12.7
.8	131.7	3.1	.8	478.9	5.5	.8	1018.1	7.9	.8	1749.3	10.3	.8	2672.5	12.7
.9	134.8	3.2	.9	484.4	5.6	.9	1026.0	8.0	.9	1759.6	10.4	.9	2685.2	12.8
6.0	138.0	3.2	14.0	490.0	5.6	22.0	1034.0	8.0	30.0	1770.0	10.4	38.0	2698.0	12.8
.1	141.2	3.2	.1	495.6	5.6	.1	1042.0	8.0	.1	1780.4	10.4	.1	2710.8	12.8
.2	144.5	3.2	.2	501.3	5.6	.2	1050.1	8.0	.2	1790.9	10.4	.2	2723.7	12.8
.3	147.7	3.3	.3	506.9	5.7	.3	1058.1	8.1	.3	1801.3	10.5	.3	2736.5	12.9
.4	151.0	3.3	.4	512.6	5.7	.4	1066.2	8.1	.4	1811.8	10.5	.4	2749.4	12.9
.5	154.4	3.3	.5	518.4	5.7	.5	1074.4	8.1	.5	1822.4	10.5	.5	2762.4	12.9
.6	157.7	3.4	.6	524.1	5.8	.6	1082.5	8.2	.6	1832.9	10.6	.6	2775.3	13.0
.7	161.1	3.4	.7	529.9	5.8	.7	1090.7	8.2	.7	1843.5	10.6	.7	2788.3	13.0
.8	164.6	3.4	.8	535.8	5.8	.8	1099.0	8.2	.8	1854.2	10.6	.8	2801.4	13.0
.9	168.0	3.5	.9	541.6	5.9	.9	1107.2	8.3	.9	1864.8	10.7	.9	2814.4	13.1
7.0	171.5	3.5	15.0	547.5	5.9	23.0	1115.5	8.3	31.0	1875.5	10.7	39.0	2827.5	13.1
.1	175.0	3.5	.1	553.4	5.9	.1	1123.8	8.3	.1	1886.2	10.7	.1	2840.6	13.1
.2	178.6	3.5	.2	559.4	5.9	.2	1132.2	8.3	.2	1897.0	10.7	.2	2853.8	13.1
.3	182.1	3.6	.3	565.3	6.0	.3	1140.5	8.4	.3	1907.7	10.8	.3	2866.9	13.2
.4	185.7	3.6	.4	571.3	6.0	.4	1148.9	8.4	.4	1918.5	10.8	.4	2880.1	13.2
.5	189.4	3.6	.5	577.4	6.0	.5	1157.4	8.4	.5	1929.4	10.8	.5	2893.4	13.2
.6	193.0	3.7	.6	583.4	6.1	.6	1165.8	8.5	.6	1940.2	10.9	.6	2906.6	13.3
.7	196.7	3.7	.7	589.5	6.1	.7	1174.3	8.5	.7	1951.1	10.9	.7	2919.9	13.3
.8	200.5	3.7	.8	595.7	6.1	.8	1182.9	8.5	.8	1962.1	10.9	.8	2933.3	13.3
.9	204.2	3.8	.9	601.8	6.2	.9	1191.4	8.6	.9	1973.0	11.0	.9	2946.6	13.4

TABLE NO. 9.

EXCAVATION AND EMBANKMENT.

AREAS.

Breadth of Base, Fifteen Feet. Slope 1 1-2 Horizontal to 1 Perpendicular.

Height.	Area in Feet.	Diff.	Height.	Area in Feet.	Diff.	Height.	Area in Feet.	Diff.	Height.	Area in Feet.	Diff.	Height.	Area in Feet.	Diff.
0.0	0.0	0.0	8.0	216.0	3.9	16.0	624.0	6.3	24.0	1224.0	8.7	32.0	2016.0	11.1
.1	1.5	1.5	.1	219.9	3.9	.1	630.3	6.3	.1	1232.7	8.7	.1	2027.1	11.1
.2	3.1	1.5	.2	223.9	3.9	.2	636.7	6.3	.2	1241.5	8.7	.2	2038.3	11.1
.3	4.6	1.6	.3	227.8	4.0	.3	643.0	6.4	.3	1250.2	8.8	.3	2049.4	11.2
.4	6.2	1.6	.4	231.8	4.0	.4	649.4	6.4	.4	1259.0	8.8	.4	2060.6	11.2
.5	7.9	1.6	.5	235.9	4.0	.5	655.9	6.4	.5	1267.9	8.8	.5	2071.9	11.2
.6	9.5	1.7	.6	239.9	4.1	.6	662.3	6.5	.6	1276.7	8.9	.6	2083.1	11.3
.7	11.2	1.7	.7	244.0	4.1	.7	668.8	6.5	.7	1285.6	8.9	.7	2094.4	11.3
.8	13.0	1.7	.8	248.2	4.1	.8	675.4	6.5	.8	1294.6	8.9	.8	2105.8	11.3
.9	14.7	1.8	.9	252.3	4.2	.9	681.9	6.6	.9	1303.5	9.0	.9	2117.1	11.4
1.0	16.5	1.8	9.0	256.5	4.2	17.0	688.5	6.6	25.0	1312.5	9.0	33.0	2128.5	11.4
.1	18.3	1.8	.1	260.7	4.2	.1	695.1	6.6	.1	1321.5	9.0	.1	2139.9	11.4
.2	20.2	1.8	.2	265.0	4.2	.2	701.8	6.6	.2	1330.6	9.0	.2	2151.4	11.4
.3	22.0	1.9	.3	269.2	4.3	.3	708.4	6.7	.3	1339.6	9.1	.3	2162.8	11.5
.4	23.9	1.9	.4	273.5	4.3	.4	715.1	6.7	.4	1348.7	9.1	.4	2174.3	11.5
.5	25.9	1.9	.5	277.9	4.3	.5	721.9	6.7	.5	1357.9	9.1	.5	2185.9	11.5
.6	27.8	2.0	.6	282.2	4.4	.6	728.6	6.8	.6	1367.0	9.2	.6	2197.4	11.6
.7	29.8	2.0	.7	286.6	4.4	.7	735.4	6.8	.7	1376.2	9.2	.7	2209.0	11.6
.8	31.9	2.0	.8	291.1	4.4	.8	742.3	6.8	.8	1385.5	9.2	.8	2220.7	11.6
.9	33.9	2.1	.9	295.5	4.5	.9	749.1	6.9	.9	1394.7	9.3	.9	2232.3	11.7
2.0	36.0	2.1	10.0	300.0	4.5	18.0	756.0	6.9	26.0	1404.0	9.3	34.0	2244.0	11.7
.1	38.1	2.1	.1	304.5	4.5	.1	762.9	6.9	.1	1413.3	9.3	.1	2255.7	11.7
.2	40.3	2.1	.2	309.1	4.5	.2	769.9	6.9	.2	1422.7	9.3	.2	2267.5	11.7
.3	42.4	2.2	.3	313.6	4.6	.3	776.8	7.0	.3	1432.0	9.4	.3	2279.2	11.8
.4	44.6	2.2	.4	318.2	4.6	.4	783.8	7.0	.4	1441.4	9.4	.4	2291.0	11.8
.5	46.9	2.2	.5	322.9	4.6	.5	790.9	7.0	.5	1450.9	9.4	.5	2302.9	11.8
.6	49.1	2.3	.6	327.5	4.7	.6	797.9	7.1	.6	1460.3	9.5	.6	2314.7	11.9
.7	51.4	2.3	.7	332.2	4.7	.7	805.0	7.1	.7	1469.8	9.5	.7	2326.6	11.9
.8	53.8	2.3	.8	337.0	4.7	.8	812.2	7.1	.8	1479.4	9.5	.8	2338.6	11.9
.9	56.1	2.4	.9	341.7	4.8	.9	819.3	7.2	.9	1488.9	9.6	.9	2350.5	12.0
3.0	58.5	2.4	11.0	346.5	4.8	19.0	826.5	7.2	27.0	1498.5	9.6	35.0	2362.5	12.0
.1	60.9	2.4	.1	351.3	4.8	.1	833.7	7.2	.1	1508.1	9.6	.1	2374.5	12.0
.2	63.4	2.4	.2	356.2	4.8	.2	841.0	7.2	.2	1517.8	9.6	.2	2386.6	12.0
.3	65.8	2.5	.3	361.0	4.9	.3	848.2	7.3	.3	1527.4	9.7	.3	2398.6	12.1
.4	68.3	2.5	.4	365.9	4.9	.4	855.5	7.3	.4	1537.1	9.7	.4	2410.7	12.1
.5	70.9	2.5	.5	370.9	4.9	.5	862.9	7.3	.5	1546.9	9.7	.5	2422.9	12.1
.6	73.4	2.6	.6	375.8	5.0	.6	870.2	7.4	.6	1556.6	9.8	.6	2435.0	12.2
.7	76.0	2.6	.7	380.8	5.0	.7	877.6	7.4	.7	1566.4	9.8	.7	2447.2	12.2
.8	78.7	2.6	.8	385.9	5.0	.8	885.1	7.4	.8	1576.3	9.8	.8	2459.5	12.2
.9	81.3	2.7	.9	390.9	5.1	.9	892.5	7.5	.9	1586.1	9.9	.9	2471.7	12.3
4.0	84.0	2.7	12.0	396.0	5.1	20.0	900.0	7.5	28.0	1596.0	9.9	36.0	2484.0	12.3
.1	86.7	2.7	.1	401.1	5.1	.1	907.5	7.5	.1	1605.9	9.9	.1	2496.3	12.3
.2	89.5	2.7	.2	406.3	5.1	.2	915.1	7.5	.2	1615.9	9.9	.2	2508.7	12.3
.3	92.2	2.8	.3	411.4	5.2	.3	922.6	7.6	.3	1625.8	10.0	.3	2521.0	12.4
.4	95.0	2.8	.4	416.6	5.2	.4	930.2	7.6	.4	1635.8	10.0	.4	2533.4	12.4
.5	97.9	2.8	.5	421.9	5.2	.5	937.9	7.6	.5	1645.9	10.0	.5	2545.9	12.4
.6	100.7	2.9	.6	427.1	5.3	.6	945.5	7.7	.6	1655.9	10.1	.6	2558.3	12.5
.7	103.6	2.9	.7	432.4	5.3	.7	953.2	7.7	.7	1666.0	10.1	.7	2570.8	12.5
.8	106.6	2.9	.8	437.8	5.3	.8	961.0	7.7	.8	1676.2	10.1	.8	2583.4	12.5
.9	109.5	3.0	.9	443.1	5.4	.9	968.7	7.8	.9	1686.3	10.2	.9	2595.9	12.6
5.0	112.5	3.0	13.0	448.5	5.4	21.0	976.5	7.8	29.0	1696.5	10.2	37.0	2608.5	12.6
.1	115.5	3.0	.1	453.9	5.4	.1	984.3	7.8	.1	1706.7	10.2	.1	2621.1	12.6
.2	118.6	3.0	.2	459.4	5.4	.2	992.2	7.8	.2	1717.0	10.2	.2	2633.8	12.6
.3	121.6	3.1	.3	464.8	5.5	.3	1000.0	7.9	.3	1727.2	10.3	.3	2646.4	12.7
.4	124.7	3.1	.4	470.3	5.5	.4	1007.9	7.9	.4	1737.5	10.3	.4	2659.1	12.7
.5	127.9	3.1	.5	475.9	5.5	.5	1015.9	7.9	.5	1747.9	10.3	.5	2671.9	12.7
.6	131.0	3.2	.6	481.4	5.6	.6	1023.8	8.0	.6	1758.2	10.4	.6	2684.6	12.8
.7	134.2	3.2	.7	487.0	5.6	.7	1031.8	8.0	.7	1768.6	10.4	.7	2697.4	12.8
.8	137.5	3.2	.8	492.7	5.6	.8	1039.9	8.0	.8	1779.1	10.4	.8	2710.3	12.8
.9	140.7	3.3	.9	498.3	5.7	.9	1047.9	8.1	.9	1789.5	10.5	.9	2723.1	12.9
6.0	144.0	3.3	14.0	504.0	5.7	22.0	1056.0	8.1	30.0	1800.0	10.5	38.0	2736.0	12.9
.1	147.3	3.3	.1	509.7	5.7	.1	1064.1	8.1	.1	1810.5	10.5	.1	2748.9	12.9
.2	150.7	3.3	.2	515.5	5.7	.2	1072.3	8.1	.2	1821.1	10.5	.2	2761.9	12.9
.3	154.0	3.4	.3	521.2	5.8	.3	1080.4	8.2	.3	1831.6	10.6	.3	2774.8	13.0
.4	157.4	3.4	.4	527.0	5.8	.4	1088.6	8.2	.4	1842.2	10.6	.4	2787.8	13.0
.5	160.9	3.4	.5	532.9	5.8	.5	1096.9	8.2	.5	1852.9	10.6	.5	2800.9	13.0
.6	164.3	3.5	.6	538.7	5.9	.6	1105.1	8.3	.6	1863.5	10.7	.6	2813.9	13.1
.7	167.8	3.5	.7	544.6	5.9	.7	1113.4	8.3	.7	1874.2	10.7	.7	2827.0	13.1
.8	171.4	3.5	.8	550.6	5.9	.8	1121.8	8.3	.8	1885.0	10.7	.8	2840.2	13.1
.9	174.9	3.6	.9	556.5	6.0	.9	1130.1	8.4	.9	1895.7	10.8	.9	2853.3	13.2
7.0	178.5	3.6	15.0	562.5	6.0	23.0	1138.5	8.4	31.0	1906.5	10.8	39.0	2866.5	13.2
.1	182.1	3.6	.1	568.5	6.0	.1	1146.9	8.4	.1	1917.3	10.8	.1	2879.7	13.2
.2	185.8	3.6	.2	574.6	6.0	.2	1155.4	8.4	.2	1928.2	10.8	.2	2893.0	13.2
.3	189.4	3.7	.3	580.6	6.1	.3	1163.8	8.5	.3	1939.0	10.9	.3	2906.2	13.3
.4	193.1	3.7	.4	586.7	6.1	.4	1172.3	8.5	.4	1949.9	10.9	.4	2919.5	13.3
.5	196.9	3.7	.5	592.9	6.1	.5	1180.9	8.5	.5	1960.9	10.9	.5	2932.9	13.3
.6	200.6	3.8	.6	599.0	6.2	.6	1189.4	8.6	.6	1971.8	11.0	.6	2946.2	13.4
.7	204.4	3.8	.7	605.2	6.2	.7	1198.0	8.6	.7	1982.8	11.0	.7	2959.6	13.4
.8	208.3	3.8	.8	611.5	6.2	.8	1206.7	8.6	.8	1993.9	11.0	.8	2973.1	13.4
.9	212.1	3.9	.9	617.7	6.3	.9	1215.3	8.7	.9	2004.9	11.1	.9	2986.5	13.5

TABLE NO. 9.

EXCAVATION AND EMBANKMENT.

AREAS.

Breadth of Base, Fifteen Feet. Slope 1 1-2 Horizontal to 1 Perpendicular.

Height.	Area in Feet.	Diff.	Height.	Area in Feet.	Diff.	Height.	Area in Feet.	Diff.	Height.	Area in Feet.	Diff.	Height.	Area in Feet.	Diff.
40.0	3000.0	13.5	48.0	4176.0	15.9	56.0	5544.0	18.3	64.0	7104.0	20.7	72.0	8856.0	23.1
.1	3013.5	13.5	.1	4191.9	15.9	.1	5562.3	18.3	.1	7124.7	20.7	.1	8879.1	23.1
.2	3027.1	13.5	.2	4207.9	15.9	.2	5580.7	18.3	.2	7145.5	20.7	.2	8902.3	23.1
.3	3040.6	13.6	.3	4223.8	16.0	.3	5599.0	18.4	.3	7166.2	20.8	.3	8925.4	23.2
.4	3054.2	13.6	.4	4239.8	16.0	.4	5617.4	18.4	.4	7187.0	20.8	.4	8948.6	23.2
.5	3067.9	13.6	.5	4255.9	16.0	.5	5635.9	18.4	.5	7207.9	20.8	.5	8971.9	23.2
.6	3081.5	13.7	.6	4271.9	16.1	.6	5654.3	18.5	.6	7228.7	20.9	.6	8995.1	23.3
.7	3095.2	13.7	.7	4288.0	16.1	.7	5672.8	18.5	.7	7249.6	20.9	.7	9018.4	23.3
.8	3109.0	13.7	.8	4304.2	16.1	.8	5691.4	18.5	.8	7270.6	20.9	.8	9041.8	23.3
.9	3122.7	13.8	.9	4320.3	16.2	.9	5709.9	18.6	.9	7291.5	21.0	.9	9065.1	23.4
41.0	3136.5	13.8	49.0	4336.5	16.2	57.0	5728.5	18.6	65.0	7312.5	21.0	73.0	9088.5	23.4
.1	3150.3	13.8	.1	4352.7	16.2	.1	5747.1	18.6	.1	7333.5	21.0	.1	9111.9	23.4
.2	3164.2	13.8	.2	4369.0	16.2	.2	5765.8	18.6	.2	7354.6	21.0	.2	9135.4	23.4
.3	3178.0	13.9	.3	4385.2	16.3	.3	5784.4	18.7	.3	7375.6	21.1	.3	9158.8	23.5
.4	3191.9	13.9	.4	4401.5	16.3	.4	5803.1	18.7	.4	7396.7	21.1	.4	9182.3	23.5
.5	3205.9	13.9	.5	4417.9	16.3	.5	5821.9	18.7	.5	7417.9	21.1	.5	9205.9	23.5
.6	3219.8	14.0	.6	4434.2	16.4	.6	5840.6	18.8	.6	7439.0	21.2	.6	9229.4	23.6
.7	3233.8	14.0	.7	4450.6	16.4	.7	5859.4	18.8	.7	7460.2	21.2	.7	9253.0	23.6
.8	3247.9	14.0	.8	4467.1	16.4	.8	5878.3	18.8	.8	7481.5	21.2	.8	9276.7	23.6
.9	3261.9	14.1	.9	4483.5	16.5	.9	5897.1	18.9	.9	7502.7	21.3	.9	9300.3	23.7
42.0	3276.0	14.1	50.0	4500.0	16.5	58.0	5916.0	18.9	66.0	7524.0	21.3	74.0	9324.0	23.7
.1	3290.1	14.1	.1	4516.5	16.5	.1	5934.9	18.9	.1	7545.3	21.3	.1	9347.7	23.7
.2	3304.3	14.1	.2	4533.1	16.5	.2	5953.9	18.9	.2	7566.7	21.3	.2	9371.5	23.7
.3	3318.4	14.2	.3	4549.6	16.6	.3	5972.8	19.0	.3	7588.0	21.4	.3	9395.2	23.8
.4	3332.6	14.2	.4	4566.2	16.6	.4	5991.8	19.0	.4	7609.4	21.4	.4	9419.0	23.8
.5	3346.9	14.2	.5	4582.9	16.6	.5	6010.9	19.0	.5	7630.9	21.4	.5	9442.9	23.8
.6	3361.1	14.3	.6	4599.5	16.7	.6	6029.9	19.1	.6	7652.3	21.5	.6	9466.7	23.9
.7	3375.4	14.3	.7	4616.2	16.7	.7	6049.0	19.1	.7	7673.8	21.5	.7	9490.6	23.9
.8	3389.8	14.3	.8	4633.0	16.7	.8	6068.2	19.1	.8	7695.4	21.5	.8	9514.6	23.9
.9	3404.1	14.4	.9	4649.7	16.8	.9	6087.3	19.2	.9	7716.9	21.6	.9	9538.5	24.0
43.0	3418.5	14.4	51.0	4666.5	16.8	59.0	6106.5	19.2	67.0	7738.5	21.6	75.0	9562.5	24.0
.1	3432.9	14.4	.1	4683.3	16.8	.1	6125.7	19.2	.1	7760.1	21.6	.1	9586.5	24.0
.2	3447.4	14.4	.2	4700.2	16.8	.2	6145.0	19.2	.2	7781.8	21.6	.2	9610.6	24.0
.3	3461.8	14.5	.3	4717.0	16.9	.3	6164.2	19.3	.3	7803.4	21.7	.3	9634.6	24.1
.4	3476.3	14.5	.4	4733.9	16.9	.4	6183.5	19.3	.4	7825.1	21.7	.4	9658.7	24.1
.5	3490.9	14.5	.5	4750.9	16.9	.5	6202.9	19.3	.5	7846.9	21.7	.5	9682.9	24.1
.6	3505.4	14.6	.6	4767.8	17.0	.6	6222.2	19.4	.6	7868.6	21.8	.6	9707.0	24.2
.7	3520.0	14.6	.7	4784.8	17.0	.7	6241.6	19.4	.7	7890.4	21.8	.7	9731.2	24.2
.8	3534.7	14.6	.8	4801.9	17.0	.8	6261.1	19.4	.8	7912.3	21.8	.8	9755.5	24.2
.9	3549.3	14.7	.9	4818.9	17.1	.9	6280.5	19.5	.9	7934.1	21.9	.9	9779.7	24.3
44.0	3564.0	14.7	52.0	4836.0	17.1	60.0	6300.0	19.5	68.0	7956.0	21.9	76.0	9804.0	24.3
.1	3578.7	14.7	.1	4853.1	17.1	.1	6319.5	19.5	.1	7977.9	21.9	.1	9828.3	24.3
.2	3593.5	14.7	.2	4870.3	17.1	.2	6339.1	19.5	.2	7999.9	21.9	.2	9852.7	24.3
.3	3608.2	14.8	.3	4887.4	17.2	.3	6358.6	19.6	.3	8021.8	22.0	.3	9877.0	24.4
.4	3623.0	14.8	.4	4904.6	17.2	.4	6378.2	19.6	.4	8043.8	22.0	.4	9901.4	24.4
.5	3637.9	14.8	.5	4921.9	17.2	.5	6397.9	19.6	.5	8065.9	22.0	.5	9925.9	24.4
.6	3652.7	14.9	.6	4939.1	17.3	.6	6417.5	19.7	.6	8087.9	22.1	.6	9950.3	24.5
.7	3667.6	14.9	.7	4956.4	17.3	.7	6437.2	19.7	.7	8110.0	22.1	.7	9974.8	24.5
.8	3682.6	14.9	.8	4973.8	17.3	.8	6457.0	19.7	.8	8132.2	22.1	.8	9999.4	24.5
.9	3697.5	15.0	.9	4991.1	17.4	.9	6476.7	19.8	.9	8154.3	22.2	.9	10023.9	24.6
45.0	3712.5	15.0	53.0	5008.5	17.4	61.0	6496.5	19.8	69.0	8176.5	22.2	77.0	10048.5	24.6
.1	3727.5	15.0	.1	5025.9	17.4	.1	6516.3	19.8	.1	8198.7	22.2	.1	10073.1	24.6
.2	3742.6	15.0	.2	5043.4	17.4	.2	6536.2	19.8	.2	8221.0	22.2	.2	10097.8	24.6
.3	3757.6	15.1	.3	5060.8	17.5	.3	6556.0	19.9	.3	8243.2	22.3	.3	10122.4	24.7
.4	3772.7	15.1	.4	5078.3	17.5	.4	6575.9	19.9	.4	8265.5	22.3	.4	10147.1	24.7
.5	3787.9	15.1	.5	5095.9	17.5	.5	6595.9	19.9	.5	8287.9	22.3	.5	10171.9	24.7
.6	3803.0	15.2	.6	5113.4	17.6	.6	6615.8	20.0	.6	8310.2	22.4	.6	10196.6	24.8
.7	3818.2	15.2	.7	5131.0	17.6	.7	6635.8	20.0	.7	8332.6	22.4	.7	10221.4	24.8
.8	3833.5	15.2	.8	5148.7	17.6	.8	6655.9	20.0	.8	8355.1	22.4	.8	10246.3	24.8
.9	3848.7	15.3	.9	5166.3	17.7	.9	6675.9	20.1	.9	8377.5	22.5	.9	10271.1	24.9
46.0	3864.0	15.3	54.0	5184.0	17.7	62.0	6696.0	20.1	70.0	8400.0	22.5	78.0	10296.0	24.9
.1	3879.3	15.3	.1	5201.7	17.7	.1	6716.1	20.1	.1	8422.5	22.5	.1	10320.9	24.9
.2	3894.7	15.3	.2	5219.5	17.7	.2	6736.3	20.1	.2	8445.1	22.5	.2	10345.9	24.9
.3	3910.0	15.4	.3	5237.2	17.8	.3	6756.4	20.2	.3	8467.6	22.6	.3	10370.8	25.0
.4	3925.4	15.4	.4	5255.0	17.8	.4	6776.6	20.2	.4	8490.2	22.6	.4	10395.8	25.0
.5	3940.9	15.4	.5	5272.9	17.8	.5	6796.9	20.2	.5	8512.9	22.6	.5	10420.9	25.0
.6	3956.3	15.5	.6	5290.7	17.9	.6	6817.1	20.3	.6	8535.5	22.7	.6	10445.9	25.1
.7	3971.8	15.5	.7	5308.6	17.9	.7	6837.4	20.3	.7	8558.2	22.7	.7	10471.0	25.1
.8	3987.4	15.5	.8	5326.6	17.9	.8	6857.8	20.3	.8	8581.0	22.7	.8	10496.2	25.1
.9	4002.9	15.6	.9	5344.5	18.0	.9	6878.1	20.4	.9	8603.7	22.8	.9	10521.3	25.2
47.0	4018.5	15.6	55.0	5362.5	18.0	63.0	6898.5	20.4	71.0	8626.5	22.8	79.0	10546.5	25.2
.1	4034.1	15.6	.1	5380.5	18.0	.1	6918.9	20.4	.1	8649.3	22.8	.1	10571.7	25.2
.2	4049.8	15.6	.2	5398.6	18.0	.2	6939.4	20.4	.2	8672.2	22.8	.2	10597.0	25.2
.3	4065.4	15.7	.3	5416.6	18.1	.3	6959.8	20.5	.3	8695.0	22.9	.3	10622.2	25.3
.4	4081.1	15.7	.4	5434.7	18.1	.4	6980.3	20.5	.4	8717.9	22.9	.4	10647.5	25.3
.5	4096.9	15.7	.5	5452.9	18.1	.5	7000.9	20.5	.5	8740.9	22.9	.5	10672.9	25.3
.6	4112.6	15.8	.6	5471.0	18.2	.6	7021.4	20.6	.6	8763.8	23.0	.6	10698.2	25.4
.7	4128.4	15.8	.7	5489.2	18.2	.7	7042.0	20.6	.7	8786.8	23.0	.7	10723.6	25.4
.8	4144.3	15.8	.8	5507.5	18.2	.8	7062.7	20.6	.8	8809.9	23.0	.8	10749.1	25.4
.9	4160.1	15.9	.9	5525.7	18.3	.9	7083.3	20.7	.9	8832.9	23.1	.9	10774.5	25.5

TABLE No. 10.

EXCAVATION AND EMBANKMENT.

AREAS.

Breadth of Base, 16 feet. Slopes, 1 1-2 Horizontal to 1 Perpendicular.

Height.	Area in Feet.	Difference.	Height.	Area in Feet.	Difference.	Height.	Area in Feet.	Difference.	Height.	Area in Feet.	Difference.	Height.	Area in Feet.	Difference.
0.0	0.0	0.0	8.0	224.0	4.0	16.0	640.0	6.4	24.0	1248.0	8.8	32.0	2048.0	11.2
1	1.6	1.6	1	228.0	4.0	1	646.4	6.4	1	1256.8	8.8	1	2059.2	11.2
2	3.3	1.6	2	232.1	4.0	2	652.9	6.4	2	1265.7	8.8	2	2070.5	11.2
3	4.9	1.7	3	236.1	4.1	3	659.3	6.5	3	1274.5	8.9	3	2081.7	11.3
4	6.6	1.7	4	240.2	4.1	4	665.8	6.5	4	1283.4	8.9	4	2093.0	11.3
5	8.4	1.7	5	244.4	4.1	5	672.4	6.5	5	1292.4	8.9	5	2104.4	11.3
6	10.1	1.8	6	248.5	4.2	6	678.9	6.6	6	1301.3	9.0	6	2115.7	11.4
7	11.9	1.8	7	252.7	4.2	7	685.5	6.6	7	1310.3	9.0	7	2127.1	11.4
8	13.8	1.8	8	257.0	4.2	8	692.2	6.6	8	1319.4	9.0	8	2138.6	11.4
9	15.6	1.9	9	261.2	4.3	9	698.8	6.7	9	1328.4	9.1	9	2150.0	11.5
1.0	17.5	1.9	9.0	265.5	4.3	17.0	705.5	6.7	25.0	1337.5	9.1	33.0	2161.5	11.5
1	19.4	1.9	1	269.8	4.3	1	712.2	6.7	1	1346.6	9.1	1	2173.0	11.5
2	21.4	1.9	2	274.2	4.3	2	719.0	6.7	2	1355.8	9.1	2	2184.6	11.5
3	23.3	2.0	3	278.5	4.4	3	725.7	6.8	3	1364.9	9.2	3	2196.1	11.6
4	25.3	2.0	4	282.9	4.4	4	732.5	6.8	4	1374.1	9.2	4	2207.7	11.6
5	27.4	2.0	5	287.4	4.4	5	739.4	6.8	5	1383.4	9.2	5	2219.4	11.6
6	29.4	2.1	6	291.8	4.5	6	746.2	6.9	6	1392.6	9.3	6	2231.0	11.7
7	31.5	2.1	7	296.3	4.5	7	753.1	6.9	7	1401.9	9.3	7	2242.7	11.7
8	33.7	2.1	8	300.9	4.5	8	760.1	6.9	8	1411.3	9.3	8	2254.5	11.7
9	35.8	2.2	9	305.4	4.6	9	767.0	7.0	9	1420.6	9.4	9	2266.2	11.8
2.0	38.0	2.2	10.0	310.0	4.6	18.0	774.0	7.0	26.0	1430.0	9.4	34.0	2278.0	11.8
1	40.2	2.2	1	314.6	4.6	1	781.0	7.0	1	1439.4	9.4	1	2289.8	11.8
2	42.5	2.2	2	319.3	4.6	2	788.1	7.0	2	1448.9	9.4	2	2301.7	11.8
3	44.7	2.3	3	323.9	4.7	3	795.1	7.1	3	1458.3	9.5	3	2313.5	11.9
4	47.0	2.3	4	328.6	4.7	4	802.2	7.1	4	1467.8	9.5	4	2325.4	11.9
5	49.4	2.3	5	333.4	4.7	5	809.4	7.1	5	1477.4	9.5	5	2337.4	11.9
6	51.7	2.4	6	338.1	4.8	6	816.5	7.2	6	1486.9	9.6	6	2349.3	12.0
7	54.1	2.4	7	342.9	4.8	7	823.7	7.2	7	1496.5	9.6	7	2361.3	12.0
8	56.6	2.4	8	347.8	4.8	8	831.0	7.2	8	1506.2	9.6	8	2373.4	12.0
9	59.0	2.5	9	352.6	4.9	9	838.2	7.3	9	1515.8	9.7	9	2385.4	12.1
3.0	61.5	2.5	11.0	357.5	4.9	19.0	845.5	7.3	27.0	1525.5	9.7	35.0	2397.5	12.1
1	64.0	2.5	1	362.4	4.9	1	852.8	7.3	1	1535.2	9.7	1	2409.6	12.1
2	66.6	2.5	2	367.4	4.9	2	860.2	7.3	2	1545.0	9.7	2	2421.8	12.1
3	69.1	2.6	3	372.3	5.0	3	867.5	7.4	3	1554.7	9.8	3	2433.9	12.2
4	71.7	2.6	4	377.3	5.0	4	874.9	7.4	4	1564.5	9.8	4	2446.1	12.2
5	74.4	2.6	5	382.4	5.0	5	882.4	7.4	5	1574.4	9.8	5	2458.4	12.2
6	77.0	2.7	6	387.4	5.1	6	889.8	7.5	6	1584.2	9.9	6	2470.6	12.3
7	79.7	2.7	7	392.5	5.1	7	897.3	7.5	7	1594.1	9.9	7	2482.9	12.3
8	82.5	2.7	8	397.7	5.1	8	904.9	7.5	8	1604.1	9.9	8	2495.3	12.3
9	85.2	2.8	9	402.8	5.2	9	912.4	7.6	9	1614.0	10.0	9	2507.6	12.4
4.0	88.0	2.8	12.0	408.0	5.2	20.0	920.0	7.6	28.0	1624.0	10.0	36.0	2520.0	12.4
1	90.8	2.8	1	413.2	5.2	1	927.6	7.6	1	1634.0	10.0	1	2532.4	12.4
2	93.7	2.8	2	418.5	5.2	2	935.3	7.6	2	1644.1	10.0	2	2544.9	12.4
3	96.5	2.9	3	423.7	5.3	3	942.9	7.7	3	1654.1	10.1	3	2557.3	12.5
4	99.4	2.9	4	429.0	5.3	4	950.6	7.7	4	1664.2	10.1	4	2569.8	12.5
5	102.4	2.9	5	434.4	5.3	5	958.4	7.7	5	1674.4	10.1	5	2582.4	12.5
6	105.3	3.0	6	439.7	5.4	6	966.1	7.8	6	1684.5	10.2	6	2594.9	12.6
7	108.3	3.0	7	445.1	5.4	7	973.9	7.8	7	1694.7	10.2	7	2607.5	12.6
8	111.4	3.0	8	450.6	5.4	8	981.8	7.8	8	1705.0	10.2	8	2620.2	12.6
9	114.4	3.1	9	456.0	5.5	9	989.6	7.9	9	1715.2	10.3	9	2632.8	12.7
5.0	117.5	3.1	13.0	461.5	5.5	21.0	997.5	7.9	29.0	1725.5	10.3	37.0	2645.5	12.7
1	120.6	3.1	1	467.0	5.5	1	1005.4	7.9	1	1735.8	10.3	1	2658.2	12.7
2	123.8	3.1	2	472.6	5.5	2	1013.4	7.9	2	1746.2	10.3	2	2671.0	12.7
3	126.9	3.2	3	478.1	5.6	3	1021.3	8.0	3	1756.5	10.4	3	2683.7	12.8
4	130.1	3.2	4	483.7	5.6	4	1029.3	8.0	4	1766.9	10.4	4	2696.5	12.8
5	133.4	3.2	5	489.4	5.6	5	1037.4	8.0	5	1777.4	10.4	5	2709.4	12.8
6	136.6	3.3	6	495.0	5.7	6	1045.4	8.1	6	1787.8	10.5	6	2722.2	12.9
7	139.9	3.3	7	500.7	5.7	7	1053.5	8.1	7	1798.3	10.5	7	2735.1	12.9
8	143.3	3.3	8	506.5	5.7	8	1061.7	8.1	8	1808.9	10.5	8	2748.1	12.9
9	146.6	3.4	9	512.2	5.8	9	1069.8	8.2	9	1819.4	10.6	9	2761.0	13.0
6.0	150.0	3.4	14.0	518.0	5.8	22.0	1078.0	8.2	30.0	1830.0	10.6	38.0	2774.0	13.0
1	153.4	3.4	1	523.8	5.8	1	1086.2	8.2	1	1840.6	10.6	1	2787.0	13.0
2	156.9	3.4	2	529.7	5.8	2	1094.5	8.2	2	1851.3	10.6	2	2800.1	13.0
3	160.3	3.5	3	535.5	5.9	3	1102.7	8.3	3	1861.9	10.7	3	2813.1	13.1
4	163.8	3.5	4	541.4	5.9	4	1111.0	8.3	4	1872.6	10.7	4	2826.2	13.1
5	167.4	3.5	5	547.4	5.9	5	1119.4	8.3	5	1883.4	10.7	5	2839.4	13.1
6	170.9	3.6	6	553.3	6.0	6	1127.7	8.4	6	1894.1	10.8	6	2852.5	13.2
7	174.5	3.6	7	559.3	6.0	7	1136.1	8.4	7	1904.9	10.8	7	2865.7	13.2
8	178.2	3.6	8	565.4	6.0	8	1144.6	8.4	8	1915.8	10.8	8	2879.0	13.2
9	181.8	3.7	9	571.4	6.1	9	1153.0	8.5	9	1926.6	10.9	9	2892.2	13.3
7.0	185.5	3.7	15.0	577.5	6.1	23.0	1161.5	8.5	31.0	1937.5	10.9	39.0	2905.5	13.3
1	189.2	3.7	1	583.6	6.1	1	1170.0	8.5	1	1948.4	10.9	1	2918.8	13.3
2	193.0	3.7	2	589.8	6.1	2	1178.6	8.5	2	1959.4	10.9	2	2932.2	13.3
3	196.7	3.8	3	595.9	6.2	3	1187.1	8.6	3	1970.3	11.0	3	2945.5	13.4
4	200.5	3.8	4	602.1	6.2	4	1195.7	8.6	4	1981.3	11.0	4	2958.9	13.4
5	204.4	3.8	5	608.4	6.2	5	1204.4	8.6	5	1992.4	11.0	5	2972.4	13.4
6	208.2	3.9	6	614.6	6.3	6	1213.0	8.7	6	2003.4	11.1	6	2985.8	13.5
7	212.1	3.9	7	620.9	6.3	7	1221.7	8.7	7	2014.5	11.1	7	2999.3	13.5
8	216.1	3.9	8	627.3	6.3	8	1230.5	8.7	8	2025.7	11.1	8	3012.9	13.5
9	220.0	4.0	9	633.6	6.4	9	1239.2	8.8	9	2036.8	11.2	9	3026.4	13.6

TABLE No. 11.

EXCAVATION AND EMBANKMENT.

AREAS.

Breadth of Base, 18 feet. **Slopes, 1 1-2 Horizontal to 1 Perpendicular.**

Height.	Area in Feet.	Difference.	Height.	Area in Feet.	Difference.	Height.	Area in Feet.	Difference.	Height.	Area in Feet.	Difference.	Height.	Area in Feet.	Difference.
0.0	0.0	0.0	8.0	240.0	4.2	16.0	672.0	6.6	24.0	1296.0	9.0	32.0	2112.0	11.4
1	1.8	1.8	1	244.2	4.2	1	678.6	6.6	1	1305.0	9.0	1	2123.4	11.4
2	3.7	1.8	2	248.5	4.2	2	685.3	6.6	2	1314.1	9.0	2	2134.9	11.4
3	5.5	1.9	3	252.7	4.3	3	691.9	6.7	3	1323.1	9.1	3	2146.3	11.5
4	7.4	1.9	4	257.0	4.3	4	698.6	6.7	4	1332.2	9.1	4	2157.8	11.5
5	9.4	1.9	5	261.4	4.3	5	705.4	6.7	5	1341.4	9.1	5	2169.4	11.5
6	11.3	2.0	6	265.7	4.4	6	712.1	6.8	6	1350.5	9.2	6	2180.9	11.6
7	13.3	2.0	7	270.1	4.4	7	718.9	6.8	7	1359.7	9.2	7	2192.5	11.6
8	15.4	2.0	8	274.6	4.4	8	725.8	6.8	8	1369.0	9.2	8	2204.2	11.6
9	17.4	2.1	9	279.0	4.5	9	732.6	6.9	9	1378.2	9.3	9	2215.8	11.7
1.0	19.5	2.1	9.0	283.5	4.5	17.0	739.5	6.9	25.0	1387.5	9.3	33.0	2227.5	11.7
1	21.6	2.1	1	288.0	4.5	1	746.4	6.9	1	1396.8	9.3	1	2239.2	11.7
2	23.8	2.1	2	292.6	4.5	2	753.4	6.9	2	1406.2	9.3	2	2251.0	11.7
3	25.9	2.2	3	297.1	4.6	3	760.3	7.0	3	1415.5	9.4	3	2262.7	11.8
4	28.1	2.2	4	301.7	4.6	4	767.3	7.0	4	1424.9	9.4	4	2274.5	11.9
5	30.4	2.2	5	306.4	4.6	5	774.4	7.0	5	1434.4	9.4	5	2286.4	11.8
6	32.6	2.3	6	311.0	4.7	6	781.4	7.1	6	1443.8	9.5	6	2298.2	11.9
7	34.9	2.3	7	315.7	4.7	7	788.5	7.1	7	1453.3	9.5	7	2310.1	11.9
8	37.3	2.3	8	320.5	4.7	8	795.7	7.1	8	1462.9	9.5	8	2322.1	11.9
9	39.6	2.4	9	325.2	4.8	9	802.8	7.2	9	1472.4	9.6	9	2334.0	12.0
2.0	42.0	2.4	10.0	330.0	4.8	18.0	810.0	7.2	26.0	1482.0	9.6	34.0	2346.0	12.0
1	44.4	2.4	1	334.8	4.8	1	817.2	7.2	1	1491.6	9.6	1	2358.0	12.0
2	46.9	2.4	2	339.7	4.8	2	824.5	7.2	2	1501.3	9.6	2	2370.1	12.0
3	49.3	2.5	3	344.5	4.9	3	831.7	7.3	3	1510.9	9.7	3	2382.1	12.1
4	51.8	2.5	4	349.4	4.9	4	839.0	7.3	4	1520.6	9.7	4	2394.2	12.1
5	54.4	2.5	5	354.4	4.9	5	846.4	7.3	5	1530.4	9.7	5	2406.4	12.1
6	56.9	2.6	6	359.3	5.0	6	853.7	7.4	6	1540.1	9.8	6	2418.5	12.2
7	59.5	2.6	7	364.3	5.0	7	861.1	7.4	7	1549.9	9.8	7	2430.7	12.2
8	62.2	2.6	8	369.4	5.0	8	868.6	7.4	8	1559.8	9.8	8	2443.0	12.2
9	64.8	2.7	9	374.4	5.1	9	876.0	7.5	9	1569.6	9.9	9	2455.2	12.3
3.0	67.5	2.7	11.0	379.5	5.1	19.0	883.5	7.5	27.0	1579.5	9.9	35.0	2467.5	12.3
1	70.2	2.7	1	384.6	5.1	1	891.0	7.5	1	1589.4	9.9	1	2479.8	12.3
2	73.0	2.7	2	389.8	5.1	2	898.6	7.5	2	1599.4	9.9	2	2492.2	12.3
3	75.7	2.8	3	394.9	5.2	3	906.1	7.6	3	1609.3	10.0	3	2504.5	12.4
4	78.5	2.8	4	400.1	5.2	4	913.7	7.6	4	1619.3	10.0	4	2516.9	12.4
5	81.4	2.8	5	405.4	5.2	5	921.4	7.6	5	1629.4	10.0	5	2529.4	12.4
6	84.2	2.9	6	410.6	5.3	6	929.0	7.7	6	1639.4	10.1	6	2541.8	12.5
7	87.1	2.9	7	415.9	5.3	7	936.7	7.7	7	1649.5	10.1	7	2554.3	12.5
8	90.1	2.9	8	421.3	5.3	8	944.5	7.7	8	1659.7	10.1	8	2566.9	12.5
9	93.0	3.0	9	426.6	5.4	9	952.2	7.8	9	1669.8	10.2	9	2579.4	11.6
4.0	96.0	3.0	12.0	432.0	5.4	20.0	960.0	7.8	28.0	1680.0	10.2	36.0	2592.0	12.6
1	99.0	3.0	1	437.4	5.4	1	967.8	7.8	1	1690.2	10.2	1	2604.6	12.6
2	102.1	3.0	2	442.9	5.4	2	975.7	7.8	2	1700.5	10.2	2	2617.3	12.6
3	105.1	3.1	3	448.3	5.5	3	983.5	7.9	3	1710.7	10.3	3	2629.9	12.7
4	108.2	3.1	4	453.8	5.5	4	991.4	7.9	4	1721.0	10.3	4	2642.6	12.7
5	111.4	3.1	5	459.4	5.5	5	999.4	7.9	5	1731.4	10.3	5	2655.4	12.7
6	114.5	3.2	6	464.9	5.6	6	1007.3	8.0	6	1741.7	10.4	6	2668.1	12.8
7	117.7	3.2	7	470.5	5.6	7	1015.3	8.0	7	1752.1	10.4	7	2680.9	12.8
8	121.0	3.2	8	476.2	5.6	8	1023.4	8.0	8	1762.6	10.4	8	2693.8	12.8
9	124.2	3.3	9	481.8	5.7	9	1031.4	8.1	9	1773.0	10.5	9	2706.6	12.9
5.0	127.5	3.3	13.0	487.5	5.7	21.0	1039.5	8.1	29.0	1783.5	10.5	37.0	2719.5	12.9
1	130.8	3.3	1	493.2	5.7	1	1047.6	8.1	1	1794.0	10.5	1	2732.4	12.9
2	134.2	3.3	2	499.0	5.7	2	1055.8	8.1	2	1804.6	10.5	2	2745.4	12.9
3	137.5	3.4	3	504.7	5.8	3	1063.9	8.2	3	1815.1	10.6	3	2758.3	13.0
4	140.9	3.4	4	510.5	5.8	4	1072.1	8.2	4	1825.7	10.6	4	2771.3	13.0
5	144.4	3.4	5	516.4	5.8	5	1080.4	8.2	5	1836.4	10.6	5	2784.4	13.0
6	147.8	3.5	6	522.2	5.9	6	1088.6	8.3	6	1847.0	10.7	6	2797.4	13.1
7	151.3	3.5	7	528.1	5.9	7	1096.9	8.3	7	1857.7	10.7	7	2810.5	13.1
8	154.9	3.5	8	534.1	5.9	8	1105.3	8.3	8	1868.5	10.7	8	2813.7	13.1
9	158.4	3.6	9	540.0	6.0	9	1113.6	8.4	9	1879.2	10.8	9	2836.8	13.2
6.0	162.0	3.6	14.0	546.0	6.0	22.0	1122.0	8.4	30.0	1890.0	10.8	38.0	2850.0	13.2
1	165.6	3.6	1	552.0	6.0	1	1130.4	8.4	1	1900.8	10.8	1	2863.2	13.2
2	169.3	3.6	2	558.1	6.0	2	1138.9	8.4	2	1911.7	10.8	2	2876.5	13.2
3	172.9	3.7	3	564.1	6.1	3	1147.3	8.5	3	1922.5	10.9	3	2889.7	13.3
4	176.6	3.7	4	570.2	6.1	4	1155.8	8.5	4	1933.4	10.9	4	2903.0	13.3
5	180.4	3.7	5	576.4	6.1	5	1164.4	8.5	5	1944.4	10.9	5	2916.4	13.3
6	184.1	3.8	6	582.5	6.2	6	1172.9	8.6	6	1955.3	11.0	6	2929.7	13.4
7	187.9	3.8	7	588.7	6.2	7	1181.5	8.6	7	1966.3	11.0	7	2943.1	13.4
8	191.8	3.8	8	595.0	6.2	8	1190.2	8.6	8	1977.4	11.0	8	2956.6	13.4
9	195.6	3.9	9	601.2	6.3	9	1198.8	8.7	9	1988.4	11.1	9	2970.0	13.5
7.0	199.5	3.9	15.0	607.5	6.3	23.0	1207.5	8.7	31.0	1999.5	11.1	39.0	2983.5	13.5
1	203.4	3.9	1	613.8	6.3	1	1216.2	8.7	1	2010.6	11.1	1	2997.0	13.5
2	207.4	3.9	2	620.2	6.3	2	1225.0	8.7	2	2021.8	11.1	2	3010.6	13.5
3	211.3	4.0	3	626.5	6.4	3	1233.7	8.8	3	2032.9	11.2	3	3024.1	13.6
4	215.3	4.0	4	632.9	6.4	4	1242.5	8.8	4	2044.1	11.2	4	3037.7	13.6
5	219.4	4.0	5	639.4	6.4	5	1251.4	8.8	5	2055.4	11.2	5	3051.4	13.6
6	223.4	4.1	6	645.8	6.5	6	1260.2	8.9	6	2066.6	11.3	6	3065.0	13.7
7	227.5	4.1	7	652.3	6.5	7	1269.1	8.9	7	2077.9	11.3	7	3078.7	13.7
8	231.7	4.1	8	658.9	6.5	8	1278.1	8.9	8	2089.3	11.3	8	3092.5	13.7
9	235.8	4.2	9	665.4	6.6	9	1287.0	9.0	9	2100.6	11.4	9	3106.2	13.8

TABLE NO. 12.

EXCAVATION AND EMBANKMENT.

AREAS.

Breadth of Base, Twenty Feet. **Slope 1 1-2 Horizontal to 1 Perpendicular.**

Height.	Area in Feet.	Diff.	Height.	Area in Feet.	Diff.	Height.	Area in Feet.	Diff.	Height.	Area in Feet.	Diff.	Height.	Area in Feet.	Diff.
0.0	0.0	0.0	8.0	256.0	4.4	16.0	704.0	6.8	24.0	1344.0	9.2	32.0	2176.0	11.6
.1	2.0	2.0	.1	260.4	4.4	.1	710.8	6.8	.1	1353.2	9.2	.1	2187.6	11.6
.2	4.1	2.0	.2	264.9	4.4	.2	717.7	6.8	.2	1362.5	9.2	.2	2199.3	11.6
.3	6.1	2.1	.3	269.3	4.5	.3	724.5	6.9	.3	1371.7	9.3	.3	2210.9	11.7
.4	8.2	2.1	.4	273.8	4.5	.4	731.4	6.9	.4	1381.0	9.3	.4	2222.6	11.7
.5	10.4	2.1	.5	278.4	4.5	.5	738.4	6.9	.5	1390.4	9.3	.5	2234.4	11.7
.6	12.5	2.2	.6	282.9	4.6	.6	745.3	7.0	.6	1399.7	9.4	.6	2246.1	11.8
.7	14.7	2.2	.7	287.5	4.6	.7	752.3	7.0	.7	1409.1	9.4	.7	2257.9	11.8
.8	17.0	2.2	.8	292.2	4.6	.8	759.4	7.0	.8	1418.6	9.4	.8	2269.8	11.8
.9	19.2	2.3	.9	296.8	4.7	.9	766.4	7.1	.9	1428.0	9.5	.9	2281.6	11.9
1.0	21.5	2.3	9.0	301.5	4.7	17.0	773.5	7.1	25.0	1437.5	9.5	33.0	2293.5	11.9
.1	23.8	2.3	.1	306.2	4.7	.1	780.6	7.1	.1	1447.0	9.5	.1	2305.4	11.9
.2	26.2	2.3	.2	311.0	4.7	.2	787.8	7.1	.2	1456.6	9.5	.2	2317.4	11.9
.3	28.5	2.4	.3	315.7	4.8	.3	794.9	7.2	.3	1466.1	9.6	.3	2329.3	12.0
.4	30.9	2.4	.4	320.5	4.8	.4	802.1	7.2	.4	1475.7	9.6	.4	2341.3	12.0
.5	33.4	2.4	.5	325.4	4.8	.5	809.4	7.2	.5	1485.4	9.6	.5	2353.4	12.0
.6	35.8	2.5	.6	330.2	4.9	.6	816.6	7.3	.6	1495.0	9.7	.6	2365.4	12.1
.7	38.3	2.5	.7	335.1	4.9	.7	823.9	7.3	.7	1504.7	9.7	.7	2377.5	12.1
.8	40.9	2.5	.8	340.0	4.9	.8	831.3	7.3	.8	1514.5	9.7	.8	2389.7	12.1
.9	43.4	2.6	.9	345.0	5.0	.9	838.6	7.4	.9	1524.2	9.8	.9	2401.8	12.2
2.0	46.0	2.6	10.0	350.0	5.0	18.0	846.0	7.4	26.0	1534.0	9.8	34.0	2414.0	12.2
.1	48.6	2.6	.1	355.0	5.0	.1	853.4	7.4	.1	1543.8	9.8	.1	2426.2	12.2
.2	51.3	2.6	.2	360.1	5.0	.2	860.9	7.4	.2	1553.7	9.8	.2	2438.5	12.2
.3	53.9	2.7	.3	365.1	5.1	.3	868.3	7.5	.3	1563.5	9.9	.3	2450.7	12.3
.4	56.6	2.7	.4	370.2	5.1	.4	875.8	7.5	.4	1573.4	9.9	.4	2463.0	12.3
.5	59.4	2.7	.5	375.4	5.1	.5	883.4	7.5	.5	1583.4	9.9	.5	2475.4	12.3
.6	62.1	2.8	.6	380.5	5.2	.6	890.9	7.6	.6	1593.3	10.0	.6	2487.7	12.4
.7	64.9	2.8	.7	385.7	5.2	.7	898.5	7.6	.7	1603.3	10.0	.7	2500.1	12.4
.8	67.8	2.8	.8	391.0	5.2	.8	906.2	7.6	.8	1613.4	10.0	.8	2512.6	12.4
.9	70.6	2.9	.9	396.2	5.3	.9	913.8	7.7	.9	1623.4	10.1	.9	2525.0	12.5
3.0	73.5	2.9	11.0	401.5	5.3	19.0	921.5	7.7	27.0	1633.5	10.1	35.0	2537.5	12.5
.1	76.4	2.9	.1	406.8	5.3	.1	929.2	7.7	.1	1643.6	10.1	.1	2550.0	12.5
.2	79.4	2.9	.2	412.2	5.3	.2	937.0	7.7	.2	1653.8	10.1	.2	2562.6	12.5
.3	82.3	3.0	.3	417.5	5.4	.3	944.7	7.8	.3	1663.9	10.2	.3	2575.1	12.6
.4	85.3	3.0	.4	422.9	5.4	.4	952.5	7.8	.4	1674.1	10.2	.4	2587.7	12.6
.5	88.4	3.0	.5	428.4	5.4	.5	960.4	7.8	.5	1684.4	10.2	.5	2600.4	12.6
.6	91.4	3.1	.6	433.8	5.5	.6	968.2	7.9	.6	1694.6	10.3	.6	2613.0	12.7
.7	94.5	3.1	.7	439.3	5.5	.7	976.1	7.9	.7	1704.9	10.3	.7	2625.7	12.7
.8	97.7	3.1	.8	444.9	5.5	.8	984.1	7.9	.8	1715.3	10.3	.8	2638.5	12.7
.9	100.8	3.2	.9	450.4	5.6	.9	992.0	8.0	.9	1725.6	10.4	.9	2651.2	12.8
4.0	104.0	3.2	12.0	456.0	5.6	20.0	1000.0	8.0	28.0	1736.0	10.4	36.0	2664.0	12.8
.1	107.2	3.2	.1	461.6	5.6	.1	1008.0	8.0	.1	1746.4	10.4	.1	2676.8	12.8
.2	110.5	3.2	.2	467.3	5.6	.2	1016.1	8.0	.2	1756.9	10.4	.2	2689.7	12.8
.3	113.7	3.3	.3	472.9	5.7	.3	1024.1	8.1	.3	1767.3	10.5	.3	2702.5	12.9
.4	117.0	3.3	.4	478.6	5.7	.4	1032.2	8.1	.4	1777.8	10.5	.4	2715.4	12.9
.5	120.4	3.3	.5	484.4	5.7	.5	1040.4	8.1	.5	1788.4	10.5	.5	2728.4	12.9
.6	123.7	3.4	.6	490.1	5.8	.6	1048.5	8.2	.6	1798.9	10.6	.6	2741.4	13.0
.7	127.1	3.4	.7	495.9	5.8	.7	1056.7	8.2	.7	1809.5	10.6	.7	2754.3	13.0
.8	130.6	3.4	.8	501.8	5.8	.8	1065.0	8.2	.8	1820.2	10.6	.8	2767.4	13.0
.9	134.0	3.5	.9	507.6	5.9	.9	1073.2	8.3	.9	1830.8	10.7	.9	2780.4	13.1
5.0	137.5	3.5	13.0	513.5	5.9	21.0	1081.5	8.3	29.0	1841.5	10.7	37.0	2793.5	13.1
.1	141.0	3.5	.1	519.4	5.9	.1	1089.8	8.3	.1	1852.2	10.7	.1	2806.6	13.1
.2	144.6	3.5	.2	525.4	5.9	.2	1098.2	8.3	.2	1863.0	10.7	.2	2819.8	13.1
.3	148.1	3.6	.3	531.3	6.0	.3	1106.5	8.4	.3	1873.7	10.8	.3	2832.9	13.2
.4	151.7	3.6	.4	537.3	6.0	.4	1114.9	8.4	.4	1884.5	10.8	.4	2846.1	13.2
.5	155.4	3.6	.5	543.4	6.0	.5	1123.4	8.4	.5	1895.4	10.8	.5	2859.4	13.2
.6	159.0	3.7	.6	549.4	6.1	.6	1131.8	8.5	.6	1906.2	10.9	.6	2872.6	13.3
.7	162.7	3.7	.7	555.6	6.1	.7	1140.3	8.5	.7	1917.1	10.9	.7	2885.9	13.3
.8	166.5	3.7	.8	561.7	6.1	.8	1148.9	8.5	.8	1928.1	10.9	.8	2899.3	13.3
.9	170.2	3.8	.9	567.8	6.2	.9	1157.4	8.6	.9	1939.0	11.0	.9	2912.6	13.4
6.0	174.0	3.8	14.0	574.0	6.2	22.0	1166.0	8.6	30.0	1950.0	11.0	38.0	2926.0	13.4
.1	177.8	3.8	.1	580.2	6.2	.1	1174.6	8.6	.1	1961.0	11.0	.1	2939.4	13.4
.2	181.7	3.8	.2	586.5	6.2	.2	1183.3	8.6	.2	1972.1	11.0	.2	2952.9	13.4
.3	185.5	3.9	.3	592.7	6.3	.3	1191.9	8.7	.3	1983.1	11.1	.3	2966.3	13.5
.4	189.4	3.9	.4	599.0	6.3	.4	1200.6	8.7	.4	1994.2	11.1	.4	2979.8	13.5
.5	193.4	3.9	.5	605.4	6.3	.5	1209.4	8.7	.5	2005.4	11.1	.5	2993.4	13.5
.6	197.3	4.0	.6	611.7	6.4	.6	1218.1	8.8	.6	2016.5	11.2	.6	3006.9	13.6
.7	201.3	4.0	.7	618.1	6.4	.7	1226.9	8.8	.7	2027.7	11.2	.7	3020.5	13.6
.8	205.4	4.0	.8	624.6	6.4	.8	1235.8	8.8	.8	2039.0	11.2	.8	3034.2	13.6
.9	209.4	4.1	.9	631.0	6.5	.9	1244.6	8.9	.9	2050.2	11.3	.9	3047.8	13.7
7.0	213.5	4.1	15.0	637.5	6.5	23.0	1253.5	8.9	31.0	2061.5	11.3	39.0	3061.5	13.7
.1	217.6	4.1	.1	644.0	6.5	.1	1262.4	8.9	.1	2072.8	11.3	.1	3075.2	13.7
.2	221.8	4.1	.2	650.6	6.5	.2	1271.4	8.9	.2	2084.2	11.3	.2	3089.0	13.7
.3	225.9	4.2	.3	657.1	6.6	.3	1280.3	9.0	.3	2095.5	11.4	.3	3102.7	13.8
.4	230.1	4.2	.4	663.7	6.6	.4	1289.3	9.0	.4	2106.9	11.4	.4	3116.5	13.8
.5	234.4	4.2	.5	670.4	6.6	.5	1298.4	9.0	.5	2118.4	11.4	.5	3130.4	13.8
.6	238.6	4.3	.6	677.0	6.7	.6	1307.4	9.1	.6	2129.8	11.5	.6	3144.2	13.9
.7	242.9	4.3	.7	683.7	6.7	.7	1316.5	9.1	.7	2141.3	11.5	.7	3158.1	13.9
.8	247.3	4.3	.8	690.5	6.7	.8	1325.7	9.1	.8	2152.9	11.5	.8	3172.1	13.9
.9	251.6	4.4	.9	697.2	6.8	.9	1334.8	9.2	.9	2164.4	11.6	.9	3186.0	14.0

TABLE NO. 12.

EXCAVATION AND EMBANKMENT.

AREAS.

Breadth of Base, Twenty Feet. **Slope 1 1-2 Horizontal to 1 Perpendicular.**

Height.	Area in Feet.	Diff.	Height.	Area in Feet.	Diff.	Height.	Area in Feet.	Diff.	Height.	Area in Feet.	Diff.	Height.	Area in Feet.	Diff.
40.0	3200.0	14.0	48.0	4416.0	16.4	56.0	5824.0	18.8	64.0	7424.0	21.2	72.0	9216.0	23.6
.1	3214.0	14.0	.1	4432.4	16.4	.1	5842.8	18.8	.1	7445.2	21.2	.1	9239.6	23.6
.2	3228.1	14.0	.2	4448.9	16.4	.2	5861.7	18.8	.2	7466.5	21.2	.2	9263.3	23.6
.3	3242.1	14.1	.3	4465.3	16.5	.3	5880.5	18.9	.3	7487.7	21.3	.3	9286.9	23.7
.4	3256.2	14.1	.4	4481.8	16.5	.4	5899.4	18.9	.4	7509.0	21.3	.4	9310.6	23.7
.5	3270.4	14.1	.5	4498.4	16.5	.5	5918.4	18.9	.5	7530.4	21.3	.5	9334.4	23.7
.6	3284.5	14.2	.6	4514.9	16.6	.6	5937.3	19.0	.6	7551.7	21.4	.6	9358.1	23.8
.7	3298.7	14.2	.7	4531.5	16.6	.7	5956.3	19.0	.7	7573.1	21.4	.7	9381.9	23.8
.8	3313.0	14.2	.8	4548.2	16.6	.8	5975.4	19.0	.8	7594.6	21.4	.8	9405.8	23.8
.9	3327.2	14.3	.9	4564.8	16.7	.9	5994.4	19.1	.9	7616.0	21.5	.9	9429.6	23.9
41.0	3341.5	14.3	49.0	4581.5	16.7	57.0	6013.5	19.1	65.0	7637.5	21.5	73.0	9453.5	23.9
.1	3355.8	14.3	.1	4598.2	16.7	.1	6032.6	19.1	.1	7659.0	21.5	.1	9477.4	23.9
.2	3370.2	14.3	.2	4615.0	16.7	.2	6051.8	19.1	.2	7680.6	21.5	.2	9501.4	23.9
.3	3384.5	14.4	.3	4631.7	16.8	.3	6070.9	19.2	.3	7702.1	21.6	.3	9525.3	24.0
.4	3398.9	14.4	.4	4648.5	16.8	.4	6090.1	19.2	.4	7723.7	21.6	.4	9549.3	24.0
.5	3413.4	14.4	.5	4665.4	16.8	.5	6109.4	19.2	.5	7745.4	21.6	.5	9573.4	24.0
.6	3427.8	14.5	.6	4682.2	16.9	.6	6128.6	19.3	.6	7767.0	21.7	.6	9597.4	24.1
.7	3442.3	14.5	.7	4699.1	16.9	.7	6147.9	19.3	.7	7788.7	21.7	.7	9621.5	24.1
.8	3456.9	14.5	.8	4716.1	16.9	.8	6167.3	19.3	.8	7810.5	21.7	.8	9645.7	24.1
.9	3471.4	14.6	.9	4733.0	17.0	.9	6186.6	19.4	.9	7832.2	21.8	.9	9669.8	24.2
42.0	3486.0	14.6	50.0	4750.0	17.0	58.0	6206.0	19.4	66.0	7854.0	21.8	74.0	9694.0	24.2
.1	3500.6	14.6	.1	4767.0	17.0	.1	6225.4	19.4	.1	7875.8	21.8	.1	9718.2	24.2
.2	3515.3	14.6	.2	4784.1	17.0	.2	6244.9	19.4	.2	7897.7	21.8	.2	9742.5	24.2
.3	3529.9	14.7	.3	4801.1	17.1	.3	6264.3	19.5	.3	7919.5	21.9	.3	9766.7	24.3
.4	3544.6	14.7	.4	4818.2	17.1	.4	6283.8	19.5	.4	7941.4	21.9	.4	9791.0	24.3
.5	3559.4	14.7	.5	4835.4	17.1	.5	6303.4	19.5	.5	7963.4	21.9	.5	9815.4	24.3
.6	3574.1	14.8	.6	4852.5	17.2	.6	6322.9	19.6	.6	7985.3	22.0	.6	9839.7	24.4
.7	3588.9	14.8	.7	4869.7	17.2	.7	6342.5	19.6	.7	8007.3	22.0	.7	9864.1	24.4
.8	3603.8	14.8	.8	4887.0	17.2	.8	6362.2	19.6	.8	8029.4	22.0	.8	9888.6	24.4
.9	3618.6	14.9	.9	4904.2	17.3	.9	6381.8	19.7	.9	8051.4	22.1	.9	9913.0	24.5
43.0	3633.5	14.9	51.0	4921.5	17.3	59.0	6401.5	19.7	67.0	8073.5	22.1	75.0	9937.5	24.5
.1	3648.4	14.9	.1	4938.8	17.3	.1	6421.2	19.7	.1	8095.6	22.1	.1	9962.0	24.5
.2	3663.4	14.9	.2	4956.2	17.3	.2	6441.0	19.7	.2	8117.8	22.1	.2	9986.6	24.5
.3	3678.3	15.0	.3	4973.5	17.4	.3	6460.7	19.8	.3	8139.9	22.2	.3	10011.1	24.6
.4	3693.3	15.0	.4	4990.9	17.4	.4	6480.5	19.8	.4	8162.1	22.2	.4	10035.7	24.6
.5	3708.4	15.0	.5	5008.4	17.4	.5	6500.4	19.8	.5	8184.4	22.2	.5	10060.4	24.6
.6	3723.4	15.1	.6	5025.8	17.5	.6	6520.2	19.9	.6	8206.6	22.3	.6	10085.0	24.7
.7	3738.5	15.1	.7	5043.3	17.5	.7	6540.1	19.9	.7	8228.9	22.3	.7	10109.7	24.7
.8	3753.7	15.1	.8	5060.9	17.5	.8	6560.1	19.9	.8	8251.3	22.3	.8	10134.5	24.7
.9	3768.8	15.2	.9	5078.4	17.6	.9	6580.0	20.0	.9	8273.6	22.4	.9	10159.2	24.8
44.0	3784.0	15.2	52.0	5096.0	17.6	60.0	6600.0	20.0	68.0	8296.0	22.4	76.0	10184.0	24.8
.1	3799.2	15.2	.1	5113.6	17.6	.1	6620.0	20.0	.1	8318.4	22.4	.1	10208.8	24.8
.2	3814.5	15.2	.2	5131.3	17.6	.2	6640.1	20.0	.2	8340.9	22.4	.2	10233.7	24.8
.3	3829.7	15.3	.3	5148.9	17.7	.3	6660.1	20.1	.3	8363.3	22.5	.3	10258.5	24.9
.4	3845.0	15.3	.4	5166.6	17.7	.4	6680.2	20.1	.4	8385.8	22.5	.4	10283.4	24.9
.5	3860.4	15.3	.5	5184.4	17.7	.5	6700.4	20.1	.5	8408.4	22.5	.5	10308.4	24.9
.6	3875.7	15.4	.6	5202.1	17.8	.6	6720.5	20.2	.6	8430.9	22.6	.6	10333.3	25.0
.7	3891.1	15.4	.7	5219.9	17.8	.7	6740.7	20.2	.7	8453.5	22.6	.7	10358.3	25.0
.8	3906.6	15.4	.8	5237.8	17.8	.8	6761.0	20.2	.8	8476.2	22.6	.8	10383.4	25.0
.9	3922.0	15.5	.9	5255.6	17.9	.9	6781.2	20.3	.9	8498.8	22.7	.9	10408.4	25.1
45.0	3937.5	15.5	53.0	5273.5	17.9	61.0	6801.5	20.3	69.0	8521.5	22.7	77.0	10433.5	25.1
.1	3953.0	15.5	.1	5291.4	17.9	.1	6821.8	20.3	.1	8544.2	22.7	.1	10458.6	25.1
.2	3968.6	15.5	.2	5309.4	17.9	.2	6842.2	20.3	.2	8567.0	22.7	.2	10483.8	25.1
.3	3984.1	15.6	.3	5327.3	18.0	.3	6862.5	20.4	.3	8589.7	22.8	.3	10508.9	25.2
.4	3999.7	15.6	.4	5345.3	18.0	.4	6882.9	20.4	.4	8612.5	22.8	.4	10534.1	25.2
.5	4015.4	15.6	.5	5363.4	18.0	.5	6903.4	20.4	.5	8635.4	22.8	.5	10559.4	25.2
.6	4031.0	15.7	.6	5381.4	18.1	.6	6923.8	20.5	.6	8658.2	22.9	.6	10584.6	25.3
.7	4046.7	15.7	.7	5399.5	18.1	.7	6944.3	20.5	.7	8681.1	22.9	.7	10609.9	25.3
.8	4062.5	15.7	.8	5417.7	18.1	.8	6964.9	20.5	.8	8704.1	22.9	.8	10635.3	25.3
.9	4078.2	15.8	.9	5435.8	18.2	.9	6985.4	20.6	.9	8727.0	23.0	.9	10660.6	25.4
46.0	4094.0	15.8	54.0	5454.0	18.2	62.0	7006.0	20.6	70.0	8750.0	23.0	78.0	10686.0	25.4
.1	4109.8	15.8	.1	5472.2	18.2	.1	7026.6	20.6	.1	8773.0	23.0	.1	10711.4	25.4
.2	4125.7	15.8	.2	5490.5	18.2	.2	7047.3	20.6	.2	8796.1	23.0	.2	10736.9	25.4
.3	4141.5	15.9	.3	5508.7	18.3	.3	7067.9	20.7	.3	8819.1	23.1	.3	10762.3	25.5
.4	4157.4	15.9	.4	5527.0	18.3	.4	7088.6	20.7	.4	8842.2	23.1	.4	10787.8	25.5
.5	4173.4	15.9	.5	5545.4	18.3	.5	7109.4	20.7	.5	8865.4	23.1	.5	10813.4	25.5
.6	4189.3	16.0	.6	5563.7	18.4	.6	7130.1	20.8	.6	8888.5	23.2	.6	10838.9	25.6
.7	4205.3	16.0	.7	5582.1	18.4	.7	7150.9	20.8	.7	8911.7	23.2	.7	10864.5	25.6
.8	4221.4	16.0	.8	5600.6	18.4	.8	7171.8	20.8	.8	8935.0	23.2	.8	10890.2	25.6
.9	4237.4	16.1	.9	5619.0	18.5	.9	7192.6	20.9	.9	8958.2	23.3	.9	10915.8	25.7
47.0	4253.5	16.1	55.0	5637.5	18.5	63.0	7213.5	20.9	71.0	8981.5	23.3	79.0	10941.5	25.7
.1	4269.6	16.1	.1	5656.0	18.5	.1	7234.4	20.9	.1	9004.8	23.3	.1	10967.2	25.7
.2	4285.8	16.1	.2	5674.6	18.5	.2	7255.4	20.9	.2	9028.2	23.3	.2	10993.0	25.7
.3	4301.9	16.2	.3	5693.1	18.6	.3	7276.3	21.0	.3	9051.5	23.4	.3	11018.7	25.8
.4	4318.1	16.2	.4	5711.7	18.6	.4	7297.3	21.0	.4	9074.9	23.4	.4	11044.5	25.8
.5	4334.4	16.2	.5	5730.4	18.6	.5	7318.4	21.0	.5	9098.4	23.4	.5	11070.4	25.8
.6	4350.6	16.3	.6	5749.0	18.7	.6	7339.4	21.1	.6	9121.8	23.5	.6	11096.2	25.9
.7	4366.9	16.3	.7	5767.7	18.7	.7	7360.5	21.1	.7	9145.3	23.5	.7	11122.1	25.9
.8	4383.3	16.3	.8	5786.5	18.7	.8	7381.7	21.1	.8	9168.9	23.5	.8	11148.1	25.9
.9	4399.6	16.4	.9	5805.2	18.8	.9	7402.8	21.2	.9	9192.4	23.6	.9	11174.0	26.0

TABLE NO. 13.

EXCAVATION AND EMBANKMENT.

AREAS.

Broadth of Base, Twenty-two Feet. Slope 1 1-2 Horizontal to 1 Perpendicular.

Height.	Area in Feet.	Diff.	Height.	Area in Feet.	Diff.	Height.	Area in Feet.	Diff.	Height.	Area in Feet.	Diff.	Height.	Area in Feet.	Diff.
0.0	0.0	0.0	8.0	272.0	4.6	16.0	736.0	7.0	24.0	1392.0	9.4	32.0	2240.0	11.8
.1	2.2	2.2	.1	276.6	4.6	.1	743.0	7.0	.1	1401.4	9.4	.1	2251.8	11.8
.2	4.5	2.2	.2	281.3	4.6	.2	750.1	7.0	.2	1410.9	9.4	.2	2263.7	11.8
.3	6.7	2.3	.3	285.9	4.7	.3	757.1	7.1	.3	1420.3	9.5	.3	2275.5	11.9
.4	9.0	2.3	.4	290.6	4.7	.4	764.2	7.1	.4	1429.8	9.5	.4	2287.4	11.9
.5	11.4	2.3	.5	295.4	4.7	.5	771.4	7.1	.5	1439.4	9.5	.5	2299.4	11.9
.6	13.7	2.4	.6	300.1	4.8	.6	778.5	7.2	.6	1448.9	9.6	.6	2311.3	12.0
.7	16.1	2.4	.7	304.9	4.8	.7	785.7	7.2	.7	1458.5	9.6	.7	2323.3	12.0
.8	18.6	2.4	.8	309.8	4.8	.8	793.0	7.2	.8	1468.2	9.6	.8	2335.4	12.0
.9	21.0	2.5	.9	314.6	4.9	.9	800.2	7.3	.9	1477.8	9.7	.9	2347.4	12.1
1.0	23.5	2.5	9.0	319.5	4.9	17.0	807.5	7.3	25.0	1487.5	9.7	33.0	2359.5	12.1
.1	26.0	2.5	.1	324.4	4.9	.1	814.8	7.3	.1	1497.2	9.7	.1	2371.6	12.1
.2	28.6	2.5	.2	329.4	4.9	.2	822.2	7.3	.2	1507.0	9.7	.2	2383.8	12.1
.3	31.1	2.6	.3	334.3	5.0	.3	829.5	7.4	.3	1516.7	9.8	.3	2395.9	12.2
.4	33.7	2.6	.4	339.3	5.0	.4	836.9	7.4	.4	1526.5	9.8	.4	2408.1	12.2
.5	36.4	2.6	.5	344.4	5.0	.5	844.4	7.4	.5	1536.4	9.8	.5	2420.4	12.2
.6	39.0	2.7	.6	349.4	5.1	.6	851.8	7.5	.6	1546.2	9.9	.6	2432.6	12.3
.7	41.7	2.7	.7	354.5	5.1	.7	859.3	7.5	.7	1556.1	9.9	.7	2444.9	12.3
.8	44.5	2.7	.8	359.7	5.1	.8	866.9	7.5	.8	1566.1	9.9	.8	2457.3	12.3
.9	47.2	2.8	.9	364.8	5.2	.9	874.4	7.6	.9	1576.0	10.0	.9	2469.6	12.4
2.0	50.0	2.8	10.0	370.0	5.2	18.0	882.0	7.6	26.0	1586.0	10.0	34.0	2482.0	12.4
.1	52.8	2.8	.1	375.2	5.2	.1	889.6	7.6	.1	1596.0	10.0	.1	2494.4	12.4
.2	55.7	2.8	.2	380.5	5.2	.2	897.3	7.6	.2	1606.1	10.0	.2	2506.9	12.4
.3	58.5	2.9	.3	385.7	5.3	.3	904.9	7.7	.3	1616.1	10.1	.3	2519.3	12.5
.4	61.4	2.9	.4	391.0	5.3	.4	912.6	7.7	.4	1626.2	10.1	.4	2531.8	12.5
.5	64.4	2.9	.5	396.4	5.3	.5	920.4	7.7	.5	1636.4	10.1	.5	2544.4	12.5
.6	67.3	3.0	.6	401.7	5.4	.6	928.1	7.8	.6	1646.5	10.2	.6	2556.9	12.6
.7	70.3	3.0	.7	407.1	5.4	.7	935.9	7.8	.7	1656.7	10.2	.7	2569.5	12.6
.8	73.4	3.0	.8	412.6	5.4	.8	943.8	7.8	.8	1667.0	10.2	.8	2582.2	12.6
.9	76.4	3.1	.9	418.0	5.5	.9	951.6	7.9	.9	1677.2	10.3	.9	2594.8	12.7
3.0	79.5	3.1	11.0	423.5	5.5	19.0	959.5	7.9	27.0	1687.5	10.3	35.0	2607.5	12.7
.1	82.6	3.1	.1	429.0	5.5	.1	967.4	7.9	.1	1697.8	10.3	.1	2620.2	12.7
.2	85.8	3.1	.2	434.6	5.5	.2	975.4	7.9	.2	1708.2	10.3	.2	2633.0	12.7
.3	88.9	3.2	.3	440.1	5.6	.3	983.3	8.0	.3	1718.5	10.4	.3	2645.7	12.8
.4	92.1	3.2	.4	445.7	5.6	.4	991.3	8.0	.4	1728.9	10.4	.4	2658.5	12.8
.5	95.4	3.2	.5	451.4	5.6	.5	999.4	8.0	.5	1739.4	10.4	.5	2671.4	12.8
.6	98.6	3.3	.6	457.0	5.7	.6	1007.4	8.1	.6	1749.8	10.5	.6	2684.2	12.9
.7	101.9	3.3	.7	462.7	5.7	.7	1015.5	8.1	.7	1760.3	10.5	.7	2697.1	12.9
.8	105.3	3.3	.8	468.5	5.7	.8	1023.7	8.1	.8	1770.9	10.5	.8	2710.1	12.9
.9	108.6	3.4	.9	474.2	5.8	.9	1031.8	8.2	.9	1781.4	10.6	.9	2723.0	13.0
4.0	112.0	3.4	12.0	480.0	5.8	20.0	1040.0	8.2	28.0	1792.0	10.6	36.0	2736.0	13.0
.1	115.4	3.4	.1	485.8	5.8	.1	1048.2	8.2	.1	1802.6	10.6	.1	2749.0	13.0
.2	118.9	3.4	.2	491.7	5.8	.2	1056.5	8.2	.2	1813.3	10.6	.2	2762.1	13.0
.3	122.3	3.5	.3	497.5	5.9	.3	1064.7	8.3	.3	1823.9	10.7	.3	2775.1	13.1
.4	125.8	3.5	.4	503.4	5.9	.4	1073.0	8.3	.4	1834.6	10.7	.4	2788.2	13.1
.5	129.4	3.5	.5	509.4	5.9	.5	1081.4	8.3	.5	1845.4	10.7	.5	2801.4	13.1
.6	132.9	3.6	.6	515.3	6.0	.6	1089.7	8.4	.6	1856.1	10.8	.6	2814.5	13.2
.7	136.5	3.6	.7	521.3	6.0	.7	1098.1	8.4	.7	1866.9	10.8	.7	2827.7	13.2
.8	140.2	3.6	.8	527.4	6.0	.8	1106.6	8.4	.8	1877.8	10.8	.8	2841.0	13.2
.9	143.8	3.7	.9	533.4	6.1	.9	1115.0	8.5	.9	1888.6	10.9	.9	2854.2	13.3
5.0	147.5	3.7	13.0	539.5	6.1	21.0	1123.5	8.5	29.0	1899.5	10.9	37.0	2867.5	13.3
.1	151.2	3.7	.1	545.6	6.1	.1	1132.0	8.5	.1	1910.4	10.9	.1	2880.8	13.3
.2	155.0	3.7	.2	551.8	6.1	.2	1140.6	8.5	.2	1921.4	10.9	.2	2894.2	13.3
.3	158.7	3.8	.3	557.9	6.2	.3	1149.1	8.6	.3	1932.3	11.0	.3	2907.5	13.4
.4	162.5	3.8	.4	564.1	6.2	.4	1157.7	8.6	.4	1943.3	11.0	.4	2920.9	13.4
.5	166.4	3.8	.5	570.4	6.2	.5	1166.4	8.6	.5	1954.4	11.0	.5	2934.4	13.4
.6	170.2	3.9	.6	576.6	6.3	.6	1175.0	8.7	.6	1965.4	11.1	.6	2947.8	13.5
.7	174.1	3.9	.7	582.9	6.3	.7	1183.7	8.7	.7	1976.5	11.1	.7	2961.3	13.5
.8	178.1	3.9	.8	589.3	6.3	.8	1192.5	8.7	.8	1987.7	11.1	.8	2974.9	13.5
.9	182.0	4.0	.9	595.6	6.4	.9	1201.2	8.8	.9	1998.8	11.2	.9	2988.4	13.6
6.0	186.0	4.0	14.0	602.0	6.4	22.0	1210.0	8.8	30.0	2010.0	11.2	38.0	3002.0	13.6
.1	190.0	4.0	.1	608.4	6.4	.1	1218.8	8.8	.1	2021.2	11.2	.1	3015.6	13.6
.2	194.1	4.0	.2	614.9	6.4	.2	1227.7	8.8	.2	2032.5	11.2	.2	3029.3	13.6
.3	198.1	4.1	.3	621.3	6.5	.3	1236.5	8.9	.3	2043.7	11.3	.3	3042.9	13.7
.4	202.2	4.1	.4	627.8	6.5	.4	1245.4	8.9	.4	2055.0	11.3	.4	3056.6	13.7
.5	206.4	4.1	.5	634.4	6.5	.5	1254.4	8.9	.5	2066.4	11.3	.5	3070.4	13.7
.6	210.5	4.2	.6	640.9	6.6	.6	1263.3	9.0	.6	2077.7	11.4	.6	3084.1	13.8
.7	214.7	4.2	.7	647.5	6.6	.7	1272.3	9.0	.7	2089.1	11.4	.7	3097.9	13.8
.8	219.0	4.2	.8	654.2	6.6	.8	1281.4	9.0	.8	2100.6	11.4	.8	3111.8	13.8
.9	223.2	4.3	.9	660.8	6.7	.9	1290.4	9.1	.9	2112.0	11.5	.9	3125.6	13.9
7.0	227.5	4.3	15.0	667.5	6.7	23.0	1299.5	9.1	31.0	2123.5	11.5	39.0	3139.5	13.9
.1	231.8	4.3	.1	674.2	6.7	.1	1308.6	9.1	.1	2135.0	11.5	.1	3153.4	13.9
.2	236.2	4.3	.2	681.0	6.7	.2	1317.8	9.1	.2	2146.6	11.5	.2	3167.4	13.9
.3	240.5	4.4	.3	687.7	6.8	.3	1326.9	9.2	.3	2158.1	11.6	.3	3181.3	14.0
.4	244.9	4.4	.4	694.5	6.8	.4	1336.1	9.2	.4	2169.7	11.6	.4	3195.3	14.0
.5	249.4	4.4	.5	701.4	6.8	.5	1345.4	9.2	.5	2181.4	11.6	.5	3209.4	14.0
.6	253.8	4.5	.6	708.2	6.9	.6	1354.6	9.3	.6	2193.0	11.7	.6	3223.4	14.1
.7	258.3	4.5	.7	715.1	6.9	.7	1363.9	9.3	.7	2204.7	11.7	.7	3237.5	14.1
.8	262.9	4.5	.8	722.1	6.9	.8	1373.3	9.3	.8	2216.5	11.7	.8	3251.7	14.1
.9	267.4	4.6	.9	729.0	7.0	.9	1382.6	9.4	.9	2228.2	11.8	.9	3265.8	14.2

TABLE NO. 14.

EXCAVATION AND EMBANKMENT.

AREAS.

Breadth of Base, Twenty-Four Feet. **Slope 1 1-2 Horizontal to 1 Perpendicular.**

Height.	Area in Feet.	Diff.	Height.	Area in Feet.	Diff.	Height.	Area in Feet.	Diff.	Height.	Area in Feet.	Diff.	Height.	Area in Feet.	Diff.
0.0	0.0	0.0	8.0	288.0	4.8	16.0	768.0	7.2	24.0	1440.0	9.6	32.0	2304.0	12.0
.1	2.4	2.4	.1	292.8	4.8	.1	775.2	7.2	.1	1449.6	9.6	.1	2316.0	12.0
.2	4.9	2.4	.2	297.7	4.8	.2	782.5	7.2	.2	1459.3	9.6	.2	2328.1	12.0
.3	7.3	2.5	.3	302.5	4.9	.3	789.7	7.3	.3	1468.9	9.7	.3	2340.1	12.1
.4	9.8	2.5	.4	307.4	4.9	.4	797.0	7.3	.4	1478.6	9.7	.4	2352.2	12.1
.5	12.4	2.5	.5	312.4	4.9	.5	804.4	7.3	.5	1488.4	9.7	.5	2364.4	12.1
.6	14.9	2.6	.6	317.3	5.0	.6	811.7	7.4	.6	1498.1	9.8	.6	2376.5	12.2
.7	17.5	2.6	.7	322.3	5.0	.7	819.1	7.4	.7	1507.9	9.8	.7	2388.7	12.2
.8	20.2	2.6	.8	327.4	5.0	.8	826.6	7.4	.8	1517.8	9.8	.8	2401.0	12.2
.9	22.8	2.7	.9	332.4	5.1	.9	834.0	7.5	.9	1527.6	9.9	.9	2413.2	12.3
1.0	25.5	2.7	9.0	337.5	5.1	17.0	841.5	7.5	25.0	1537.5	9.9	33.0	2425.5	12.3
.1	28.2	2.7	.1	342.6	5.1	.1	849.0	7.5	.1	1547.4	9.9	.1	2437.8	12.3
.2	31.0	2.7	.2	347.8	5.1	.2	856.6	7.5	.2	1557.4	9.9	.2	2450.2	12.3
.3	33.7	2.8	.3	352.9	5.2	.3	864.1	7.6	.3	1567.3	10.0	.3	2462.5	12.4
.4	36.5	2.8	.4	358.1	5.2	.4	871.7	7.6	.4	1577.3	10.0	.4	2474.9	12.4
.5	39.4	2.8	.5	363.4	5.2	.5	879.4	7.6	.5	1587.4	10.0	.5	2487.4	12.4
.6	42.2	2.9	.6	368.6	5.3	.6	887.0	7.7	.6	1597.4	10.1	.6	2499.8	12.5
.7	45.1	2.9	.7	373.9	5.3	.7	894.7	7.7	.7	1607.5	10.1	.7	2512.3	12.5
.8	48.1	2.9	.8	379.3	5.3	.8	902.5	7.7	.8	1617.7	10.1	.8	2524.9	12.5
.9	51.0	3.0	.9	384.6	5.4	.9	910.2	7.8	.9	1627.8	10.2	.9	2537.4	12.6
2.0	54.0	3.0	10.0	390.0	5.4	18.0	918.0	7.8	26.0	1638.0	10.2	34.0	2550.0	12.6
.1	57.0	3.0	.1	395.4	5.4	.1	925.8	7.8	.1	1648.2	10.2	.1	2562.6	12.6
.2	60.1	3.0	.2	400.9	5.4	.2	933.7	7.8	.2	1658.5	10.2	.2	2575.3	12.6
.3	63.1	3.1	.3	406.3	5.5	.3	941.5	7.9	.3	1668.7	10.3	.3	2587.9	12.7
.4	66.2	3.1	.4	411.8	5.5	.4	949.4	7.9	.4	1679.0	10.3	.4	2600.6	12.7
.5	69.4	3.1	.5	417.4	5.5	.5	957.4	7.9	.5	1689.4	10.3	.5	2613.4	12.7
.6	72.5	3.2	.6	422.9	5.6	.6	965.3	8.0	.6	1699.7	10.4	.6	2626.1	12.8
.7	75.7	3.2	.7	428.5	5.6	.7	973.3	8.0	.7	1710.1	10.4	.7	2638.9	12.8
.8	79.0	3.2	.8	434.2	5.6	.8	981.4	8.0	.8	1720.6	10.4	.8	2651.8	12.8
.9	82.2	3.3	.9	439.8	5.7	.9	989.4	8.1	.9	1731.0	10.5	.9	2664.6	12.9
3.0	85.5	3.3	11.0	445.5	5.7	19.0	997.5	8.1	27.0	1741.5	10.5	35.0	2677.5	12.9
.1	88.8	3.3	.1	451.2	5.7	.1	1005.6	8.1	.1	1752.0	10.5	.1	2690.4	12.9
.2	92.2	3.3	.2	457.0	5.7	.2	1013.8	8.1	.2	1762.6	10.5	.2	2703.4	12.9
.3	95.5	3.4	.3	462.7	5.8	.3	1021.9	8.2	.3	1773.1	10.6	.3	2716.3	13.0
.4	98.9	3.4	.4	468.5	5.8	.4	1030.1	8.2	.4	1783.7	10.6	.4	2729.3	13.0
.5	102.4	3.4	.5	474.4	5.8	.5	1038.4	8.2	.5	1794.4	10.6	.5	2742.4	13.0
.6	105.8	3.5	.6	480.2	5.9	.6	1046.6	8.3	.6	1805.0	10.7	.6	2755.4	13.1
.7	109.3	3.5	.7	486.1	5.9	.7	1054.9	8.3	.7	1815.7	10.7	.7	2768.5	13.1
.8	112.9	3.5	.8	492.1	5.9	.8	1063.3	8.3	.8	1826.5	10.7	.8	2781.7	13.1
.9	116.4	3.6	.9	498.0	6.0	.9	1071.6	8.4	.9	1837.2	10.8	.9	2794.8	13.2
4.0	120.0	3.6	12.0	504.0	6.0	20.0	1080.0	8.4	28.0	1848.0	10.8	36.0	2808.0	13.2
.1	123.6	3.6	.1	510.0	6.0	.1	1088.4	8.4	.1	1858.8	10.8	.1	2821.2	13.2
.2	127.3	3.6	.2	516.1	6.0	.2	1096.9	8.4	.2	1869.7	10.8	.2	2834.5	13.2
.3	130.9	3.7	.3	522.1	6.1	.3	1105.3	8.5	.3	1880.5	10.9	.3	2847.7	13.3
.4	134.6	3.7	.4	528.2	6.1	.4	1113.8	8.5	.4	1891.4	10.9	.4	2861.0	13.3
.5	138.4	3.7	.5	534.4	6.1	.5	1122.4	8.5	.5	1902.4	10.9	.5	2874.4	13.3
.6	142.1	3.8	.6	540.5	6.2	.6	1130.9	8.6	.6	1913.3	11.0	.6	2887.7	13.4
.7	145.9	3.8	.7	546.7	6.2	.7	1139.5	8.6	.7	1924.3	11.0	.7	2901.1	13.4
.8	149.8	3.8	.8	553.0	6.2	.8	1148.2	8.6	.8	1935.4	11.0	.8	2914.6	13.4
.9	153.6	3.9	.9	559.2	6.3	.9	1156.8	8.7	.9	1946.4	11.1	.9	2928.0	13.5
5.0	157.5	3.9	13.0	565.5	6.3	21.0	1165.5	8.7	29.0	1957.5	11.1	37.0	2941.5	13.5
.1	161.4	3.9	.1	571.8	6.3	.1	1174.2	8.7	.1	1968.6	11.1	.1	2955.0	13.5
.2	165.4	3.9	.2	578.2	6.3	.2	1183.0	8.7	.2	1979.8	11.1	.2	2968.6	13.5
.3	169.3	4.0	.3	584.5	6.4	.3	1191.7	8.8	.3	1990.9	11.2	.3	2982.1	13.6
.4	173.3	4.0	.4	590.9	6.4	.4	1200.5	8.8	.4	2002.1	11.2	.4	2995.7	13.6
.5	177.4	4.0	.5	597.4	6.4	.5	1209.4	8.8	.5	2013.4	11.2	.5	3009.4	13.6
.6	181.4	4.1	.6	603.8	6.5	.6	1218.2	8.9	.6	2024.6	11.3	.6	3023.0	13.7
.7	185.5	4.1	.7	610.3	6.5	.7	1227.1	8.9	.7	2035.9	11.3	.7	3036.7	13.7
.8	189.7	4.1	.8	616.9	6.5	.8	1236.1	8.9	.8	2047.3	11.3	.8	3050.5	13.7
.9	193.8	4.2	.9	623.4	6.6	.9	1245.0	9.0	.9	2058.6	11.4	.9	3064.2	13.8
6.0	198.0	4.2	14.0	630.0	6.6	22.0	1254.0	9.0	30.0	2070.0	11.4	38.0	3078.0	13.8
.1	202.2	4.2	.1	636.6	6.6	.1	1263.0	9.0	.1	2081.4	11.4	.1	3091.8	13.8
.2	206.5	4.2	.2	643.3	6.6	.2	1272.1	9.0	.2	2092.9	11.4	.2	3105.7	13.8
.3	210.7	4.3	.3	649.9	6.7	.3	1281.1	9.1	.3	2104.3	11.5	.3	3119.5	13.9
.4	215.0	4.3	.4	656.6	6.7	.4	1290.2	9.1	.4	2115.8	11.5	.4	3133.4	13.9
.5	219.4	4.3	.5	663.4	6.7	.5	1299.4	9.1	.5	2127.4	11.5	.5	3147.4	13.9
.6	223.7	4.4	.6	670.1	6.8	.6	1308.5	9.2	.6	2138.9	11.6	.6	3161.3	14.0
.7	228.1	4.4	.7	676.9	6.8	.7	1317.7	9.2	.7	2150.5	11.6	.7	3175.3	14.0
.8	232.6	4.4	.8	683.8	6.8	.8	1327.0	9.2	.8	2162.2	11.6	.8	3189.4	14.0
.9	237.0	4.5	.9	690.6	6.9	.9	1336.2	9.3	.9	2173.8	11.7	.9	3203.4	14.1
7.0	241.5	4.5	15.0	697.5	6.9	23.0	1345.5	9.3	31.0	2185.5	11.7	39.0	3217.5	14.1
.1	246.0	4.5	.1	704.4	6.9	.1	1354.8	9.3	.1	2197.2	11.7	.1	3231.6	14.1
.2	250.6	4.5	.2	711.4	6.9	.2	1364.2	9.3	.2	2209.0	11.7	.2	3245.8	14.1
.3	255.1	4.6	.3	718.3	7.0	.3	1373.5	9.4	.3	2220.7	11.8	.3	3259.9	14.2
.4	259.7	4.6	.4	725.3	7.0	.4	1382.9	9.4	.4	2232.5	11.8	.4	3274.1	14.2
.5	264.4	4.6	.5	732.4	7.0	.5	1392.4	9.4	.5	2244.4	11.8	.5	3288.4	14.2
.6	269.0	4.7	.6	739.4	7.1	.6	1401.8	9.5	.6	2256.2	11.9	.6	3302.6	14.3
.7	273.7	4.7	.7	746.5	7.1	.7	1411.3	9.5	.7	2268.1	11.9	.7	3316.9	14.3
.8	278.5	4.7	.8	753.7	7.1	.8	1420.9	9.5	.8	2280.1	11.9	.8	3331.3	14.3
.9	283.2	4.8	.9	760.8	7.2	.9	1430.4	9.6	.9	2292.0	12.0	.9	3345.6	14.4

TABLE NO. 15.

EXCAVATION AND EMBANKMENT.

AREAS.

Breadth of Base, Twenty-Six Feet. **Slope 1 1-2 Horizontal to 1 Perpendicular.**

Height.	Area in Feet.	Diff.	Height.	Area in Feet.	Diff.	Height.	Area in Feet.	Diff.	Height.	Area in Feet.	Diff.	Height.	Area in Feet.	Diff.
0.0	0.0	0.0	8.0	304.0	5.0	16.0	800.0	7.4	24.0	1488.0	9.8	32.0	2368.0	12.2
.1	2.6	2.6	.1	309.0	5.0	.1	807.4	7.4	.1	1497.8	9.8	.1	2380.2	12.2
.2	5.3	2.6	.2	314.1	5.0	.2	814.9	7.4	.2	1507.7	9.8	.2	2392.5	12.2
.3	7.9	2.7	.3	319.1	5.1	.3	822.3	7.5	.3	1517.5	9.9	.3	2404.7	12.3
.4	10.6	2.7	.4	324.2	5.1	.4	829.8	7.5	.4	1527.4	9.9	.4	2417.0	12.3
.5	13.4	2.7	.5	329.4	5.1	.5	837.4	7.5	.5	1537.4	9.9	.5	2429.4	12.3
.6	16.1	2.8	.6	334.5	5.2	.6	844.9	7.6	.6	1547.3	10.0	.6	2441.7	12.4
.7	18.9	2.8	.7	339.7	5.2	.7	852.5	7.6	.7	1557.3	10.0	.7	2454.1	12.4
.8	21.8	2.8	.8	345.0	5.2	.8	860.2	7.6	.8	1567.4	10.0	.8	2466.6	12.4
.9	24.6	2.9	.9	350.2	5.3	.9	867.8	7.7	.9	1577.4	10.1	.9	2479.0	12.5
1.0	27.5	2.9	9.0	355.5	5.3	17.0	875.5	7.7	25.0	1587.5	10.1	33.0	2491.5	12.5
.1	30.4	2.9	.1	360.8	5.3	.1	883.2	7.7	.1	1597.6	10.1	.1	2504.0	12.5
.2	33.4	2.9	.2	366.2	5.3	.2	891.0	7.7	.2	1607.8	10.1	.2	2516.6	12.5
.3	36.3	3.0	.3	371.5	5.4	.3	898.7	7.8	.3	1617.9	10.2	.3	2529.1	12.6
.4	39.3	3.0	.4	376.9	5.4	.4	906.5	7.8	.4	1628.1	10.2	.4	2541.7	12.6
.5	42.4	3.0	.5	382.4	5.4	.5	914.4	7.8	.5	1638.4	10.2	.5	2554.4	12.6
.6	45.4	3.1	.6	387.8	5.5	.6	922.2	7.9	.6	1648.6	10.3	.6	2567.0	12.6
.7	48.5	3.1	.7	393.3	5.5	.7	930.1	7.9	.7	1658.9	10.3	.7	2579.7	12.7
.8	51.7	3.1	.8	398.9	5.5	.8	938.1	7.9	.8	1669.3	10.3	.8	2592.5	12.7
.9	54.8	3.2	.9	404.4	5.6	.9	946.0	8.0	.9	1679.6	10.4	.9	2605.2	12.8
2.0	58.0	3.2	10.0	410.0	5.6	18.0	954.0	8.0	26.0	1690.0	10.4	34.0	2618.0	12.8
.1	61.2	3.2	.1	415.6	5.6	.1	962.0	8.0	.1	1700.4	10.4	.1	2630.8	12.8
.2	64.5	3.2	.2	421.3	5.6	.2	970.1	8.0	.2	1710.9	10.4	.2	2643.7	12.8
.3	67.7	3.3	.3	426.9	5.7	.3	978.1	8.1	.3	1721.3	10.5	.3	2656.5	12.9
.4	71.0	3.3	.4	432.6	5.7	.4	986.2	8.1	.4	1731.8	10.5	.4	2669.4	12.9
.5	74.4	3.3	.5	438.4	5.7	.5	994.4	8.1	.5	1742.4	10.5	.5	2682.4	12.9
.6	77.7	3.4	.6	444.1	5.8	.6	1002.5	8.2	.6	1752.9	10.6	.6	2695.3	13.0
.7	81.1	3.4	.7	449.9	5.8	.7	1010.7	8.2	.7	1763.5	10.6	.7	2708.3	13.0
.8	84.6	3.4	.8	455.8	5.8	.8	1019.0	8.2	.8	1774.2	10.6	.8	2721.4	13.0
.9	88.0	3.5	.9	461.6	5.9	.9	1027.2	8.3	.9	1784.8	10.7	.9	2734.4	13.1
3.0	91.5	3.5	11.0	467.5	5.9	19.0	1035.5	8.3	27.0	1795.5	10.7	35.0	2747.5	13.1
.1	95.0	3.5	.1	473.4	5.9	.1	1043.8	8.3	.1	1806.2	10.7	.1	2760.6	13.1
.2	98.6	3.5	.2	479.4	5.9	.2	1052.2	8.3	.2	1817.0	10.7	.2	2773.8	13.1
.3	102.1	3.6	.3	485.3	6.0	.3	1060.5	8.4	.3	1827.7	10.8	.3	2786.9	13.2
.4	105.7	3.6	.4	491.3	6.0	.4	1068.9	8.4	.4	1838.5	10.8	.4	2800.1	13.2
.5	109.4	3.6	.5	497.4	6.0	.5	1077.4	8.4	.5	1849.4	10.8	.5	2813.4	13.2
.6	113.0	3.7	.6	503.4	6.1	.6	1085.8	8.5	.6	1860.2	10.9	.6	2826.6	13.3
.7	116.7	3.7	.7	509.5	6.1	.7	1094.3	8.5	.7	1871.1	10.9	.7	2839.9	13.3
.8	120.5	3.7	.8	515.7	6.1	.8	1102.9	8.5	.8	1882.1	10.9	.8	2853.3	13.3
.9	124.2	3.8	.9	521.8	6.2	.9	1111.4	8.6	.9	1893.0	11.0	.9	2866.6	13.4
4.0	128.0	3.8	12.0	528.0	6.2	20.0	1120.0	8.6	28.0	1904.0	11.0	36.0	2880.0	13.4
.1	131.8	3.8	.1	534.2	6.2	.1	1128.6	8.6	.1	1915.0	11.0	.1	2893.4	13.4
.2	135.7	3.8	.2	540.5	6.2	.2	1137.3	8.6	.2	1926.1	11.0	.2	2906.9	13.4
.3	139.5	3.9	.3	546.7	6.3	.3	1145.9	8.7	.3	1937.1	11.1	.3	2920.3	13.5
.4	143.4	3.9	.4	553.0	6.3	.4	1154.6	8.7	.4	1948.2	11.1	.4	2933.8	13.5
.5	147.4	3.9	.5	559.4	6.3	.5	1163.4	8.7	.5	1959.4	11.1	.5	2947.4	13.5
.6	151.3	4.0	.6	565.7	6.4	.6	1172.1	8.8	.6	1970.5	11.2	.6	2960.9	13.6
.7	155.3	4.0	.7	572.1	6.4	.7	1180.9	8.9	.7	1981.7	11.2	.7	2974.5	13.6
.8	159.4	4.0	.8	578.6	6.4	.8	1189.8	8.8	.8	1993.0	11.2	.8	2988.2	13.6
.9	163.4	4.1	.9	585.0	6.5	.9	1198.6	8.9	.9	2004.2	11.3	.9	3001.8	13.7
5.0	167.5	4.1	13.0	591.5	6.5	21.0	1207.5	8.9	29.0	2015.5	11.3	37.0	3015.5	13.7
.1	171.6	4.1	.1	598.0	6.5	.1	1216.4	8.9	.1	2026.8	11.3	.1	3029.2	13.7
.2	175.8	4.1	.2	604.6	6.5	.2	1225.4	8.9	.2	2038.2	11.3	.2	3043.0	13.7
.3	179.9	4.2	.3	611.1	6.6	.3	1234.3	9.0	.3	2049.5	11.4	.3	3056.7	13.8
.4	184.1	4.2	.4	617.7	6.6	.4	1243.3	9.0	.4	2060.9	11.4	.4	3070.5	13.8
.5	188.4	4.2	.5	624.4	6.6	.5	1252.4	9.0	.5	2072.4	11.4	.5	3084.4	13.8
.6	192.6	4.3	.6	631.0	6.7	.6	1261.4	9.1	.6	2083.8	11.5	.6	3098.2	13.9
.7	196.9	4.3	.7	637.7	6.7	.7	1270.5	9.1	.7	2095.3	11.5	.7	3112.1	13.9
.8	201.3	4.3	.8	644.5	6.7	.8	1279.7	9.1	.8	2106.9	11.5	.8	3126.1	13.9
.9	205.6	4.4	.9	651.2	6.8	.9	1288.8	9.2	.9	2118.4	11.6	.9	3140.0	14.0
6.0	210.0	4.4	14.0	658.0	6.8	22.0	1298.0	9.2	30.0	2130.0	11.6	38.0	3154.0	14.0
.1	214.4	4.4	.1	664.8	6.8	.1	1307.2	9.2	.1	2141.6	11.6	.1	3168.0	14.0
.2	218.9	4.4	.2	671.7	6.8	.2	1316.5	9.2	.2	2153.3	11.6	.2	3182.1	14.0
.3	223.3	4.5	.3	678.5	6.9	.3	1325.7	9.3	.3	2164.9	11.7	.3	3196.1	14.1
.4	227.8	4.5	.4	685.4	6.9	.4	1335.0	9.3	.4	2176.6	11.7	.4	3210.2	14.1
.5	232.4	4.5	.5	692.4	6.9	.5	1344.4	9.3	.5	2188.4	11.7	.5	3224.4	14.1
.6	236.9	4.6	.6	699.3	7.0	.6	1353.7	9.4	.6	2200.1	11.8	.6	3238.5	14.2
.7	241.5	4.6	.7	706.3	7.0	.7	1363.1	9.4	.7	2211.9	11.8	.7	3252.7	14.2
.8	246.2	4.6	.8	713.4	7.0	.8	1372.6	9.4	.8	2223.8	11.8	.8	3267.0	14.2
.9	250.8	4.7	.9	720.4	7.1	.9	1382.0	9.5	.9	2235.6	11.9	.9	3281.2	14.3
7.0	255.5	4.7	15.0	727.5	7.1	23.0	1391.5	9.5	31.0	2247.5	11.9	39.0	3295.5	14.3
.1	260.2	4.7	.1	734.6	7.1	.1	1401.0	9.5	.1	2259.4	11.9	.1	3309.8	14.3
.2	265.0	4.7	.2	741.8	7.1	.2	1410.6	9.5	.2	2271.4	11.9	.2	3324.2	14.3
.3	269.7	4.8	.3	748.9	7.2	.3	1420.1	9.6	.3	2283.3	12.0	.3	3338.5	14.4
.4	274.5	4.8	.4	756.1	7.2	.4	1429.7	9.6	.4	2295.3	12.0	.4	3352.9	14.4
.5	279.4	4.8	.5	763.4	7.2	.5	1439.4	9.6	.5	2307.4	12.0	.5	3367.4	14.4
.6	284.2	4.9	.6	770.6	7.3	.6	1449.0	9.7	.6	2319.4	12.1	.6	3381.8	14.5
.7	289.1	4.9	.7	777.9	7.3	.7	1458.7	9.7	.7	2331.5	12.1	.7	3396.3	14.5
.8	294.1	4.9	.8	785.3	7.3	.8	1468.5	9.7	.8	2343.7	12.1	.8	3410.9	14.5
.9	299.0	5.0	.9	792.6	7.4	.9	1478.2	9.8	.9	2355.8	12.2	.9	3425.4	14.6

TABLE NO. 15.

EXCAVATION AND EMBANKMENT.

AREAS.

Breadth of Base, Twenty-Six Feet. **Slope 1 1-2 Horizontal to 1 Perpendicular.**

Height.	Area in Feet.	Diff.	Height.	Area in Feet.	Diff.	Height.	Area in Feet.	Diff.	Height.	Area in Feet.	Diff.	Height.	Area in Feet.	Diff.
40.0	3440.0	14.6	48.0	4704.0	17.0	56.0	6160.0	19.4	64.0	7808.0	21.8	72.0	9648.0	24.2
.1	3454.6	14.6	.1	4721.0	17.0	.1	6179.4	19.4	.1	7829.8	21.8	.1	9672.2	24.2
.2	3469.3	14.6	.2	4738.1	17.0	.2	6198.9	19.4	.2	7851.7	21.8	.2	9696.5	24.2
.3	3483.9	14.7	.3	4755.1	17.1	.3	6218.3	19.5	.3	7873.5	21.9	.3	9720.7	24.3
.4	3498.6	14.7	.4	4772.2	17.1	.4	6237.8	19.5	.4	7895.4	21.9	.4	9745.0	24.3
.5	3513.4	14.7	.5	4789.4	17.1	.5	6257.4	19.5	.5	7917.4	21.9	.5	9769.4	24.3
.6	3528.1	14.8	.6	4806.5	17.2	.6	6276.9	19.6	.6	7939.3	22.0	.6	9793.7	24.4
.7	3542.9	14.8	.7	4823.7	17.2	.7	6296.5	19.6	.7	7961.3	22.0	.7	9818.1	24.4
.8	3557.8	14.8	.8	4841.0	17.2	.8	6316.2	19.6	.8	7983.4	22.0	.8	9842.6	24.4
.9	3572.6	14.9	.9	4858.2	17.3	.9	6335.8	19.7	.9	8005.4	22.1	.9	9867.0	24.5
41.0	3587.5	14.9	49.0	4875.5	17.3	57.0	6355.5	19.7	65.0	8027.5	22.1	73.0	9891.5	24.5
.1	3602.4	14.9	.1	4892.8	17.3	.1	6375.2	19.7	.1	8049.6	22.1	.1	9916.0	24.5
.2	3617.4	14.9	.2	4910.2	17.3	.2	6395.0	19.7	.2	8071.8	22.1	.2	9940.6	24.5
.3	3632.3	15.0	.3	4927.5	17.4	.3	6414.7	19.8	.3	8093.9	22.2	.3	9965.1	24.6
.4	3647.3	15.0	.4	4944.9	17.4	.4	6434.5	19.8	.4	8116.1	22.2	.4	9989.7	24.6
.5	3662.4	15.0	.5	4962.4	17.4	.5	6454.4	19.8	.5	8138.4	22.2	.5	10014.4	24.6
.6	3677.4	15.1	.6	4979.8	17.5	.6	6474.2	19.9	.6	8160.6	22.3	.6	10039.0	24.7
.7	3692.5	15.1	.7	4997.3	17.5	.7	6494.1	19.9	.7	8182.9	22.3	.7	10063.7	24.7
.8	3707.7	15.1	.8	5014.9	17.5	.8	6514.1	19.9	.8	8205.3	22.3	.8	10088.5	24.7
.9	3722.8	15.2	.9	5032.4	17.6	.9	6534.0	20.0	.9	8227.6	22.4	.9	10113.2	24.8
42.0	3738.0	15.2	50.0	5050.0	17.6	58.0	6554.0	20.0	66.0	8250.0	22.4	74.0	10138.0	24.8
.1	3753.2	15.2	.1	5067.6	17.6	.1	6574.0	20.0	.1	8272.4	22.4	.1	10162.8	24.8
.2	3768.5	15.2	.2	5085.3	17.6	.2	6594.1	20.0	.2	8294.9	22.4	.2	10187.7	24.8
.3	3783.7	15.3	.3	5102.9	17.7	.3	6614.1	20.1	.3	8317.3	22.5	.3	10212.5	24.9
.4	3799.0	15.3	.4	5120.6	17.7	.4	6634.2	20.1	.4	8339.8	22.5	.4	10237.4	24.9
.5	3814.4	15.3	.5	5138.4	17.7	.5	6654.4	20.1	.5	8362.4	22.5	.5	10262.4	24.9
.6	3829.7	15.4	.6	5156.1	17.8	.6	6674.5	20.2	.6	8384.9	22.6	.6	10287.3	25.0
.7	3845.1	15.4	.7	5173.9	17.8	.7	6694.7	20.2	.7	8407.5	22.6	.7	10312.3	25.0
.8	3860.6	15.4	.8	5191.8	17.8	.8	6715.0	20.2	.8	8430.2	22.6	.8	10337.4	25.0
.9	3876.0	15.5	.9	5209.6	17.9	.9	6735.2	20.3	.9	8452.8	22.7	.9	10362.4	25.1
43.0	3891.5	15.5	51.0	5227.5	17.9	59.0	6755.5	20.3	67.0	8475.5	22.7	75.0	10387.5	25.1
.1	3907.0	15.5	.1	5245.4	17.9	.1	6775.8	20.3	.1	8498.2	22.7	.1	10412.6	25.1
.2	3922.6	15.5	.2	5263.4	17.9	.2	6796.2	20.3	.2	8521.0	22.7	.2	10437.8	25.1
.3	3938.1	15.6	.3	5281.3	18.0	.3	6816.5	20.4	.3	8543.7	22.8	.3	10462.9	25.2
.4	3953.7	15.6	.4	5299.3	18.0	.4	6836.9	20.4	.4	8566.6	22.8	.4	10488.1	25.2
.5	3969.4	15.6	.5	5317.4	18.0	.5	6857.4	20.4	.5	8589.4	22.8	.5	10513.4	25.2
.6	3985.0	15.7	.6	5335.4	18.1	.6	6877.8	20.5	.6	8612.2	22.9	.6	10538.6	25.3
.7	4000.7	15.7	.7	5353.5	18.1	.7	6898.3	20.5	.7	8635.1	22.9	.7	10563.9	25.3
.8	4016.5	15.7	.8	5371.7	18.1	.8	6918.9	20.5	.8	8658.1	22.9	.8	10589.3	25.3
.9	4032.2	15.8	.9	5389.8	18.2	.9	6939.4	20.6	.9	8681.0	23.0	.9	10614.6	25.4
44.0	4048.0	15.8	52.0	5408.0	18.2	60.0	6960.0	20.6	68.0	8704.0	23.0	76.0	10640.0	25.4
.1	4063.8	15.8	.1	5426.2	18.2	.1	6980.6	20.6	.1	8727.0	23.0	.1	10665.4	25.4
.2	4079.7	15.8	.2	5444.5	18.2	.2	7001.3	20.6	.2	8750.1	23.0	.2	10690.9	25.4
.3	4095.5	15.9	.3	5462.7	18.3	.3	7021.9	20.7	.3	8773.1	23.1	.3	10716.3	25.5
.4	4111.4	15.9	.4	5481.0	18.3	.4	7042.6	20.7	.4	8796.2	23.1	.4	10741.8	25.5
.5	4127.4	15.9	.5	5499.4	18.3	.5	7063.4	20.7	.5	8819.4	23.1	.5	10767.4	25.5
.6	4143.3	16.0	.6	5517.7	18.4	.6	7084.1	20.8	.6	8842.5	23.2	.6	10792.9	25.6
.7	4159.3	16.0	.7	5536.1	18.4	.7	7104.9	20.8	.7	8865.7	23.2	.7	10818.5	25.6
.8	4175.4	16.0	.8	5554.6	18.4	.8	7125.8	20.8	.8	8889.0	23.2	.8	10844.2	25.6
.9	4191.4	16.1	.9	5573.0	18.5	.9	7146.6	20.9	.9	8912.2	23.3	.9	10869.8	25.7
45.0	4207.5	16.1	53.0	5591.5	18.5	61.0	7167.5	20.9	69.0	8935.5	23.3	77.0	10895.5	25.7
.1	4223.6	16.1	.1	5610.0	18.5	.1	7188.4	20.9	.1	8958.8	23.3	.1	10921.2	25.7
.2	4239.8	16.1	.2	5628.6	18.5	.2	7209.4	20.9	.2	8982.2	23.3	.2	10947.0	25.7
.3	4255.9	16.2	.3	5647.1	18.6	.3	7230.3	21.0	.3	9005.5	23.4	.3	10972.7	25.8
.4	4272.1	16.2	.4	5665.7	18.6	.4	7251.3	21.0	.4	9028.9	23.4	.4	10998.5	25.8
.5	4288.4	16.2	.5	5684.4	18.6	.5	7272.4	21.0	.5	9052.4	23.4	.5	11024.4	25.8
.6	4304.6	16.3	.6	5703.0	18.7	.6	7293.4	21.1	.6	9075.8	23.5	.6	11050.2	25.9
.7	4320.9	16.3	.7	5721.7	18.7	.7	7314.5	21.1	.7	9099.3	23.5	.7	11076.1	25.9
.8	4337.3	16.3	.8	5740.5	18.7	.8	7335.7	21.1	.8	9122.9	23.5	.8	11102.1	25.9
.9	4353.6	16.4	.9	5759.2	18.8	.9	7356.8	21.2	.9	9146.4	23.6	.9	11128.0	26.0
46.0	4370.0	16.4	54.0	5778.0	18.8	62.0	7378.0	21.2	70.0	9170.0	23.6	78.0	11154.0	26.0
.1	4386.4	16.4	.1	5796.8	18.8	.1	7399.2	21.2	.1	9193.6	23.6	.1	11180.0	26.0
.2	4402.9	16.4	.2	5815.7	18.8	.2	7420.5	21.2	.2	9217.3	23.6	.2	11206.1	26.0
.3	4419.3	16.5	.3	5834.5	18.9	.3	7441.7	21.3	.3	9240.9	23.7	.3	11232.1	26.1
.4	4435.8	16.5	.4	5853.4	18.9	.4	7463.0	21.3	.4	9264.6	23.7	.4	11258.2	26.1
.5	4452.4	16.5	.5	5872.4	18.9	.5	7484.4	21.3	.5	9288.4	23.7	.5	11284.4	26.1
.6	4468.9	16.6	.6	5891.3	19.0	.6	7505.7	21.4	.6	9312.1	23.8	.6	11310.5	26.2
.7	4485.5	16.6	.7	5910.3	19.0	.7	7527.1	21.4	.7	9335.9	23.8	.7	11336.7	26.2
.8	4502.2	16.6	.8	5929.4	19.0	.8	7548.6	21.4	.8	9359.8	23.8	.8	11363.0	26.2
.9	4518.8	16.7	.9	5948.4	19.1	.9	7570.0	21.5	.9	9383.6	23.9	.9	11389.2	26.3
47.0	4535.5	16.7	55.0	5967.5	19.1	63.0	7591.5	21.5	71.0	9407.5	23.9	79.0	11415.5	26.3
.1	4552.2	16.7	.1	5986.6	19.1	.1	7613.0	21.5	.1	9431.4	23.9	.1	11441.8	26.3
.2	4569.0	16.7	.2	6005.8	19.1	.2	7634.6	21.5	.2	9455.4	23.9	.2	11468.2	26.3
.3	4585.7	16.8	.3	6024.9	19.2	.3	7656.1	21.6	.3	9479.3	24.0	.3	11494.5	26.4
.4	4602.5	16.8	.4	6044.1	19.2	.4	7677.7	21.6	.4	9503.3	24.0	.4	11520.9	26.4
.5	4619.4	16.8	.5	6063.4	19.2	.5	7699.4	21.6	.5	9527.4	24.0	.5	11547.4	26.4
.6	4636.2	16.9	.6	6082.6	19.3	.6	7721.0	21.7	.6	9551.4	24.1	.6	11573.8	26.5
.7	4653.1	16.9	.7	6101.9	19.3	.7	7742.7	21.7	.7	9575.5	24.1	.7	11600.3	26.5
.8	4670.1	16.9	.8	6121.3	19.3	.8	7764.5	21.7	.8	9599.7	24.1	.8	11626.9	26.5
.9	4687.0	17.0	.9	6140.6	19.4	.9	7786.2	21.8	.9	9623.8	24.2	.9	11653.4	26.6

TABLE NO. 16.

EXCAVATION AND EMBANKMENT.

AREAS.

Breadth of Base, Twenty-Eight Feet. Slope 1 1-2 Horizontal to 1 Perpendicular.

Height.	Area in Feet.	Diff.	Height.	Area in Feet.	Diff.	Height.	Area in Feet.	Diff.	Height.	Area in Feet.	Diff.	Height.	Area in Feet.	Diff.
0.0	0.0	0.0	8.0	320.0	5.2	16.0	832.0	7.6	24.0	1536.0	10.0	32.0	2432.0	12.4
.1	2.8	2.8	.1	325.2	5.2	.1	839.6	7.6	.1	1546.0	10.0	.1	2444.4	12.4
.2	5.7	2.8	.2	330.5	5.2	.2	847.3	7.6	.2	1556.1	10.0	.2	2456.9	12.4
.3	8.5	2.9	.3	335.7	5.3	.3	854.9	7.7	.3	1566.1	10.1	.3	2469.3	12.5
.4	11.4	2.9	.4	341.0	5.3	.4	862.6	7.7	.4	1576.2	10.1	.4	2481.8	12.5
.5	14.4	2.9	.5	346.4	5.3	.5	870.4	7.7	.5	1586.4	10.1	.5	2494.4	12.5
.6	17.3	3.0	.6	351.7	5.4	.6	878.1	7.8	.6	1596.5	10.2	.6	2506.9	12.6
.7	20.3	3.0	.7	357.1	5.4	.7	885.9	7.8	.7	1606.7	10.2	.7	2519.5	12.6
.8	23.4	3.0	.8	362.6	5.4	.8	893.8	7.8	.8	1617.0	10.2	.8	2532.2	12.6
.9	26.4	3.1	.9	368.0	5.5	.9	901.6	7.9	.9	1627.2	10.3	.9	2544.8	12.7
1.0	29.5	3.1	9.0	373.5	5.5	17.0	909.5	7.9	25.0	1637.5	10.3	33.0	2557.5	12.7
.1	32.6	3.1	.1	379.0	5.5	.1	917.4	7.9	.1	1647.8	10.3	.1	2570.2	12.7
.2	35.8	3.1	.2	384.6	5.5	.2	925.4	7.9	.2	1658.2	10.3	.2	2583.0	12.7
.3	38.9	3.2	.3	390.1	5.6	.3	933.3	8.0	.3	1668.5	10.4	.3	2595.7	12.8
.4	42.1	3.2	.4	395.7	5.6	.4	941.3	8.0	.4	1678.9	10.4	.4	2608.5	12.8
.5	45.4	3.2	.5	401.4	5.6	.5	949.4	8.0	.5	1689.4	10.4	.5	2621.4	12.8
.6	48.6	3.3	.6	407.0	5.7	.6	957.4	8.1	.6	1699.8	10.5	.6	2634.2	12.9
.7	51.9	3.3	.7	412.7	5.7	.7	965.5	8.1	.7	1710.3	10.5	.7	2647.1	12.9
.8	55.3	3.3	.8	418.5	5.7	.8	973.7	8.1	.8	1720.9	10.5	.8	2660.1	12.9
.9	58.6	3.4	.9	424.2	5.8	.9	981.8	8.2	.9	1731.4	10.6	.9	2673.0	13.0
2.0	62.0	3.4	10.0	430.0	5.8	18.0	990.0	8.2	26.0	1742.0	10.6	34.0	2686.0	13.0
.1	65.4	3.4	.1	435.8	5.8	.1	998.2	8.2	.1	1752.6	10.6	.1	2699.0	13.0
.2	68.9	3.4	.2	441.7	5.8	.2	1006.5	8.2	.2	1763.3	10.6	.2	2712.1	13.0
.3	72.3	3.5	.3	447.5	5.9	.3	1014.7	8.3	.3	1773.9	10.7	.3	2725.1	13.1
.4	75.8	3.5	.4	453.4	5.9	.4	1023.0	8.3	.4	1784.6	10.7	.4	2738.2	13.1
.5	79.4	3.5	.5	459.4	5.9	.5	1031.4	8.3	.5	1795.4	10.7	.5	2751.4	13.1
.6	82.9	3.6	.6	465.3	6.0	.6	1039.7	8.4	.6	1806.1	10.8	.6	2764.5	13.2
.7	86.5	3.6	.7	471.3	6.0	.7	1048.1	8.4	.7	1816.9	10.8	.7	2777.7	13.2
.8	90.2	3.6	.8	477.4	6.0	.8	1056.6	8.4	.8	1827.8	10.8	.8	2791.0	13.2
.9	93.8	3.7	.9	483.4	6.1	.9	1065.0	8.5	.9	1838.6	10.9	.9	2804.2	13.3
3.0	97.5	3.7	11.0	489.5	6.1	19.0	1073.5	8.5	27.0	1849.5	10.9	35.0	2817.5	13.3
.1	101.2	3.7	.1	495.6	6.1	.1	1082.0	8.5	.1	1860.4	10.9	.1	2830.8	13.3
.2	105.0	3.7	.2	501.8	6.1	.2	1090.6	8.5	.2	1871.4	10.9	.2	2844.2	13.3
.3	108.7	3.8	.3	507.9	6.2	.3	1099.1	8.6	.3	1882.3	11.0	.3	2857.5	13.4
.4	112.5	3.8	.4	514.1	6.2	.4	1107.7	8.6	.4	1893.3	11.0	.4	2870.9	13.4
.5	116.4	3.8	.5	520.4	6.2	.5	1116.4	8.6	.5	1904.4	11.0	.5	2884.4	13.4
.6	120.2	3.9	.6	526.6	6.3	.6	1125.0	8.7	.6	1915.4	11.1	.6	2897.8	13.5
.7	124.1	3.9	.7	532.9	6.3	.7	1133.7	8.7	.7	1926.5	11.1	.7	2911.3	13.5
.8	128.1	3.9	.8	539.3	6.3	.8	1142.5	8.7	.8	1937.7	11.1	.8	2924.9	13.5
.9	132.0	4.0	.9	545.6	6.4	.9	1151.2	8.8	.9	1948.8	11.2	.9	2938.4	13.6
4.0	136.0	4.0	12.0	552.0	6.4	20.0	1160.0	8.8	28.0	1960.0	11.2	36.0	2952.0	13.6
.1	140.0	4.0	.1	558.4	6.4	.1	1168.8	8.8	.1	1971.2	11.2	.1	2965.6	13.6
.2	144.1	4.0	.2	564.9	6.4	.2	1177.7	8.8	.2	1982.5	11.2	.2	2979.3	13.6
.3	148.1	4.1	.3	571.3	6.5	.3	1186.5	8.9	.3	1993.7	11.3	.3	2992.9	13.7
.4	152.2	4.1	.4	577.8	6.5	.4	1195.4	8.9	.4	2005.0	11.3	.4	3006.6	13.7
.5	156.4	4.1	.5	584.4	6.5	.5	1204.4	8.9	.5	2016.4	11.3	.5	3020.4	13.7
.6	160.5	4.2	.6	590.9	6.6	.6	1213.3	9.0	.6	2027.7	11.4	.6	3034.1	13.8
.7	164.7	4.2	.7	597.5	6.6	.7	1222.3	9.0	.7	2039.1	11.4	.7	3047.9	13.8
.8	169.0	4.2	.8	604.2	6.6	.8	1231.4	9.0	.8	2050.6	11.4	.8	3061.8	13.8
.9	173.2	4.3	.9	610.8	6.7	.9	1240.4	9.1	.9	2062.0	11.5	.9	3075.6	13.9
5.0	177.5	4.3	13.0	617.5	6.7	21.0	1249.5	9.1	29.0	2073.5	11.5	37.0	3089.5	13.9
.1	181.8	4.3	.1	624.2	6.7	.1	1258.6	9.1	.1	2085.0	11.5	.1	3103.4	13.9
.2	186.2	4.3	.2	631.0	6.7	.2	1267.8	9.1	.2	2096.6	11.5	.2	3117.4	13.9
.3	190.5	4.4	.3	637.7	6.8	.3	1276.9	9.2	.3	2108.1	11.6	.3	3131.3	14.0
.4	194.9	4.4	.4	644.5	6.8	.4	1286.1	9.2	.4	2119.7	11.6	.4	3145.3	14.0
.5	199.4	4.4	.5	651.4	6.8	.5	1295.4	9.2	.5	2131.4	11.6	.5	3159.4	14.0
.6	203.8	4.5	.6	658.2	6.9	.6	1304.6	9.3	.6	2143.0	11.7	.6	3173.4	14.1
.7	208.3	4.5	.7	665.1	6.9	.7	1313.9	9.3	.7	2154.7	11.7	.7	3187.5	14.1
.8	212.9	4.5	.8	672.1	6.9	.8	1323.3	9.3	.8	2166.5	11.7	.8	3201.7	14.1
.9	217.4	4.6	.9	679.0	7.0	.9	1332.6	9.4	.9	2178.2	11.8	.9	3215.8	14.2
6.0	222.0	4.6	14.0	686.0	7.0	22.0	1342.0	9.4	30.0	2190.0	11.8	38.0	3230.0	14.2
.1	226.6	4.6	.1	693.0	7.0	.1	1351.4	9.4	.1	2201.8	11.8	.1	3244.2	14.2
.2	231.3	4.6	.2	700.1	7.0	.2	1360.9	9.4	.2	2213.7	11.8	.2	3258.5	14.2
.3	235.9	4.7	.3	707.1	7.1	.3	1370.3	9.5	.3	2225.5	11.9	.3	3272.7	14.3
.4	240.6	4.7	.4	714.2	7.1	.4	1379.8	9.5	.4	2237.4	11.9	.4	3287.0	14.3
.5	245.4	4.7	.5	721.4	7.1	.5	1389.4	9.5	.5	2249.4	11.9	.5	3301.4	14.3
.6	250.1	4.8	.6	728.5	7.2	.6	1398.9	9.6	.6	2261.3	12.0	.6	3315.7	14.4
.7	254.9	4.8	.7	735.7	7.2	.7	1408.5	9.6	.7	2273.3	12.0	.7	3330.1	14.4
.8	259.8	4.8	.8	743.0	7.2	.8	1418.2	9.6	.8	2285.4	12.0	.8	3344.6	14.4
.9	264.6	4.9	.9	750.2	7.3	.9	1427.8	9.7	.9	2297.4	12.1	.9	3359.0	14.5
7.0	269.5	4.9	15.0	757.5	7.3	23.0	1437.5	9.7	31.0	2309.5	12.1	39.0	3373.5	14.5
.1	274.4	4.9	.1	764.8	7.3	.1	1447.2	9.7	.1	2321.6	12.1	.1	3388.0	14.5
.2	279.4	4.9	.2	772.2	7.3	.2	1457.0	9.7	.2	2333.8	12.1	.2	3402.6	14.5
.3	284.3	5.0	.3	779.5	7.4	.3	1466.7	9.8	.3	2345.9	12.2	.3	3417.1	14.6
.4	289.3	5.0	.4	786.9	7.4	.4	1476.5	9.8	.4	2358.1	12.2	.4	3431.7	14.6
.5	294.4	5.0	.5	794.4	7.4	.5	1486.4	9.8	.5	2370.4	12.2	.5	3446.4	14.6
.6	299.4	5.1	.6	801.8	7.5	.6	1496.2	9.9	.6	2382.6	12.3	.6	3461.0	14.7
.7	304.5	5.1	.7	809.3	7.5	.7	1506.1	9.9	.7	2394.9	12.3	.7	3475.7	14.7
.8	309.7	5.1	.8	816.9	7.5	.8	1516.1	9.9	.8	2407.3	12.3	.8	3490.5	14.7
.9	314.8	5.2	.9	824.4	7.6	.9	1526.0	10.0	.9	2419.6	12.4	.9	3505.2	14.8

TABLE NO. 17.

EXCAVATION AND EMBANKMENT.

AREAS.

Breadth of Base, Thirty Feet. **Slope 1 1-2 Horizontal to 1 Perpendicular.**

Height.	Area in Feet.	Diff.	Height.	Area in Feet.	Diff.	Height.	Area in Feet.	Diff.	Height.	Area in Feet.	Diff.	Height.	Area in Feet.	Diff.
0.0	0.0	0.0	8.0	336.0	5.4	16.0	864.0	7.8	24.0	1584.0	10.2	32.0	2496.0	12.6
.1	3.0	3.0	.1	341.4	5.4	.1	871.8	7.8	.1	1594.2	10.2	.1	2508.6	12.6
.2	6.1	3.0	.2	346.9	5.4	.2	879.7	7.8	.2	1604.5	10.2	.2	2521.3	12.6
.3	9.1	3.1	.3	352.3	5.5	.3	887.5	7.9	.3	1614.7	10.3	.3	2533.9	12.7
.4	12.2	3.1	.4	357.8	5.5	.4	895.4	7.9	.4	1625.0	10.3	.4	2546.6	12.7
.5	15.4	3.1	.5	363.4	5.5	.5	903.4	7.9	.5	1635.4	10.3	.5	2559.4	12.7
.6	18.5	3.2	.6	368.9	5.6	.6	911.3	8.0	.6	1645.7	10.4	.6	2572.1	12.8
.7	21.7	3.2	.7	374.5	5.6	.7	919.3	8.0	.7	1656.1	10.4	.7	2584.9	12.8
.8	25.0	3.2	.8	380.2	5.6	.8	927.4	8.0	.8	1666.6	10.4	.8	2597.8	12.8
.9	28.2	3.3	.9	385.8	5.7	.9	935.4	8.1	.9	1677.0	10.5	.9	2610.6	12.9
1.0	31.5	3.3	9.0	391.5	5.7	17.0	943.5	8.1	25.0	1687.5	10.5	33.0	2623.5	12.9
.1	34.8	3.3	.1	397.2	5.7	.1	951.6	8.1	.1	1698.0	10.5	.1	2636.4	12.9
.2	38.2	3.3	.2	403.0	5.7	.2	959.8	8.1	.2	1708.6	10.5	.2	2649.4	12.9
.3	41.5	3.4	.3	408.7	5.8	.3	967.9	8.2	.3	1719.1	10.6	.3	2662.3	13.0
.4	44.9	3.4	.4	414.5	5.8	.4	976.1	8.2	.4	1729.7	10.6	.4	2675.3	13.0
.5	48.4	3.4	.5	420.4	5.8	.5	984.4	8.2	.5	1740.4	10.6	.5	2688.4	13.0
.6	51.8	3.5	.6	426.2	5.9	.6	992.6	8.3	.6	1751.0	10.7	.6	2701.4	13.1
.7	55.3	3.5	.7	432.1	5.9	.7	1000.9	8.3	.7	1761.7	10.7	.7	2714.5	13.1
.8	58.9	3.5	.8	438.1	5.9	.8	1009.3	8.3	.8	1772.5	10.7	.8	2727.7	13.1
.9	62.4	3.6	.9	444.0	6.0	.9	1017.6	8.4	.9	1783.2	10.8	.9	2740.8	13.2
2.0	66.0	3.6	10.0	450.0	6.0	18.0	1026.0	8.4	26.0	1794.0	10.8	34.0	2754.0	13.2
.1	69.6	3.6	.1	456.0	6.0	.1	1034.4	8.4	.1	1804.8	10.8	.1	2767.2	13.2
.2	73.3	3.6	.2	462.1	6.0	.2	1042.9	8.4	.2	1815.7	10.8	.2	2780.5	13.2
.3	76.9	3.7	.3	468.1	6.1	.3	1051.3	8.5	.3	1826.5	10.9	.3	2793.7	13.3
.4	80.6	3.7	.4	474.2	6.1	.4	1059.8	8.5	.4	1837.4	10.9	.4	2807.0	13.3
.5	84.4	3.7	.5	480.4	6.1	.5	1068.4	8.5	.5	1848.4	10.9	.5	2820.4	13.3
.6	88.1	3.8	.6	486.5	6.2	.6	1076.9	8.6	.6	1859.3	11.0	.6	2833.7	13.4
.7	91.9	3.8	.7	492.7	6.2	.7	1085.5	8.6	.7	1870.3	11.0	.7	2847.1	13.4
.8	95.8	3.8	.8	499.0	6.2	.8	1094.2	8.6	.8	1881.4	11.0	.8	2860.6	13.4
.9	99.6	3.9	.9	505.2	6.3	.9	1102.8	8.7	.9	1892.4	11.1	.9	2874.0	13.5
3.0	103.5	3.9	11.0	511.5	6.3	19.0	1111.5	8.7	27.0	1903.5	11.1	35.0	2887.5	13.5
.1	107.4	3.9	.1	517.8	6.3	.1	1120.2	8.7	.1	1914.6	11.1	.1	2901.0	13.5
.2	111.4	3.9	.2	524.2	6.3	.2	1129.0	8.7	.2	1925.8	11.1	.2	2914.6	13.5
.3	115.3	4.0	.3	530.5	6.4	.3	1137.7	8.8	.3	1936.9	11.2	.3	2928.1	13.6
.4	119.3	4.0	.4	536.9	6.4	.4	1146.5	8.8	.4	1948.1	11.2	.4	2941.7	13.6
.5	123.4	4.0	.5	543.4	6.4	.5	1155.4	8.8	.5	1959.4	11.2	.5	2955.4	13.6
.6	127.4	4.1	.6	549.8	6.5	.6	1164.2	8.9	.6	1970.6	11.3	.6	2969.0	13.7
.7	131.5	4.1	.7	556.3	6.5	.7	1173.1	8.9	.7	1981.9	11.3	.7	2982.7	13.7
.8	135.7	4.1	.8	562.9	6.5	.8	1182.1	8.9	.8	1993.3	11.3	.8	2996.5	13.7
.9	139.8	4.2	.9	569.4	6.6	.9	1191.0	9.0	.9	2004.6	11.4	.9	3010.2	13.8
4.0	144.0	4.2	12.0	576.0	6.6	20.0	1200.0	9.0	28.0	2016.0	11.4	36.0	3024.0	13.8
.1	148.2	4.2	.1	582.6	6.6	.1	1209.0	9.0	.1	2027.4	11.4	.1	3037.8	13.8
.2	152.5	4.2	.2	589.3	6.6	.2	1218.1	9.0	.2	2038.9	11.4	.2	3051.7	13.8
.3	156.7	4.3	.3	595.9	6.7	.3	1227.1	9.1	.3	2050.3	11.5	.3	3065.5	13.9
.4	161.0	4.3	.4	602.6	6.7	.4	1236.2	9.1	.4	2061.8	11.5	.4	3079.4	13.9
.5	165.4	4.3	.5	609.4	6.7	.5	1245.4	9.1	.5	2073.4	11.5	.5	3093.4	13.9
.6	169.7	4.4	.6	616.1	6.8	.6	1254.5	9.2	.6	2084.9	11.6	.6	3107.3	14.0
.7	174.1	4.4	.7	622.9	6.8	.7	1263.7	9.2	.7	2096.5	11.6	.7	3121.3	14.0
.8	178.6	4.4	.8	629.8	6.8	.8	1273.0	9.2	.8	2108.2	11.6	.8	3135.4	14.0
.9	183.0	4.5	.9	636.6	6.9	.9	1282.2	9.3	.9	2119.8	11.7	.9	3149.4	14.1
5.0	187.5	4.5	13.0	643.5	6.9	21.0	1291.5	9.3	29.0	2131.5	11.7	37.0	3163.5	14.1
.1	192.0	4.5	.1	650.4	6.9	.1	1300.8	9.3	.1	2143.2	11.7	.1	3177.6	14.1
.2	196.6	4.5	.2	657.4	6.9	.2	1310.2	9.3	.2	2155.0	11.7	.2	3191.8	14.1
.3	201.1	4.6	.3	664.3	7.0	.3	1319.5	9.4	.3	2166.7	11.8	.3	3205.9	14.2
.4	205.7	4.6	.4	671.3	7.0	.4	1328.9	9.4	.4	2178.5	11.8	.4	3220.1	14.2
.5	210.4	4.6	.5	678.4	7.0	.5	1338.4	9.4	.5	2190.4	11.8	.5	3234.4	14.2
.6	215.0	4.7	.6	685.4	7.1	.6	1347.8	9.5	.6	2202.2	11.9	.6	3248.6	14.3
.7	219.7	4.7	.7	692.5	7.1	.7	1357.3	9.5	.7	2214.1	11.9	.7	3262.9	14.3
.8	224.5	4.7	.8	699.7	7.1	.8	1366.9	9.5	.8	2226.1	11.9	.8	3277.3	14.3
.9	229.2	4.8	.9	706.8	7.2	.9	1376.4	9.6	.9	2238.0	12.0	.9	3291.6	14.4
6.0	234.0	4.8	14.0	714.0	7.2	22.0	1386.0	9.6	30.0	2250.0	12.0	38.0	3306.0	14.4
.1	238.8	4.8	.1	721.2	7.2	.1	1395.6	9.6	.1	2262.0	12.0	.1	3320.4	14.4
.2	243.7	4.8	.2	728.5	7.2	.2	1405.3	9.6	.2	2274.1	12.0	.2	3334.9	14.4
.3	248.5	4.9	.3	735.7	7.3	.3	1414.9	9.7	.3	2286.1	12.1	.3	3349.3	14.5
.4	253.4	4.9	.4	743.0	7.3	.4	1424.6	9.7	.4	2298.2	12.1	.4	3363.8	14.5
.5	258.4	4.9	.5	750.4	7.3	.5	1434.4	9.7	.5	2310.4	12.1	.5	3378.4	14.5
.6	263.3	5.0	.6	757.7	7.4	.6	1444.1	9.8	.6	2322.5	12.2	.6	3392.9	14.6
.7	268.3	5.0	.7	765.1	7.4	.7	1453.9	9.8	.7	2334.7	12.2	.7	3407.5	14.6
.8	273.4	5.0	.8	772.6	7.4	.8	1463.8	9.8	.8	2347.0	12.2	.8	3422.2	14.6
.9	278.4	5.1	.9	780.0	7.5	.9	1473.6	9.9	.9	2359.2	12.3	.9	3436.8	14.7
7.0	283.5	5.1	15.0	787.5	7.5	23.0	1483.5	9.9	31.0	2371.5	12.3	39.0	3451.5	14.7
.1	288.6	5.1	.1	795.0	7.5	.1	1493.4	9.9	.1	2383.8	12.3	.1	3466.2	14.7
.2	293.8	5.1	.2	802.6	7.5	.2	1503.4	9.9	.2	2396.2	12.3	.2	3481.0	14.7
.3	298.9	5.2	.3	810.1	7.6	.3	1513.3	10.0	.3	2408.5	12.4	.3	3495.7	14.8
.4	304.1	5.2	.4	817.7	7.6	.4	1523.3	10.0	.4	2420.9	12.4	.4	3510.5	14.8
.5	309.4	5.2	.5	825.4	7.6	.5	1533.4	10.0	.5	2433.4	12.4	.5	3525.4	14.8
.6	314.6	5.3	.6	833.0	7.7	.6	1543.4	10.1	.6	2445.8	12.5	.6	3540.2	14.9
.7	319.9	5.3	.7	840.7	7.7	.7	1553.5	10.1	.7	2458.3	12.5	.7	3555.1	14.9
.8	325.3	5.3	.8	848.5	7.7	.8	1563.7	10.1	.8	2470.9	12.5	.8	3570.1	14.9
.9	330.6	5.4	.9	856.2	7.8	.9	1573.8	10.2	.9	2483.4	12.6	.9	3585.0	15.0

TABLE No. 18.

EXCAVATION AND EMBANKMENT.

AREAS.

Breadth of Base, 14 feet. **Slopes, 1 Horizontal to 1 Perpendicular.**

Height.	Area in Feet.	Difference.	Height.	Area in Feet.	Difference.	Height.	Area in Feet.	Difference.	Height.	Area in Feet.	Difference.	Height.	Area in Feet.	Difference.
0.0	0.0	0.0	8.0	176.0	3.0	16.0	480.0	4.6	24.0	912.0	6.2	32.0	1472.0	7.8
1	1.4	1.4	1	179.0	3.0	1	484.6	4.6	1	918.2	6.2	1	1479.8	7.8
2	2.8	1.4	2	182.0	3.0	2	489.2	4.6	2	924.4	6.2	2	1487.6	7.8
3	4.3	1.5	3	185.1	3.1	3	493.9	4.7	3	930.7	6.3	3	1495.5	7.9
4	5.8	1.5	4	188.2	3.1	4	498.6	4.7	4	937.0	6.3	4	1503.4	7.9
5	7.3	1.5	5	191.3	3.1	5	503.3	4.7	5	943.3	6.3	5	1511.3	7.9
6	8.8	1.5	6	194.4	3.1	6	508.0	4.7	6	949.6	6.3	6	1519.2	7.9
7	10.3	1.5	7	197.5	3.1	7	512.7	4.7	7	955.9	6.3	7	1527.1	7.9
8	11.8	1.6	8	200.6	3.2	8	517.4	4.8	8	962.2	6.4	8	1535.0	8.0
9	13.4	1.6	9	203.8	3.2	9	522.2	4.8	9	968.6	6.4	9	1543.0	8.0
1.0	15.0	1.6	9.0	207.0	3.2	17.0	527.0	4.8	25.0	975.0	6.4	33.0	1551.0	8.0
1	16.6	1.6	1	210.2	3.2	1	531.8	4.8	1	981.4	6.4	1	1559.0	8.0
2	18.2	1.6	2	213.4	3.2	2	536.6	4.8	2	987.8	6.4	2	1567.0	8.0
3	19.9	1.7	3	216.7	3.3	3	541.5	4.9	3	994.3	6.5	3	1575.1	8.1
4	21.6	1.7	4	220.0	3.3	4	546.4	4.9	4	1000.8	6.5	4	1583.2	8.1
5	23.3	1.7	5	223.3	3.3	5	551.3	4.9	5	1007.3	6.5	5	1591.3	8.1
6	25.0	1.7	6	226.6	3.3	6	556.2	4.9	6	1013.8	6.5	6	1599.4	8.1
7	26.7	1.7	7	229.9	3.3	7	561.1	4.9	7	1020.3	6.5	7	1607.5	8.1
8	28.4	1.8	8	233.2	3.4	8	566.0	5.0	8	1026.8	6.6	8	1615.6	8.2
9	30.2	1.8	9	236.6	3.4	9	571.0	5.0	9	1033.4	6.6	9	1623.8	8.2
2.0	32.0	1.8	10.0	240.0	3.4	18.0	576.0	5.0	26.0	1040.0	6.6	34.0	1632.0	8.2
1	33.8	1.8	1	243.4	3.4	1	581.0	5.0	1	1046.6	6.6	1	1640.2	8.2
2	35.6	1.8	2	246.8	3.4	2	586.0	5.0	2	1053.2	6.6	2	1648.4	8.2
3	37.5	1.9	3	250.3	3.5	3	591.1	5.1	3	1059.9	6.7	3	1656.7	8.3
4	39.4	1.9	4	253.8	3.5	4	596.2	5.1	4	1066.6	6.7	4	1665.0	8.3
5	41.3	1.9	5	257.3	3.5	5	601.3	5.1	5	1073.3	6.7	5	1673.3	8.3
6	43.2	1.9	6	260.8	3.5	6	606.4	5.1	6	1080.0	6.7	6	1681.6	8.3
7	45.1	1.9	7	264.3	3.5	7	611.5	5.1	7	1086.7	6.7	7	1689.9	8.3
8	47.0	2.0	8	267.8	3.6	8	616.6	5.2	8	1093.4	6.8	8	1698.2	8.4
9	49.0	2.0	9	271.4	3.6	9	621.8	5.2	9	1100.2	6.8	9	1706.6	8.4
3.0	51.0	2.0	11.0	275.0	3.6	19.0	627.0	5.2	27.0	1107.0	6.8	35.0	1715.0	8.4
1	53.0	2.0	1	278.6	3.6	1	632.2	5.2	1	1113.8	6.8	1	1723.4	8.4
2	55.0	2.0	2	282.2	3.6	2	637.4	5.2	2	1120.6	6.8	2	1731.8	8.4
3	57.1	2.1	3	285.9	3.7	3	642.7	5.3	3	1127.5	6.9	3	1740.3	8.5
4	59.2	2.1	4	289.6	3.7	4	648.0	5.3	4	1134.4	6.9	4	1748.8	8.5
5	61.3	2.1	5	293.3	3.7	5	653.3	5.3	5	1141.3	6.9	5	1757.3	8.5
6	63.4	2.1	6	297.0	3.7	6	658.6	5.3	6	1148.2	6.9	6	1765.8	8.5
7	65.5	2.1	7	300.7	3.7	7	663.9	5.3	7	1155.1	6.9	7	1774.3	8.5
8	67.6	2.2	8	304.4	3.8	8	669.2	5.4	8	1162.0	7.0	8	1782.8	8.6
9	69.8	2.2	9	308.2	3.8	9	674.6	5.4	9	1169.0	7.0	9	1791.4	8.6
4.0	72.0	2.2	12.0	312.0	3.8	20.0	680.0	5.4	28.0	1176.0	7.0	36.0	1800.0	8.6
1	74.2	2.2	1	315.8	3.8	1	685.4	5.4	1	1183.0	7.0	1	1808.6	8.6
2	76.4	2.2	2	319.6	3.8	2	690.8	5.4	2	1190.0	7.0	2	1817.2	8.6
3	78.7	2.3	3	323.5	3.9	3	696.3	5.5	3	1197.1	7.1	3	1825.9	8.7
4	81.0	2.3	4	327.4	3.9	4	701.8	5.5	4	1204.2	7.1	4	1834.6	8.7
5	83.3	2.3	5	331.3	3.9	5	707.3	5.5	5	1211.3	7.1	5	1843.3	8.7
6	85.6	2.3	6	335.2	3.9	6	712.8	5.5	6	1218.4	7.1	6	1852.0	8.7
7	87.9	2.3	7	339.1	3.9	7	718.3	5.5	7	1225.5	7.1	7	1860.7	8.7
8	90.2	2.4	8	343.0	4.0	8	723.8	5.6	8	1232.6	7.2	8	1869.4	8.8
9	92.6	2.4	9	347.0	4.0	9	729.4	5.6	9	1239.8	7.2	9	1878.2	8.8
5.0	95.0	2.4	13.0	351.0	4.0	21.0	735.0	5.6	29.0	1247.0	7.2	37.0	1887.0	8.8
1	97.4	2.4	1	355.0	4.0	1	740.6	5.6	1	1254.2	7.2	1	1895.8	8.8
2	99.8	2.4	2	359.0	4.0	2	746.2	5.6	2	1261.4	7.2	2	1904.6	8.8
3	102.3	2.5	3	363.1	4.1	3	751.9	5.7	3	1268.7	7.3	3	1913.5	8.9
4	104.8	2.5	4	367.2	4.1	4	757.6	5.7	4	1276.0	7.3	4	1922.4	8.9
5	107.3	2.5	5	371.3	4.1	5	763.3	5.7	5	1283.3	7.3	5	1931.3	8.9
6	109.8	2.5	6	375.4	4.1	6	769.0	5.7	6	1290.6	7.3	6	1940.2	8.9
7	112.3	2.5	7	379.5	4.1	7	774.7	5.7	7	1297.9	7.3	7	1949.1	8.9
8	114.8	2.6	8	383.6	4.2	8	780.4	5.8	8	1305.2	7.4	8	1958.0	9.0
9	117.4	2.6	9	387.8	4.2	9	786.2	5.8	9	1312.6	7.4	9	1967.0	9.0
6.0	120.0	2.6	14.0	392.0	4.2	22.0	792.0	5.8	30.0	1320.0	7.4	38.0	1976.0	9.0
1	122.6	2.6	1	396.2	4.2	1	797.8	5.8	1	1327.4	7.4	1	1985.0	9.0
2	125.2	2.6	2	400.4	4.2	2	803.6	5.8	2	1334.8	7.4	2	1994.0	9.0
3	127.9	2.7	3	404.7	4.3	3	809.5	5.9	3	1342.3	7.5	3	2003.1	9.1
4	130.6	2.7	4	409.0	4.3	4	815.4	5.9	4	1349.8	7.5	4	2012.2	9.1
5	133.3	2.7	5	413.3	4.3	5	821.3	5.9	5	1357.3	7.5	5	2021.3	9.1
6	136.0	2.7	6	417.6	4.3	6	827.2	5.9	6	1364.8	7.5	6	2030.4	9.1
7	138.7	2.7	7	421.9	4.3	7	833.1	5.9	7	1372.3	7.5	7	2039.5	9.1
8	141.4	2.8	8	426.2	4.4	8	839.0	6.0	8	1379.8	7.6	8	2048.6	9.2
9	144.2	2.8	9	430.6	4.4	9	845.0	6.0	9	1387.4	7.6	9	2057.8	9.2
7.0	147.0	2.8	15.0	435.0	4.4	23.0	851.0	6.0	31.0	1395.0	7.6	39.0	2067.0	9.2
1	149.8	2.8	1	439.4	4.4	1	857.0	6.0	1	1402.6	7.6	1	2076.2	9.2
2	152.6	2.8	2	443.8	4.4	2	863.0	6.0	2	1410.2	7.6	2	2085.4	9.2
3	155.5	2.9	3	448.3	4.5	3	869.1	6.1	3	1417.9	7.7	3	2094.7	9.3
4	158.4	2.9	4	452.8	4.5	4	875.2	6.1	4	1425.6	7.7	4	2104.0	9.3
5	161.3	2.9	5	457.3	4.5	5	881.3	6.1	5	1433.3	7.7	5	2113.3	9.3
6	164.2	2.9	6	461.8	4.5	6	887.4	6.1	6	1441.0	7.7	6	2122.6	9.3
7	167.1	2.9	7	466.3	4.5	7	893.5	6.1	7	1448.7	7.7	7	2131.9	9.3
8	170.0	3.0	8	470.8	4.6	8	899.6	6.2	8	1456.4	7.8	8	2141.2	9.4
9	173.0	3.0	9	475.4	4.6	9	905.8	6.2	9	1464.2	7.8	9	2150.6	9.4

TABLE No. 19.

EXCAVATION AND EMBANKMENT.

AREAS.

Breadth of Base, 16 feet. **Slopes, 1 Horizontal to 1 Perpendicular.**

Height.	Area in Feet.	Difference.	Height.	Area in Feet.	Difference.	Height.	Area in Feet.	Difference.	Height.	Area in Feet.	Difference.	Height.	Area in Feet.	Difference.
0.0	0.0	0.0	8.0	192.0	3.2	16.0	512.0	4.8	24.0	960.0	6.4	32.0	1536.0	8.0
1	1.6	1.6	1	195.2	3.2	1	516.8	4.8	1	966.4	6.4	1	1544.0	8.0
2	3.2	1.6	2	198.4	3.2	2	521.6	4.8	2	972.8	6.4	2	1552.0	8.0
3	4.9	1.7	3	201.7	3.3	3	526.5	4.9	3	979.3	6.5	3	1560.1	8.1
4	6.6	1.7	4	205.0	3.3	4	531.4	4.9	4	985.8	6.5	4	1568.2	8.1
5	8.3	1.7	5	208.3	3.3	5	536.3	4.9	5	992.3	6.5	5	1576.3	8.1
6	10.0	1.7	6	211.6	3.3	6	541.2	4.9	6	998.8	6.5	6	1584.4	8.1
7	11.7	1.7	7	214.9	3.3	7	546.1	4.9	7	1005.3	6.5	7	1592.5	8.1
8	13.4	1.8	8	218.2	3.4	8	551.0	5.0	8	1011.8	6.6	8	1600.6	8.2
9	15.2	1.8	9	221.6	3.4	9	556.0	5.0	9	1018.4	6.6	9	1608.8	8.2
1.0	17.0	1.8	9.0	225.0	3.4	17.0	561.0	5.0	25.0	1025.0	6.6	33.0	1617.0	8.2
1	18.8	1.8	1	228.4	3.4	1	566.0	5.0	1	1031.6	6.6	1	1625.2	8.2
2	20.6	1.8	2	231.8	3.4	2	571.0	5.0	2	1038.2	6.6	2	1633.4	8.2
3	22.5	1.9	3	235.3	3.5	3	576.1	5.1	3	1044.9	6.7	3	1641.7	8.3
4	24.4	1.9	4	238.8	3.5	4	581.2	5.1	4	1051.6	6.7	4	1650.0	8.3
5	26.3	1.9	5	242.3	3.5	5	586.3	5.1	5	1058.3	6.7	5	1658.3	8.3
6	28.2	1.9	6	245.8	3.5	6	591.4	5.1	6	1065.0	6.7	6	1666.6	8.3
7	30.1	1.9	7	249.3	3.5	7	596.5	5.1	7	1071.7	6.7	7	1674.9	8.3
8	32.0	2.0	8	252.8	3.6	8	601.6	5.2	8	1078.4	6.8	8	1683.2	8.4
9	34.0	2.0	9	256.4	3.6	9	606.8	5.2	9	1085.2	6.8	9	1691.6	8.4
2.0	36.0	2.0	10.0	260.0	3.6	18.0	612.0	5.2	26.0	1092.0	6.8	34.0	1700.0	8.4
1	38.0	2.0	1	263.6	3.6	1	617.2	5.2	1	1098.8	6.8	1	1708.4	8.4
2	40.0	2.0	2	267.2	3.6	2	622.4	5.2	2	1105.6	6.8	2	1716.8	8.4
3	42.1	2.1	3	270.9	3.7	3	627.7	5.3	3	1112.5	6.9	3	1725.3	8.5
4	44.2	2.1	4	274.6	3.7	4	633.0	5.3	4	1119.4	6.9	4	1733.8	8.5
5	46.3	2.1	5	278.3	3.7	5	638.3	5.3	5	1126.3	6.9	5	1742.3	8.5
6	48.4	2.1	6	282.0	3.7	6	643.6	5.3	6	1133.2	6.9	6	1750.8	8.5
7	50.5	2.1	7	285.7	3.7	7	648.9	5.3	7	1140.1	6 9	7	1759.3	8.5
8	52.6	2.2	8	289.4	3.8	8	654.2	5.4	8	1147.0	7.0	8	1767.8	8.6
9	54.8	2.2	9	293.2	3.8	9	659.6	5.4	9	1154.0	7.0	9	1776.4	8.6
3.0	57.0	2.2	11.0	297.0	3.8	19.0	665.0	5.4	27.0	1161.0	7.0	35.0	1785.0	8.6
1	59.2	2.2	1	300.8	3.8	1	670.4	5.4	1	1168.0	7.0	1	1793.6	8.6
2	61.4	2.2	2	304.6	3.8	2	675.8	5.4	2	1175.0	7.0	2	1802.2	8.6
3	63.7	2.3	3	308.5	3.9	3	681.3	5.5	3	1182.1	7.1	3	1810.9	8.7
4	66.0	2.3	4	312.4	3.9	4	686.8	5.5	4	1189.2	7 1	4	1819.6	8.7
5	68 3	2.3	5	316.3	3.9	5	692.3	5.5	5	1196.3	7.1	5	1828.3	8.7
6	70.6	2.3	6	320.2	3.9	6	697.8	5.5	6	1203.4	7.1	6	1837.0	8.7
7	72.9	2.3	7	324.1	3.9	7	703.3	5.5	7	1210.5	7.1	7	1845.7	8.7
8	75.2	2.4	8	328.0	4.0	8	708.8	5.6	8	1217.6	7.2	8	1854.4	8.8
9	77.6	2.4	9	332.0	4.0	9	714.4	5.6	9	1224.8	7 2	9	1863.2	8.8
4.0	80.0	2.4	12.0	336.0	4.0	20.0	720.0	5.6	28.0	1232.0	7.2	36.0	1872.0	8.8
1	82.4	2.4	1	340.0	4.0	1	725.6	5.6	1	1239.2	7.2	1	1880.8	8.8
2	84.8	2.4	2	344.0	4.0	2	731.2	5.6	2	1246.4	7.2	2	1889.6	8.8
3	87.3	2.5	3	348.1	4.1	3	736.9	5.7	3	1253.7	7.3	3	1898.5	8.9
4	89.8	2.5	4	352.2	4.1	4	742.6	5.7	4	1261.0	7.3	4	1907.4	8.9
5	92.3	2.5	5	356.3	4.1	5	748.3	5.7	5	1268.3	7.3	5	1916.3	8.9
6	94.8	2.5	6	360.4	4.1	6	754.0	5.7	6	1275.6	7.3	6	1925.2	8.9
7	97.3	2.5	7	364.5	4.1	7	759.7	5.7	7	1282.9	7.3	7	1934.1	8.9
8	99.8	2.6	8	368.6	4.2	8	765.4	5.8	8	1290.2	7.4	8	1943.0	9.0
9	102.4	2.6	9	372.8	4.2	9	771.2	5.8	9	1297.6	7.4	9	1952.0	9.0
5.0	105.0	2.6	13.0	377.0	4.2	21.0	777.0	5.8	29.0	1305.0	7.4	37.0	1961.0	9.0
1	107.6	2.6	1	381.2	4.2	1	782.8	5.8	1	1312.4	7.4	1	1970.0	9.0
2	110.2	2.6	2	385.4	4.2	2	788.6	5.8	2	1319.8	7.4	2	1979.0	9.0
3	112.9	2.7	3	389.7	4.3	3	794.5	5.9	3	1327.3	7.5	3	1988.1	9.1
4	115.6	2.7	4	394.0	4.3	4	800.4	5.9	4	1334.8	7.5	4	1997.2	9.1
5	118.3	2.7	5	398.3	4.3	5	806.3	5.9	5	1342.3	7.5	5	2006.3	9.1
6	121.0	2.7	6	402.6	4.3	6	812.2	5.9	6	1349.8	7.5	6	2015.4	9.1
7	123.7	2.7	7	406.9	4.3	7	818.1	5.9	7	1357.3	7.5	7	2024.5	9.1
8	126.4	2.8	8	411.2	4.4	8	824.0	6.0	8	1364.8	7.6	8	2033.6	9.2
9	129.2	2.8	9	415.6	4.4	9	830.0	6.0	9	1372.4	7.6	9	2042.8	9.2
6.0	132.0	2.8	14.0	420.0	4.4	22.0	836.0	6.0	30.0	1380.0	7.6	38.0	2052.0	9.2
1	134.8	2.8	1	424.4	4.4	1	842.0	6.0	1	1387.6	7.6	1	2061.2	9.2
2	137.6	2.8	2	428.8	4.4	2	848.0	6.0	2	1395.2	7.6	2	2070.4	9.2
3	140.5	2.9	3	433.3	4.5	3	854.1	6.1	3	1402.9	7.7	3	2079.7	9.3
4	143.4	2.9	4	437.8	4.5	4	860.2	6.1	4	1410.6	7.7	4	2089.0	9.3
5	146.3	2.9	5	442.3	4.5	5	866.3	6.1	5	1418.3	7.7	5	2098.3	9.3
6	149.2	2.9	6	446.8	4.5	6	872.4	6.1	6	1426.0	7.7	6	2107.6	9.3
7	152.1	2.9	7	451.3	4.5	7	878.5	6.1	7	1433.7	7.7	7	2116.9	9.3
8	155.0	3.0	8	455.8	4.6	8	884.6	6.2	8	1441.4	7.8	8	2126.2	9.4
9	158.0	3.0	9	460.4	4.6	9	890.8	6.2	9	1449.2	7.8	9	2135.6	9.4
7.0	161.0	3.0	15.0	465.0	4.6	23.0	897.0	6.2	31.0	1457.0	7.8	39.0	2145.0	9.4
1	164.0	3.0	1	469.6	4.6	1	903.2	6.2	1	1464.8	7.8	1	2154.4	9.4
2	167.0	3.0	2	474.2	4.6	2	909.4	6.2	2	1472.6	7.8	2	2163.8	9.4
3	170.1	3.1	3	478.9	4.7	3	915.7	6.3	3	1480.5	7.9	3	2173.3	9.5
4	173.2	3.1	4	483.6	4.7	4	922.0	6.3	4	1488.4	7.9	4	2182.8	9.5
5	176.3	3.1	5	488.3	4.7	5	928.3	6.3	5	1496.3	7.9	5	2192.3	9.5
6	179.4	3.1	6	493.0	4.7	6	934.6	6.3	6	1504.2	7.9	6	2201.8	9.5
7	182.5	3.1	7	497.7	4.7	7	940.9	6.3	7	1512.1	7.9	7	2211.3	9.5
8	185.6	3.2	8	502.4	4.8	8	947.2	6.4	8	1520.0	8.0	8	2220.8	9.6
9	188.8	3.2	9	507.2	4.8	9	953.6	6.4	9	1528.0	8.0	9	2230.4	9.6

TABLE No. 20

EXCAVATION AND EMBANKMENT.

AREAS.

Breadth of Base, 18 feet. **Slopes, 1 Horizontal to 1 Perpendicular.**

Height.	Area in Feet.	Difference.	Height.	Area in Feet.	Difference.	Height.	Area in Feet.	Difference.	Height.	Area in Feet.	Difference.	Height.	Area in Feet.	Difference.
0.0	0.0	0.0	8.0	208.0	3.4	16.0	544.0	5.0	24.0	1008.0	6.6	32.0	1600.0	8.2
1	1.8	1.8	1	211.4	3.4	1	549.0	5.0	1	1014.6	6.6	1	1608.2	8.2
2	3.6	1.8	2	214.8	3.4	2	554.0	5.0	2	1021.2	6.6	2	1616.4	8.2
3	5.5	1.9	3	218.3	3.5	3	559.1	5.1	3	1027.9	6.7	3	1624.7	8.3
4	7.4	1.9	4	221.8	3.5	4	564.2	5.1	4	1034.6	6.7	4	1633.0	8.3
5	9.3	1.9	5	225.3	3.5	5	569.3	5.1	5	1041.3	6.7	5	1641.3	8.3
6	11.2	1.9	6	228.8	3.5	6	574.4	5.1	6	1048.0	6.7	6	1649.6	8.3
7	13.1	1.9	7	232.3	3.5	7	579.5	5.1	7	1054.7	6.7	7	1657.9	8.3
8	15.0	2.0	8	235.8	3.6	8	584.6	5.2	8	1061.4	6.8	8	1666.2	8.4
9	17.0	2.0	9	239.4	3.6	9	589.8	5.2	9	1068.2	6.8	9	1674.6	8.4
1.0	19.0	2.0	9.0	243.0	3.6	17.0	595.0	5.2	25.0	1075.0	6.8	33.0	1683.0	8.4
1	21.0	2.0	1	246.6	3.6	1	600.2	5.2	1	1081.8	6.8	1	1691.4	8.4
2	23.0	2.0	2	250.2	3.6	2	605.4	5.2	2	1088.6	6.8	2	1699.8	8.4
3	25.1	2.1	3	253.9	3.7	3	610.7	5.3	3	1095.5	6.9	3	1708.3	8.5
4	27.2	2.1	4	257.6	3.7	4	616.0	5.3	4	1102.4	6.9	4	1716.8	8.5
5	29.3	2.1	5	261.3	3.7	5	621.3	5.3	5	1109.3	6.9	5	1725.3	8.5
6	31.4	2.1	6	265.0	3.7	6	626.6	5.3	6	1116.2	6.9	6	1733.8	8.5
7	33.5	2.1	7	268.7	3.7	7	631.9	5.3	7	1123.1	6.9	7	1742.3	8.5
8	35.6	2.2	8	272.4	3.8	8	637.2	5.4	8	1130.0	7.0	8	1750.8	8.6
9	37.8	2.2	9	276.2	3.8	9	642.6	5.4	9	1137.0	7.0	9	1759.4	8.6
2.0	40.0	2.2	10.0	280.0	3.8	18.0	648.0	5.4	26.0	1144.0	7.0	34.0	1768.0	8.6
1	42.2	2.2	1	283.8	3.8	1	653.4	5.4	1	1151.0	7.0	1	1776.6	8.6
2	44.4	2.2	2	287.6	3.8	2	658.8	5.4	2	1158.0	7.0	2	1785.2	8.6
3	46.7	2.3	3	291.5	3.9	3	664.3	5.5	3	1165.1	7.1	3	1793.9	8.7
4	49.0	2.3	4	295.4	3.9	4	669.8	5.5	4	1172.2	7.1	4	1802.6	8.7
5	51.3	2.3	5	299.3	3.9	5	675.3	5.5	5	1179.3	7.1	5	1811.3	8.7
6	53.6	2.3	6	303.2	3.9	6	680.8	5.5	6	1186.4	7.1	6	1820.0	8.7
7	55.9	2.3	7	307.1	3.9	7	686.3	5.5	7	1193.5	7.1	7	1828.7	8.7
8	58.2	2.4	8	311.0	4.0	8	691.8	5.6	8	1200.6	7.2	8	1837.4	8.8
9	60.6	2.4	9	315.0	4.0	9	697.4	5.6	9	1207.8	7.2	9	1846.2	8.8
3.0	63.0	2.4	11.0	319.0	4.0	19.0	703.0	5.6	27.0	1215.0	7.2	35.0	1855.0	8.8
1	65.4	2.4	1	323.0	4.0	1	708.6	5.6	1	1222.2	7.2	1	1863.8	8.8
2	67.8	2.4	2	327.0	4.0	2	714.2	5.6	2	1229.4	7.2	2	1872.6	8.8
3	70.3	2.5	3	331.1	4.1	3	719.9	5.7	3	1236.7	7.3	3	1881.5	8.9
4	72.8	2.5	4	335.2	4.1	4	725.6	5.7	4	1244.0	7.3	4	1890.4	8.9
5	75.3	2.5	5	339.3	4.1	5	731.3	5.7	5	1251.3	7.3	5	1899.3	8.9
6	77.8	2.5	6	343.4	4.1	6	737.0	5.7	6	1258.6	7.3	6	1908.2	8.9
7	80.3	2.5	7	347.5	4.1	7	742.7	5.7	7	1265.9	7.3	7	1917.1	8.9
8	82.8	2.6	8	351.6	4.2	8	748.4	5.8	8	1273.2	7.4	8	1926.0	9.0
9	85.4	2.6	9	355.8	4.2	9	754.2	5.8	9	1280.6	7.4	9	1935.0	9.0
4.0	88.0	2.6	12.0	360.0	4.2	20.0	760.0	5.8	28.0	1288.0	7.4	36.0	1944.0	9.0
1	90.6	2.6	1	364.2	4.2	1	765.8	5.8	1	1295.4	7.4	1	1953.0	9.0
2	93.2	2.6	2	368.4	4.2	2	771.6	5.8	2	1302.8	7.4	2	1962.0	9.0
3	95.9	2.7	3	372.7	4.3	3	777.5	5.9	3	1310.3	7.5	3	1971.1	9.1
4	98.6	2.7	4	377.0	4.3	4	783.4	5.9	4	1317.8	7.5	4	1980.2	9.1
5	101.3	2.7	5	381.3	4.3	5	789.3	5.9	5	1325.3	7.5	5	1989.3	9.1
6	104.0	2.7	6	385.6	4.3	6	795.2	5.9	6	1332.8	7.5	6	1998.4	9.1
7	106.7	2.7	7	389.9	4.3	7	801.1	5.9	7	1340.3	7.5	7	2007.5	9.1
8	109.4	2.8	8	394.2	4.4	8	807.0	6.0	8	1347.8	7.6	8	2016.6	9.2
9	112.2	2.8	9	398.6	4.4	9	813.0	6.0	9	1355.4	7.6	9	2025.8	9.2
5.0	115.0	2.8	13.0	403.0	4.4	21.0	819.0	6.0	29.0	1363.0	7.6	37.0	2035.0	9.2
1	117.8	2.8	1	407.4	4.4	1	825.0	6.0	1	1370.6	7.6	1	2044.2	9.2
2	120.6	2.8	2	411.8	4.4	2	831.0	6.0	2	1378.2	7.6	2	2053.4	9.2
3	123.5	2.9	3	416.3	4.5	3	837.1	6.1	3	1385.9	7.7	3	2062.7	9.3
4	126.4	2.9	4	420.8	4.5	4	843.2	6.1	4	1393.6	7.7	4	2072.0	9.3
5	129.3	2.9	5	425.3	4.5	5	849.3	6.1	5	1401.3	7.7	5	2081.3	9.3
6	132.2	2.9	6	429.8	4.5	6	855.4	6.1	6	1409.0	7.7	6	2090.6	9.3
7	135.1	2.9	7	434.3	4.5	7	861.5	6.1	7	1416.7	7.7	7	2099.9	9.3
8	138.0	3.0	8	438.8	4.6	8	867.6	6.2	8	1424.4	7.8	8	2109.2	9.4
9	141.0	3.0	9	443.4	4.6	9	873.8	6.2	9	1432.2	7.8	9	2118.6	9.4
6.0	144.0	3.0	14.0	448.0	4.6	22.0	880.0	6.2	30.0	1440.0	7.8	38.0	2128.0	9.4
1	147.0	3.0	1	452.6	4.6	1	886.2	6.2	1	1447.8	7.8	1	2137.4	9.4
2	150.0	3.0	2	457.2	4.6	2	892.4	6.2	2	1455.6	7.8	2	2146.8	9.4
3	153.1	3.1	3	461.9	4.7	3	898.7	6.3	3	1463.5	7.9	3	2156.3	9.5
4	156.2	3.1	4	466.6	4.7	4	905.0	6.3	4	1471.4	7.9	4	2165.8	9.5
5	159.3	3.1	5	471.3	4.7	5	911.3	6.3	5	1479.3	7.9	5	2175.3	9.5
6	162.4	3.1	6	476.0	4.7	6	917.6	6.3	6	1487.2	7.9	6	2184.8	9.5
7	165.5	3.1	7	480.7	4.7	7	923.9	6.3	7	1495.1	7.9	7	2194.3	9.5
8	168.6	3.2	8	485.4	4.8	8	930.2	6.4	8	1503.0	8.0	8	2203.8	9.6
9	171.8	3.2	9	490.2	4.8	9	936.6	6.4	9	1511.0	8.0	9	2213.4	9.6
7.0	175.0	3.2	15.0	495.0	4.8	23.0	943.0	6.4	31.0	1519.0	8.0	39.0	2223.0	9.6
1	178.2	3.2	1	499.8	4.8	1	949.4	6.4	1	1527.0	8.0	1	2232.6	9.6
2	181.4	3.2	2	504.6	4.8	2	955.8	6.4	2	1535.0	8.0	2	2242.2	9.6
3	184.7	3.3	3	509.5	4.9	3	962.3	6.5	3	1543.1	8.1	3	2251.9	9.7
4	188.0	3.3	4	514.4	4.9	4	968.8	6.5	4	1551.2	8.1	4	2261.6	9.7
5	191.3	3.3	5	519.3	4.9	5	875.3	6.5	5	1559.3	8.1	5	2271.3	9.7
6	194.6	3.3	6	524.2	4.9	6	981.8	6.5	6	1567.4	8.1	6	2281.0	9.7
7	197.9	3.3	7	529.1	4.9	7	988.3	6.5	7	1575.5	8.1	7	2290.7	9.7
8	201.2	3.4	8	534.0	5.0	8	994.8	6.6	8	1583.6	8.2	8	2300.4	9.8
9	204.6	3.4	9	539.0	5.0	9	1001.4	6.6	9	1591.8	8.2	9	2310.2	9.8

TABLE NO. 21.

EXCAVATION AND EMBANKMENT.

AREAS.

Breadth of Base, Twenty Feet. Slope 1 Horizontal to 1 Perpendicular.

Height.	Area in Feet.	Diff.	Height.	Area in Feet.	Diff.	Height.	Area in Feet.	Diff.	Height.	Area in Feet.	Diff.	Height.	Area in Feet.	Diff.
0.0	0.0	0.0	8.0	224.0	3.6	16.0	576.0	5.2	24.0	1056.0	6.8	32.0	1664.0	8.4
.1	2.0	2.0	.1	227.6	3.6	.1	581.2	5.2	.1	1062.8	6.8	.1	1672.4	8.4
.2	4.0	2.1	.2	231.2	3.6	.2	586.4	5.2	.2	1069.6	6.8	.2	1680.8	8.5
.3	6.1	2.1	.3	234.9	3.7	.3	591.7	5.3	.3	1076.5	6.9	.3	1689.3	8.5
.4	8.2	2.1	.4	238.6	3.7	.4	597.0	5.3	.4	1083.4	6.9	.4	1697.8	8.5
.5	10.3	2.1	.5	242.3	3.7	.5	602.3	5.3	.5	1090.3	6.9	.5	1706.3	8.5
.6	12.4	2.1	.6	246.0	3.7	.6	607.6	5.3	.6	1097.2	6.9	.6	1714.8	8.5
.7	14.5	2.2	.7	249.7	3.7	.7	612.9	5.3	.7	1104.1	6.9	.7	1723.3	8.6
.8	16.6	2.2	.8	253.4	3.8	.8	618.2	5.4	.8	1111.0	7.0	.8	1731.8	8.6
.9	18.8	2.2	.9	257.2	3.8	.9	623.6	5.4	.9	1118.0	7.0	.9	1740.4	8.6
1.0	21.0	2.2	9.0	261.0	3.8	17.0	629.0	5.4	25.0	1125.0	7.0	33.0	1749.0	8.6
.1	23.2	2.2	.1	264.8	3.8	.1	634.4	5.4	.1	1132.0	7.0	.1	1757.6	8.6
.2	25.4	2.3	.2	268.6	3.8	.2	639.8	5.4	.2	1139.0	7.0	.2	1766.2	8.6
.3	27.7	2.3	.3	272.5	3.9	.3	645.3	5.5	.3	1146.1	7.1	.3	1774.9	8.7
.4	30.0	2.3	.4	276.4	3.9	.4	650.8	5.5	.4	1153.2	7.1	.4	1783.6	8.7
.5	32.3	2.3	.5	280.3	3.9	.5	656.3	5.5	.5	1160.3	7.1	.5	1792.3	8.7
.6	34.6	2.3	.6	284.2	3.9	.6	661.8	5.5	.6	1167.4	7.1	.6	1801.0	8.7
.7	36.9	2.4	.7	288.1	3.9	.7	667.3	5.5	.7	1174.5	7.1	.7	1809.7	8.7
.8	39.2	2.4	.8	292.0	4.0	.8	672.8	5.6	.8	1181.6	7.2	.8	1818.4	8.8
.9	41.6	2.4	.9	296.0	4.0	.9	678.4	5.6	.9	1188.8	7.2	.9	1827.2	8.8
2.0	44.0	2.4	10.0	300.0	4.0	18.0	684.0	5.6	26.0	1196.0	7.2	34.0	1836.0	8.8
.1	46.4	2.4	.1	304.0	4.0	.1	689.6	5.6	.1	1203.2	7.2	.1	1844.8	8.8
.2	48.8	2.5	.2	308.0	4.1	.2	695.2	5.6	.2	1210.4	7.2	.2	1853.6	8.8
.3	51.3	2.5	.3	312.1	4.1	.3	700.9	5.7	.3	1217.7	7.3	.3	1862.5	8.9
.4	53.8	2.5	.4	316.2	4.1	.4	706.6	5.7	.4	1225.0	7.3	.4	1871.4	8.9
.5	56.3	2.5	.5	320.3	4.1	.5	712.3	5.7	.5	1232.3	7.3	.5	1880.3	8.9
.6	58.8	2.5	.6	324.4	4.1	.6	718.0	5.7	.6	1239.6	7.3	.6	1889.2	8.9
.7	61.3	2.6	.7	328.5	4.2	.7	723.7	5.7	.7	1246.9	7.3	.7	1898.1	8.9
.8	63.8	2.6	.8	332.6	4.2	.8	729.4	5.8	.8	1254.2	7.4	.8	1907.0	9.0
.9	66.4	2.6	.9	336.8	4.2	.9	735.2	5.8	.9	1261.6	7.4	.9	1916.0	9.0
3.0	69.0	2.6	11.0	341.0	4.2	19.0	741.0	5.8	27.0	1269.0	7.4	35.0	1925.0	9.0
.1	71.6	2.6	.1	345.2	4.2	.1	746.8	5.8	.1	1276.4	7.4	.1	1934.0	9.0
.2	74.2	2.6	.2	349.4	4.3	.2	752.6	5.8	.2	1283.8	7.4	.2	1943.0	9.0
.3	76.9	2.7	.3	353.7	4.3	.3	758.5	5.9	.3	1291.3	7.5	.3	1952.1	9.1
.4	79.6	2.7	.4	358.0	4.3	.4	764.4	5.9	.4	1298.8	7.5	.4	1961.2	9.1
.5	82.3	2.7	.5	362.3	4.3	.5	770.3	5.9	.5	1306.3	7.5	.5	1970.3	9.1
.6	85.0	2.7	.6	366.6	4.3	.6	776.2	5.9	.6	1313.8	7.5	.6	1979.4	9.1
.7	87.7	2.7	.7	370.9	4.4	.7	782.1	5.9	.7	1321.3	7.5	.7	1988.5	9.1
.8	90.4	2.8	.8	375.2	4.4	.8	788.0	6.0	.8	1328.8	7.6	.8	1997.6	9.2
.9	93.2	2.8	.9	379.6	4.4	.9	794.0	6.0	.9	1336.4	7.6	.9	2006.8	9.2
4.0	96.0	2.8	12.0	384.0	4.4	20.0	800.0	6.0	28.0	1344.0	7.6	36.0	2016.0	9.2
.1	98.8	2.8	.1	388.4	4.4	.1	806.0	6.0	.1	1351.6	7.6	.1	2025.2	9.2
.2	101.6	2.8	.2	392.8	4.5	.2	812.0	6.1	.2	1359.2	7.6	.2	2034.4	9.2
.3	104.5	2.9	.3	397.3	4.5	.3	818.1	6.1	.3	1366.9	7.7	.3	2043.7	9.3
.4	107.4	2.9	.4	401.8	4.5	.4	824.2	6.1	.4	1374.6	7.7	.4	2053.0	9.3
.5	110.3	2.9	.5	406.3	4.5	.5	830.3	6.1	.5	1382.3	7.7	.5	2062.3	9.3
.6	113.2	2.9	.6	410.8	4.5	.6	836.4	6.1	.6	1390.0	7.7	.6	2071.6	9.3
.7	116.1	2.9	.7	415.3	4.6	.7	842.5	6.2	.7	1397.7	7.7	.7	2080.9	9.3
.8	119.0	3.0	.8	419.8	4.6	.8	848.6	6.2	.8	1405.4	7.8	.8	2090.2	9.4
.9	122.0	3.0	.9	424.4	4.6	.9	854.8	6.2	.9	1413.2	7.8	.9	2099.6	9.4
5.0	125.0	3.0	13.0	429.0	4.6	21.0	861.0	6.2	29.0	1421.0	7.8	37.0	2109.0	9.4
.1	128.0	3.0	.1	433.6	4.6	.1	867.2	6.2	.1	1428.8	7.8	.1	2118.4	9.4
.2	131.0	3.0	.2	438.2	4.6	.2	873.4	6.3	.2	1436.6	7.8	.2	2127.8	9.4
.3	134.1	3.1	.3	442.9	4.7	.3	879.7	6.3	.3	1444.5	7.9	.3	2137.3	9.5
.4	137.2	3.1	.4	447.6	4.7	.4	886.0	6.3	.4	1452.4	7.9	.4	2146.8	9.5
.5	140.3	3.1	.5	452.3	4.7	.5	892.3	6.3	.5	1460.3	7.9	.5	2156.3	9.5
.6	143.4	3.1	.6	457.0	4.7	.6	898.6	6.3	.6	1468.2	7.9	.6	2165.8	9.5
.7	146.5	3.1	.7	461.7	4.7	.7	904.9	6.4	.7	1476.1	7.9	.7	2175.3	9.5
.8	149.6	3.2	.8	466.4	4.8	.8	911.2	6.4	.8	1484.0	8.0	.8	2184.8	9.6
.9	152.8	3.2	.9	471.2	4.8	.9	917.6	6.4	.9	1492.0	8.0	.9	2194.4	9.6
6.0	156.0	3.2	14.0	476.0	4.8	22.0	924.0	6.4	30.0	1500.0	8.0	38.0	2204.0	9.6
.1	159.2	3.2	.1	480.8	4.8	.1	930.4	6.4	.1	1508.0	8.0	.1	2213.6	9.6
.2	162.4	3.2	.2	485.6	4.8	.2	936.8	6.5	.2	1516.0	8.1	.2	2223.2	9.6
.3	165.7	3.3	.3	490.5	4.9	.3	943.3	6.5	.3	1524.1	8.1	.3	2232.9	9.7
.4	169.0	3.3	.4	495.4	4.9	.4	949.8	6.5	.4	1532.2	8.1	.4	2242.6	9.7
.5	172.3	3.3	.5	500.3	4.9	.5	956.3	6.5	.5	1540.3	8.1	.5	2252.3	9.7
.6	175.6	3.3	.6	505.2	4.9	.6	962.8	6.5	.6	1548.4	8.1	.6	2262.0	9.7
.7	178.9	3.3	.7	510.1	4.9	.7	969.3	6.6	.7	1556.5	8.2	.7	2271.7	9.7
.8	182.2	3.4	.8	515.0	5.0	.8	975.8	6.6	.8	1564.6	8.2	.8	2281.4	9.8
.9	185.6	3.4	.9	520.0	5.0	.9	982.4	6.6	.9	1572.8	8.2	.9	2291.2	9.8
7.0	189.0	3.4	15.0	525.0	5.0	23.0	989.0	6.6	31.0	1581.0	8.2	39.0	2301.0	9.8
.1	192.4	3.4	.1	530.0	5.0	.1	995.6	6.6	.1	1589.2	8.2	.1	2310.8	9.8
.2	195.8	3.4	.2	535.0	5.0	.2	1002.2	6.6	.2	1597.4	8.3	.2	2320.6	9.8
.3	199.3	3.5	.3	540.1	5.1	.3	1008.9	6.7	.3	1605.7	8.3	.3	2330.5	9.9
.4	202.8	3.5	.4	545.2	5.1	.4	1015.6	6.7	.4	1614.0	8.3	.4	2340.4	9.9
.5	206.3	3.5	.5	550.3	5.1	.5	1022.3	6.7	.5	1622.3	8.3	.5	2350.3	9.9
.6	209.8	3.5	.6	555.4	5.1	.6	1029.0	6.7	.6	1630.6	8.3	.6	2360.2	9.9
.7	213.3	3.5	.7	560.5	5.1	.7	1035.7	6.7	.7	1638.9	8.4	.7	2370.1	9.9
.8	216.8	3.6	.8	565.6	5.2	.8	1042.4	6.8	.8	1647.2	8.4	.8	2380.0	10.0
.9	220.4	3.6	.9	570.8	5.2	.9	1049.2	6.8	.9	1655.6	8.4	.9	2390.0	10.0

TABLE No. 22.

EXCAVATION AND EMBANKMENT.

AREAS.

Breadth of Base, 22 feet. **Slopes, 1 Horizontal to 1 Perpendicular.**

Height.	Area in Feet.	Difference.	Height.	Area in Feet.	Difference.	Height.	Area in Feet.	Difference.	Height.	Area in Feet.	Difference.	Height.	Area in Feet.	Difference.
0.0	0.0	0.0	8.0	240.0	3.8	16.0	608.0	5.4	24.0	1104.0	7.0	32.0	1728.0	8.6
1	2.2	2.2	1	243.8	3.8	1	613.4	5.4	1	1111.0	7.0	1	1736.6	8.6
2	4.4	2.2	2	247.6	3.8	2	618.8	5.4	2	1118.0	7.0	2	1745.2	8.6
3	6.7	2.3	3	251.5	3.9	3	624.3	5.5	3	1125.1	7.1	3	1753.9	8.7
4	9.0	2.3	4	255.4	3.9	4	629.8	5.5	4	1132.2	7.1	4	1762.6	8.7
5	11.3	2.3	5	259.3	3.9	5	635.3	5.5	5	1139.3	7.1	5	1771.3	8.7
6	13.6	2.3	6	263.2	3.9	6	640.8	5.5	6	1146.4	7.1	6	1780.0	8.7
7	15.9	2.3	7	267.1	3.9	7	646.3	5.5	7	1153.5	7.1	7	1788.7	8.7
8	18.2	2.4	8	271.0	4.0	8	651.8	5.6	8	1160.6	7.2	8	1797.4	8.8
9	20.6	2.4	9	275.0	4.0	9	657.4	5.6	9	1167.8	7.2	9	1806.2	8.8
1.0	23.0	2.4	9.0	279.0	4.0	17.0	663.0	5.6	25.0	1175.0	7.2	33.0	1815.0	8.8
1	25.4	2.4	1	283.0	4.0	1	668.6	5.6	1	1182.2	7.2	1	1823.8	8.8
2	27.8	2.4	2	287.0	4.0	2	674.2	5.6	2	1189.4	7.2	2	1832.6	8.8
3	30.3	2.5	3	291.1	4.1	3	679.9	5.7	3	1196.7	7.3	3	1841.5	8.9
4	32.8	2.5	4	295.2	4.1	4	685.6	5.7	4	1204.0	7.3	4	1850.4	8.9
5	35.3	2.5	5	299.3	4.1	5	691.3	5.7	5	1211.3	7.3	5	1859.3	8.9
6	37.8	2.5	6	303.4	4.1	6	697.0	5.7	6	1218.6	7.3	6	1868.2	8.9
7	40.3	2.5	7	307.5	4.1	7	702.7	5.7	7	1225.9	7.3	7	1877.1	8.9
8	42.8	2.6	8	311.6	4.2	8	708.4	5.8	8	1233.2	7.4	8	1886.0	9.0
9	45.4	2.6	9	315.8	4.2	9	714.2	5.8	9	1240.6	7.4	9	1895.0	9.0
2.0	48.0	2.6	10.0	320.0	4.2	18.0	720.0	5.8	26.0	1248.0	7.4	34.0	1904.0	9.0
1	50.6	2.6	1	324.2	4.2	1	725.8	5.8	1	1255.4	7.4	1	1913.0	9.0
2	53.2	2.6	2	328.4	4.2	2	731.6	5.8	2	1262.8	7.4	2	1922.0	9.0
3	55.9	2.7	3	332.7	4.3	3	737.5	5.9	3	1270.3	7.5	3	1931.1	9.1
4	58.6	2.7	4	337.0	4.3	4	743.4	5.9	4	1277.8	7.5	4	1940.2	9.1
5	61.3	2.7	5	341.3	4.3	5	749.3	5.9	5	1285.3	7.5	5	1949.3	9.1
6	64.0	2.7	6	345.6	4.3	6	755.2	5.9	6	1292.8	7.5	6	1958.4	9.1
7	66.7	2.7	7	349.9	4.3	7	761.1	5.9	7	1300.3	7.5	7	1967.5	9.1
8	69.4	2.8	8	354.2	4.4	8	767.0	6.0	8	1307.8	7.6	8	1976.6	9.2
9	72.2	2.8	9	358.6	4.4	9	773.0	6.0	9	1315.4	7.6	9	1985.8	9.2
3.0	75.0	2.8	11.0	363.0	4.4	19.0	779.0	6.0	27.0	1323.0	7.6	35.0	1995.0	9.2
1	77.8	2.8	1	367.4	4.4	1	785.0	6.0	1	1330.6	7.6	1	2004.2	9.2
2	80.6	2.8	2	371.8	4.4	2	791.0	6.0	2	1338.2	7.6	2	2013.4	9.2
3	83.5	2.9	3	376.3	4.5	3	797.1	6.1	3	1345.9	7.7	3	2022.7	9.3
4	86.4	2.9	4	380.8	4.5	4	803.2	6.1	4	1353.6	7.7	4	2032.0	9.3
5	89.3	2.9	5	385.3	4.5	5	809.3	6.1	5	1361.3	7.7	5	2041.3	9.3
6	92.2	2.9	6	389.8	4.5	6	815.4	6.1	6	1369.0	7.7	6	2050.6	9.3
7	95.1	2.9	7	394.3	4.5	7	821.5	6.1	7	1376.7	7.7	7	2059.9	9.3
8	98.0	3.0	8	398.8	4.6	8	827.6	6.2	8	1384.4	7.8	8	2069.2	9.4
9	101.0	3.0	9	403.4	4.6	9	833.8	6.2	9	1392.2	7.8	9	2078.6	9.4
4.0	104.0	3.0	12.0	408.0	4.6	20.0	840.0	6.2	28.0	1400.0	7.8	36.0	2088.0	9.4
1	107.0	3.0	1	412.6	4.6	1	846.2	6.2	1	1407.8	7.8	1	2097.4	9.4
2	110.0	3.0	2	417.2	4.6	2	852.4	6.2	2	1415.6	7.8	2	2106.8	9.4
3	113.1	3.1	3	421.9	4.7	3	858.7	6.3	3	1423.5	7.9	3	2116.3	9.5
4	116.2	3.1	4	426.6	4.7	4	865.0	6.3	4	1431.4	7.9	4	2125.8	9.5
5	119.3	3.1	5	431.3	4.7	5	871.3	6.3	5	1439.3	7.9	5	2135.3	9.5
6	122.4	3.1	6	436.0	4.7	6	877.6	6.3	6	1447.2	7.9	6	2144.8	9.5
7	125.5	3.1	7	440.7	4.7	7	883.9	6.3	7	1455.1	7.9	7	2154.3	9.5
8	128.6	3.2	8	445.4	4.8	8	890.2	6.4	8	1463.0	8.0	8	2163.8	9.6
9	131.8	3.2	9	450.2	4.8	9	896.6	6.4	9	1471.0	8.0	9	2173.4	9.6
5.0	135.0	3.2	13.0	455.0	4.8	21.0	903.0	6.4	29.0	1479.0	8.0	37.0	2183.0	9.6
1	138.2	3.2	1	459.8	4.8	1	909.4	6.4	1	1487.0	8.0	1	2192.6	9.6
2	141.4	3.2	2	464.6	4.8	2	915.8	6.4	2	1495.0	8.0	2	2202.2	9.6
3	144.7	3.3	3	469.5	4.9	3	922.3	6.5	3	1503.1	8.1	3	2211.9	9.7
4	148.0	3.3	4	474.4	4.9	4	928.8	6.5	4	1511.2	8.1	4	2221.6	9.7
5	151.3	3.3	5	479.3	4.9	5	935.3	6.5	5	1519.3	8.1	5	2231.3	9.7
6	154.6	3.3	6	484.2	4.9	6	941.8	6.5	6	1527.4	8.1	6	2241.0	9.7
7	157.9	3.3	7	489.1	4.9	7	948.3	6.5	7	1535.5	8.1	7	2250.7	9.7
8	161.2	3.4	8	494.0	5.0	8	954.8	6.6	8	1543.6	8.2	8	2260.4	9.8
9	164.6	3.4	9	499.0	5.0	9	961.4	6.6	9	1551.8	8.2	9	2270.2	9.8
6.0	168.0	3.4	14.0	504.0	5.0	22.0	968.0	6.6	30.0	1560.0	8.2	38.0	2280.0	9.8
1	171.4	3.4	1	509.0	5.0	1	974.6	6.6	1	1568.2	8.2	1	2289.8	9.8
2	174.8	3.4	2	514.0	5.0	2	981.2	6.6	2	1576.4	8.2	2	2299.6	9.8
3	178.3	3.5	3	519.1	5.1	3	987.9	6.7	3	1584.7	8.3	3	2309.5	9.9
4	181.8	3.5	4	524.2	5.1	4	994.6	6.7	4	1593.0	8.3	4	2319.4	9.9
5	185.3	3.5	5	529.3	5.1	5	1001.3	6.7	5	1601.3	8.3	5	2329.3	9.9
6	188.8	3.5	6	534.4	5.1	6	1008.0	6.7	6	1609.6	8.3	6	2339.2	9.9
7	192.3	3.5	7	539.5	5.1	7	1014.7	6.7	7	1617.9	8.3	7	2349.1	9.9
8	195.8	3.6	8	544.6	5.2	8	1021.4	6.8	8	1626.2	8.4	8	2359.0	10.0
9	199.4	3.6	9	549.8	5.2	9	1028.2	6.8	9	1634.6	8.4	9	2369.0	10.0
7.0	203.0	3.6	15.0	555.0	5.2	23.0	1035.0	6.8	31.0	1643.0	8.4	39.0	2379.0	10.0
1	206.6	3.6	1	560.2	5.2	1	1041.8	6.8	1	1651.4	8.4	1	2389.0	10.0
2	210.2	3.6	2	565.4	5.2	2	1048.6	6.8	2	1659.8	8.4	2	2399.0	10.0
3	213.9	3.7	3	570.7	5.3	3	1055.5	6.9	3	1668.3	8.5	3	2409.1	10.1
4	217.6	3.7	4	576.0	5.3	4	1062.4	6.9	4	1676.8	8.5	4	2419.2	10.1
5	221.3	3.7	5	581.3	5.3	5	1069.3	6.9	5	1685.3	8.5	5	2429.3	10.1
6	225.0	3.7	6	586.6	5.3	6	1076.2	6.9	6	1693.8	8.5	6	2439.4	10.1
7	228.7	3.7	7	591.9	5.3	7	1083.1	6.9	7	1702.3	8.5	7	2449.5	10.1
8	232.4	3.8	8	597.2	5.4	8	1090.0	7.0	8	1710.8	8.6	8	2459.6	10.2
9	236.2	3.8	9	602.6	5.4	9	1097.0	7.0	9	1719.4	8.6	9	2469.8	10.2

TABLE No. 23.

EXCAVATION AND EMBANKMENT.

AREAS.

Breadth of Base, 24 feet. **Slopes, 1 Horizontal to 1 Perpendicular.**

Height.	Area in Feet.	Difference.	Height.	Area in Feet.	Difference.	Height.	Area in Feet.	Difference.	Height.	Area in Feet.	Difference.	Height.	Area in Feet.	Difference.
0.0	0.0	0.0	8.0	256.0	4.0	16.0	640.0	5.6	24.0	1152.0	7.2	32.0	1792.0	8.8
1	2.4	2.4	1	260.0	4.0	1	645.6	5.6	1	1159.2	7.2	1	1800.8	8.8
2	4.8	2.4	2	264.0	4.0	2	651.2	5.6	2	1166.4	7.2	2	1809.6	8.8
3	7.3	2.5	3	268.1	4.1	3	656.9	5.7	3	1173.7	7.3	3	1818.5	8.9
4	9.8	2.5	4	272.2	4.1	4	662.6	5.7	4	1181.0	7.3	4	1827.4	8.9
5	12.3	2.5	5	276.3	4.1	5	668.3	5.7	5	1188.3	7.3	5	1836.3	8.9
6	14.8	2.5	6	280.4	4.1	6	674.0	5.7	6	1195.6	7.3	6	1845.2	8.9
7	17.3	2.5	7	284.5	4.1	7	679.7	5.7	7	1202.9	7.3	7	1854.1	8.9
8	19.8	2.6	8	288.6	4.2	8	685.4	5.8	8	1210.2	7.4	8	1863.0	9.0
9	22.4	2.6	9	292.8	4.2	9	691.2	5.8	9	1217.6	7.4	9	1872.0	9.0
1.0	25.0	2.6	9.0	297.0	4.2	17.0	697.0	5.8	25.0	1225.0	7.4	33.0	1881.0	9.0
1	27.6	2.6	1	301.2	4.2	1	702.8	5.8	1	1232.4	7.4	1	1890.0	9.0
2	30.2	2.6	2	305.4	4.2	2	708.6	5.8	2	1239.8	7.4	2	1899.0	9.0
3	32.9	2.7	3	309.7	4.3	3	714.5	5.9	3	1247.3	7.5	3	1908.1	9.1
4	35.6	2.7	4	314.0	4.3	4	720.4	5.9	4	1254.8	7.5	4	1917.2	9.1
5	38.3	2.7	5	318.3	4.3	5	726.3	5.9	5	1262.3	7.5	5	1926.3	9.1
6	41.0	2.7	6	322.6	4.3	6	732.2	5.9	6	1269.8	7.5	6	1935.4	9.1
7	43.7	2.7	7	326.9	4.3	7	738.1	5.9	7	1277.3	7.5	7	1944.5	9.1
8	46.4	2.8	8	331.2	4.4	8	744.0	6.0	8	1284.8	7.6	8	1953.6	9.2
9	49.2	2.8	9	335.6	4.4	9	750.0	6.0	9	1292.4	7.6	9	1962.8	9.2
2.0	52.0	2.8	10.0	340.0	4.4	18.0	756.0	6.0	26.0	1300.0	7.6	34.0	1972.0	9.2
1	54.8	2.8	1	344.4	4.4	1	762.0	6.0	1	1307.6	7.6	1	1981.2	9.2
2	57.6	2.8	2	348.8	4.4	2	768.0	6.0	2	1315.2	7.6	2	1990.4	9.2
3	60.5	2.9	3	353.3	4.5	3	774.1	6.1	3	1322.9	7.7	3	1999.7	9.3
4	63.4	2.9	4	357.8	4.5	4	780.2	6.1	4	1330.6	7.7	4	2009.0	9.3
5	66.3	2.9	5	362.3	4.5	5	786.3	6.1	5	1338.3	7.7	5	2018.3	9.3
6	69.2	2.9	6	366.8	4.5	6	792.4	6.1	6	1346.0	7.7	6	2027.6	9.3
7	72.1	2.9	7	371.3	4.5	7	798.5	6.1	7	1353.7	7.7	7	2036.9	9.3
8	75.0	3.0	8	375.8	4.6	8	804.6	6.2	8	1361.4	7.8	8	2046.2	9.4
9	78.0	3.0	9	380.4	4.6	9	810.8	6.2	9	1369.2	7.8	9	2055.6	9.4
3.0	81.0	3.0	11.0	385.0	4.6	19.0	817.0	6.2	27.0	1377.0	7.8	35.0	2065.0	9.4
1	84.0	3.0	1	389.6	4.6	1	823.2	6.2	1	1384.8	7.8	1	2074.4	9.4
2	87.0	3.0	2	394.2	4.6	2	829.4	6.2	2	1392.6	7.8	2	2083.8	9.4
3	90.1	3.1	3	398.9	4.7	3	835.7	6.3	3	1400.5	7.9	3	2093.3	9.5
4	93.2	3.1	4	403.6	4.7	4	842.0	6.3	4	1408.4	7.9	4	2102.8	9.5
5	96.3	3.1	5	408.3	4.7	5	848.3	6.3	5	1416.3	7.9	5	2112.3	9.5
6	99.4	3.1	6	413.0	4.7	6	854.6	6.3	6	1424.2	7.9	6	2121.8	9.5
7	102.5	3.1	7	417.7	4.7	7	860.9	6.3	7	1432.1	7.9	7	2131.3	9.5
8	105.6	3.2	8	422.4	4.8	8	867.2	6.4	8	1440.0	8.0	8	2140.8	9.6
9	108.8	3.2	9	427.2	4.8	9	873.6	6.4	9	1448.0	8.0	9	2150.4	9.6
4.0	112.0	3.2	12.0	432.0	4.8	20.0	880.0	6.4	28.0	1456.0	8.0	36.0	2160.0	9.6
1	115.2	3.2	1	436.8	4.8	1	886.4	6.4	1	1464.0	8.0	1	2169.6	9.6
2	118.4	3.2	2	441.6	4.8	2	892.8	6.4	2	1472.0	8.0	2	2179.2	9.6
3	121.7	3.3	3	446.5	4.9	3	899.3	6.5	3	1480.1	8.1	3	2188.9	9.7
4	125.0	3.3	4	451.4	4.9	4	905.8	6.5	4	1488.2	8.1	4	2198.6	9.7
5	128.3	3.3	5	456.3	4.9	5	912.3	6.5	5	1496.3	8.1	5	2208.3	9.7
6	131.6	3.3	6	461.2	4.9	6	918.8	6.5	6	1504.4	8.1	6	2218.0	9.7
7	134.9	3.3	7	466.1	4.9	7	925.3	6.5	7	1512.5	8.1	7	2227.7	9.7
8	138.2	3.4	8	471.0	5.0	8	931.8	6.6	8	1520.6	8.2	8	2237.4	9.8
9	141.6	3.4	9	476.0	5.0	9	938.4	6.6	9	1528.8	8.2	9	2247.2	9.8
5.0	145.0	3.4	13.0	481.0	5.0	21.0	945.0	6.6	29.0	1537.0	8.2	37.0	2257.0	9.8
1	148.4	3.4	1	486.0	5.0	1	951.6	6.6	1	1545.2	8.2	1	2266.8	9.8
2	151.8	3.4	2	491.0	5.0	2	958.2	6.6	2	1553.4	8.2	2	2276.6	9.8
3	155.3	3.5	3	496.1	5.1	3	964.9	6.7	3	1561.7	8.3	3	2286.5	9.9
4	158.8	3.5	4	501.2	5.1	4	971.6	6.7	4	1570.0	8.3	4	2296.4	9.9
5	162.3	3.5	5	506.3	5.1	5	978.3	6.7	5	1578.3	8.3	5	2306.3	9.9
6	165.8	3.5	6	511.4	5.1	6	985.0	6.7	6	1586.6	8.3	6	2316.2	9.9
7	169.3	3.5	7	516.5	5.1	7	991.7	6.7	7	1594.9	8.3	7	2326.1	9.9
8	172.8	3.6	8	521.6	5.2	8	998.4	6.8	8	1603.2	8.4	8	2336.0	10.0
9	176.4	3.6	9	526.8	5.2	9	1005.2	6.8	9	1611.6	8.4	9	2346.0	10.0
6.0	180.0	3.6	14.0	532.0	5.2	22.0	1012.0	6.8	30.0	1620.0	8.4	38.0	2356.0	10.0
1	183.6	3.6	1	537.2	5.2	1	1018.8	6.8	1	1628.4	8.4	1	2366.0	10.0
2	187.2	3.6	2	542.4	5.2	2	1025.6	6.8	2	1636.8	8.4	2	2376.0	10.0
3	190.9	3.7	3	547.7	5.3	3	1032.5	6.9	3	1645.3	8.5	3	2386.1	10.1
4	194.6	3.7	4	553.0	5.3	4	1039.4	6.9	4	1653.8	8.5	4	2396.2	10.1
5	198.3	3.7	5	558.3	5.3	5	1046.3	6.9	5	1662.3	8.5	5	2406.3	10.1
6	202.0	3.7	6	563.6	5.3	6	1053.2	6.9	6	1670.8	8.5	6	2416.4	10.1
7	205.7	3.7	7	568.9	5.3	7	1060.1	6.9	7	1679.3	8.5	7	3426.5	10.1
8	209.4	3.8	8	574.2	5.4	8	1067.0	7.0	8	1687.8	8.6	8	2436.6	10.2
9	213.2	3.8	9	579.6	5.4	9	1074.0	7.0	9	1696.4	8.6	9	2436.8	10.2
7.0	217.0	3.8	15.0	585.0	5.4	23.0	1081.0	7.0	31.0	1705.0	8.6	39.0	2457.0	10.2
1	220.8	3.8	1	590.4	5.4	1	1088.0	7.0	1	1713.6	8.6	1	2467.2	10.2
2	224.6	3.8	2	595.8	5.4	2	1095.0	7.0	2	1722.2	8.6	2	2477.4	10.2
3	228.5	3.9	3	601.3	5.5	3	1102.1	7.1	3	1730.9	8.7	3	2487.7	10.3
4	232.4	3.9	4	606.8	5.5	4	1109.2	7.1	4	1739.6	8.7	4	2498.0	10.3
5	236.3	3.9	5	612.3	5.5	5	1116.3	7.1	5	1748.3	8.7	5	2508.3	10.3
6	240.2	3.9	6	617.8	5.5	6	1123.4	7.1	6	1757.0	8.7	6	2518.6	10.3
7	244.1	3.9	7	623.3	5.5	7	1130.5	7.1	7	1765.7	8.7	7	2528.9	10.3
8	248.0	4.0	8	628.8	5.6	8	1137.6	7.2	8	1774.4	8.8	8	2539.2	10.4
9	252.0	4.0	9	634.4	5.6	9	1144.8	7.2	9	1783.2	8.8	9	2549.6	10.4

TABLE NO. 24.

EXCAVATION AND EMBANKMENT.

AREAS.

Breadth of Base, Twenty-Six Feet. Slope 1 Horizontal to 1 Perpendicular.

Height.	Area in Feet.	Diff.	Height.	Area in Feet.	Diff.	Height.	Area in Feet.	Diff.	Height.	Area in Feet.	Diff.	Height.	Area in Feet.	Diff.
0.0	0.0	0.0	8.0	272.0	4.2	16.0	672.0	5.8	24.0	1200.0	7.4	32.0	1856.0	9.0
.1	2.6	2.6	.1	276.2	4.2	.1	677.8	5.8	.1	1207.4	7.4	.1	1865.0	9.0
.2	5.2	2.6	.2	280.4	4.2	.2	683.6	5.8	.2	1214.8	7.4	.2	1874.0	9.0
.3	7.9	2.7	.3	284.7	4.3	.3	689.5	5.9	.3	1222.3	7.5	.3	1883.1	9.1
.4	10.6	2.7	.4	289.0	4.3	.4	695.4	5.9	.4	1229.8	7.5	.4	1892.2	9.1
.5	13.3	2.7	.5	293.3	4.3	.5	701.3	5.9	.5	1237.3	7.5	.5	1901.3	9.1
.6	16.0	2.7	.6	297.6	4.3	.6	707.2	5.9	.6	1244.8	7.5	.6	1910.4	9.1
.7	18.7	2.7	.7	301.9	4.3	.7	713.1	5.9	.7	1252.3	7.5	.7	1919.5	9.1
.8	21.4	2.8	.8	306.2	4.4	.8	719.0	6.0	.8	1259.8	7.6	.8	1928.6	9.2
.9	24.2	2.8	.9	310.6	4.4	.9	725.0	6.0	.9	1267.4	7.6	.9	1937.8	9.2
1.0	27.0	2.8	9.0	315.0	4.4	17.0	731.0	6.0	25.0	1275.0	7.6	33.0	1947.0	9.2
.1	29.8	2.8	.1	319.4	4.4	.1	737.0	6.0	.1	1282.6	7.6	.1	1956.2	9.2
.2	32.6	2.8	.2	323.8	4.4	.2	743.0	6.0	.2	1290.2	7.6	.2	1965.4	9.2
.3	35.5	2.9	.3	328.3	4.5	.3	749.1	6.1	.3	1297.9	7.7	.3	1974.7	9.3
.4	38.4	2.9	.4	332.8	4.5	.4	755.2	6.1	.4	1305.6	7.7	.4	1984.0	9.3
.5	41.3	2.9	.5	337.3	4.5	.5	761.3	6.1	.5	1313.3	7.7	.5	1993.3	9.3
.6	44.2	2.9	.6	341.8	4.5	.6	767.4	6.1	.6	1321.0	7.7	.6	2002.6	9.3
.7	47.1	2.9	.7	346.3	4.5	.7	773.5	6.1	.7	1328.7	7.7	.7	2011.9	9.3
.8	50.0	3.0	.8	350.8	4.6	.8	779.6	6.2	.8	1336.4	7.8	.8	2021.2	9.4
.9	53.0	3.0	.9	355.4	4.6	.9	785.8	6.2	.9	1344.2	7.8	.9	2030.6	9.4
2.0	56.0	3.0	10.0	360.0	4.6	18.0	792.0	6.2	26.0	1352.0	7.8	34.0	2040.0	9.4
.1	59.0	3.0	.1	364.6	4.6	.1	798.2	6.2	.1	1359.8	7.8	.1	2049.4	9.4
.2	62.0	3.0	.2	369.2	4.6	.2	804.4	6.2	.2	1367.6	7.8	.2	2058.8	9.4
.3	65.1	3.1	.3	373.9	4.7	.3	810.7	6.3	.3	1375.5	7.9	.3	2068.3	9.5
.4	68.2	3.1	.4	378.6	4.7	.4	817.0	6.3	.4	1383.4	7.9	.4	2077.8	9.5
.5	71.3	3.1	.5	383.3	4.7	.5	823.3	6.3	.5	1391.3	7.9	.5	2087.3	9.5
.6	74.4	3.1	.6	388.0	4.7	.6	829.6	6.3	.6	1399.2	7.9	.6	2096.8	9.5
.7	77.5	3.1	.7	392.7	4.7	.7	835.9	6.3	.7	1407.1	7.9	.7	2106.3	9.5
.8	80.6	3.2	.8	397.4	4.8	.8	842.2	6.4	.8	1415.0	8.0	.8	2115.8	9.6
.9	83.8	3.2	.9	402.2	4.8	.9	848.6	6.4	.9	1423.0	8.0	.9	2125.4	9.6
3.0	87.0	3.2	11.0	407.0	4.8	19.0	855.0	6.4	27.0	1431.0	8.0	35.0	2135.0	9.6
.1	90.2	3.2	.1	411.8	4.8	.1	861.4	6.4	.1	1439.0	8.0	.1	2144.6	9.6
.2	93.4	3.2	.2	416.6	4.8	.2	867.8	6.4	.2	1447.0	8.0	.2	2154.2	9.6
.3	96.7	3.3	.3	421.5	4.9	.3	874.3	6.5	.3	1455.1	8.1	.3	2163.9	9.7
.4	100.0	3.3	.4	426.4	4.9	.4	880.8	6.5	.4	1463.2	8.1	.4	2173.6	9.7
.5	103.3	3.3	.5	431.3	4.9	.5	887.3	6.5	.5	1471.3	8.1	.5	2183.3	9.7
.6	106.6	3.3	.6	436.2	4.9	.6	893.8	6.5	.6	1479.4	8.1	.6	2193.0	9.7
.7	109.9	3.3	.7	441.1	4.9	.7	900.3	6.5	.7	1487.5	8.1	.7	2202.7	9.7
.8	113.2	3.4	.8	446.0	5.0	.8	906.8	6.6	.8	1495.6	8.2	.8	2212.4	9.8
.9	116.6	3.4	.9	451.0	5.0	.9	913.4	6.6	.9	1503.8	8.2	.9	2222.2	9.8
4.0	120.0	3.4	12.0	456.0	5.0	20.0	920.0	6.6	28.0	1512.0	8.2	36.0	2232.0	9.8
.1	123.4	3.4	.1	461.0	5.0	.1	926.6	6.6	.1	1520.2	8.2	.1	2241.8	9.8
.2	126.8	3.4	.2	466.0	5.0	.2	933.2	6.6	.2	1528.4	8.2	.2	2251.6	9.8
.3	130.3	3.5	.3	471.1	5.1	.3	939.9	6.7	.3	1536.7	8.3	.3	2261.5	9.9
.4	133.8	3.5	.4	476.2	5.1	.4	946.6	6.7	.4	1545.0	8.3	.4	2271.4	9.9
.5	137.3	3.5	.5	481.3	5.1	.5	953.3	6.7	.5	1553.3	8.3	.5	2281.3	9.9
.6	140.8	3.5	.6	486.4	5.1	.6	960.0	6.7	.6	1561.6	8.3	.6	2291.2	9.9
.7	144.3	3.5	.7	491.5	5.1	.7	966.7	6.7	.7	1569.9	8.3	.7	2301.1	9.9
.8	147.8	3.6	.8	496.6	5.2	.8	973.4	6.8	.8	1578.2	8.4	.8	2311.0	10.0
.9	151.4	3.6	.9	501.8	5.2	.9	980.2	6.8	.9	1586.6	8.4	.9	2321.0	10.0
5.0	155.0	3.6	13.0	507.0	5.2	21.0	987.0	6.8	29.0	1595.0	8.4	37.0	2331.0	10.0
.1	158.6	3.6	.1	512.2	5.2	.1	993.8	6.8	.1	1603.4	8.4	.1	2341.0	10.0
.2	162.2	3.6	.2	517.4	5.2	.2	1000.6	6.8	.2	1611.8	8.4	.2	2351.0	10.0
.3	165.9	3.7	.3	522.7	5.3	.3	1007.5	6.9	.3	1620.3	8.5	.3	2361.1	10.1
.4	169.6	3.7	.4	528.0	5.3	.4	1014.4	6.9	.4	1628.8	8.5	.4	2371.2	10.1
.5	173.3	3.7	.5	533.3	5.3	.5	1021.3	6.9	.5	1637.3	8.5	.5	2381.3	10.1
.6	177.0	3.7	.6	538.6	5.3	.6	1028.2	6.9	.6	1645.8	8.5	.6	2391.4	10.1
.7	180.7	3.7	.7	543.9	5.3	.7	1035.1	6.9	.7	1654.3	8.5	.7	2401.5	10.1
.8	184.4	3.8	.8	549.2	5.4	.8	1042.0	7.0	.8	1662.8	8.6	.8	2411.6	10.2
.9	188.2	3.8	.9	554.6	5.4	.9	1049.0	7.0	.9	1671.4	8.6	.9	2421.8	10.2
6.0	192.0	3.8	14.0	560.0	5.4	22.0	1056.0	7.0	30.0	1680.0	8.6	38.0	2432.0	10.2
.1	195.8	3.8	.1	565.4	5.4	.1	1063.0	7.0	.1	1688.6	8.6	.1	2442.2	10.2
.2	199.6	3.8	.2	570.8	5.4	.2	1070.0	7.0	.2	1697.2	8.6	.2	2452.4	10.2
.3	203.5	3.9	.3	576.3	5.5	.3	1077.1	7.1	.3	1705.9	8.7	.3	2462.7	10.3
.4	207.4	3.9	.4	581.8	5.5	.4	1084.2	7.1	.4	1714.6	8.7	.4	2473.0	10.3
.5	211.3	3.9	.5	587.3	5.5	.5	1091.3	7.1	.5	1723.3	8.7	.5	2483.3	10.3
.6	215.2	3.9	.6	592.8	5.5	.6	1098.4	7.1	.6	1732.0	8.7	.6	2493.6	10.3
.7	219.1	3.9	.7	598.3	5.5	.7	1105.5	7.1	.7	1740.7	8.7	.7	2503.9	10.3
.8	223.0	4.0	.8	603.8	5.6	.8	1112.6	7.2	.8	1749.4	8.8	.8	2514.2	10.4
.9	227.0	4.0	.9	609.4	5.6	.9	1119.8	7.2	.9	1758.2	8.8	.9	2524.6	10.4
7.0	231.0	4.0	15.0	615.0	5.6	23.0	1127.0	7.2	31.0	1767.0	8.8	39.0	2535.0	10.4
.1	235.0	4.0	.1	620.6	5.6	.1	1134.2	7.2	.1	1775.8	8.8	.1	2545.4	10.4
.2	239.0	4.0	.2	626.2	5.6	.2	1141.4	7.2	.2	1784.6	8.8	.2	2555.8	10.4
.3	243.1	4.1	.3	631.9	5.7	.3	1148.7	7.3	.3	1793.5	8.9	.3	2566.3	10.5
.4	247.2	4.1	.4	637.6	5.7	.4	1156.0	7.3	.4	1802.4	8.9	.4	2576.8	10.5
.5	251.3	4.1	.5	643.3	5.7	.5	1163.3	7.3	.5	1811.3	8.9	.5	2587.3	10.5
.6	255.4	4.1	.6	649.0	5.7	.6	1170.6	7.3	.6	1820.2	8.9	.6	2597.8	10.5
.7	259.5	4.1	.7	654.7	5.7	.7	1177.9	7.3	.7	1829.1	8.9	.7	2608.3	10.5
.8	263.6	4.2	.8	660.4	5.8	.8	1185.2	7.4	.8	1838.0	9.0	.8	2618.8	10.6
.9	267.8	4.2	.9	666.2	5.8	.9	1192.6	7.4	.9	1847.0	9.0	.9	2629.4	10.6

TABLE No. 25.

EXCAVATION AND EMBANKMENT.

AREAS.

Breadth of Base, 28 feet. Slopes, 1 Horizontal to 1 Perpendicular.

Height.	Area in Feet.	Difference.	Height.	Area in Feet.	Difference.	Height.	Area in Feet.	Difference.	Height.	Area in Feet.	Difference.	Height.	Area in Feet.	Difference.
0.0	0.0	0.0	8.0	288.0	4.4	16.0	704.0	6.0	24.0	1248.0	7.6	32.0	1920.0	9.2
1	2.8	2.8	1	292.4	4.4	1	710.0	6.0	1	1255.6	7.6	1	1929.2	9.2
2	5.6	2.8	2	296.8	4.4	2	716.0	6.0	2	1263.2	7.6	2	1938.4	9.2
3	8.5	2.9	3	301.3	4.5	3	722.1	6.1	3	1270.9	7.7	3	1947.7	9.3
4	11.4	2.9	4	305.8	4.5	4	728.2	6.1	4	1278.6	7.7	4	1957.0	9.3
5	14.3	2.9	5	310.3	4.5	5	734.3	6.1	5	1286.3	7.7	5	1966.3	9.3
6	17.2	2.9	6	314.8	4.5	6	740.4	6.1	6	1294.0	7.7	6	1975.6	9.3
7	20.1	2.9	7	319.3	4.5	7	746.5	6.1	7	1301.7	7.7	7	1984.9	9.3
8	23.0	3.0	8	323.8	4.6	8	752.6	6.2	8	1309.4	7.8	8	1994.2	9.4
9	26.0	3.0	9	328.4	4.6	9	758.8	6.2	9	1317.2	7.8	9	2003.6	9.4
1.0	29.0	3.0	9.0	333.0	4.6	17.0	765.0	6.2	25.0	1325.0	7.8	33.0	2013.0	9.4
1	32.0	3.0	1	337.6	4.6	1	771.2	6.2	1	1332.8	7.8	1	2022.4	9.4
2	35.0	3.0	2	342.2	4.6	2	777.4	6.2	2	1340.6	7.8	2	2031.8	9.4
3	38.1	3.1	3	346.9	4.7	3	783.7	6.3	3	1348.5	7.9	3	2041.3	9.5
4	41.2	3.1	4	351.6	4.7	4	790.0	6.3	4	1356.4	7.9	4	2050.8	9.5
5	44.3	3.1	5	356.3	4.7	5	796.3	6.3	5	1364.3	7.9	5	2060.3	9.5
6	47.4	3.1	6	361.0	4.7	6	802.6	6.3	6	1372.2	7.9	6	2069.8	9.5
7	50.5	3.1	7	365.7	4.7	7	808.9	6.3	7	1380.1	7.9	7	2079.3	9.5
8	53.6	3.2	8	370.4	4.8	8	815.2	6.4	8	1388.0	8.0	8	2088.8	9.6
9	56.8	3.2	9	375.2	4.8	9	821.6	6.4	9	1396.0	8.0	9	2098.4	9.6
2.0	60.0	3.2	10.0	380.0	4.8	18.0	828.0	6.4	26.0	1404.0	8.0	34.0	2108.0	9.6
1	63.2	3.2	1	384.8	4.8	1	834.4	6.4	1	1412.0	8.0	1	2117.6	9.6
2	66.4	3.2	2	389.6	4.8	2	840.8	6.4	2	1420.0	8.0	2	2127.2	9.6
3	69.7	3.3	3	394.5	4.9	3	847.3	6.5	3	1428.1	8.1	3	2136.9	9.7
4	73.0	3.3	4	399.4	4.9	4	853.8	6.5	4	1436.2	8.1	4	2146.6	9.7
5	76.3	3.3	5	404.3	4.9	5	860.3	6.5	5	1444.3	8.1	5	2156.3	9.7
6	79.6	3.3	6	409.2	4.9	6	866.8	6.5	6	1452.4	8.1	6	2166.0	9.7
7	82.9	3.3	7	414.1	4.9	7	873.3	6.5	7	1460.5	8.1	7	2175.7	9.7
8	86.2	3.4	8	419.0	5.0	8	879.8	6.6	8	1468.6	8.2	8	2185.4	9.8
9	89.6	3.4	9	424.0	5.0	9	886.4	6.6	9	1476.8	8.2	9	2195.2	9.8
3.0	93.0	3.4	11.0	429.0	5.0	19.0	893.0	6.6	27.0	1485.0	8.2	35.0	2205.0	9.8
1	96.4	3.4	1	434.0	5.0	1	899.6	6.6	1	1493.2	8.2	1	2214.8	9.8
2	99.8	3.4	2	439.0	5.0	2	906.2	6.6	2	1501.4	8.2	2	2224.6	9.8
3	103.3	3.5	3	444.1	5.1	3	912.9	6.7	3	1509.7	8.3	3	2234.5	9.9
4	106.8	3.5	4	449.2	5.1	4	919.6	6.7	4	1518.0	8.3	4	2244.4	9.9
5	110.3	3.5	5	454.3	5.1	5	926.3	6.7	5	1526.3	8.3	5	2254.3	9.9
6	113.8	3.5	6	459.4	5.1	6	933.0	6.7	6	1534.6	8.3	6	2264.2	9.9
7	117.3	3.5	7	464.5	5.1	7	939.7	6.7	7	1542.9	8.3	7	2274.1	9.9
8	120.8	3.6	8	469.6	5.2	8	946.4	6.8	8	1551.2	8.4	8	2284.0	10.0
9	124.4	3.6	9	474.8	5.2	9	953.2	6.8	9	1559.6	8.4	9	2294.0	10.0
4.0	128.0	3.6	12.0	480.0	5.2	20.0	960.0	6.8	28.0	1568.0	8.4	36.0	2304.0	10.0
1	131.6	3.6	1	485.2	5.2	1	966.8	6.8	1	1576.4	8.4	1	2314.0	10.0
2	135.2	3.6	2	490.4	5.2	2	973.6	6.8	2	1584.8	8.4	2	2324.0	10.0
3	138.9	3.7	3	495.7	5.3	3	980.5	6.9	3	1593.3	8.5	3	2334.1	10.1
4	142.6	3.7	4	501.0	5.3	4	987.4	6.9	4	1601.8	8.5	4	2344.2	10.1
5	146.3	3.7	5	506.3	5.3	5	994.3	6.9	5	1610.3	8.5	5	2354.3	10.1
6	150.0	3.7	6	511.6	5.3	6	1001.2	6.9	6	1618.8	8.5	6	2364.4	10.1
7	153.7	3.7	7	516.9	5.3	7	1008.1	6.9	7	1627.3	8.5	7	2374.5	10.1
8	157.4	3.8	8	522.2	5.4	8	1015.0	7.0	8	1635.8	8.6	8	2384.6	10.2
9	161.2	3.8	9	527.6	5.4	9	1022.0	7.0	9	1644.4	8.6	9	2394.8	10.2
5.0	165.0	3.8	13.0	533.0	5.4	21.0	1029.0	7.0	29.0	1653.0	8.6	37.0	2405.0	10.2
1	168.8	3.8	1	538.4	5.4	1	1036.0	7.0	1	1661.6	8.6	1	2415.2	10.2
2	172.6	3.8	2	543.8	5.4	2	1043.0	7.0	2	1670.2	8.6	2	2425.4	10.2
3	176.5	3.9	3	549.3	5.5	3	1050.1	7.1	3	1678.9	8.7	3	2435.7	10.3
4	180.4	3.9	4	554.8	5.5	4	1057.2	7.1	4	1687.6	8.7	4	2446.0	10.3
5	184.3	3.9	5	560.3	5.5	5	1064.3	7.1	5	1696.3	8.7	5	2456.3	10.3
6	188.2	3.9	6	565.8	5.5	6	1071.4	7.1	6	1705.0	8.7	6	2466.6	10.3
7	192.1	3.9	7	571.3	5.5	7	1078.5	7.1	7	1713.7	8.7	7	2476.9	10.3
8	196.0	4.0	8	576.8	5.6	8	1085.6	7.2	8	1722.4	8.8	8	2487.2	10.4
9	200.0	4.0	9	582.4	5.6	9	1092.8	7.2	9	1731.2	8.8	9	2497.6	10.4
6.0	204.0	4.0	14.0	588.0	5.6	22.0	1100.0	7.2	30.0	1740.0	8.8	38.0	2508.0	10.4
1	208.0	4.0	1	593.6	5.6	1	1107.2	7.2	1	1748.8	8.8	1	2518.4	10.4
2	212.0	4.0	2	599.2	5.6	2	1114.4	7.2	2	1757.6	8.8	2	2528.8	10.4
3	216.1	4.1	3	604.9	5.7	3	1121.7	7.3	3	1766.5	8.9	3	2539.3	10.5
4	220.2	4.1	4	610.6	5.7	4	1129.0	7.3	4	1775.4	8.9	4	2549.8	10.5
5	224.3	4.1	5	616.3	5.7	5	1136.3	7.3	5	1784.3	8.9	5	2560.3	10.5
6	228.4	4.1	6	622.0	5.7	6	1143.6	7.3	6	1793.2	8.9	6	2570.8	10.5
7	232.5	4.1	7	627.7	5.7	7	1150.9	7.3	7	1802.1	8.9	7	2581.3	10.5
8	236.6	4.2	8	633.4	5.8	8	1158.2	7.4	8	1811.0	9.0	8	2591.8	10.6
9	240.8	4.2	9	639.2	5.8	9	1165.6	7.4	9	1820.0	9.0	9	2602.4	10.6
7.0	245.0	4.2	15.0	645.0	5.8	23.0	1173.0	7.4	31.0	1829.0	9.0	39.0	2613.0	10.6
1	249.2	4.2	1	650.8	5.8	1	1180.4	7.4	1	1838.0	9.0	1	2623.6	10.6
2	253.4	4.2	2	656.6	5.8	2	1187.8	7.4	2	1847.0	9.0	2	2634.2	10.6
3	257.7	4.3	3	662.5	5.9	3	1195.3	7.5	3	1856.1	9.1	3	2644.9	10.7
4	262.0	4.3	4	668.4	5.9	4	1202.8	7.5	4	1865.2	9.1	4	2655.6	10.7
5	266.3	4.3	5	674.3	5.9	5	1210.3	7.5	5	1874.3	9.1	5	2666.3	10.7
6	270.6	4.3	6	680.2	5.9	6	1217.8	7.5	6	1883.4	9.1	6	2677.0	10.7
7	274.9	4.3	7	686.1	5.9	7	1225.3	7.5	7	1892.5	9.1	7	2687.7	10.7
8	279.2	4.4	8	692.0	6.0	8	1232.8	7.6	8	1901.6	9.2	8	2698.4	10.8
9	283.6	4.4	9	698.0	6.0	9	1240.4	7.6	9	1910.8	9.2	9	2709.2	10.8

TABLE NO. 26.

EXCAVATION AND EMBANKMENT.

AREAS.

Breadth of Base, Thirty Feet. **Slope 1 Horizontal to 1 Perpendicular.**

Height.	Area in Feet.	Diff.	Height.	Area in Feet.	Diff.	Height.	Area in Feet.	Diff.	Height.	Area in Feet.	Diff.	Height.	Area in Feet.	Diff.
0.0	0.0	0.0	8.0	304.0	4.6	16.0	736.0	6.2	24.0	1296.0	7.8	32.0	1984.0	9.4
.1	3.0	3.0	.1	308.6	4.6	.1	742.2	6.2	.1	1303.8	7.8	.1	1993.4	9.4
.2	6.0	3.0	.2	313.2	4.6	.2	748.4	6.2	.2	1311.6	7.8	.2	2002.8	9.4
.3	9.1	3.1	.3	317.9	4.7	.3	754.7	6.3	.3	1319.5	7.9	.3	2012.3	9.5
.4	12.2	3.1	.4	322.6	4.7	.4	761.0	6.3	.4	1327.4	7.9	.4	2021.8	9.5
.5	15.3	3.1	.5	327.3	4.7	.5	767.3	6.3	.5	1335.3	7.9	.5	2031.3	9.5
.6	18.4	3.1	.6	332.0	4.7	.6	773.6	6.3	.6	1343.2	7.9	.6	2040.8	9.5
.7	21.5	3.1	.7	336.7	4.7	.7	779.9	6.3	.7	1351.1	7.9	.7	2050.3	9.5
.8	24.6	3.2	.8	341.4	4.8	.8	786.2	6.4	.8	1359.0	8.0	.8	2059.8	9.6
.9	27.8	3.2	.9	346.2	4.8	.9	792.6	6.4	.9	1367.0	8.0	.9	2069.4	9.6
1.0	31.0	3.2	9.0	351.0	4.8	17.0	799.0	6.4	25.0	1375.0	8.0	33.0	2079.0	9.6
.1	34.2	3.2	.1	355.8	4.8	.1	805.4	6.4	.1	1383.0	8.0	.1	2088.6	9.6
.2	37.4	3.2	.2	360.6	4.8	.2	811.8	6.4	.2	1391.0	8.0	.2	2098.2	9.6
.3	40.7	3.3	.3	365.5	4.9	.3	818.3	6.5	.3	1399.1	8.1	.3	2107.9	9.7
.4	44.0	3.3	.4	370.4	4.9	.4	824.8	6.5	.4	1407.2	8.1	.4	2117.6	9.7
.5	47.3	3.3	.5	375.3	4.9	.5	831.3	6.5	.5	1415.3	8.1	.5	2127.3	9.7
.6	50.6	3.3	.6	380.2	4.9	.6	837.8	6.5	.6	1423.4	8.1	.6	2137.0	9.7
.7	53.9	3.3	.7	385.1	4.9	.7	844.3	6.5	.7	1431.5	8.1	.7	2146.7	9.7
.8	57.2	3.4	.8	390.0	5.0	.8	850.8	6.6	.8	1439.6	8.2	.8	2156.4	9.8
.9	60.6	3.4	.9	395.0	5.0	.9	857.4	6.6	.9	1447.8	8.2	.9	2166.2	9.8
2.0	64.0	3.4	10.0	400.0	5.0	18.0	864.0	6.6	26.0	1456.0	8.2	34.0	2176.0	9.8
.1	67.4	3.4	.1	405.0	5.0	.1	870.6	6.6	.1	1464.2	8.2	.1	2185.8	9.8
.2	70.8	3.4	.2	410.0	5.0	.2	877.2	6.6	.2	1472.4	8.2	.2	2195.6	9.8
.3	74.3	3.5	.3	415.1	5.1	.3	883.9	6.7	.3	1480.7	8.3	.3	2205.5	9.9
.4	77.8	3.5	.4	420.2	5.1	.4	890.6	6.7	.4	1489.0	8.3	.4	2215.4	9.9
.5	81.3	3.5	.5	425.3	5.1	.5	897.3	6.7	.5	1497.3	8.3	.5	2225.3	9.9
.6	84.8	3.5	.6	430.4	5.1	.6	904.0	6.7	.6	1505.6	8.3	.6	2235.2	9.9
.7	88.3	3.5	.7	435.5	5.1	.7	910.7	6.7	.7	1513.9	8.3	.7	2245.1	9.9
.8	91.8	3.6	.8	440.6	5.2	.8	917.4	6.8	.8	1522.2	8.4	.8	2255.0	10.0
.9	95.4	3.6	.9	445.8	5.2	.9	924.2	6.8	.9	1530.6	8.4	.9	2265.0	10.0
3.0	99.0	3.6	11.0	451.0	5.2	19.0	931.0	6.8	27.0	1539.0	8.4	35.0	2275.0	10.0
.1	102.6	3.6	.1	456.2	5.2	.1	937.8	6.8	.1	1547.4	8.4	.1	2285.0	10.0
.2	106.3	3.6	.2	461.4	5.2	.2	944.6	6.8	.2	1555.8	8.4	.2	2295.0	10.0
.3	109.9	3.7	.3	466.7	5.3	.3	951.5	6.9	.3	1564.3	8.5	.3	2305.1	10.1
.4	113.6	3.7	.4	472.0	5.3	.4	958.4	6.9	.4	1572.8	8.5	.4	2315.2	10.1
.5	117.3	3.7	.5	477.3	5.3	.5	965.3	6.9	.5	1581.3	8.5	.5	2325.3	10.1
.6	121.0	3.7	.6	482.6	5.3	.6	972.2	6.9	.6	1589.8	8.5	.6	2335.4	10.1
.7	124.7	3.7	.7	487.9	5.3	.7	979.1	6.9	.7	1598.3	8.5	.7	2345.5	10.1
.8	128.4	3.8	.8	493.2	5.4	.8	986.0	7.0	.8	1606.8	8.6	.8	2355.6	10.2
.9	132.2	3.8	.9	498.6	5.4	.9	993.0	7.0	.9	1615.4	8.6	.9	2365.8	10.2
4.0	136.0	3.8	12.0	504.0	5.4	20.0	1000.0	7.0	28.0	1624.0	8.6	36.0	2376.0	10.2
.1	139.8	3.8	.1	509.4	5.4	.1	1007.0	7.0	.1	1632.6	8.6	.1	2386.2	10.2
.2	143.6	3.8	.2	514.8	5.4	.2	1014.0	7.0	.2	1641.2	8.6	.2	2396.4	10.2
.3	147.4	3.9	.3	520.3	5.5	.3	1021.1	7.1	.3	1649.9	8.7	.3	2406.7	10.3
.4	151.4	3.9	.4	525.8	5.5	.4	1028.2	7.1	.4	1658.6	8.7	.4	2417.0	10.3
.5	155.3	3.9	.5	531.3	5.5	.5	1035.3	7.1	.5	1667.3	8.7	.5	2427.3	10.3
.6	159.2	3.9	.6	536.8	5.5	.6	1042.4	7.1	.6	1676.0	8.7	.6	2437.6	10.3
.7	163.1	3.9	.7	542.3	5.5	.7	1049.5	7.1	.7	1684.7	8.7	.7	2447.9	10.3
.8	167.0	4.0	.8	547.8	5.6	.8	1056.6	7.2	.8	1693.4	8.8	.8	2458.2	10.4
.9	171.0	4.0	.9	553.4	5.6	.9	1063.8	7.2	.9	1702.2	8.8	.9	2468.6	10.4
5.0	175.0	4.0	13.0	559.0	5.6	21.0	1071.0	7.2	29.0	1711.0	8.8	37.0	2479.0	10.4
.1	179.0	4.0	.1	564.6	5.6	.1	1078.2	7.2	.1	1719.8	8.8	.1	2489.4	10.4
.2	183.0	4.0	.2	570.2	5.6	.2	1085.4	7.2	.2	1728.6	8.8	.2	2499.8	10.4
.3	187.1	4.1	.3	575.9	5.7	.3	1092.7	7.3	.3	1737.5	8.9	.3	2510.3	10.5
.4	191.2	4.1	.4	581.6	5.7	.4	1100.0	7.3	.4	1746.4	8.9	.4	2520.8	10.5
.5	195.3	4.1	.5	587.3	5.7	.5	1107.3	7.3	.5	1755.3	8.9	.5	2531.3	10.5
.6	199.4	4.1	.6	593.0	5.7	.6	1114.6	7.3	.6	1764.2	8.9	.6	2541.8	10.5
.7	203.5	4.1	.7	598.7	5.7	.7	1121.9	7.3	.7	1773.1	8.9	.7	2552.3	10.5
.8	207.6	4.2	.8	604.4	5.8	.8	1129.2	7.4	.8	1782.0	9.0	.8	2562.8	10.6
.9	211.8	4.2	.9	610.2	5.8	.9	1136.6	7.4	.9	1791.0	9.0	.9	2573.4	10.6
6.0	216.0	4.2	14.0	616.0	5.8	22.0	1144.0	7.4	30.0	1800.0	9.0	38.0	2584.0	10.6
.1	220.2	4.2	.1	621.8	5.8	.1	1151.4	7.4	.1	1809.0	9.0	.1	2594.6	10.6
.2	224.4	4.2	.2	627.6	5.8	.2	1158.8	7.4	.2	1818.0	9.0	.2	2605.2	10.6
.3	228.7	4.3	.3	633.5	5.9	.3	1166.3	7.5	.3	1827.1	9.1	.3	2615.9	10.7
.4	233.0	4.3	.4	639.4	5.9	.4	1173.8	7.5	.4	1836.2	9.1	.4	2626.6	10.7
.5	237.3	4.3	.5	645.3	5.9	.5	1181.3	7.5	.5	1845.3	9.1	.5	2637.3	10.7
.6	241.6	4.3	.6	651.2	5.9	.6	1188.8	7.5	.6	1854.4	9.1	.6	2648.0	10.7
.7	245.9	4.3	.7	657.1	5.9	.7	1196.3	7.5	.7	1863.5	9.1	.7	2658.7	10.7
.8	250.2	4.4	.8	663.0	6.0	.8	1203.8	7.6	.8	1872.6	9.2	.8	2669.4	10.8
.9	254.6	4.4	.9	669.0	6.0	.9	1211.4	7.6	.9	1881.8	9.2	.9	2680.2	10.8
7.0	259.0	4.4	15.0	675.0	6.0	23.0	1219.0	7.6	31.0	1891.0	9.2	39.0	2691.0	10.8
.1	263.4	4.4	.1	681.0	6.0	.1	1226.6	7.6	.1	1900.2	9.2	.1	2701.8	10.8
.2	267.8	4.4	.2	687.0	6.0	.2	1234.2	7.6	.2	1909.4	9.2	.2	2712.6	10.8
.3	272.3	4.5	.3	693.1	6.1	.3	1241.9	7.7	.3	1918.7	9.3	.3	2723.5	10.9
.4	276.8	4.5	.4	699.2	6.1	.4	1249.6	7.7	.4	1928.0	9.3	.4	2734.4	10.9
.5	281.3	4.5	.5	705.3	6.1	.5	1257.3	7.7	.5	1937.3	9.3	.5	2745.3	10.9
.6	285.8	4.5	.6	711.4	6.1	.6	1265.0	7.7	.6	1946.6	9.3	.6	2756.2	10.9
.7	290.3	4.5	.7	717.5	6.1	.7	1272.7	7.7	.7	1955.9	9.3	.7	2767.1	10.9
.8	294.8	4.6	.8	723.6	6.2	.8	1280.4	7.8	.8	1965.2	9.4	.8	2778.0	11.0
.9	299.4	4.6	.9	729.8	6.2	.9	1288.2	7.8	.9	1974.6	9.4	.9	2789.0	11.0

TABLE NO. 27.

EXCAVATION AND EMBANKMENT.

AREAS.

Breadth of Base, Twenty Feet. Slope 1-2 Horizontal to 1 Perpendicular.

Height.	Area in Feet.	Diff.	Height.	Area in Feet.	Diff.	Height.	Area in Feet.	Diff.	Height.	Area in Feet.	Diff.	Height.	Area in Feet.	Diff.
0.0	0.0	2.0	8.0	192.0	2.8	16.0	448.0	3.6	24.0	768.0	4.4	32.0	1152.0	5.2
.1	2.0	2.0	.1	194.8	2.8	.1	451.6	3.6	.1	772.4	4.4	.1	1157.2	5.2
.2	4.0	2.0	.2	197.6	2.8	.2	455.2	3.6	.2	776.8	4.4	.2	1162.4	5.2
.3	6.0	2.0	.3	200.4	2.8	.3	458.8	3.6	.3	781.2	4.4	.3	1167.6	5.2
.4	8.1	2.0	.4	203.3	2.8	.4	462.5	3.6	.4	785.7	4.4	.4	1172.9	5.2
.5	10.1	2.0	.5	206.1	2.8	.5	466.1	3.6	.5	790.1	4.4	.5	1178.1	5.2
.6	12.2	2.1	.6	209.0	2.9	.6	469.8	3.7	.6	794.6	4.5	.6	1183.4	5.3
.7	14.2	2.1	.7	211.8	2.9	.7	473.4	3.7	.7	799.0	4.5	.7	1188.6	5.3
.8	16.3	2.1	.8	214.7	2.9	.8	477.1	3.7	.8	803.5	4.5	.8	1193.9	5.3
.9	18.4	2.1	.9	217.6	2.9	.9	480.8	3.7	.9	808.0	4.5	.9	1199.2	5.3
1.0	20.5	2.1	9.0	220.5	2.9	17.0	484.5	3.7	25.0	812.5	4.5	33.0	1204.5	5.3
.1	22.6	2.1	.1	223.4	2.9	.1	488.2	3.7	.1	817.0	4.5	.1	1209.8	5.3
.2	24.7	2.1	.2	226.3	2.9	.2	491.9	3.7	.2	821.5	4.5	.2	1215.1	5.3
.3	26.8	2.1	.3	229.2	2.9	.3	495.6	3.7	.3	826.0	4.5	.3	1220.4	5.3
.4	29.0	2.1	.4	232.2	2.9	.4	499.4	3.7	.4	830.6	4.5	.4	1225.8	5.3
.5	31.1	2.1	.5	235.1	2.9	.5	503.1	3.7	.5	835.1	4.5	.5	1231.1	5.3
.6	33.3	2.2	.6	238.1	3.0	.6	506.9	3.8	.6	839.7	4.6	.6	1236.5	5.4
.7	35.4	2.2	.7	241.0	3.0	.7	510.6	3.8	.7	844.2	4.6	.7	1241.8	5.4
.8	37.6	2.2	.8	244.0	3.0	.8	514.4	3.8	.8	848.8	4.6	.8	1247.2	5.4
.9	39.8	2.2	.9	247.0	3.0	.9	518.2	3.8	.9	853.4	4.6	.9	1252.6	5.4
2.0	42.0	2.2	10.0	250.0	3.0	18.0	522.0	3.8	26.0	858.0	4.6	34.0	1258.0	5.4
.1	44.2	2.2	.1	253.0	3.0	.1	525.8	3.8	.1	862.6	4.6	.1	1263.4	5.4
.2	46.4	2.2	.2	256.0	3.0	.2	529.6	3.8	.2	867.2	4.6	.2	1268.8	5.4
.3	48.6	2.2	.3	259.0	3.0	.3	533.4	3.8	.3	871.8	4.6	.3	1274.2	5.4
.4	50.9	2.2	.4	262.1	3.0	.4	537.3	3.8	.4	876.5	4.6	.4	1279.7	5.4
.5	53.1	2.2	.5	265.1	3.0	.5	541.1	3.8	.5	881.1	4.6	.5	1285.1	5.4
.6	55.4	2.3	.6	268.2	3.1	.6	545.0	3.9	.6	885.8	4.7	.6	1290.6	5.5
.7	57.6	2.3	.7	271.2	3.1	.7	548.8	3.9	.7	890.4	4.7	.7	1296.0	5.5
.8	59.9	2.3	.8	274.3	3.1	.8	552.7	3.9	.8	895.1	4.7	.8	1301.5	5.5
.9	62.2	2.3	.9	277.4	3.1	.9	556.6	3.9	.9	899.8	4.7	.9	1307.0	5.5
3.0	64.5	2.3	11.0	280.5	3.1	19.0	560.5	3.9	27.0	904.5	4.7	35.0	1312.5	5.5
.1	66.8	2.3	.1	283.6	3.1	.1	564.4	3.9	.1	909.2	4.7	.1	1318.0	5.5
.2	69.1	2.3	.2	286.7	3.1	.2	568.3	3.9	.2	913.9	4.7	.2	1323.5	5.5
.3	71.4	2.3	.3	289.8	3.1	.3	572.2	3.9	.3	918.6	4.7	.3	1329.0	5.5
.4	73.8	2.3	.4	293.0	3.1	.4	576.2	3.9	.4	923.4	4.7	.4	1334.6	5.5
.5	76.1	2.3	.5	296.1	3.1	.5	580.1	3.9	.5	928.1	4.7	.5	1340.1	5.5
.6	78.5	2.4	.6	299.3	3.2	.6	584.1	4.0	.6	932.9	4.8	.6	1345.7	5.6
.7	80.8	2.4	.7	302.4	3.2	.7	588.0	4.0	.7	937.6	4.8	.7	1351.2	5.6
.8	83.2	2.4	.8	305.6	3.2	.8	592.0	4.0	.8	942.4	4.8	.8	1356.8	5.6
.9	85.6	2.4	.9	308.8	3.2	.9	596.0	4.0	.9	947.2	4.8	.9	1362.4	5.6
4.0	88.0	2.4	12.0	312.0	3.2	20.0	600.0	4.0	28.0	952.0	4.8	36.0	1368.0	5.6
.1	90.4	2.4	.1	315.2	3.2	.1	604.0	4.0	.1	956.8	4.8	.1	1373.6	5.6
.2	92.8	2.4	.2	318.4	3.2	.2	608.0	4.0	.2	961.6	4.8	.2	1379.2	5.6
.3	95.2	2.4	.3	321.6	3.2	.3	612.0	4.0	.3	966.4	4.8	.3	1384.8	5.6
.4	97.7	2.4	.4	324.9	3.2	.4	616.1	4.0	.4	971.3	4.8	.4	1390.5	5.6
.5	100.1	2.4	.5	328.1	3.2	.5	620.1	4.0	.5	976.1	4.8	.5	1396.1	5.6
.6	102.6	2.5	.6	331.4	3.3	.6	624.2	4.1	.6	981.0	4.9	.6	1401.8	5.7
.7	105.0	2.5	.7	334.6	3.3	.7	628.2	4.1	.7	985.8	4.9	.7	1407.4	5.7
.8	107.5	2.5	.8	337.9	3.3	.8	632.3	4.1	.8	990.7	4.9	.8	1413.1	5.7
.9	110.0	2.5	.9	341.2	3.3	.9	636.4	4.1	.9	995.6	4.9	.9	1418.8	5.7
5.0	112.5	2.5	13.0	344.5	3.3	21.0	640.5	4.1	29.0	1000.5	4.9	37.0	1424.5	5.7
.1	115.0	2.5	.1	347.8	3.3	.1	644.6	4.1	.1	1005.4	4.9	.1	1430.2	5.7
.2	117.5	2.5	.2	351.1	3.3	.2	648.7	4.1	.2	1010.3	4.9	.2	1435.9	5.7
.3	120.0	2.5	.3	354.4	3.3	.3	652.8	4.1	.3	1015.2	4.9	.3	1441.6	5.7
.4	122.6	2.5	.4	357.8	3.3	.4	657.0	4.1	.4	1020.2	4.9	.4	1447.4	5.7
.5	125.1	2.5	.5	361.1	3.3	.5	661.1	4.1	.5	1025.1	4.9	.5	1453.1	5.7
.6	127.7	2.6	.6	364.5	3.4	.6	665.3	4.2	.6	1030.1	5.0	.6	1458.9	5.8
.7	130.2	2.6	.7	367.8	3.4	.7	669.4	4.2	.7	1035.0	5.0	.7	1464.6	5.8
.8	132.8	2.6	.8	371.2	3.4	.8	673.6	4.2	.8	1040.0	5.0	.8	1470.4	5.8
.9	135.4	2.6	.9	374.6	3.4	.9	677.8	4.2	.9	1045.0	5.0	.9	1476.2	5.8
6.0	138.0	2.6	14.0	378.0	3.4	22.0	682.0	4.2	30.0	1050.0	5.0	38.0	1482.0	5.8
.1	140.6	2.6	.1	381.4	3.4	.1	686.2	4.2	.1	1055.0	5.0	.1	1487.8	5.8
.2	143.2	2.6	.2	384.8	3.4	.2	690.4	4.2	.2	1060.0	5.0	.2	1493.6	5.8
.3	145.8	2.6	.3	388.2	3.4	.3	694.6	4.2	.3	1065.0	5.0	.3	1499.4	5.8
.4	148.5	2.6	.4	391.7	3.4	.4	698.9	4.2	.4	1070.1	5.0	.4	1505.3	5.8
.5	151.1	2.6	.5	395.1	3.4	.5	703.1	4.2	.5	1075.1	5.0	.5	1511.1	5.8
.6	153.8	2.7	.6	398.6	3.5	.6	707.4	4.3	.6	1080.2	5.1	.6	1517.0	5.9
.7	156.4	2.7	.7	402.0	3.5	.7	711.6	4.3	.7	1085.2	5.1	.7	1522.8	5.9
.8	159.1	2.7	.8	405.5	3.5	.8	715.9	4.3	.8	1090.3	5.1	.8	1528.7	5.9
.9	161.8	2.7	.9	409.0	3.5	.9	720.2	4.3	.9	1095.4	5.1	.9	1534.6	5.9
7.0	164.5	2.7	15.0	412.5	3.5	23.0	724.5	4.3	31.0	1100.5	5.1	39.0	1540.5	5.9
.1	167.2	2.7	.1	416.0	3.5	.1	728.8	4.3	.1	1105.6	5.1	.1	1546.4	5.9
.2	169.9	2.7	.2	419.5	3.5	.2	733.1	4.3	.2	1110.7	5.1	.2	1552.3	5.9
.3	172.6	2.7	.3	423.0	3.5	.3	737.4	4.3	.3	1115.8	5.1	.3	1558.2	5.9
.4	175.4	2.7	.4	426.6	3.5	.4	741.8	4.3	.4	1121.0	5.1	.4	1564.2	5.9
.5	178.1	2.7	.5	430.1	3.5	.5	746.1	4.3	.5	1126.1	5.1	.5	1570.1	5.9
.6	180.9	2.8	.6	433.7	3.6	.6	750.5	4.4	.6	1131.3	5.2	.6	1576.1	6.0
.7	183.6	2.8	.7	437.2	3.6	.7	754.8	4.4	.7	1136.4	5.2	.7	1582.0	6.0
.8	186.4	2.8	.8	440.8	3.6	.8	759.2	4.4	.8	1141.6	5.2	.8	1588.0	6.0
.9	189.2	2.8	.9	444.4	3.6	.9	763.6	4.4	.9	1146.8	5.2	.9	1594.0	6.0

TABLE NO. 28.

EXCAVATION AND EMBANKMENT.

AREAS.

Breadth of Base, Twenty-Six Feet. Slope 1-2 Horizontal to 1 Perpendicular.

Height.	Area in Feet.	Diff.	Height.	Area in Feet.	Diff.	Height.	Area in Feet.	Diff.	Height.	Area in Feet.	Diff.	Height.	Area in Feet.	Diff.
0.0	0.0	0.0	8.0	240.0	3.4	16.0	544.0	4.2	24.0	912.0	5.0	32.0	1344.0	5.8
.1	2.6	2.6	.1	243.4	3.4	.1	548.2	4.2	.1	917.0	5.0	.1	1349.8	5.8
.2	5.2	2.6	.2	246.8	3.4	.2	552.4	4.2	.2	922.0	5.0	.2	1355.6	5.8
.3	7.8	2.6	.3	250.2	3.4	.3	556.6	4.2	.3	927.0	5.0	.3	1361.4	5.8
.4	10.5	2.6	.4	253.7	3.4	.4	560.9	4.2	.4	932.1	5.0	.4	1367.3	5.8
.5	13.1	2.6	.5	257.1	3.4	.5	565.1	4.2	.5	937.1	5.0	.5	1373.1	5.8
.6	15.8	2.7	.6	260.6	3.5	.6	569.4	4.3	.6	942.2	5.1	.6	1379.0	5.9
.7	18.4	2.7	.7	264.0	3.5	.7	573.6	4.3	.7	947.2	5.1	.7	1384.8	5.9
.8	21.1	2.7	.8	267.5	3.5	.8	577.9	4.3	.8	952.3	5.1	.8	1390.7	5.9
.9	23.8	2.7	.9	271.0	3.5	.9	582.2	4.3	.9	957.4	5.1	.9	1396.6	5.9
1.0	26.5	2.7	9.0	274.5	3.5	17.0	586.5	4.3	25.0	962.5	5.1	33.0	1402.5	5.9
.1	29.2	2.7	.1	278.0	3.5	.1	590.8	4.3	.1	967.6	5.1	.1	1408.4	5.9
.2	31.9	2.7	.2	281.5	3.5	.2	595.1	4.3	.2	972.7	5.1	.2	1414.3	5.9
.3	34.6	2.7	.3	285.0	3.5	.3	599.4	4.3	.3	977.8	5.1	.3	1420.2	5.9
.4	37.4	2.7	.4	288.6	3.5	.4	603.8	4.3	.4	983.0	5.1	.4	1426.2	5.9
.5	40.1	2.7	.5	292.1	3.5	.5	608.1	4.3	.5	988.1	5.1	.5	1432.1	5.9
.6	42.9	2.8	.6	295.7	3.6	.6	612.5	4.4	.6	993.3	5.2	.6	1438.1	6.0
.7	45.6	2.8	.7	299.2	3.6	.7	616.8	4.4	.7	998.4	5.2	.7	1444.0	6.0
.8	48.4	2.8	.8	302.8	3.6	.8	621.2	4.4	.8	1003.6	5.2	.8	1450.0	6.0
.9	51.2	2.8	.9	306.4	3.6	.9	625.6	4.4	.9	1008.8	5.2	.9	1456.0	6.0
2.0	54.0	2.8	10.0	310.0	3.6	18.0	630.0	4.4	26.0	1014.0	5.2	34.0	1462.0	6.0
.1	56.8	2.8	.1	313.6	3.6	.1	634.4	4.4	.1	1019.2	5.2	.1	1468.0	6.0
.2	59.6	2.8	.2	317.2	3.6	.2	638.8	4.4	.2	1024.4	5.2	.2	1474.0	6.0
.3	62.4	2.8	.3	320.8	3.6	.3	643.2	4.4	.3	1029.6	5.2	.3	1480.0	6.0
.4	65.3	2.8	.4	324.5	3.6	.4	647.7	4.4	.4	1034.9	5.2	.4	1486.1	6.0
.5	68.1	2.8	.5	328.1	3.6	.5	652.1	4.4	.5	1040.1	5.2	.5	1492.1	6.0
.6	71.0	2.9	.6	331.8	3.7	.6	656.6	4.5	.6	1045.4	5.3	.6	1498.2	6.1
.7	73.8	2.9	.7	335.4	3.7	.7	661.0	4.5	.7	1050.6	5.3	.7	1504.2	6.1
.8	76.7	2.9	.8	339.1	3.7	.8	665.5	4.5	.8	1055.9	5.3	.8	1510.3	6.1
.9	79.6	2.9	.9	342.8	3.7	.9	670.0	4.5	.9	1061.2	5.3	.9	1516.4	6.1
3.0	82.5	2.9	11.0	346.5	3.7	19.0	674.5	4.5	27.0	1066.5	5.3	35.0	1522.5	6.1
.1	85.4	2.9	.1	350.2	3.7	.1	679.0	4.5	.1	1071.8	5.3	.1	1528.6	6.1
.2	88.3	2.9	.2	353.9	3.7	.2	683.5	4.5	.2	1077.1	5.3	.2	1534.7	6.1
.3	91.2	2.9	.3	357.6	3.7	.3	688.0	4.5	.3	1082.4	5.3	.3	1540.8	6.1
.4	94.2	2.9	.4	361.4	3.7	.4	692.6	4.5	.4	1087.8	5.3	.4	1547.0	6.1
.5	97.1	2.9	.5	365.1	3.7	.5	697.1	4.5	.5	1093.1	5.3	.5	1553.1	6.1
.6	100.1	3.0	.6	368.9	3.8	.6	701.7	4.6	.6	1098.5	5.4	.6	1559.3	6.2
.7	103.0	3.0	.7	372.6	3.8	.7	706.2	4.6	.7	1103.8	5.4	.7	1565.4	6.2
.8	106.0	3.0	.8	376.4	3.8	.8	710.8	4.6	.8	1109.2	5.4	.8	1571.6	6.2
.9	109.0	3.0	.9	380.2	3.8	.9	715.4	4.6	.9	1114.6	5.4	.9	1577.8	6.2
4.0	112.0	3.0	12.0	384.0	3.8	20.0	720.0	4.6	28.0	1120.0	5.4	36.0	1584.0	6.2
.1	115.0	3.0	.1	387.8	3.8	.1	724.6	4.6	.1	1125.4	5.4	.1	1590.2	6.2
.2	118.0	3.0	.2	391.6	3.8	.2	729.2	4.6	.2	1130.8	5.4	.2	1596.4	6.2
.3	121.0	3.0	.3	395.4	3.8	.3	733.8	4.6	.3	1136.2	5.4	.3	1602.6	6.2
.4	124.1	3.0	.4	399.3	3.8	.4	738.5	4.6	.4	1141.7	5.4	.4	1608.9	6.2
.5	127.1	3.0	.5	403.1	3.8	.5	743.1	4.6	.5	1147.1	5.4	.5	1615.1	6.2
.6	130.2	3.1	.6	407.0	3.9	.6	747.8	4.7	.6	1152.6	5.5	.6	1621.4	6.3
.7	133.2	3.1	.7	410.8	3.9	.7	752.4	4.7	.7	1158.0	5.5	.7	1627.6	6.3
.8	136.3	3.1	.8	414.7	3.9	.8	757.1	4.7	.8	1163.5	5.5	.8	1633.9	6.3
.9	139.4	3.1	.9	418.6	3.9	.9	761.8	4.7	.9	1169.0	5.5	.9	1640.2	6.3
5.0	142.5	3.1	13.0	422.5	3.9	21.0	766.5	4.7	29.0	1174.5	5.5	37.0	1646.5	6.3
.1	145.6	3.1	.1	426.4	3.9	.1	771.2	4.7	.1	1180.0	5.5	.1	1652.8	6.3
.2	148.7	3.1	.2	430.3	3.9	.2	775.9	4.7	.2	1185.5	5.5	.2	1659.1	6.3
.3	151.8	3.1	.3	434.2	3.9	.3	780.7	4.7	.3	1191.0	5.5	.3	1665.4	6.3
.4	155.0	3.1	.4	438.2	3.9	.4	785.4	4.7	.4	1196.6	5.5	.4	1671.8	6.3
.5	158.1	3.1	.5	442.1	3.9	.5	790.1	4.7	.5	1202.1	5.5	.5	1678.1	6.3
.6	161.3	3.2	.6	446.1	4.0	.6	794.9	4.8	.6	1207.7	5.6	.6	1684.5	6.4
.7	164.4	3.2	.7	450.0	4.0	.7	799.7	4.8	.7	1213.2	5.6	.7	1690.8	6.4
.8	167.6	3.2	.8	454.0	4.0	.8	804.4	4.8	.8	1218.8	5.6	.8	1697.2	6.4
.9	170.8	3.2	.9	458.0	4.0	.9	809.2	4.8	.9	1224.4	5.6	.9	1703.6	6.4
6.0	174.0	3.2	14.0	462.0	4.0	22.0	814.0	4.8	30.0	1230.0	5.6	38.0	1710.0	6.4
.1	177.2	3.2	.1	466.0	4.0	.1	818.8	4.8	.1	1235.6	5.6	.1	1716.4	6.4
.2	180.4	3.2	.2	470.0	4.0	.2	823.6	4.8	.2	1241.2	5.6	.2	1722.8	6.4
.3	183.6	3.2	.3	474.0	4.0	.3	828.4	4.8	.3	1246.8	5.6	.3	1729.2	6.4
.4	186.9	3.2	.4	478.1	4.0	.4	833.3	4.8	.4	1252.5	5.6	.4	1735.7	6.4
.5	190.1	3.2	.5	482.1	4.0	.5	838.1	4.8	.5	1258.1	5.6	.5	1742.1	6.4
.6	193.4	3.3	.6	486.2	4.1	.6	843.0	4.9	.6	1263.8	5.7	.6	1748.6	6.5
.7	196.6	3.3	.7	490.2	4.1	.7	847.8	4.9	.7	1269.4	5.7	.7	1755.0	6.5
.8	199.9	3.3	.8	494.3	4.1	.8	852.7	4.9	.8	1275.1	5.7	.8	1761.5	6.5
.9	203.2	3.3	.9	498.4	4.1	.9	857.6	4.9	.9	1280.8	5.7	.9	1768.0	6.5
7.0	206.5	3.3	15.0	502.5	4.1	23.0	862.5	4.9	31.0	1286.5	5.7	39.0	1774.5	6.5
.1	209.8	3.3	.1	506.6	4.1	.1	867.4	4.9	.1	1292.2	5.7	.1	1781.0	6.5
.2	213.1	3.3	.2	510.7	4.1	.2	872.3	4.9	.2	1297.9	5.7	.2	1787.5	6.5
.3	216.4	3.3	.3	514.8	4.1	.3	877.2	4.9	.3	1303.6	5.7	.3	1794.0	6.5
.4	219.8	3.3	.4	519.0	4.1	.4	882.2	4.9	.4	1309.4	5.7	.4	1800.6	6.5
.5	223.1	3.3	.5	523.1	4.1	.5	887.1	4.9	.5	1315.1	5.7	.5	1807.1	6.5
.6	226.5	3.4	.6	527.3	4.2	.6	892.1	5.0	.6	1320.9	5.8	.6	1813.7	6.6
.7	229.8	3.4	.7	531.4	4.2	.7	897.0	5.0	.7	1326.6	5.8	.7	1820.2	6.6
.8	233.2	3.4	.8	535.6	4.2	.8	902.0	5.0	.8	1332.4	5.8	.8	1826.8	6.6
.9	236.6	3.4	.9	539.8	4.2	.9	907.0	5.0	.9	1338.2	5.8	.9	1833.4	6.6

TABLE NO. 29.

EXCAVATION AND EMBANKMENT.

AREAS.

Breadth of Base, Thirty Feet. **Slope 1-2 Horizontal to 1 Perpendicular.**

Height.	Area in Feet.	Diff.	Height.	Area in Feet.	Diff.	Height.	Area in Feet.	Diff.	Height.	Area in Feet.	Diff.	Height.	Area in Feet.	Diff.
0.0	0.0	0.0	8.0	272.0	3.8	16.0	608.0	4.6	24.0	1008.0	5.4	32.0	1472.0	6.2
.1	3.0	3.0	.1	275.8	3.8	.1	612.6	4.6	.1	1013.4	5.4	.1	1478.2	6.2
.2	6.0	3.0	.2	279.6	3.8	.2	617.2	4.6	.2	1018.8	5.4	.2	1484.4	6.2
.3	9.0	3.0	.3	283.4	3.8	.3	621.8	4.6	.3	1024.2	5.4	.3	1490.6	6.2
.4	12.1	3.0	.4	287.3	3.8	.4	626.5	4.6	.4	1029.7	5.4	.4	1496.9	6.2
.5	15.1	3.0	.5	291.1	3.8	.5	631.1	4.6	.5	1035.1	5.4	.5	1503.1	6.2
.6	18.2	3.1	.6	295.0	3.9	.6	635.8	4.7	.6	1040.6	5.5	.6	1509.4	6.3
.7	21.2	3.1	.7	298.8	3.9	.7	640.4	4.7	.7	1046.0	5.5	.7	1515.6	6.3
.8	24.3	3.1	.8	302.7	3.9	.8	645.1	4.7	.8	1051.5	5.5	.8	1521.9	6.3
.9	27.4	3.1	.9	306.6	3.9	.9	649.8	4.7	.9	1057.0	5.5	.9	1528.2	6.3
1.0	30.5	3.1	9.0	310.5	3.9	17.0	654.5	4.7	25.0	1062.5	5.5	33.0	1534.5	6.3
.1	33.6	3.1	.1	314.4	3.9	.1	659.2	4.7	.1	1068.0	5.5	.1	1540.8	6.3
.2	36.7	3.1	.2	318.3	3.9	.2	663.9	4.7	.2	1073.5	5.5	.2	1547.1	6.3
.3	39.8	3.1	.3	322.2	3.9	.3	668.6	4.7	.3	1079.0	5.5	.3	1553.4	6.3
.4	43.0	3.1	.4	326.2	3.9	.4	673.4	4.7	.4	1084.6	5.5	.4	1559.8	6.3
.5	46.1	3.1	.5	330.1	3.9	.5	678.1	4.7	.5	1090.1	5.5	.5	1566.1	6.3
.6	49.3	3.2	.6	334.1	4.0	.6	682.9	4.8	.6	1095.7	5.6	.6	1572.5	6.4
.7	52.4	3.2	.7	338.0	4.0	.7	687.6	4.8	.7	1101.2	5.6	.7	1578.8	6.4
.8	55.6	3.2	.8	342.0	4.0	.8	692.4	4.8	.8	1106.8	5.6	.8	1585.2	6.4
.9	58.8	3.2	.9	346.0	4.0	.9	697.2	4.8	.9	1112.4	5.6	.9	1591.6	6.4
2.0	62.0	3.2	10.0	350.0	4.0	18.0	702.0	4.8	26.0	1118.0	5.6	34.0	1598.0	6.4
.1	65.2	3.2	.1	354.0	4.0	.1	706.8	4.8	.1	1123.6	5.6	.1	1604.4	6.4
.2	68.4	3.2	.2	358.0	4.0	.2	711.6	4.8	.2	1129.2	5.6	.2	1610.8	6.4
.3	71.6	3.2	.3	362.0	4.0	.3	716.4	4.8	.3	1134.8	5.6	.3	1617.2	6.4
.4	74.9	3.2	.4	366.1	4.0	.4	721.3	4.8	.4	1140.5	5.6	.4	1623.7	6.4
.5	78.1	3.2	.5	370.1	4.0	.5	726.1	4.8	.5	1146.1	5.6	.5	1630.1	6.4
.6	81.4	3.3	.6	374.2	4.1	.6	731.0	4.9	.6	1151.8	5.7	.6	1636.6	6.5
.7	84.6	3.3	.7	378.2	4.1	.7	735.8	4.9	.7	1157.4	5.7	.7	1643.0	6.5
.8	87.9	3.3	.8	382.3	4.1	.8	740.7	4.9	.8	1163.1	5.7	.8	1649.5	6.5
.9	91.2	3.3	.9	386.4	4.1	.9	745.6	4.9	.9	1168.8	5.7	.9	1656.0	6.5
3.0	94.5	3.3	11.0	390.5	4.1	19.0	750.5	4.9	27.0	1174.5	5.7	35.0	1662.5	6.5
.1	97.8	3.3	.1	394.6	4.1	.1	755.4	4.9	.1	1180.2	5.7	.1	1669.0	6.5
.2	101.1	3.3	.2	398.7	4.1	.2	760.3	4.9	.2	1185.9	5.7	.2	1675.5	6.5
.3	104.4	3.3	.3	402.8	4.1	.3	765.2	4.9	.3	1191.6	5.7	.3	1682.0	6.5
.4	107.8	3.3	.4	407.0	4.1	.4	770.2	4.9	.4	1197.4	5.7	.4	1688.6	6.5
.5	111.1	3.3	.5	411.1	4.1	.5	775.1	4.9	.5	1203.1	5.7	.5	1695.1	6.5
.6	114.5	3.4	.6	415.3	4.2	.6	780.1	5.0	.6	1208.9	5.8	.6	1701.7	6.6
.7	117.8	3.4	.7	419.4	4.2	.7	785.0	5.0	.7	1214.6	5.8	.7	1708.2	6.6
.8	121.2	3.4	.8	423.6	4.2	.8	790.0	5.0	.8	1220.4	5.8	.8	1714.8	6.6
.9	124.6	3.4	.9	427.8	4.2	.9	795.0	5.0	.9	1226.2	5.8	.9	1721.4	6.6
4.0	128.0	3.4	12.0	432.0	4.2	20.0	800.0	5.0	28.0	1232.0	5.8	36.0	1728.0	6.6
.1	131.4	3.4	.1	436.2	4.2	.1	805.0	5.0	.1	1237.8	5.8	.1	1734.6	6.6
.2	134.8	3.4	.2	440.4	4.2	.2	810.0	5.0	.2	1243.6	5.8	.2	1741.2	6.6
.3	138.2	3.4	.3	444.6	4.2	.3	815.0	5.0	.3	1249.4	5.8	.3	1747.8	6.6
.4	141.7	3.4	.4	448.9	4.2	.4	820.1	5.0	.4	1255.3	5.8	.4	1754.5	6.6
.5	145.1	3.4	.5	453.1	4.2	.5	825.1	5.0	.5	1261.1	5.8	.5	1761.1	6.6
.6	148.6	3.5	.6	457.4	4.3	.6	830.2	5.1	.6	1267.0	5.9	.6	1767.8	6.7
.7	152.0	3.5	.7	461.6	4.3	.7	835.2	5.1	.7	1272.8	5.9	.7	1774.4	6.7
.8	155.5	3.5	.8	465.9	4.3	.8	840.3	5.1	.8	1278.7	5.9	.8	1781.1	6.7
.9	159.0	3.5	.9	470.2	4.3	.9	845.4	5.1	.9	1284.6	5.9	.9	1787.8	6.7
5.0	162.5	3.5	13.0	474.5	4.3	21.0	850.5	5.1	29.0	1290.5	5.9	37.0	1794.5	6.7
.1	166.0	3.5	.1	478.8	4.3	.1	855.6	5.1	.1	1296.4	5.9	.1	1801.2	6.7
.2	169.5	3.5	.2	483.1	4.3	.2	860.7	5.1	.2	1302.3	5.9	.2	1807.9	6.7
.3	173.0	3.5	.3	487.4	4.3	.3	865.8	5.1	.3	1308.2	5.9	.3	1814.6	6.7
.4	176.6	3.5	.4	491.8	4.3	.4	871.0	5.1	.4	1314.2	5.9	.4	1821.4	6.7
.5	180.1	3.5	.5	496.1	4.3	.5	876.1	5.1	.5	1320.1	5.9	.5	1828.1	6.7
.6	183.7	3.6	.6	500.5	4.4	.6	881.3	5.2	.6	1326.1	6.0	.6	1834.9	6.8
.7	187.2	3.6	.7	504.8	4.4	.7	886.4	5.2	.7	1332.0	6.0	.7	1841.6	6.8
.8	190.8	3.6	.8	509.2	4.4	.8	891.6	5.2	.8	1338.0	6.0	.8	1848.4	6.8
.9	194.4	3.6	.9	513.6	4.4	.9	896.8	5.2	.9	1344.0	6.0	.9	1855.2	6.8
6.0	198.0	3.6	14.0	518.0	4.4	22.0	902.0	5.2	30.0	1350.0	6.0	38.0	1862.0	6.8
.1	201.6	3.6	.1	522.4	4.4	.1	907.2	5.2	.1	1356.0	6.0	.1	1868.8	6.8
.2	205.2	3.6	.2	526.8	4.4	.2	912.4	5.2	.2	1362.0	6.0	.2	1875.6	6.8
.3	208.8	3.6	.3	531.2	4.4	.3	917.6	5.2	.3	1368.0	6.0	.3	1882.4	6.8
.4	212.5	3.6	.4	535.7	4.4	.4	922.9	5.2	.4	1374.1	6.0	.4	1889.3	6.8
.5	216.1	3.6	.5	540.1	4.4	.5	928.1	5.2	.5	1380.1	6.0	.5	1896.1	6.8
.6	219.8	3.7	.6	544.6	4.5	.6	933.4	5.3	.6	1386.2	6.1	.6	1903.0	6.9
.7	223.4	3.7	.7	549.0	4.5	.7	938.6	5.3	.7	1392.2	6.1	.7	1909.8	6.9
.8	227.1	3.7	.8	553.5	4.5	.8	943.9	5.3	.8	1398.3	6.1	.8	1916.7	6.9
.9	230.8	3.7	.9	558.0	4.5	.9	949.2	5.3	.9	1404.4	6.1	.9	1923.6	6.9
7.0	234.5	3.7	15.0	562.5	4.5	23.0	954.5	5.3	31.0	1410.5	6.1	39.0	1930.5	6.9
.1	238.2	3.7	.1	567.0	4.5	.1	959.8	5.3	.1	1416.6	6.1	.1	1937.4	6.9
.2	241.9	3.7	.2	571.5	4.5	.2	965.1	5.3	.2	1422.7	6.1	.2	1944.3	6.9
.3	245.6	3.7	.3	576.0	4.5	.3	970.4	5.3	.3	1428.8	6.1	.3	1951.2	6.9
.4	249.4	3.7	.4	580.6	4.5	.4	975.8	5.3	.4	1435.0	6.1	.4	1958.2	6.9
.5	253.1	3.7	.5	585.1	4.5	.5	981.1	5.3	.5	1441.1	6.1	.5	1965.1	6.9
.6	256.9	3.8	.6	589.7	4.6	.6	986.5	5.4	.6	1447.3	6.2	.6	1972.1	7.0
.7	260.6	3.8	.7	594.2	4.6	.7	991.8	5.4	.7	1453.4	6.2	.7	1979.0	7.0
.8	264.4	3.8	.8	598.8	4.6	.8	997.2	5.4	.8	1459.6	6.2	.8	1986.0	7.0
.9	268.2	3.8	.9	603.4	4.6	.9	1002.6	5.4	.9	1465.8	6.2	.9	1993.0	7.0

TABLE No. 30.

EXCAVATION AND EMBANKMENT.

AREAS.

Breadth of Base, 15 feet. **Slopes, 1-4 Horizontal to 1 Perpendicular.**

Height.	Area in Feet.	Difference.	Height.	Area in Feet.	Difference.	Height.	Area in Feet.	Difference.	Height.	Area in Feet.	Difference.	Height.	Area in Feet.	Difference.
0.0	0.0	1.5	8.0	136.0	1.9	16.0	304.0	2.3	24.0	504.0	2.7	32.0	736.0	3.1
1	1.5	1.5	1	137.9	1.9	1	306.3	2.3	1	506.7	2.7	1	739.1	3.1
2	3.0	1.5	2	139.8	1.9	2	308.6	2.3	2	509.4	2.7	2	742.2	3.1
3	4.5	1.5	3	141.7	1.9	3	310.9	2.3	3	512.1	2.7	3	745.3	3.1
4	6.0	1.5	4	143.6	1.9	4	313.2	2.3	4	514.8	2.7	4	748.4	3.1
5	7.6	1.5	5	145.6	1.9	5	315.6	2.3	5	517.6	2.7	5	751.6	3.1
6	9.1	1.5	6	147.5	1.9	6	317.9	2.3	6	520.3	2.7	6	754.7	3.1
7	10.6	1.5	7	149.4	1.9	7	320.2	2.3	7	523.0	2.7	7	757.8	3.1
8	12.2	1.5	8	151.4	1.9	8	322.6	2.3	8	525.8	2.7	8	761.0	3.1
9	13.7	1.5	9	153.3	1.9	9	324.9	2.3	9	528.5	2.7	9	764.1	3.1
1.0	15.3	1.5	9.0	155.3	1.9	17.0	327.3	2.3	25.0	531.3	2.7	33.0	767.3	3.1
1	16.8	1.6	1	157.2	2.0	1	329.6	2.4	1	534.0	2.8	1	770.4	3.2
2	18.4	1.6	2	159.2	2.0	2	332.0	2.4	2	536.8	2.8	2	773.6	3.2
3	19.9	1.6	3	161.1	2.0	3	334.3	2.4	3	539.5	2.8	3	776.7	3.2
4	21.5	1.6	4	163.1	2.0	4	336.7	2.4	4	542.3	2.8	4	779.9	3.2
5	23.1	1.6	5	165.1	2.0	5	339.1	2.4	5	545.1	2.8	5	783.1	3.2
6	24.6	1.6	6	167.0	2.0	6	341.4	2.4	6	547.8	2.8	6	786.2	3.2
7	26.2	1.6	7	169.0	2.0	7	343.8	2.4	7	550.6	2.8	7	789.4	3.2
8	27.8	1.6	8	171.0	2.0	8	346.2	2.4	8	553.4	2.8	8	792.6	3.2
9	29.4	1.6	9	173.0	2.0	9	348.6	2.4	9	556.2	2.8	9	795.8	3.2
2.0	31.0	1.6	10.0	175.0	2.0	18.0	351.0	2.4	26.0	559.0	2.8	34.0	799.0	3.2
1	32.6	1.6	1	177.0	2.0	1	353.4	2.4	1	561.8	2.8	1	802.2	3.2
2	34.2	1.6	2	179.0	2.0	2	355.8	2.4	2	564.6	2.8	2	805.4	3.2
3	35.8	1.6	3	181.0	2.0	3	358.2	2.4	3	567.4	2.8	3	808.6	3.2
4	37.4	1.6	4	183.0	2.0	4	360.6	2.4	4	570.2	2.8	4	811.8	3.2
5	39.1	1.6	5	185.1	2.0	5	363.1	2.4	5	573.1	2.8	5	815.1	3.2
6	40.7	1.6	6	187.1	2.0	6	365.5	2.4	6	575.9	2.8	6	818.3	3.2
7	42.3	1.6	7	189.1	2.0	7	367.9	2.4	7	578.7	2.8	7	821.5	3.2
8	44.0	1.6	8	191.2	2.0	8	370.4	2.4	8	581.6	2.8	8	824.8	3.2
9	45.6	1.6	9	193.2	2.0	9	372.8	2.4	9	584.4	2.8	9	828.0	3.2
3.0	47.3	1.6	11.0	195.3	2.0	19.0	375.3	2.4	27.0	587.3	2.8	35.0	831.3	3.2
1	48.9	1.7	1	197.3	2.1	1	377.7	2.5	1	590.1	2.9	1	834.5	3.3
2	50.6	1.7	2	199.4	2.1	2	380.2	2.5	2	593.0	2.9	2	837.8	3.3
3	52.2	1.7	3	201.4	2.1	3	382.6	2.5	3	595.8	2.9	3	841.0	3.3
4	53.9	1.7	4	203.5	2.1	4	385.1	2.5	4	598.7	2.9	4	844.3	3.3
5	55.6	1.7	5	205.6	2.1	5	387.6	2.5	5	601.6	2.9	5	847.6	3.3
6	57.2	1.7	6	207.6	2.1	6	390.0	2.5	6	604.4	2.9	6	850.8	3.3
7	58.9	1.7	7	209.7	2.1	7	392.5	2.5	7	607.3	2.9	7	854.1	3.3
8	60.6	1.7	8	211.8	2.1	8	395.0	2.5	8	610.2	2.9	8	857.4	3.3
9	62.3	1.7	9	213.9	2.1	9	397.5	2.5	9	613.1	2.9	9	860.7	3.3
4.0	64.0	1.7	12.0	216.0	2.1	20.0	400.0	2.5	28.0	616.0	2.9	36.0	861.0	3.3
1	65.7	1.7	1	218.1	2.1	1	402.5	2.5	1	618.9	2.9	1	867.3	3.3
2	67.4	1.7	2	220.2	2.1	2	405.0	2.5	2	621.8	2.9	2	870.6	3.3
3	69.1	1.7	3	222.3	2.1	3	407.5	2.5	3	624.7	2.9	3	873.9	3.3
4	70.8	1.7	4	224.4	2.1	4	410.0	2.5	4	627.6	2.9	4	877.2	3.3
5	72.6	1.7	5	226.6	2.1	5	412.6	2.5	5	630.6	2.9	5	880.6	3.3
6	74.3	1.7	6	228.7	2.1	6	415.1	2.5	6	633.5	2.9	6	883.9	3.3
7	76.0	1.7	7	230.8	2.1	7	417.6	2.5	7	636.4	2.9	7	887.2	3.3
8	77.8	1.7	8	233.0	2.1	8	420.2	2.5	8	639.4	2.9	8	890.6	3.3
9	79.5	1.7	9	235.1	2.1	9	422.7	2.5	9	642.3	2.9	9	893.9	3.3
5.0	81.3	1.7	13.0	237.3	2.1	21.0	425.3	2.5	29.0	645.3	2.9	37.0	897.3	3.3
1	83.0	1.8	1	239.4	2.2	1	427.8	2.6	1	648.2	3.0	1	900.6	3.4
2	84.8	1.8	2	241.6	2.2	2	430.4	2.6	2	651.2	3.0	2	904.0	3.4
3	86.5	1.8	3	243.7	2.2	3	432.9	2.6	3	654.1	3.0	3	907.3	3.4
4	88.3	1.8	4	245.9	2.2	4	435.5	2.6	4	657.1	3.0	4	910.7	3.4
5	90.1	1.8	5	248.1	2.2	5	438.1	2.6	5	660.1	3.0	5	914.1	3.4
6	91.8	1.8	6	250.2	2.2	6	440.6	2.6	6	663.0	3.0	6	917.4	3.4
7	93.6	1.8	7	252.4	2.2	7	443.2	2.6	7	666.0	3.0	7	920.8	3.4
8	95.4	1.8	8	254.6	2.2	8	445.8	2.6	8	669.0	3.0	8	934.2	3.4
9	97.2	1.8	9	256.8	2.2	9	448.4	2.6	9	672.0	3.0	9	927.6	3.4
6.0	99.0	1.8	14.0	259.0	2.2	22.0	451.0	2.6	30.0	675.0	3.0	38.0	931.0	3.4
1	100.8	1.8	1	261.2	2.2	1	453.6	2.6	1	678.0	3.0	1	934.4	3.4
2	102.6	1.8	2	263.4	2.2	2	456.2	2.6	2	681.0	3.0	2	937.8	3.4
3	104.4	1.8	3	265.6	2.2	3	458.8	2.6	3	684.0	3.0	3	941.2	3.4
4	106.2	1.8	4	267.8	2.2	4	461.4	2.6	4	687.0	3.0	4	944.6	3.4
5	108.1	1.8	5	270.1	2.2	5	464.1	2.6	5	690.1	3.0	5	948.1	3.4
6	109.9	1.8	6	272.3	2.2	6	466.7	2.6	6	693.1	3.0	6	951.5	3.4
7	111.7	1.8	7	274.5	2.2	7	469.3	2.6	7	696.1	3.0	7	954.9	3.4
8	113.6	1.8	8	276.8	2.2	8	472.0	2.6	8	699.2	3.0	8	958.4	3.4
9	115.4	1.8	9	279.0	2.2	9	474.6	2.6	9	702.2	3.0	9	961.8	3.4
7.0	117.3	1.8	15.0	281.3	2.2	23.0	477.3	2.6	31.0	705.3	3.0	39.0	965.3	3.4
1	119.1	1.9	1	283.5	2.3	1	479.9	2.7	1	708.3	3.1	1	968.7	3.5
2	121.0	1.9	2	285.8	2.3	2	482.6	2.7	2	711.4	3.1	2	972.2	3.5
3	122.8	1.9	3	288.0	2.3	3	485.2	2.7	3	714.4	3.1	3	975.6	3.5
4	124.7	1.9	4	290.3	2.3	4	487.9	2.7	4	717.5	3.1	4	979.1	3.5
5	126.6	1.9	5	292.6	2.3	5	490.6	2.7	5	720.6	3.1	5	982.6	3.5
6	128.4	1.9	6	294.8	2.3	6	493.2	2.7	6	723.6	3.1	6	986.0	3.5
7	130.3	1.9	7	297.1	2.3	7	495.9	2.7	7	726.7	3.1	7	989.5	3.5
8	132.2	1.9	8	299.4	2.3	8	498.6	2.7	8	729.8	3.1	8	903.0	3.5
9	134.1	1.9	9	301.7	2.3	9	501.3	2.7	9	732.9	3.1	9	996.5	3.5

TABLE No. 31.

EXCAVATION AND EMBANKMENT.

AREAS.

Breadth of Base, 18 feet. **Slopes, 1-4 Horizontal to 1 Perpendicular.**

Height.	Area in Feet.	Difference.	Height.	Area in Feet.	Difference.	Height.	Area in Feet.	Difference.	Height.	Area in Feet.	Difference.	Height.	Area in Feet.	Difference.
0.0	0.0	0.0	8.0	160.0	2.2	16.0	352.0	2.6	24.0	576.0	3.0	32.0	832.0	3.4
1	1.8	1.8	1	162.2	2.2	1	354.6	2.6	1	579.0	3.0	1	835.4	3.4
2	3.6	1.8	2	164.4	2.2	2	357.2	2.6	2	582.0	3.0	2	838.8	3.4
3	5.4	1.8	3	166.6	2.2	3	359.8	2.6	3	585.0	3.0	3	842.2	3.4
4	7.2	1.8	4	168.8	2.2	4	362.4	2.6	4	588.0	3.0	4	845.6	3.4
5	9.1	1.8	5	171.1	2.2	5	365.1	2.6	5	591.1	3.0	5	849.1	3.4
6	10.9	1.8	6	173.3	2.2	6	367.7	2.6	6	594.1	3.0	6	852.5	3.4
7	12.7	1.8	7	175.5	2.2	7	370.3	2.6	7	597.1	3.0	7	855.9	3.4
8	14.6	1.8	8	177.8	2.2	8	373.0	2.6	8	600.2	3.0	8	859.4	3.4
9	16.4	1.8	9	180.0	2.2	9	375.6	2.6	9	603.2	3.0	9	862.8	3.4
1.0	18.3	1.8	9.0	182.3	2.2	17.0	378.3	2.6	25.0	606.3	3.0	33.0	866.3	3.4
1	20.1	1.9	1	184.5	2.3	1	380.9	2.7	1	609.3	3.1	1	869.7	3.5
2	22.0	1.9	2	186.8	2.3	2	383.6	2.7	2	612.4	3.1	2	873.2	3.5
3	23.8	1.9	3	189.0	2.3	3	386.2	2.7	3	615.4	3.1	3	876.6	3.5
4	25.7	1.9	4	191.3	2.3	4	388.9	2.7	4	618.5	3.1	4	880.1	3.5
5	27.6	1.9	5	193.6	2.3	5	391.6	2.7	5	621.6	3.1	5	883.6	3.5
6	29.4	1.9	6	195.8	2.3	6	394.2	2.7	6	624.6	3.1	6	887.0	3.5
7	31.3	1.9	7	198.1	2.3	7	396.9	2.7	7	627.7	3.1	7	890.5	3.5
8	33.2	1.9	8	200.4	2.3	8	399.6	2.7	8	630.8	3.1	8	894.0	3.5
9	35.1	1.9	9	202.7	2.3	9	402.3	2.7	9	633.9	3.1	9	897.5	3.5
2.0	37.0	1.9	10.0	205.0	2.3	18.0	405.0	2.7	26.0	637.0	3.1	34.0	901.0	3.5
1	38.9	1.9	1	207.3	2.3	1	407.7	2.7	1	640.1	3.1	1	904.5	3.5
2	40.8	1.9	2	209.6	2.3	2	410.4	2.7	2	643.2	3.1	2	908.0	3.5
3	42.7	1.9	3	211.9	2.3	3	413.1	2.7	3	646.3	3.1	3	911.5	3.5
4	44.6	1.9	4	214.2	2.3	4	415.8	2.7	4	649.4	3.1	4	915.0	3.5
5	46.6	1.9	5	216.6	2.3	5	418.6	2.7	5	652.6	3.1	5	918.6	3.5
6	48.5	1.9	6	218.9	2.3	6	421.3	2.7	6	655.7	3.1	6	922.1	3.5
7	50.4	1.9	7	221.2	2.3	7	424.0	2.7	7	658.8	3.1	7	925.6	3.5
8	52.4	1.9	8	223.6	2.3	8	426.8	2.7	8	662.0	3.1	8	929.2	3.5
9	54.3	1.9	9	225.9	2.3	9	429.5	2.7	9	665.1	3.1	9	932.7	3.5
3.0	56.3	1.9	11.0	228.3	2.3	19.0	432.3	2.7	27.0	668.3	3.1	35.0	936.3	3.5
1	58.2	2.0	1	230.6	2.4	1	435.0	2.8	1	671.4	3.2	1	939.8	3.6
2	60.2	2.0	2	233.0	2.4	2	437.8	2.8	2	674.6	3.2	2	943.4	3.6
3	62.1	2.0	3	235.3	2.4	3	440.5	2.8	3	677.7	3.2	3	946.9	3.6
4	64.1	2.0	4	237.7	2.4	4	443.3	2.8	4	680.9	3.2	4	950.5	3.6
5	66.1	2.0	5	240.1	2.4	5	446.1	2.8	5	684.1	3.2	5	954.1	3.6
6	68.0	2.0	6	242.4	2.4	6	448.8	2.8	6	687.2	3.2	6	957.6	3.6
7	70.0	2.0	7	244.8	2.4	7	451.6	2.8	7	690.4	3.2	7	961.2	3.6
8	72.0	2.0	8	247.2	2.4	8	454.4	2.8	8	693.6	3.2	8	964.8	3.6
9	74.0	2.0	9	249.6	2.4	9	457.2	2.8	9	696.8	3.2	9	968.4	3.6
4.0	76.0	2.0	12.0	252.0	2.4	20.0	460.0	2.8	28.0	700.0	3.2	36.0	972.0	3.6
1	78.0	2.0	1	254.4	2.4	1	462.8	2.8	1	703.2	3.2	1	975.6	3.6
2	80.0	2.0	2	256.8	2.4	2	465.6	2.8	2	706.4	3.2	2	979.2	3.6
3	82.0	2.0	3	259.2	2.4	3	468.4	2.8	3	709.6	3.2	3	982.8	3.6
4	84.0	2.0	4	261.6	2.4	4	471.2	2.8	4	712.8	3.2	4	986.4	3.6
5	86.1	2.0	5	264.1	2.4	5	474.1	2.8	5	716.1	3.2	5	990.1	3.6
6	88.1	2.0	6	266.5	2.4	6	476.9	2.8	6	719.3	3.2	6	993.7	3.6
7	90.1	2.0	7	268.9	2.4	7	479.7	2.8	7	722.5	3.2	7	997.3	3.6
8	92.2	2.0	8	271.4	2.4	8	482.6	2.8	8	725.8	3.2	8	1001.0	3.6
9	94.2	2.0	9	273.8	2.4	9	485.4	2.8	9	729.0	3.2	9	1004.6	3.6
5.0	96.3	2.0	13.0	276.3	2.4	21.0	488.3	2.8	29.0	732.3	3.2	37.0	1008.3	3.6
1	98.3	2.1	1	278.7	2.5	1	491.1	2.9	1	735.5	3.3	1	1011.9	3.7
2	100.4	2.1	2	281.2	2.5	2	494.0	2.9	2	738.8	3.3	2	1015.6	3.7
3	102.4	2.1	3	283.6	2.5	3	496.8	2.9	3	742.0	3.3	3	1019.2	3.7
4	104.5	2.1	4	286.1	2.5	4	499.7	2.9	4	745.3	3.3	4	1022.9	3.7
5	106.6	2.1	5	288.6	2.5	5	502.6	2.9	5	748.6	3.3	5	1026.6	3.7
6	108.6	2.1	6	291.0	2.5	6	505.4	2.9	6	751.8	3.3	6	1030.2	3.7
7	110.7	2.1	7	293.5	2.5	7	508.3	2.9	7	755.1	3.3	7	1033.9	3.7
8	112.8	2.1	8	296.0	2.5	8	511.2	2.9	8	758.4	3.3	8	1037.6	3.7
9	114.9	2.1	9	298.5	2.5	9	514.1	2.9	9	761.7	3.3	9	1041.3	3.7
6.0	117.0	2.1	14.0	301.0	2.5	22.0	517.0	2.9	30.0	765.0	3.3	38.0	1045.0	3.7
1	119.1	2.1	1	303.5	2.5	1	519.9	2.9	1	768.3	3.3	1	1048.7	3.7
2	121.2	2.1	2	306.0	2.5	2	522.8	2.9	2	771.6	3.3	2	1052.4	3.7
3	123.3	2.1	3	308.5	2.5	3	525.7	2.9	3	774.9	3.3	3	1056.1	3.7
4	125.4	2.1	4	311.0	2.5	4	528.6	2.9	4	778.2	3.3	4	1059.8	3.7
5	127.6	2.1	5	313.6	2.5	5	531.6	2.9	5	781.6	3.3	5	1063.6	3.7
6	129.7	2.1	6	316.1	2.5	6	534.5	2.9	6	784.9	3.3	6	1067.3	3.7
7	131.8	2.1	7	318.6	2.5	7	537.4	2.9	7	788.2	3.3	7	1071.0	3.7
8	134.0	2.1	8	321.2	2.5	8	540.4	2.9	8	791.6	3.3	8	1074.8	3.7
9	136.1	2.1	9	323.7	2.5	9	543.3	2.9	9	794.9	3.3	9	1078.5	3.7
7.0	138.3	2.1	15.0	326.3	2.5	23.0	546.3	2.9	31.0	798.3	3.3	39.0	1082.3	3.7
1	140.4	2.2	1	328.8	2.6	1	549.2	3.0	1	801.6	3.4	1	1086.0	3.8
2	142.6	2.2	2	331.4	2.6	2	552.2	3.0	2	805.0	3.4	2	1089.8	3.8
3	144.7	2.2	3	333.9	2.6	3	555.1	3.0	3	808.3	3.4	3	1093.5	3.8
4	146.9	2.2	4	336.5	2.6	4	558.1	3.0	4	811.7	3.4	4	1097.3	3.8
5	149.1	2.2	5	339.1	2.6	5	561.1	3.0	5	815.1	3.4	5	1101.1	3.8
6	151.2	2.2	6	341.6	2.6	6	564.0	3.0	6	818.4	3.4	6	1104.8	3.8
7	153.4	2.2	7	344.2	2.6	7	567.0	3.0	7	821.8	3.4	7	1108.6	3.8
8	155.6	2.2	8	346.8	2.6	8	570.0	3.0	8	825.2	3.4	8	1112.4	3.8
9	157.8	2.2	9	349.4	2.6	9	573.0	3.0	9	828.6	3.4	9	1116.2	3.8

TABLE No. 32.

EXCAVATION AND EMBANKMENT.

AREAS.

Breadth of Base, 20 feet. Slopes, 1-4 Horizontal to 1 Perpendicular.

Height.	Area in Feet.	Difference.	Height.	Area in Feet.	Difference.	Height.	Area in Feet.	Difference.	Height.	Area in Feet.	Difference.	Height.	Area in Feet.	Difference.
0.0	0.0	0.0	8.0	176.0	2.4	16.0	384.0	2.8	24.0	624.0	3.2	32.0	896.0	3.6
1	2.0	2.0	1	178.4	2.4	1	386.8	2.8	1	627.2	3.2	1	899.6	3.6
2	4.0	2.0	2	180.8	2.4	2	389.6	2.8	2	630.4	3.2	2	903.2	3.6
3	6.0	2.0	3	183.2	2.4	3	392.4	2.8	3	633.6	3.2	3	906.8	3.6
4	8.0	2.0	4	185.6	2.4	4	395.2	2.8	4	636.8	3.2	4	910.4	3.6
5	10.1	2.0	5	188.1	2.4	5	398.1	2.8	5	640.1	3.2	5	914.1	3.6
6	12.1	2.0	6	190.5	2.4	6	400.9	2.8	6	643.3	3.2	6	917.7	3.6
7	14.1	2.0	7	192.9	2.4	7	403.7	2.8	7	646.5	3.2	7	921.3	3.6
8	16.2	2.0	8	195.4	2.4	8	406.6	2.8	8	649.8	3.2	8	925.0	3.6
9	18.2	2.0	9	197.8	2.4	9	409.4	2.8	9	653.0	3.2	9	928.6	3.6
1.0	20.3	2.0	9.0	200.3	2.4	17.0	412.3	2.8	25.0	656.3	3.2	33.0	932.3	3.6
1	22.3	2.1	1	202.7	2.5	1	415.1	2.9	1	659.5	3.3	1	935.9	3.7
2	24.4	2.1	2	205.2	2.5	2	418.0	2.9	2	662.8	3.3	2	939.6	3.7
3	26.4	2.1	3	207.6	2.5	3	420.8	2.9	3	666.0	3.3	3	943.2	3.7
4	28.5	2.1	4	210.1	2.5	4	423.7	2.9	4	669.3	3.3	4	946.9	3.7
5	30.6	2.1	5	212.6	2.5	5	426.6	2.9	5	671.6	3.3	5	950.6	3.7
6	32.6	2.1	6	215.0	2.5	6	429.4	2.9	6	675.8	3.3	6	954.2	3.7
7	34.7	2.1	7	217.5	2.5	7	432.3	2.9	7	679.1	3.3	7	957.9	3.7
8	36.8	2.1	8	220.0	2.5	8	435.2	2.9	8	682.4	3.3	8	961.6	3.7
9	38.9	2.1	9	222.5	2.5	9	438.1	2.9	9	685.7	3.3	9	965.3	3.7
2.0	41.0	2.1	10.0	225.0	2.5	18.0	441.0	2.9	26.0	689.0	3.3	34.0	969.0	3.7
1	43.1	2.1	1	227.5	2.5	1	443.9	2.9	1	692.3	3.3	1	972.7	3.7
2	45.2	2.1	2	230.0	2.5	2	446.8	2.9	2	695.6	3.3	2	976.4	3.7
3	47.3	2.1	3	232.5	2.5	3	449.7	2.9	3	698.9	3.3	3	980.1	3.7
4	49.4	2.1	4	235.0	2.5	4	452.6	2.9	4	702.2	3.3	4	983.8	3.7
5	51.6	2.1	5	237.6	2.5	5	455.6	2.9	5	705.6	3.3	5	987.6	3.7
6	53.7	2.1	6	240.1	2.5	6	458.5	2.9	6	708.9	3.3	6	991.3	3.7
7	55.8	2.1	7	242.6	2.5	7	461.4	2.9	7	712.2	3.3	7	995.0	3.7
8	58.0	2.1	8	245.2	2.5	8	464.4	2.9	8	715.6	3.3	8	998.8	3.7
9	60.1	2.1	9	247.7	2.5	9	467.3	2.9	9	718.9	3.3	9	1002.5	3.7
3.0	62.3	2.1	11.0	250.3	2.5	19.0	470.3	2.9	27.0	722.3	3.3	35.0	1006.3	3.7
1	64.4	2.2	1	252.8	2.6	1	473.2	3.0	1	725.6	3.4	1	1010.0	3.8
2	66.6	2.2	2	255.4	2.6	2	476.2	3.0	2	729.0	3.4	2	1013.8	3.8
3	68.7	2.2	3	257.9	2.6	3	479.1	3.0	3	732.3	3.4	3	1017.5	3.8
4	70.9	2.2	4	260.5	2.6	4	482.1	3.0	4	735.7	3.4	4	1021.3	3.8
5	73.1	2.2	5	263.1	2.6	5	485.1	3.0	5	739.1	3.4	5	1025.1	3.8
6	75.2	2.2	6	265.6	2.6	6	488.0	3.0	6	742.4	3.4	6	1028.8	3.8
7	77.4	2.2	7	268.2	2.6	7	491.0	3.0	7	745.8	3.4	7	1032.6	3.8
8	79.6	2.2	8	270.8	2.6	8	494.0	3.0	8	749.2	3.4	8	1036.4	3.8
9	81.8	2.2	9	273.4	2.6	9	497.0	3.0	9	752.6	3.4	9	1040.2	3.8
4.0	84.0	2.2	12.0	276.0	2.6	20.0	500.0	3.0	28.0	756.0	3.4	36.0	1044.0	3.8
1	86.2	2.2	1	278.6	2.6	1	503.0	3.0	1	759.4	3.4	1	1047.8	3.8
2	88.4	2.2	2	281.2	2.6	2	506.0	3.0	2	762.8	3.4	2	1051.6	3.8
3	90.6	2.2	3	283.8	2.6	3	509.0	3.0	3	766.2	3.4	3	1055.4	3.8
4	92.8	2.2	4	286.4	2.6	4	512.0	3.0	4	769.6	3.4	4	1059.2	3.8
5	95.1	2.2	5	289.1	2.6	5	515.1	3.0	5	773.1	3.4	5	1063.1	3.8
6	97.3	2.2	6	291.7	2.6	6	518.1	3.0	6	776.5	3.4	6	1066.9	3.8
7	99.5	2.2	7	294.3	2.6	7	521.1	3.0	7	779.9	3.4	7	1070.7	3.8
8	101.8	2.2	8	297.0	2.6	8	524.2	3.0	8	783.4	3.4	8	1074.6	3.8
9	104.0	2.2	9	299.6	2.6	9	527.2	3.0	9	786.8	3.4	9	1078.4	3.8
5.0	106.3	2.2	13.0	302.3	2.6	21.0	530.3	3.0	29.0	790.3	3.4	37.0	1082.3	3.8
1	108.5	2.3	1	304.9	2.7	1	533.3	3.1	1	793.7	3.5	1	1086.1	3.9
2	110.8	2.3	2	307.6	2.7	2	536.4	3.1	2	797.2	3.5	2	1090.0	3.9
3	113.0	2.3	3	310.2	2.7	3	539.4	3.1	3	800.6	3.5	3	1093.8	3.9
4	115.3	2.3	4	312.9	2.7	4	542.5	3.1	4	804.1	3.5	4	1097.7	3.9
5	117.6	2.3	5	315.6	2.7	5	545.6	3.1	5	807.6	3.5	5	1101.6	3.9
6	119.8	2.3	6	318.2	2.7	6	548.6	3.1	6	811.0	3.5	6	1105.4	3.9
7	122.1	2.3	7	320.9	2.7	7	551.7	3.1	7	814.5	3.5	7	1109.3	3.9
8	124.4	2.3	8	323.6	2.7	8	554.8	3.1	8	818.0	3.5	8	1113.2	3.9
9	126.7	2.3	9	326.3	2.7	9	557.9	3.1	9	821.5	3.5	9	1117.1	3.9
6.0	129.0	2.3	14.0	329.0	2.7	22.0	561.0	3.1	30.0	825.0	3.5	38.0	1121.0	3.9
1	131.3	2.3	1	331.7	2.7	1	564.1	3.1	1	828.5	3.5	1	1124.9	3.9
2	133.6	2.3	2	334.4	2.7	2	567.2	3.1	2	832.0	3.5	2	1128.8	3.9
3	135.9	2.3	3	337.1	2.7	3	570.3	3.1	3	835.5	3.5	3	1132.7	3.9
4	138.2	2.3	4	339.8	2.7	4	573.4	3.1	4	839.0	3.5	4	1136.6	3.9
5	140.6	2.3	5	342.6	2.7	5	576.6	3.1	5	842.6	3.5	5	1140.6	3.9
6	142.9	2.3	6	345.3	2.7	6	579.7	3.1	6	846.1	3.5	6	1144.5	3.9
7	145.2	2.3	7	348.0	2.7	7	582.8	3.1	7	849.6	3.5	7	1148.4	3.9
8	147.6	2.3	8	350.8	2.7	8	586.0	3.1	8	853.2	3.5	8	1152.4	3.9
9	149.9	2.3	9	353.5	2.7	9	589.1	3.1	9	856.7	3.5	9	1156.3	3.9
7.0	152.3	2.3	15.0	356.3	2.7	23.0	592.3	3.1	31.0	860.3	3.5	39.0	1160.3	3.9
1	154.6	2.4	1	359.0	2.8	1	595.4	3.2	1	863.8	3.6	1	1164.2	4.0
2	157.0	2.4	2	361.8	2.8	2	598.6	3.2	2	867.4	3.6	2	1168.2	4.0
3	159.3	2.4	3	364.5	2.8	3	601.7	3.2	3	870.9	3.6	3	1172.1	4.0
4	161.7	2.4	4	367.3	2.8	4	604.9	3.2	4	874.5	3.6	4	1176.1	4.0
5	164.1	2.4	5	370.1	2.8	5	608.1	3.2	5	878.1	3.6	5	1180.1	4.0
6	166.4	2.4	6	372.8	2.8	6	611.2	3.2	6	881.6	3.6	6	1184.0	4.0
7	168.8	2.4	7	375.6	2.8	7	614.4	3.2	7	885.2	3.6	7	1188.0	4.0
8	171.2	2.4	8	378.4	2.8	8	617.6	3.2	8	888.8	3.6	8	1192.0	4.0
9	173.6	2.4	9	381.2	2.8	9	620.8	3.2	9	892.4	3.6	9	1196.0	4.0

TABLE NO. 1.

EXCAVATION AND EMBANKMENT.

AREAS OF TRIANGLES.

Slope 1 1-2 Horizontal to 1 Perpendicular.

Height.	Area in Feet.	Height.	Area in Feet.	Height	Area in Feet.	Height.	Area in Feet.	Height.	Area in Feet.	Height.	Area in Feet.	Height.	Area in Feet.
0.0	0.0	8.0	48.0	16.0	192.0	24.0	432.0	32.0	768.0	40.0	1200.0	48.0	1728.0
.1	0.0	.1	49.2	.1	194.4	.1	435.6	.1	772.8	.1	1206.0	.1	1735.2
.2	0.0	.2	50.4	.2	196.8	.2	439.2	.2	777.6	.2	1212.0	.2	1742.4
.3	0.1	.3	51.7	.3	199.3	.3	442.9	.3	782.5	.3	1218.1	.3	1749.7
.4	0.1	.4	52.9	.4	201.7	.4	446.5	.4	787.3	.4	1224.1	.4	1756.9
.5	0.2	.5	54.2	.5	204.2	.5	450.2	.5	792.2	.5	1230.2	.5	1764.2
.6	0.3	.6	55.5	.6	206.7	.6	453.9	.6	797.1	.6	1236.3	.6	1771.5
.7	0.4	.7	56.8	.7	209.2	.7	457.6	.7	802.0	.7	1242.4	.7	1778.8
.8	0.5	.8	58.1	.8	211.7	.8	461.3	.8	806.9	.8	1248.5	.8	1786.1
.9	0.6	.9	59.4	.9	214.2	.9	465.0	.9	811.8	.9	1254.6	.9	1793.4
1.0	0.8	9.0	60.8	17.0	216.8	25.0	468.8	33.0	816.8	41.0	1260.8	49.0	1800.8
.1	0.9	.1	62.1	.1	219.3	.1	472.5	.1	821.7	.1	1266.9	.1	1808.1
.2	1.1	.2	63.5	.2	221.9	.2	476.3	.2	826.7	.2	1273.1	.2	1815.5
.3	1.3	.3	64.9	.3	224.5	.3	480.1	.3	831.7	.3	1279.3	.3	1822.9
.4	1.5	.4	66.3	.4	227.1	.4	483.9	.4	836.7	.4	1285.5	.4	1830.3
.5	1.7	.5	67.7	.5	229.7	.5	487.7	.5	841.7	.5	1291.7	.5	1837.7
.6	1.9	.6	69.1	.6	232.3	.6	491.5	.6	846.7	.6	1297.9	.6	1845.1
.7	2.2	.7	70.6	.7	235.0	.7	495.4	.7	851.8	.7	1304.2	.7	1852.6
.8	2.4	.8	72.0	.8	237.6	.8	499.2	.8	856.8	.8	1310.4	.8	1860.0
.9	2.7	.9	73.5	.9	240.3	.9	503.1	.9	861.9	.9	1316.7	.9	1867.5
2.0	3.0	10.0	75.0	18.0	243.0	26.0	507.0	34.0	867.0	42.0	1323.0	50.0	1875.0
.1	3.3	.1	76.5	.1	245.7	.1	510.9	.1	872.1	.1	1329.3	.1	1882.5
.2	3.6	.2	78.0	.2	248.4	.2	514.8	.2	877.2	.2	1335.6	.2	1890.0
.3	4.0	.3	79.6	.3	251.2	.3	518.8	.3	882.4	.3	1342.0	.3	1897.6
.4	4.3	.4	81.1	.4	253.9	.4	522.7	.4	887.5	.4	1348.3	.4	1905.1
.5	4.7	.5	82.7	.5	256.7	.5	526.7	.5	892.7	.5	1354.7	.5	1912.7
.6	5.1	.6	84.3	.6	259.5	.6	530.7	.6	897.9	.6	1361.1	.6	1920.3
.7	5.5	.7	85.9	.7	262.3	.7	534.7	.7	903.1	.7	1367.5	.7	1927.9
.8	5.9	.8	87.5	.8	265.1	.8	538.7	.8	908.3	.8	1373.9	.8	1935.5
.9	6.3	.9	89.1	.9	267.9	.9	542.7	.9	913.5	.9	1380.3	.9	1943.1
3.0	6.8	11.0	90.8	19.0	270.8	27.0	546.8	35.0	918.8	43.0	1386.8	51.0	1950.8
.1	7.2	.1	92.4	.1	273.6	.1	550.8	.1	924.0	.1	1393.2	.1	1958.4
.2	7.7	.2	94.1	.2	276.5	.2	554.9	.2	929.3	.2	1399.7	.2	1966.1
.3	8.2	.3	95.8	.3	279.4	.3	559.0	.3	934.6	.3	1406.2	.3	1973.8
.4	8.7	.4	97.5	.4	282.3	.4	563.1	.4	939.9	.4	1412.7	.4	1981.5
.5	9.2	.5	99.2	.5	285.2	.5	567.2	.5	945.2	.5	1419.2	.5	1989.2
.6	9.7	.6	100.9	.6	288.1	.6	571.3	.6	950.5	.6	1425.7	.6	1996.9
.7	10.3	.7	102.7	.7	291.1	.7	575.5	.7	955.9	.7	1432.3	.7	2004.7
.8	10.8	.8	104.4	.8	294.0	.8	579.6	.8	961.2	.8	1438.8	.8	2012.4
.9	11.4	.9	106.2	.9	297.0	.9	583.8	.9	966.6	.9	1445.4	.9	2020.2
4.0	12.0	12.0	108.0	20.0	300.0	28.0	588.0	36.0	972.0	44.0	1452.0	52.0	2028.0
.1	12.6	.1	109.8	.1	303.0	.1	592.2	.1	977.4	.1	1458.6	.1	2035.8
.2	13.2	.2	111.6	.2	306.0	.2	596.4	.2	982.8	.2	1465.2	.2	2043.6
.3	13.9	.3	113.5	.3	309.1	.3	600.7	.3	988.3	.3	1471.9	.3	2051.5
.4	14.5	.4	115.3	.4	312.1	.4	604.9	.4	993.7	.4	1478.5	.4	2059.3
.5	15.2	.5	117.2	.5	315.2	.5	609.2	.5	999.2	.5	1485.2	.5	2067.2
.6	15.9	.6	119.1	.6	318.3	.6	613.5	.6	1004.7	.6	1491.9	.6	2075.1
.7	16.6	.7	121.0	.7	321.4	.7	617.8	.7	1010.2	.7	1498.6	.7	2083.0
.8	17.3	.8	122.9	.8	324.5	.8	622.1	.8	1015.7	.8	1505.3	.8	2090.9
.9	18.0	.9	124.8	.9	327.6	.9	626.4	.9	1021.2	.9	1512.0	.9	2098.8
5.0	18.8	13.0	126.8	21.0	330.8	29.0	630.8	37.0	1026.8	45.0	1518.8	53.0	2106.8
.1	19.5	.1	128.7	.1	333.9	.1	635.1	.1	1032.3	.1	1525.5	.1	2114.7
.2	20.3	.2	130.7	.2	337.1	.2	639.5	.2	1037.9	.2	1532.3	.2	2122.7
.3	21.1	.3	132.7	.3	340.3	.3	643.9	.3	1043.5	.3	1539.1	.3	2130.7
.4	21.9	.4	134.7	.4	343.5	.4	648.3	.4	1049.1	.4	1545.9	.4	2138.7
.5	22.7	.5	136.7	.5	346.7	.5	652.7	.5	1054.7	.5	1552.7	.5	2146.7
.6	23.5	.6	138.7	.6	349.9	.6	657.1	.6	1060.3	.6	1559.5	.6	2154.7
.7	24.4	.7	140.8	.7	353.2	.7	661.6	.7	1066.0	.7	1566.4	.7	2162.8
.8	25.2	.8	142.8	.8	356.4	.8	666.0	.8	1071.6	.8	1573.2	.8	2170.8
.9	26.1	.9	144.9	.9	359.7	.9	670.5	.9	1077.3	.9	1580.1	.9	2178.9
6.0	27.0	14.0	147.0	22.0	363.0	30.0	675.0	38.0	1083.0	46.0	1587.0	54.0	2187.0
.1	27.9	.1	149.1	.1	366.3	.1	679.5	.1	1088.7	.1	1593.9	.1	2195.1
.2	28.8	.2	151.2	.2	369.6	.2	684.0	.2	1094.4	.2	1600.8	.2	2203.2
.3	29.8	.3	153.4	.3	373.0	.3	688.6	.3	1100.2	.3	1607.8	.3	2211.4
.4	30.7	.4	155.5	.4	376.3	.4	693.1	.4	1105.9	.4	1614.7	.4	2219.5
.5	31.7	.5	157.7	.5	379.7	.5	697.7	.5	1111.7	.5	1621.7	.5	2227.7
.6	32.7	.6	159.9	.6	383.1	.6	702.3	.6	1117.5	.6	1628.7	.6	2235.9
.7	33.7	.7	162.1	.7	386.5	.7	706.9	.7	1123.3	.7	1635.7	.7	2244.1
.8	34.7	.8	164.3	.8	389.9	.8	711.5	.8	1129.1	.8	1642.7	.8	2252.3
.9	35.7	.9	166.5	.9	393.3	.9	716.1	.9	1134.9	.9	1649.7	.9	2260.5
7.0	36.8	15.0	168.8	23.0	396.8	31.0	720.8	39.0	1140.8	47.0	1656.8	55.0	2268.8
.1	37.8	.1	171.0	.1	400.2	.1	725.4	.1	1146.6	.1	1663.8	.1	2277.0
.2	38.9	.2	173.3	.2	403.7	.2	730.1	.2	1152.5	.2	1670.9	.2	2285.3
.3	40.0	.3	175.6	.3	407.2	.3	734.8	.3	1158.4	.3	1678.0	.3	2293.6
.4	41.1	.4	177.9	.4	410.7	.4	739.5	.4	1164.3	.4	1685.1	.4	2301.9
.5	42.2	.5	180.2	.5	414.2	.5	744.2	.5	1170.2	.5	1692.2	.5	2310.2
.6	43.3	.6	182.5	.6	417.7	.6	748.9	.6	1176.1	.6	1699.3	.6	2318.5
.7	44.5	.7	184.9	.7	421.3	.7	753.7	.7	1182.1	.7	1706.5	.7	2326.9
.8	45.6	.8	187.2	.8	424.8	.8	758.4	.8	1188.0	.8	1713.6	.8	2335.2
.9	46.8	.9	189.6	.9	428.4	.9	763.2	.9	1194.0	.9	1720.8	.9	2343.6

TABLE NO. 2.

EXCAVATION AND EMBANKMENT.

AREAS OF TRIANGLES.

Slope 1 Horizontal to 1 Perpendicular.

Height.	Area in Feet.	Height.	Area in Feet.	Height.	Area in Feet.	Height.	Area in Feet.	Height.	Area in Feet.	Height.	Area in Feet.	Height.	Area in Feet.	Height.	Area in Feet.
0.0	0.0	8.0	32.0	16.0	128.0	24.0	288.0	32.0	512.0	40.0	800.0	48.0	1152.0
.1	0.0	.1	32.8	.1	129.6	.1	290.4	.1	515.2	.1	804.0	.1	1156.8
.2	0.0	.2	33.6	.2	131.2	.2	292.8	.2	518.4	.2	808.0	.2	1161.6
.3	0.0	.3	34.4	.3	132.8	.3	295.2	.3	521.6	.3	812.0	.3	1166.4
.4	0.1	.4	35.3	.4	134.5	.4	297.7	.4	524.9	.4	816.1	.4	1171.3
.5	0.1	.5	36.1	.5	136.1	.5	300.1	.5	528.1	.5	820.1	.5	1176.1
.6	0.2	.6	37.0	.6	137.8	.6	302.6	.6	531.4	.6	824.2	.6	1181.0
.7	0.2	.7	37.8	.7	139.4	.7	305.0	.7	534.6	.7	828.2	.7	1185.8
.8	0.3	.8	38.7	.8	141.1	.8	307.5	.8	537.9	.8	832.3	.8	1190.7
.9	0.4	.9	39.6	.9	142.8	.9	310.0	.9	541.2	.9	836.4	.9	1195.6
1.0	0.5	9.0	40.5	17.0	144.5	25.0	312.5	33.0	544.5	41.0	840.5	49.0	1200.5
.1	0.6	.1	41.4	.1	146.2	.1	315.0	.1	547.8	.1	844.6	.1	1205.4
.2	0.7	.2	42.3	.2	147.9	.2	317.5	.2	551.1	.2	848.7	.2	1210.3
.3	0.8	.3	43.2	.3	149.6	.3	320.0	.3	554.4	.3	852.8	.3	1215.2
.4	1.0	.4	44.2	.4	151.4	.4	322.6	.4	557.8	.4	857.0	.4	1220.2
.5	1.1	.5	45.1	.5	153.1	.5	325.1	.5	561.1	.5	861.1	.5	1225.1
.6	1.2	.6	46.1	.6	154.9	.6	327.7	.6	564.5	.6	865.3	.6	1230.1
.7	1.4	.7	47.0	.7	156.6	.7	330.2	.7	567.8	.7	869.4	.7	1235.0
.8	1.6	.8	48.0	.8	158.4	.8	332.8	.8	571.2	.8	873.6	.8	1240.0
.9	1.8	.9	49.0	.9	160.2	.9	335.4	.9	574.6	.9	877.8	.9	1245.0
2.0	2.0	10.0	50.0	18.0	162.0	26.0	338.0	34.0	578.0	42.0	882.0	50.0	1250.0
.1	2.2	.1	51.0	.1	163.8	.1	340.6	.1	581.4	.1	886.2	.1	1255.0
.2	2.4	.2	52.0	.2	165.6	.2	343.2	.2	584.8	.2	890.4	.2	1260.0
.3	2.6	.3	53.0	.3	167.4	.3	345.8	.3	588.2	.3	894.6	.3	1265.0
.4	2.9	.4	54.1	.4	169.3	.4	348.5	.4	591.7	.4	898.9	.4	1270.1
.5	3.1	.5	55.1	.5	171.1	.5	351.1	.5	595.1	.5	903.1	.5	1275.1
.6	3.4	.6	56.2	.6	173.0	.6	353.8	.6	598.6	.6	907.4	.6	1280.2
.7	3.6	.7	57.2	.7	174.8	.7	356.4	.7	602.0	.7	911.6	.7	1285.2
.8	3.9	.8	58.3	.8	176.7	.8	359.1	.8	605.5	.8	915.9	.8	1290.3
.9	4.2	.9	59.4	.9	178.6	.9	361.8	.9	609.0	.9	920.2	.9	1295.4
3.0	4.5	11.0	60.5	19.0	180.5	27.0	364.5	35.0	612.5	43.0	924.5	51.0	1300.5
.1	4.8	.1	61.6	.1	182.4	.1	367.2	.1	616.0	.1	928.8	.1	1305.6
.2	5.1	.2	62.7	.2	184.3	.2	369.9	.2	619.5	.2	933.1	.2	1310.7
.3	5.4	.3	63.8	.3	186.2	.3	372.6	.3	623.0	.3	937.4	.3	1315.8
.4	5.8	.4	65.0	.4	188.2	.4	375.4	.4	626.6	.4	941.8	.4	1321.0
.5	6.1	.5	66.1	.5	190.1	.5	378.1	.5	630.1	.5	946.1	.5	1326.1
.6	6.5	.6	67.3	.6	192.1	.6	380.9	.6	633.7	.6	950.5	.6	1331.3
.7	6.8	.7	68.4	.7	194.0	.7	383.6	.7	637.2	.7	954.8	.7	1336.4
.8	7.2	.8	69.6	.8	196.0	.8	386.4	.8	640.8	.8	959.2	.8	1341.6
.9	7.6	.9	70.8	.9	198.0	.9	389.2	.9	644.4	.9	963.6	.9	1346.8
4.0	8.0	12.0	72.0	20.0	200.0	28.0	392.0	36.0	648.0	44.0	968.0	52.0	1352.0
.1	8.4	.1	73.2	.1	202.0	.1	394.8	.1	651.6	.1	972.4	.1	1357.2
.2	8.8	.2	74.4	.2	204.0	.2	397.6	.2	655.2	.2	976.8	.2	1362.4
.3	9.2	.3	75.6	.3	206.0	.3	400.4	.3	658.8	.3	981.2	.3	1367.6
.4	9.7	.4	76.9	.4	208.1	.4	403.2	.4	662.5	.4	985.7	.4	1372.9
.5	10.1	.5	78.1	.5	210.1	.5	406.1	.5	666.1	.5	990.1	.5	1378.1
.6	10.6	.6	79.4	.6	212.2	.6	409.0	.6	669.8	.6	994.6	.6	1383.4
.7	11.0	.7	80.6	.7	214.2	.7	411.8	.7	673.4	.7	999.0	.7	1388.6
.8	11.5	.8	81.9	.8	216.3	.8	414.7	.8	677.1	.8	1003.5	.8	1393.9
.9	12.0	.9	83.2	.9	218.4	.9	417.6	.9	680.8	.9	1008.0	.9	1399.2
5.0	12.5	13.0	84.5	21.0	220.5	29.0	420.5	37.0	684.5	45.0	1012.5	53.0	1404.5
.1	13.0	.1	85.8	.1	222.6	.1	423.4	.1	688.2	.1	1017.0	.1	1409.8
.2	13.5	.2	87.1	.2	224.7	.2	426.3	.2	691.9	.2	1021.5	.2	1415.1
.3	14.0	.3	88.4	.3	226.8	.3	429.2	.3	695.6	.3	1026.0	.3	1420.4
.4	14.6	.4	89.8	.4	229.0	.4	432.2	.4	699.4	.4	1030.6	.4	1425.8
.5	15.1	.5	91.1	.5	231.1	.5	435.1	.5	703.1	.5	1035.1	.5	1431.1
.6	15.7	.6	92.5	.6	233.3	.6	438.1	.6	706.9	.6	1039.7	.6	1436.5
.7	16.2	.7	93.8	.7	235.4	.7	441.0	.7	710.6	.7	1044.2	.7	1441.8
.8	16.8	.8	95.2	.8	237.6	.8	444.0	.8	714.4	.8	1048.8	.8	1447.2
.9	17.4	.9	96.6	.9	239.8	.9	447.0	.9	718.2	.9	1053.4	.9	1452.6
6.0	18.0	14.0	98.0	22.0	242.0	30.0	450.0	38.0	722.0	46.0	1058.0	54.0	1458.0
.1	18.6	.1	99.4	.1	244.2	.1	453.0	.1	725.8	.1	1062.6	.1	1463.4
.2	19.2	.2	100.8	.2	246.4	.2	456.0	.2	729.6	.2	1067.2	.2	1468.8
.3	19.8	.3	102.2	.3	248.6	.3	459.0	.3	733.4	.3	1071.8	.3	1474.2
.4	20.5	.4	103.7	.4	250.9	.4	462.1	.4	737.3	.4	1076.5	.4	1479.7
.5	21.1	.5	105.1	.5	253.1	.5	465.1	.5	741.1	.5	1081.1	.5	1485.1
.6	21.8	.6	106.6	.6	255.4	.6	468.2	.6	745.0	.6	1085.8	.6	1490.6
.7	22.4	.7	108.0	.7	257.6	.7	471.2	.7	748.8	.7	1090.4	.7	1496.0
.8	23.1	.8	109.5	.8	259.9	.8	474.3	.8	752.7	.8	1095.1	.8	1501.5
.9	23.8	.9	111.0	.9	262.2	.9	477.4	.9	756.6	.9	1099.8	.9	1507.0
7.0	24.5	15.0	112.5	23.0	264.5	31.0	480.5	39.0	760.5	47.0	1104.5	55.0	1512.5
.1	25.2	.1	114.0	.1	266.8	.1	483.6	.1	764.4	.1	1109.2	.1	1518.0
.2	25.9	.2	115.5	.2	269.1	.2	486.7	.2	768.3	.2	1113.9	.2	1523.5
.3	26.6	.3	117.0	.3	271.4	.3	489.8	.3	772.2	.3	1118.6	.3	1529.0
.4	27.4	.4	118.6	.4	273.8	.4	493.0	.4	776.2	.4	1123.4	.4	1534.6
.5	28.1	.5	120.1	.5	276.1	.5	496.1	.5	780.1	.5	1128.1	.5	1540.1
.6	28.9	.6	121.7	.6	278.5	.6	499.3	.6	784.1	.6	1132.9	.6	1545.7
.7	29.6	.7	123.2	.7	280.8	.7	502.4	.7	788.0	.7	1137.6	.7	1551.2
.8	30.4	.8	124.8	.8	283.2	.8	505.6	.8	792.0	.8	1142.4	.8	1556.8
.9	31.2	.9	126.4	.9	285.6	.9	508.8	.9	796.0	.9	1147.2	.9	1562.4

TABLE NO. 3.

EXCAVATION AND EMBANKMENT.

AREAS OF TRIANGLES.

Slope 1-2 Horizontal to 1 Perpendicular.

Height.	Area in Feet.	Height.	Area in Feet.	Height.	Area in Feet.	Height.	Area in Feet.	Height.	Area in Feet.	Height.	Area in Feet.	Height.	Area in Feet.
0.0	0.0	8.0	16.0	16.0	64.0	24.0	144.0	32.0	256.0	40.0	400.0	48.0	576.0
.1	0.0	.1	16.4	.1	64.8	.1	145.2	.1	257.6	.1	402.0	.1	578.4
.2	0.0	.2	16.8	.2	65.6	.2	146.4	.2	259.2	.2	404.0	.2	580.8
.3	0.0	.3	17.2	.3	66.4	.3	147.6	.3	260.8	.3	406.0	.3	583.2
.4	0.0	.4	17.6	.4	67.2	.4	148.8	.4	262.4	.4	408.0	.4	585.6
.5	0.1	.5	18.1	.5	68.1	.5	150.1	.5	264.1	.5	410.1	.5	588.1
.6	0.1	.6	18.5	.6	68.9	.6	151.3	.6	265.7	.6	412.1	.6	590.5
.7	0.1	.7	18.9	.7	69.7	.7	152.5	.7	267.3	.7	414.1	.7	592.9
.8	0.2	.8	19.4	.8	70.6	.8	153.8	.8	269.0	.8	416.2	.8	595.4
.9	0.2	.9	19.8	.9	71.4	.9	155.0	.9	270.6	.9	418.2	.9	597.8
1.0	0.3	9.0	20.3	17.0	72.3	25.0	156.3	33.0	272.3	41.0	420.3	49.0	600.3
.1	0.3	.1	20.7	.1	73.1	.1	157.5	.1	273.9	.1	422.3	.1	602.7
.2	0.4	.2	21.2	.2	74.0	.2	158.8	.2	275.6	.2	424.4	.2	605.2
.3	0.4	.3	21.6	.3	74.8	.3	160.0	.3	277.2	.3	426.4	.3	607.6
.4	0.5	.4	22.1	.4	75.7	.4	161.3	.4	278.9	.4	428.5	.4	610.1
.5	0.6	.5	22.6	.5	76.6	.5	162.6	.5	280.6	.5	430.6	.5	612.6
.6	0.6	.6	23.0	.6	77.4	.6	163.8	.6	282.2	.6	432.6	.6	615.0
.7	0.7	.7	23.5	.7	78.3	.7	165.1	.7	283.9	.7	434.7	.7	617.5
.8	0.8	.8	24.0	.8	79.2	.8	166.4	.8	285.6	.8	436.8	.8	620.0
.9	0.9	.9	24.5	.9	80.1	.9	167.7	.9	287.3	.9	438.9	.9	622.5
2.0	1.0	10.0	25.0	18.0	81.0	26.0	169.0	34.0	289.0	42.0	441.0	50.0	625.0
.1	1.1	.1	25.5	.1	81.9	.1	170.3	.1	290.7	.1	443.1	.1	627.5
.2	1.2	.2	26.0	.2	82.8	.2	171.6	.2	292.4	.2	445.2	.2	630.0
.3	1.3	.3	26.5	.3	83.7	.3	172.9	.3	294.1	.3	447.3	.3	632.5
.4	1.4	.4	27.0	.4	84.6	.4	174.2	.4	295.8	.4	449.4	.4	635.0
.5	1.6	.5	27.6	.5	85.6	.5	175.6	.5	297.6	.5	451.6	.5	637.6
.6	1.7	.6	28.1	.6	86.5	.6	176.9	.6	299.3	.6	453.7	.6	640.1
.7	1.8	.7	28.6	.7	87.4	.7	178.2	.7	301.0	.7	455.8	.7	642.6
.8	2.0	.8	29.2	.8	88.4	.8	179.6	.8	302.8	.8	458.0	.8	645.2
.9	2.1	.9	29.7	.9	89.3	.9	180.9	.9	304.5	.9	460.1	.9	647.7
3.0	2.3	11.0	30.3	19.0	90.3	27.0	182.3	35.0	306.3	43.0	462.3	51.0	650.3
.1	2.4	.1	30.8	.1	91.2	.1	183.6	.1	308.0	.1	464.4	.1	652.8
.2	2.6	.2	31.4	.2	92.2	.2	185.0	.2	309.8	.2	466.6	.2	655.4
.3	2.7	.3	31.9	.3	93.1	.3	186.3	.3	311.5	.3	468.7	.3	657.9
.4	2.9	.4	32.5	.4	94.1	.4	187.7	.4	313.3	.4	470.9	.4	660.5
.5	3.1	.5	33.1	.5	95.1	.5	189.1	.5	315.1	.5	473.1	.5	663.1
.6	3.2	.6	33.6	.6	96.0	.6	190.4	.6	316.8	.6	475.2	.6	665.6
.7	3.4	.7	34.2	.7	97.0	.7	191.8	.7	318.6	.7	477.4	.7	668.2
.8	3.6	.8	34.8	.8	98.0	.8	193.2	.8	320.4	.8	479.6	.8	670.8
.9	3.8	.9	35.4	.9	99.0	.9	194.6	.9	322.2	.9	481.8	.9	673.4
4.0	4.0	12.0	36.0	20.0	100.0	28.0	196.0	36.0	324.0	44.0	484.0	52.0	676.0
.1	4.2	.1	36.6	.1	101.0	.1	197.4	.1	325.8	.1	486.2	.1	678.6
.2	4.4	.2	37.2	.2	102.0	.2	198.8	.2	327.6	.2	488.4	.2	681.2
.3	4.6	.3	37.8	.3	103.0	.3	200.2	.3	329.4	.3	490.6	.3	683.8
.4	4.8	.4	38.4	.4	104.0	.4	201.6	.4	331.2	.4	492.8	.4	686.4
.5	5.1	.5	39.1	.5	105.1	.5	203.1	.5	333.1	.5	495.1	.5	689.1
.6	5.3	.6	39.7	.6	106.1	.6	204.5	.6	334.9	.6	497.3	.6	691.7
.7	5.5	.7	40.3	.7	107.1	.7	205.9	.7	336.7	.7	499.5	.7	694.3
.8	5.8	.8	41.0	.8	108.2	.8	207.4	.8	338.6	.8	501.8	.8	697.0
.9	6.0	.9	41.6	.9	109.2	.9	208.8	.9	340.4	.9	504.0	.9	699.6
5.0	6.3	13.0	42.3	21.0	110.3	29.0	210.3	37.0	342.3	45.0	506.3	53.0	702.3
.1	6.5	.1	42.9	.1	111.3	.1	211.7	.1	344.1	.1	508.5	.1	704.9
.2	6.8	.2	43.6	.2	112.4	.2	213.2	.2	346.0	.2	510.8	.2	707.6
.3	7.0	.3	44.2	.3	113.4	.3	214.6	.3	347.8	.3	513.0	.3	710.2
.4	7.3	.4	44.9	.4	114.5	.4	216.1	.4	349.7	.4	515.3	.4	712.9
.5	7.6	.5	45.6	.5	115.6	.5	217.6	.5	351.6	.5	517.6	.5	715.6
.6	7.8	.6	46.2	.6	116.6	.6	219.0	.6	353.4	.6	519.8	.6	718.2
.7	8.1	.7	46.9	.7	117.7	.7	220.5	.7	355.3	.7	522.1	.7	720.9
.8	8.4	.8	47.6	.8	118.8	.8	222.0	.8	357.2	.8	524.4	.8	723.6
.9	8.7	.9	48.3	.9	119.9	.9	223.5	.9	359.1	.9	526.7	.9	726.3
6.0	9.0	14.0	49.0	22.0	121.0	30.0	225.0	38.0	361.0	46.0	529.0	54.0	729.0
.1	9.3	.1	49.7	.1	122.1	.1	226.5	.1	362.9	.1	531.3	.1	731.7
.2	9.6	.2	50.4	.2	123.2	.2	228.0	.2	364.8	.2	533.6	.2	734.4
.3	9.9	.3	51.1	.3	124.3	.3	229.5	.3	366.7	.3	535.9	.3	737.1
.4	10.2	.4	51.8	.4	125.4	.4	231.0	.4	368.6	.4	538.2	.4	739.8
.5	10.6	.5	52.6	.5	126.6	.5	232.6	.5	370.6	.5	540.6	.5	742.6
.6	10.9	.6	53.3	.6	127.7	.6	234.1	.6	372.5	.6	542.9	.6	745.3
.7	11.2	.7	54.0	.7	128.8	.7	235.6	.7	374.4	.7	545.2	.7	748.0
.8	11.6	.8	54.8	.8	130.0	.8	237.2	.8	376.4	.8	547.6	.8	750.8
.9	11.9	.9	55.5	.9	131.1	.9	238.7	.9	378.3	.9	549.9	.9	753.5
7.0	12.3	15.0	56.3	23.0	132.3	31.0	240.3	39.0	380.3	47.0	552.3	55.0	756.3
.1	12.6	.1	57.0	.1	133.4	.1	241.8	.1	382.2	.1	554.6	.1	759.0
.2	13.0	.2	57.8	.2	134.6	.2	243.4	.2	384.2	.2	557.0	.2	761.8
.3	13.3	.3	58.5	.3	135.7	.3	244.9	.3	386.1	.3	559.3	.3	764.5
.4	13.7	.4	59.3	.4	136.9	.4	246.5	.4	388.1	.4	561.7	.4	767.3
.5	14.1	.5	60.1	.5	138.1	.5	248.1	.5	390.1	.5	564.1	.5	770.1
.6	14.4	.6	60.8	.6	139.2	.6	249.6	.6	392.0	.6	566.4	.6	772.8
.7	14.8	.7	61.6	.7	140.4	.7	251.2	.7	394.0	.7	568.8	.7	775.6
.8	15.2	.8	62.4	.8	141.6	.8	252.8	.8	396.0	.8	571.2	.8	778.4
.9	15.6	.9	63.2	.9	142.8	.9	254.4	.9	398.0	.9	573.6	.9	781.2

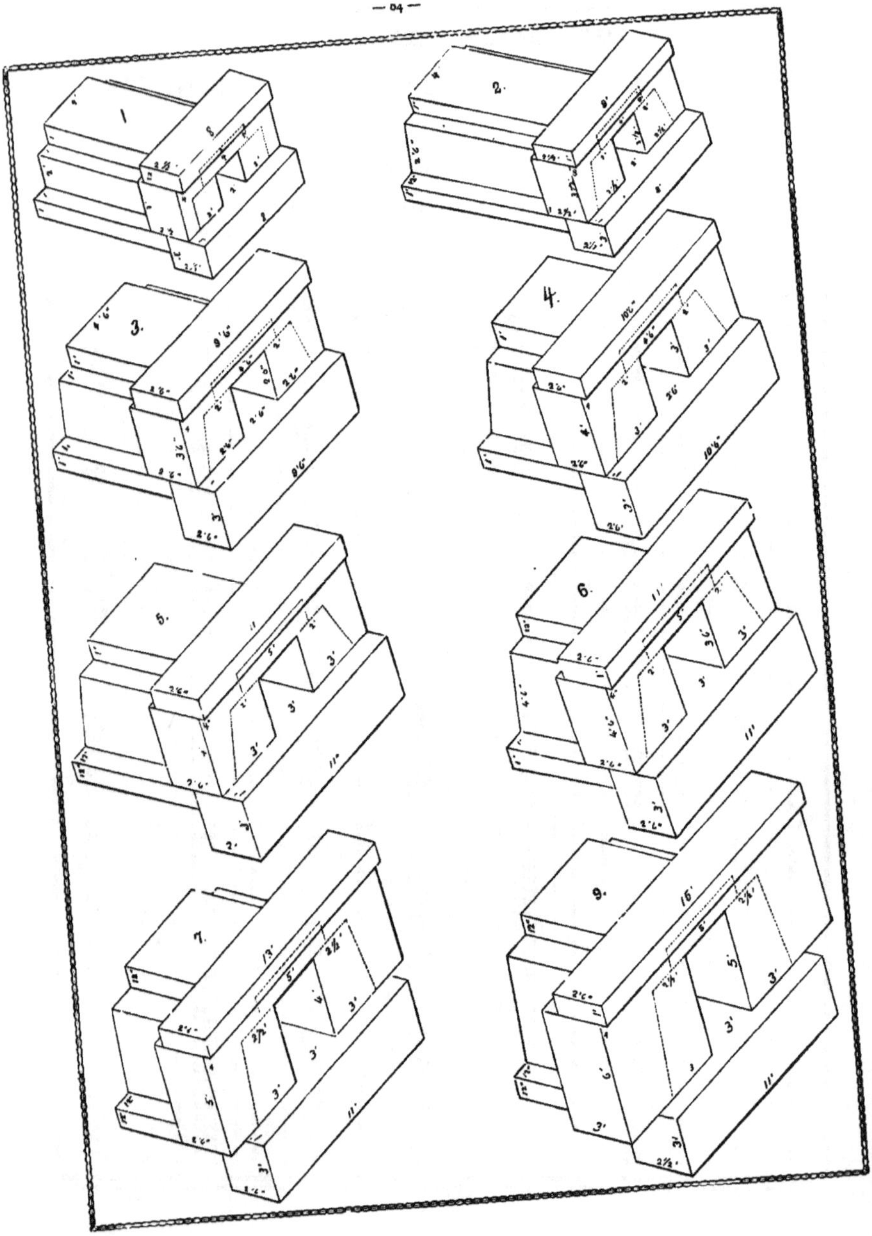
1
2.
3.
9'6"
4.
10'6"
5.
11'
6.
11'
7.
13'
9.
15'

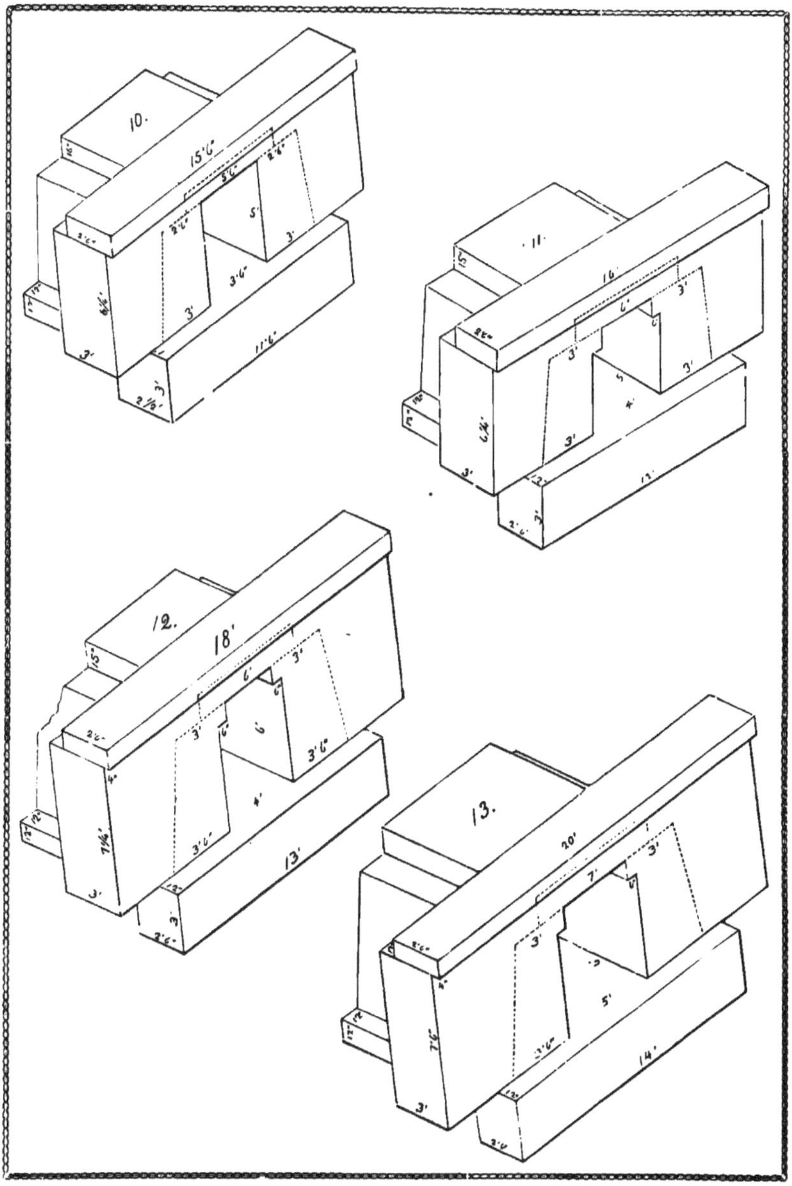
10.
15'6"
11'6"
11.
16'
12.
18'
13'
13.
20'
14'

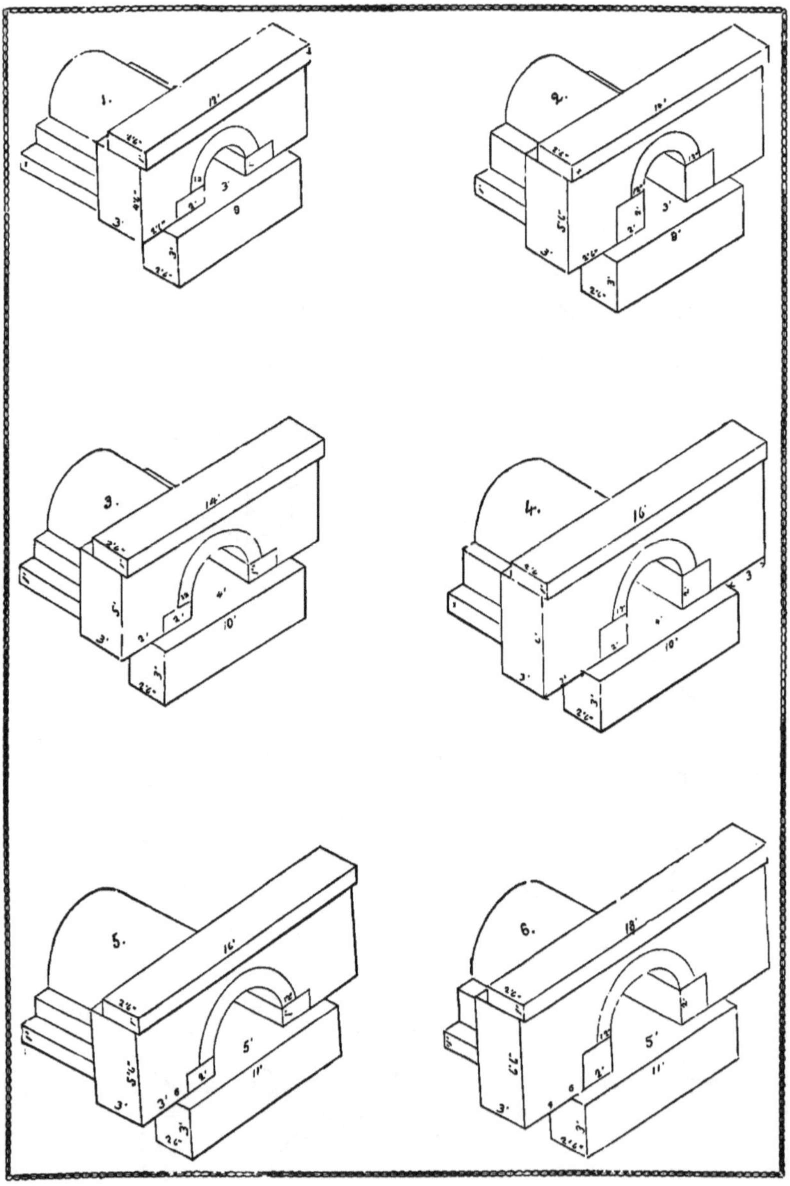
1.
2.
3.
14'
4.
16'
5.
16'
6.
18'

TABLE FOR BOX CULVERTS.

No. of Culvert....		1	2	3	4	5	6	7	8	9	10	11	12	13
Size of Culvert....		2X2	2X2½	2½X2½	2½X3	3X3	3X3½	3X4	3X4½	3X5	3½X5	4X5	4X6	5X6
Cub. Yds. in Coping, W'gs and Sunk Walls.......		6.51852	7.7500	8.084815	9.28889	9.703704	9.981481	12.111111	11.981481	17.111111	17.981481	18.290298	22.111111	26.259259
Cub. Yds. in 1 Lineal Foot of Culvert............		0.748780	0.805118	0.965185	1.111111	1.148148	1.210741	1.407407	1.509259	1.611111	1.699074	1.777778	2.129630	2.277778
Area Water-way..		4.0	5.0	6.25	7.5	9.0	10.5	12.0	13.5	15.0	17.5	20.0	24.0	30.0
Base, Fifteen Feet. HEIGHT.	Length of Culvert	Cub. Yds.	Cub. Yds.	Cub. Yds.	Cub. Yds.	Cub. Yds.	Cub. Yds.	Cub. Yds.	Cub. Yds.	Cub. Yds.	Cub. Yds.	Cub. Yds.	Cub. Yds.	Cub. Yds.
0	15	17.6	21.2	22.1	26.1	26.9	28.6	33.2	37.6	41.3	43.5	45.0	55.1	60.4
⅓	16	18.4	22.1	23.0	27.2	28.1	29.8	34.6	39.1	42.9	45.2	46.7	57.2	62.7
⅔	17	19.1	23.0	24.0	28.3	29.2	31.1	36.0	40.6	44.5	46.9	48.5	59.3	65.0
1	18	19.9	23.9	24.9	29.4	30.4	32.3	37.4	42.1	46.1	48.6	50.3	61.4	67.3
⅓	19	20.6	24.8	[illegible]	30.5	31.5	33.6	38.9	43.7	47.7	50.3	52.1	63.6	69.5
⅔	20	21.3	25.7	26.8	31.6	32.7	34.8	40.3	45.2	49.3	52.0	53.9	65.7	71.8
2	21	22.1	26.6	27.7	32.7	33.8	36.0	41.7	46.7	50.9	53.7	55.6	67.8	74.1
⅓	22	22.8	27.5	28.6	33.8	35.0	37.3	43.1	48.2	52.6	55.4	57.4	70.0	76.4
⅔	23	23.6	28.4	29.6	35.0	36.1	38.5	44.5	49.7	54.2	57.1	59.2	72.1	78.7
3	24	24.3	29.3	30.5	36.1	37.3	39.8	45.9	51.2	55.8	58.8	61.0	74.2	80.9
⅓	25	25.0	30.2	31.4	37.2	38.4	41.0	47.3	52.7	57.4	60.5	62.7	76.4	83.2
⅔	26	25.8	31.1	32.4	38.3	39.6	42.2	48.7	54.2	59.0	62.2	64.5	78.5	85.5
4	27	26.6	32.0	33.3	39.4	40.8	43.5	50.1	55.7	60.6	63.9	66.3	80.6	87.8
⅓	28	27.3	32.9	34.2	40.5	41.8	44.7	51.5	57.2	62.2	65.6	68.1	82.7	90.0
⅔	29	28.0	33.8	35.2	41.6	43.0	46.0	52.9	58.7	63.8	67.3	69.9	84.9	92.3
5	30	28.7	34.7	36.1	42.7	44.1	47.2	54.3	60.3	65.4	69.0	71.6	87.0	94.6
⅓	31	29.5	35.6	37.0	43.8	45.3	48.4	55.7	61.8	67.1	70.7	73.4	89.1	96.9
⅔	32	30.2	36.5	38.0	45.0	46.4	49.7	57.1	63.3	68.7	72.4	75.2	91.2	99.1
6	33	31.0	37.4	38.9	46.1	47.6	50.9	58.6	64.8	70.3	74.1	77.0	93.4	101.4
⅓	34	31.7	38.3	39.9	47.2	48.7	52.2	60.0	66.3	71.9	75.7	78.7	95.5	103.7
⅔	35	32.4	39.2	40.8	48.3	49.9	53.4	61.4	67.8	73.5	77.4	80.5	97.6	106.0
7	36	33.2	40.1	41.7	49.4	51.0	54.6	62.8	69.3	75.1	79.1	82.3	99.8	108.3
⅓	37	33.9	41.0	42.7	50.5	52.2	55.9	64.2	70.8	76.7	80.8	84.1	101.9	110.5
⅔	38	34.7	41.9	43.6	51.6	53.3	57.1	65.6	72.3	78.3	82.5	85.9	104.0	112.8
8	39	35.4	42.8	44.5	52.7	54.5	58.4	67.0	73.8	79.9	84.2	87.6	106.2	115.1
⅓	40	36.1	43.7	45.5	53.8	55.6	59.6	68.4	75.4	81.6	85.9	89.4	108.3	117.4
⅔	41	36.9	44.6	46.4	55.0	56.8	60.9	69.8	76.9	83.2	87.6	91.2	110.4	119.6
9	42	37.6	45.5	47.3	56.1	57.9	62.1	71.2	78.4	84.8	89.3	93.0	112.6	121.9
⅓	43	38.4	46.4	48.3	57.2	59.1	63.3	72.6	79.9	86.4	91.0	94.7	114.7	121.2
⅔	44	39.1	47.3	49.2	58.3	60.2	64.6	74.0	81.4	88.0	92.7	96.5	116.8	126.5
10	45	39.9	48.2	50.1	59.4	61.4	65.8	75.4	82.9	89.6	94.4	98.3	118.9	128.8
⅓	46	40.6	49.1	51.1	60.5	62.5	67.1	76.9	84.4	91.2	96.1	100.1	121.1	131.0
⅔	47	41.3	50.0	52.0	61.6	63.7	68.3	78.3	85.9	92.8	97.8	101.9	123.2	133.3
11	48	42.1	50.9	52.9	62.7	64.8	69.5	79.7	87.4	94.4	99.5	103.6	125.3	135.6
⅓	49	42.8	51.8	53.9	63.8	66.0	70.8	81.1	88.9	96.1	101.2	105.4	127.5	137.9
⅔	50	43.6	52.7	54.8	65.0	67.1	72.0	82.5	90.4	97.7	102.9	107.2	129.6	140.1
12	51	44.3	53.6	55.8	66.1	68.3	73.3	83.9	92.0	99.3	104.6	109.0	131.7	142.4
⅓	52	45.0	54.5	56.7	67.2	69.4	74.5	85.3	93.5	100.9	106.3	110.7	133.9	144.7
⅔	53	45.8	55.4	57.6	68.3	70.6	75.7	86.7	95.0	102.5	108.0	112.5	136.0	147.0
13	54	46.5	56.2	58.6	69.4	71.7	77.0	88.1	96.5	104.1	109.7	114.3	138.1	149.3
⅓	55	47.3	57.1	59.5	70.5	72.8	78.2	89.5	98.0	105.7	111.4	116.1	140.2	151.5
⅔	56	48.0	58.0	60.4	71.6	74.0	79.5	90.9	99.5	107.3	113.1	117.9	142.4	153.8
14	57	48.7	58.9	61.4	72.7	75.1	80.7	92.3	101.0	108.9	114.8	119.6	144.5	156.1
⅓	58	49.5	59.8	62.3	73.8	76.3	81.9	93.7	102.5	110.6	116.5	121.4	146.6	158.4
⅔	59	50.2	60.7	63.2	75.0	77.4	83.2	95.1	104.0	112.2	118.2	123.2	148.8	160.6
15	60	51.0	61.6	64.2	76.1	78.5	84.4	96.6	105.5	113.8	119.9	125.0	150.9	162.9
⅓	61	51.7	62.5	[illegible]	77.2	79.7	85.7	98.0	107.0	115.4	121.6	126.7	153.0	165.2
⅔	62	52.4	63.4	66.0	78.3	80.9	86.9	99.4	108.6	117.0	123.3	128.5	155.1	167.5
16	63	53.2	64.3	67.0	79.4	82.0	88.1	100.8	110.1	118.6	125.0	130.3	157.3	169.8
⅓	64	53.9	65.2	67.9	80.5	83.2	89.4	102.2	111.6	120.2	126.7	132.1	159.4	172.0
⅔	65	54.7	66.1	68.8	81.6	84.3	90.6	103.6	113.1	121.8	128.4	133.9	161.5	174.3
17	66	55.4	67.0	69.8	82.7	85.5	91.9	105.0	114.6	123.4	130.1	135.6	163.7	176.6
⅓	67	56.1	67.9	70.7	83.8	86.6	93.1	106.4	116.1	125.1	131.8	137.4	165.8	178.9
⅔	68	56.9	68.8	71.7	85.0	87.8	94.4	107.8	117.6	126.7	133.5	139.2	167.9	181.1
18	69	57.6	69.7	72.6	86.1	88.9	95.6	109.2	119.1	128.3	135.2	141.0	170.1	183.4
⅓	70	58.4	70.6	73.5	87.2	90.1	96.8	110.6	120.6	129.9	136.9	142.7	172.1	185.7
⅔	71	59.1	71.5	74.5	88.3	91.2	98.1	112.0	122.1	131.5	138.6	144.5	174.3	188.0
19	72	59.9	72.4	75.4	89.4	92.4	99.3	113.4	123.6	133.1	140.3	146.3	176.4	190.3
⅓	73	60.6	73.3	76.3	90.5	93.5	100.6	114.9	125.2	134.7	142.0	148.1	178.6	192.5
⅔	74	61.3	74.2	77.3	91.6	94.7	101.8	116.3	126.7	136.3	143.7	149.9	180.7	194.8
20	75	62.1	75.1	78.2	92.7	95.8	103.0	117.7	128.2	137.9	145.4	151.6	182.8	197.2
⅓	76	62.8	76.0	79.1	93.8	96.9	104.3	119.1	129.7	139.6	147.1	153.4	185.0	199.4
⅔	77	63.6	76.9	80.0	95.0	98.1	105.5	120.5	131.2	141.2	148.8	155.2	187.1	201.6
21	78	64.3	77.8	81.0	96.1	99.3	106.8	121.9	132.7	142.8	150.5	157.0	189.2	203.9
⅓	79	65.0	78.7	81.9	97.2	100.4	108.0	123.3	134.2	144.4	152.2	158.7	191.4	206.2
⅔	80	65.8	79.6	82.9	98.3	101.6	109.2	124.7	135.7	146.0	153.9	160.5	193.5	208.5
22	81	66.5	80.5	83.8	99.4	102.7	110.5	126.1	137.2	147.6	155.6	162.3	195.6	210.8
⅓	82	67.3	81.4	84.7	100.5	103.8	111.7	127.5	138.7	149.2	157.3	164.1	197.7	213.0
⅔	83	68.0	82.3	85.7	101.6	105.0	113.0	128.9	140.2	150.8	159.0	165.9	199.9	215.3
23	84	68.7	83.2	86.6	102.7	106.1	114.2	130.3	141.8	152.4	160.7	167.6	202.0	217.6
⅓	85	69.5	84.1	87.6	103.8	107.3	115.4	131.7	143.3	154.1	162.4	169.4	204.1	219.9
⅔	86	70.2	85.0	88.5	104.9	108.4	116.7	133.1	144.8	155.7	164.1	171.2	206.3	222.2

TABLE FOR BOX CULVERTS.

(CONTINUED.)

No. of Culvert....		1	2	3	4	5	6	7	8	9	10	11	12	13
Size of Culvert....		2X2	2X2½	2½X2½	2½X3	3X3	3X3½	3X4	3X4½	3X5	3½X5	4X5	4X6	5X6
Cub. Yds. in Coping, Wings and Sunk Walls......		8.51852	7.7500	8.664815	9.25926	9.768701	9.981481	12.111111	14.981481	17.111111	17.981481	18.296296	25.111111	26.259259
Cub. Yds. in 1 Lineal Foot of Culvert..........		0.740740	0.898148	0.985185	1.111111	1.148148	1.240741	1.407407	1.569259	1.611111	1.666074	1.777778	2.129630	2.277778
Area Water-Way..		4.0	5.0	6.25	7.5	9.0	10.5	12.0	13.5	15.0	17.5	20.0	24.0	30.0
Base, Fifteen Feet. HEIGHT.	Length of Culvert.	Cub. Yds.	Cub. Yds.	Cub. Yds.	Cub. Yds.	Cub. Yds.	Cub. Yds.	Cub. Yds.	Cub. Yds.	Cub. Yds.	Cub. Yds.	Cub. Yds.	Cub. Yds.	Cub. Yds.
24	87	71.0	85.9	89.4	106.1	109.6	117.9	134.6	146.3	157.3	165.8	173.0	208.4	224.4
⅓	88	71.7	86.8	90.4	107.2	110.7	119.2	136.0	147.8	158.9	167.5	174.7	210.5	226.7
⅔	89	72.4	87.7	91.3	108.3	111.9	120.4	137.4	149.3	160.5	169.2	176.5	212.6	229.0
25	90	73.2	88.6	92.2	109.4	113.0	121.6	138.8	150.8	162.1	170.9	178.3	214.8	231.3
⅓	91	73.9	89.5	93.2	110.5	114.2	122.9	140.2	152.3	163.7	172.6	180.1	216.9	233.5
⅔	92	74.7	90.4	94.1	111.6	115.3	124.1	141.6	153.8	165.3	174.3	181.9	219.0	235.8
26	93	75.4	91.3	95.0	112.7	116.5	125.4	143.0	155.3	166.9	176.0	183.6	221.2	238.1
⅓	94	76.1	92.2	96.0	113.8	117.6	126.6	144.4	156.8	168.6	177.7	185.4	223.3	240.4
⅔	95	76.9	93.1	96.9	115.0	118.8	127.9	145.8	158.4	170.2	179.4	187.2	225.4	242.6
27	96	77.6	94.0	97.8	116.1	119.9	129.1	147.2	159.9	171.8	181.1	189.0	227.6	244.9
⅓	97	78.4	94.9	98.8	117.2	121.1	130.3	148.6	161.4	173.4	182.8	190.7	229.7	247.2
⅔	98	79.1	95.8	99.7	118.3	122.2	131.6	150.0	162.9	175.0	184.5	192.5	231.8	249.5
28	99	79.9	96.7	100.6	119.4	123.4	132.8	151.4	164.4	176.6	186.2	194.3	233.9	251.8
⅓	100	80.6	97.6	101.6	120.5	124.5	134.1	152.9	165.9	178.2	187.9	196.1	236.1	254.0
⅔	101	81.3	98.5	102.5	121.6	125.7	135.3	154.3	167.4	179.8	189.6	197.9	238.2	256.3
29	102	82.1	99.4	103.4	122.7	126.8	136.5	155.7	168.9	181.4	191.3	199.6	240.3	258.6
⅓	103	82.8	100.3	104.4	123.8	128.0	137.8	157.1	170.4	183.1	193.0	201.4	242.5	260.9
⅔	104	83.6	101.1	105.3	125.0	129.1	139.0	158.5	171.9	184.7	194.7	203.2	244.6	263.1
30	105	84.3	102.1	106.3	126.1	130.3	140.3	159.9	173.4	186.3	196.4	204.9	246.7	265.4
⅓	106	85.0	103.0	107.2	127.2	131.4	141.5	161.3	175.0	187.9	198.1	206.7	248.8	267.7
⅔	107	85.8	103.9	108.1	128.3	132.5	142.7	162.7	176.5	189.5	199.8	208.5	251.0	270.0
31	108	86.5	104.7	109.1	129.4	133.7	144.0	164.1	178.0	191.1	201.5	210.3	253.1	272.3
⅓	109	87.3	105.6	110.0	130.5	134.8	145.2	165.5	179.5	192.7	203.2	212.1	255.2	274.5
⅔	110	88.0	106.5	110.9	131.6	136.0	146.5	166.9	181.0	194.3	204.9	213.9	257.4	276.8
32	111	88.7	107.4	111.9	132.7	137.1	147.7	168.3	182.5	195.9	206.6	215.6	259.5	279.1
⅓	112	89.5	108.3	112.8	133.8	138.3	148.9	169.7	184.0	197.6	208.3	217.4	261.6	281.4
⅔	113	90.2	109.2	113.7	135.0	139.4	150.2	171.1	185.5	199.2	210.0	219.2	263.8	283.6
33	114	91.0	110.1	114.7	136.1	140.6	151.4	172.6	187.0	200.8	211.6	221.0	265.9	285.9
⅓	115	91.7	111.0	115.6	137.2	141.7	152.7	174.0	188.5	202.4	213.4	222.7	268.0	288.2
⅔	116	92.4	111.9	116.5	138.3	142.9	153.9	175.4	190.1	204.0	215.1	224.5	270.1	290.5
34	117	93.2	112.8	117.5	139.4	144.0	155.1	176.8	191.6	205.6	216.8	226.3	272.3	292.8
⅓	118	93.9	113.7	118.4	140.5	145.2	156.4	178.2	193.1	207.2	218.5	228.1	274.4	295.0
⅔	119	94.7	114.6	119.3	141.6	146.3	157.6	179.6	194.6	208.8	220.2	229.9	276.5	297.3
35	120	95.4	115.5	120.3	142.7	147.5	158.9	181.0	196.1	210.4	221.9	231.6	278.7	299.6
⅓	121	96.1	116.4	121.2	143.8	148.6	160.1	182.4	197.6	212.1	223.6	233.4	280.8	301.9
⅔	122	96.9	117.3	122.2	145.0	149.8	161.4	183.8	199.1	213.7	225.3	235.2	282.9	304.1
36	123	97.6	118.2	123.1	146.1	150.9	162.6	185.2	200.6	215.3	227.0	237.0	285.1	306.4
⅓	124	98.4	119.1	124.0	147.2	152.1	163.8	186.6	202.1	216.9	228.7	238.7	287.2	308.7
⅔	125	99.1	120.0	125.0	148.3	153.2	165.1	188.0	203.6	218.5	230.4	240.5	289.3	311.0
37	126	99.9	120.9	125.9	149.4	154.4	166.3	189.4	205.1	220.1	232.1	242.3	291.4	313.3
⅓	127	100.6	121.8	126.8	150.5	155.5	167.6	190.9	206.7	221.7	233.8	244.1	293.6	315.5
⅔	128	101.3	122.7	127.8	151.6	156.7	168.8	192.3	208.2	223.3	235.5	245.9	295.7	317.8
38	129	102.1	123.6	128.7	152.7	157.8	170.0	193.7	209.7	224.9	237.2	247.6	297.8	320.1
⅓	130	102.8	124.5	129.6	153.8	159.0	171.3	195.1	211.2	226.6	238.9	249.4	300.0	322.4
⅔	131	103.6	125.4	130.6	154.9	160.1	172.5	196.5	212.7	228.2	240.6	251.2	302.1	324.6
39	132	104.3	126.3	131.5	156.1	161.3	173.8	197.9	214.2	229.8	242.3	253.0	304.2	326.9
⅓	133	105.0	127.2	132.4	157.2	162.4	175.0	199.3	215.7	231.4	244.0	254.7	306.4	329.2
⅔	134	105.8	128.1	133.4	158.3	163.6	176.2	200.7	217.2	233.0	245.7	256.5	308.5	331.5
40	135	106.5	129.0	134.3	159.4	164.7	177.5	202.1	218.7	234.6	247.4	258.3	310.6	333.8
⅓	136	107.3	129.9	135.2	160.5	165.9	178.7	203.5	220.2	236.2	249.1	260.1	312.7	336.0
⅔	137	108.0	130.8	136.2	161.6	167.0	180.0	204.9	221.7	237.8	250.8	261.9	314.9	338.3
41	138	108.7	131.7	137.1	162.7	168.1	181.2	206.3	223.3	239.4	252.5	263.6	317.0	340.6
⅓	139	109.5	132.6	138.0	163.8	169.3	182.4	207.7	224.8	241.0	254.2	265.4	319.1	342.9
⅔	140	110.2	133.5	139.0	164.9	170.4	183.7	209.1	226.3	242.7	255.9	267.2	321.3	345.1
42	141	111.0	134.4	139.9	166.1	171.6	184.9	210.6	227.8	244.3	257.6	269.0	323.4	347.4
⅓	142	111.7	135.3	140.9	167.2	172.7	186.2	212.0	229.3	245.9	259.2	270.7	325.5	349.7
⅔	143	112.4	136.2	141.8	168.3	173.9	187.4	213.4	230.8	247.5	261.0	272.5	327.6	352.0
43	144	113.2	137.1	142.7	169.4	175.0	188.6	214.8	232.3	249.1	262.6	274.3	329.8	354.3
⅓	145	113.9	138.0	143.7	170.5	176.2	189.9	216.2	233.8	250.7	264.3	276.1	331.9	356.5
⅔	146	114.7	138.9	144.6	171.6	177.3	191.1	217.6	235.3	252.3	266.0	277.9	334.0	358.8
44	147	115.4	139.8	145.5	172.7	178.5	192.4	219.0	236.8	253.9	267.7	279.6	336.2	361.1
⅓	148	116.1	140.7	146.5	173.8	179.6	193.6	220.4	238.4	255.6	269.4	281.4	338.3	363.4
⅔	149	116.9	141.6	147.4	174.9	180.8	194.9	221.8	239.9	257.2	271.1	283.2	340.4	365.6
45	150	117.6	142.5	148.3	176.1	181.9	196.1	223.2	241.4	258.8	272.8	285.0	342.6	367.9
⅓	151	118.4	143.4	149.3	177.2	183.1	197.3	224.6	242.9	260.4	274.5	286.7	344.7	370.2
⅔	152	119.1	144.3	150.2	178.3	184.2	198.6	226.0	244.4	262.0	276.2	288.5	346.8	372.5
46	153	119.8	145.2	151.4	179.4	185.4	199.8	227.4	245.9	263.6	277.9	290.3	348.9	374.8
⅓	154	120.6	146.1	152.1	180.5	186.5	201.1	228.9	247.4	265.2	279.6	292.1	351.1	377.0
⅔	155	121.3	147.0	153.0	181.6	187.7	202.3	230.3	248.9	266.8	281.3	293.9	353.2	379.3
47	156	122.1	147.9	154.0	182.7	188.8	203.5	231.7	250.4	268.4	283.0	295.6	355.3	381.6
⅓	157	122.8	148.8	154.9	183.8	190.0	204.8	233.1	251.9	270.1	284.7	297.4	357.5	383.9
⅔	158	123.6	149.7	155.8	184.9	191.1	206.0	234.5	253.4	271.7	286.4	299.2	359.6	386.1

TABLE FOR BOX CULVERTS.

(CONTINUED.)

No. of Culvert....		1	2	3	4	5	6	7	8	9	10	11	12	13
Size of Culvert....		2X2	2X2½	2½X2½	2½X3	3X3	3X3½	3X4	3X4½	3X5	3½X5	4X5	4X6	5X6
Cub Yds. in Coping, W'gs and Parapet Walls........		6.51852	7.3500	8.081815	8.90000	9.189701	9.981481	12.111111	14.981481	17.111111	17.981481	18.296296	23.111111	29.259259
Cub. Yds. in 1 Lineal Foot of Culvert...........		0.740740	0.898148	0.925185	1.111111	1.118148	1.240741	1.407407	1.509259	1.611111	1.699074	1.777778	2.129630	2.277778
Area Water-way..		4.0	5.0	6.25	7.5	9.0	10.5	12.0	13.5	15.0	17.5	19.0	23.0	29.0
Base, Fifteen Feet. HEIGHT.	Length of Culvert	Cub. Yds.	Cub. Yds.	Cub. Yds.	Cub. Yds.	Cub. Yds.	Cub. Yds.	Cub. Yds.	Cub. Yds.	Cub. Yds.	Cub. Yds.	Cub. Yds.	Cub. Yds.	Cub. Yds.
48	159	124.3	150.6	156.7	186.1	192.3	207.3	235.9	255.0	273.3	288.1	301.0	361.7	388.4
⅓	160	125.0	151.5	157.7	187.2	193.4	208.5	237.3	256.5	274.9	289.8	302.7	363.9	390.7
⅔	161	125.8	152.4	158.6	188.3	194.6	209.7	238.7	258.0	276.5	291.5	304.5	366.0	393.0
49	162	126.5	153.3	159.6	189.4	195.7	211.0	240.1	259.5	278.1	293.2	306.3	368.1	395.3
⅓	163	127.3	154.1	160.5	190.5	196.9	212.2	241.5	261.0	279.7	294.9	308.1	370.2	397.5
⅔	164	128.0	155.0	161.4	191.6	198.0	213.4	242.9	262.5	281.3	296.6	309.9	372.4	399.8
50	165	128.7	155.9	162.4	192.7	199.1	214.7	244.3	264.0	282.9	298.3	311.6	374.5	402.1
⅓	166	129.5	156.8	163.3	193.8	200.3	216.0	245.7	265.5	284.6	300.0	313.4	376.6	404.4
⅔	167	130.2	157.7	164.2	194.9	201.4	217.2	247.1	267.0	286.2	301.7	315.2	378.8	406.6
51	168	131.0	158.6	165.2	196.1	202.6	218.4	248.6	268.5	287.8	303.4	317.0	380.9	408.9
⅓	169	131.7	159.5	166.1	197.2	203.7	219.7	250.0	270.0	289.4	305.1	318.7	383.0	411.2
⅔	170	132.4	160.4	167.0	198.2	204.9	220.9	251.4	271.6	291.0	306.8	320.5	385.1	413.5
52	171	133.2	161.3	168.0	199.4	206.0	222.1	252.8	273.1	292.6	308.5	322.3	387.3	415.8
⅓	172	133.9	162.2	168.9	200.5	207.2	223.4	254.2	274.6	294.2	310.2	324.1	389.4	418.0
⅔	173	134.7	163.1	169.9	201.6	208.3	224.6	255.6	276.1	295.8	311.9	325.9	391.5	420.3
53	174	135.4	164.0	170.8	202.7	209.5	225.9	257.0	277.6	297.4	313.6	327.6	393.7	422.6
⅓	175	136.1	164.9	171.7	203.8	210.6	227.1	258.4	279.1	299.1	315.3	329.4	395.8	424.9
⅔	176	136.9	165.8	172.7	204.9	211.8	228.4	259.8	280.6	300.7	317.0	331.2	397.9	427.1
54	177	137.6	166.7	173.6	206.1	212.9	229.6	261.2	282.1	302.3	318.7	333.0	400.1	429.4
⅓	178	138.4	167.6	174.5	207.2	214.1	230.8	262.6	283.6	303.9	320.4	334.7	402.2	431.7
⅔	179	139.1	168.5	175.5	208.3	215.2	232.1	264.0	285.1	305.5	322.1	336.5	404.3	434.0
55	180	139.9	169.4	176.4	209.4	216.4	233.3	265.4	286.6	307.1	323.8	338.3	406.4	436.3
⅓	181	140.6	170.3	177.3	210.5	217.5	234.6	266.9	288.2	308.7	325.5	340.1	408.6	438.5
⅔	182	141.3	171.2	178.3	211.6	218.7	235.8	268.3	289.7	310.3	327.2	341.9	410.7	440.8
56	183	142.1	172.1	179.2	212.7	219.8	237.0	269.7	291.2	311.9	328.9	343.6	412.8	443.1
⅓	184	142.8	173.0	180.1	213.8	221.0	238.3	271.1	292.7	313.6	330.6	345.4	415.0	445.4
⅔	185	143.6	173.9	181.1	214.9	222.1	239.5	272.5	294.2	315.2	332.3	347.2	417.1	447.6
57	186	144.3	174.8	182.0	216.1	223.3	240.8	273.9	295.7	316.8	334.0	349.1	419.2	449.9
⅓	187	145.0	175.7	182.9	217.2	224.4	242.0	275.3	297.2	318.4	335.7	350.7	421.4	452.2
⅔	188	145.8	176.6	183.9	218.3	225.6	243.2	276.7	298.7	320.0	337.4	352.5	423.5	454.5
58	189	146.5	177.5	184.8	219.4	226.7	244.5	278.1	300.2	321.6	339.1	354.3	425.6	456.8
⅓	190	147.3	178.4	185.7	220.5	227.9	245.7	279.5	301.7	323.2	340.8	356.1	427.7	459.0
⅔	191	148.0	179.3	186.7	221.6	229.0	247.0	280.9	303.2	324.8	342.5	357.9	429.9	461.3
59	192	148.7	180.2	187.6	222.7	230.1	248.2	282.3	304.8	326.4	344.2	359.6	432.0	463.6
⅓	193	149.5	181.1	188.6	223.8	231.3	249.4	283.7	306.3	328.1	345.9	361.4	434.1	465.9
⅔	194	150.2	182.0	189.5	224.9	232.4	250.7	285.1	307.8	329.7	347.6	363.2	436.3	468.1
60	195	151.0	182.9	190.4	226.1	233.6	251.9	286.6	309.3	331.3	349.3	365.0	438.9	470.4
⅓	196	151.7	183.8	191.4	227.2	234.7	253.2	288.0	310.8	332.9	351.0	366.7	440.5	472.7
⅔	197	152.4	184.7	192.3	228.3	235.9	254.4	289.4	312.3	334.5	352.7	368.5	442.6	475.0
61	198	153.2	185.6	193.2	229.4	237.0	255.6	290.8	313.8	336.1	354.4	370.3	444.8	477.3
⅓	199	153.9	186.5	194.2	230.5	238.2	256.9	292.2	315.3	337.7	356.1	372.1	446.9	479.5
⅔	200	154.7	187.4	195.1	231.6	239.3	258.1	293.6	316.8	339.3	357.8	373.9	449.0	481.8
62	201	155.4	188.3	196.0	232.7	240.5	259.4	295.0	318.3	340.9	359.5	375.6	451.2	484.1
⅓	202	156.1	189.2	197.0	233.8	241.6	260.6	296.4	319.6	342.6	361.2	377.4	453.3	486.4
⅔	203	156.9	190.1	197.9	234.9	242.8	261.9	297.8	321.4	344.2	362.9	379.2	455.4	488.6
63	204	157.6	191.0	198.8	236.1	243.9	263.1	299.2	322.9	345.8	364.6	381.0	457.6	490.9
⅓	205	158.4	191.9	199.8	237.2	245.1	264.3	300.6	324.4	347.4	366.3	382.7	459.7	493.2
⅔	206	159.1	192.8	200.7	238.3	246.2	265.6	302.0	325.9	349.0	368.0	384.5	461.8	495.5
64	207	159.8	193.7	201.6	239.4	247.4	266.8	303.4	327.4	350.6	369.7	386.3	463.9	497.8
⅓	208	160.6	194.6	202.6	240.5	248.5	268.1	304.9	328.9	352.2	371.4	388.1	466.1	500.0
⅔	209	161.3	195.5	203.5	241.6	249.7	269.3	306.3	330.4	353.8	373.1	389.9	468.2	502.3
65	210	162.1	196.4	204.5	242.7	250.8	270.5	307.7	331.9	355.4	374.8	391.6	470.3	504.6
⅓	211	162.8	197.3	205.4	243.8	252.0	271.8	309.1	333.4	357.1	376.5	393.4	472.5	506.9
⅔	212	163.6	198.2	206.3	244.9	253.1	273.0	310.5	334.9	358.7	378.2	395.2	474.6	509.1
66	213	164.3	199.1	207.3	246.1	254.3	274.3	311.9	336.5	360.3	379.9	397.0	476.7	511.4
⅓	214	165.0	200.0	208.2	247.2	255.4	275.5	313.3	338.0	361.9	381.6	398.7	478.9	513.7
⅔	215	165.8	200.9	209.1	248.3	256.6	276.7	314.7	339.5	363.5	383.3	400.5	481.0	516.0
67	216	166.5	201.7	210.1	249.4	257.7	278.0	316.1	341.0	365.1	385.0	402.3	483.1	518.3
⅓	217	167.3	202.6	211.0	250.5	258.9	279.2	317.5	342.5	366.7	386.7	404.1	485.2	520.5
⅔	218	168.0	203.5	211.9	251.6	260.0	280.5	318.9	344.0	368.3	388.4	405.9	487.4	522.8
68	219	168.7	204.4	212.9	252.7	261.1	281.7	320.3	345.5	369.9	390.1	407.6	489.5	525.1
⅓	220	169.5	205.3	213.8	253.8	262.3	282.9	321.7	347.0	371.6	391.8	409.4	491.6	527.4

TABLE FOR ARCH CULVERTS.

No. of Culvert....		1	2	3	4	5	6	7	8	9	10	11	12	13
Size of Culvert....		3X1	3X2	4X1	4X2	5X1	5X2	6X2	6X3	8X2	8X3	8X4	10X2	10X4
Arch—Cub. Yds. per Lineal Foot..........		0.2327	0.2327	0.2909	0.2909	0.3490	0.3490	0.4072	0.4072	0.5727	0.5727	0.5727	0.8845	0.8845
Walls—Cubic Yards per Lineal Foot..........		0.1481	0.2962	0.1481	0.2962	0.1481	0.2962	0.3704	0.5556	0.3704	0.5556	0.9630	0.4444	1.0370
Paving—Cubic Yards per Lineal Foot..........		0.3333	0.3333	0.3704	0.3704	0.4074	0.4074	0.4444	0.5185	0.5555	0.5926	0.6666	0.6666	0.7407
Coping—Cub. Yds...		2.2222	2.5926	2.5926	2.9700	2.9630	3.3333	3.7030	4.0740	4.4444	4.8148	5.1850	5.5555	5.9260
Wings—Cub. Yds....		8.2624	11.8182	10.6262	11.6380	13.2800	17.7240	20.6270	25.5280	24.6010	40.2283	45.1460	47.2753	50.2810
Sunk W'ls—Cub. Yds.		1.0000	4.0000	4.4444	4.4444	4.8889	4.8889	5.3333	6.2220	6.6666	7.1111	8.0000	8.0000	8.8888
Area Water-way..		6.534	9.531	10.293	14.283	14.817	19.817	26.137	32.137	41.133	49.133	57.133	59.270	79.270
Cub. Yds. in 1 Lineal Foot of Culvert..........		0.7141	0.8622	0.8084	0.9575	0.9045	1.0526	1.2220	1.3923	1.5096	1.8219	2.3023	1.9935	2.8822
Base, Fifteen Feet. HEIGHT	Length of Culvert	Cub. Yds.	Cub. Yds.	Cub. Yds.	Cub. Yds.	Cub. Yds.	Cub. Yds.	Cub. Yds.	Cub. Yds.	Cub. Yds.	Cub. Yds.	Cub. Yds.	Cub. Yds.	Cub. Yds.
0	15	25.2	31.3	29.8	36.4	34.7	41.7	48.0	59.7	63.7	81.1	95.9	90.8	114.2
⅓	16	25.9	32.2	30.6	37.4	35.6	42.8	49.2	61.3	65.3	83.1	98.2	92.8	116.7
⅔	17	26.6	33.1	31.4	38.3	36.5	43.8	50.4	62.9	66.9	85.0	100.5	94.8	119.4
1	18	27.3	33.9	32.2	39.3	37.4	44.9	51.7	64.5	68.5	86.9	102.8	96.7	122.1
⅓	19	28.1	34.8	33.1	40.2	38.3	45.9	52.9	66.1	70.1	88.9	105.1	98.7	124.7
⅔	20	28.8	35.7	33.9	41.2	39.2	47.0	54.1	67.7	71.7	90.8	107.4	100.7	127.4
2	21	29.5	36.5	34.7	42.2	40.1	48.1	55.3	69.3	73.3	92.7	109.7	102.7	130.1
⅓	22	30.2	37.4	35.5	43.1	41.0	49.1	56.6	70.9	74.9	94.7	112.0	104.7	132.7
⅔	23	30.9	38.2	36.3	44.1	41.9	50.2	57.8	72.4	76.5	96.6	114.3	106.7	135.4
3	24	31.6	39.1	37.1	45.0	42.8	51.2	59.0	74.0	78.1	98.5	116.6	108.7	138.0
⅓	25	32.3	40.0	37.9	46.0	43.7	52.3	60.2	75.6	79.7	100.5	118.9	110.7	140.7
⅔	26	33.1	40.8	38.7	46.9	44.6	53.3	61.4	77.2	81.3	102.4	121.2	112.7	143.4
4	27	33.8	41.7	39.5	47.9	45.6	54.4	62.7	78.8	82.9	104.3	123.5	114.7	146.0
⅓	28	34.5	42.6	40.3	48.9	46.5	55.4	63.9	80.4	84.5	106.3	125.8	116.7	148.7
⅔	29	35.2	43.4	41.1	49.8	47.4	56.5	65.1	82.0	86.1	108.2	128.1	118.7	151.3
5	30	35.9	44.3	42.0	50.8	48.3	57.5	66.3	83.6	87.7	110.1	130.4	120.7	154.0
⅓	31	36.6	45.1	42.8	51.7	49.2	58.6	67.6	85.2	89.3	112.1	132.7	122.7	156.7
⅔	32	37.3	46.0	43.6	52.7	50.1	59.6	68.8	86.8	90.9	114.0	135.0	124.7	159.3
6	33	38.1	46.9	44.4	53.6	51.0	60.7	70.0	88.4	92.5	115.9	137.3	126.7	162.0
⅓	34	38.8	47.7	45.2	54.6	51.9	61.7	71.2	90.0	94.1	117.8	139.6	128.7	164.7
⅔	35	39.5	48.6	46.0	55.6	52.8	62.8	72.4	91.6	95.7	119.8	141.9	130.7	167.3
7	36	40.2	49.5	46.8	56.5	53.7	63.8	73.7	93.1	97.3	121.7	144.2	132.7	170.0
⅓	37	40.9	50.3	47.6	57.5	54.6	64.9	74.9	94.7	98.9	123.6	146.5	134.7	172.6
⅔	38	41.6	51.2	48.4	58.4	55.5	65.9	76.1	96.3	100.5	125.6	148.8	136.7	175.3
8	39	42.3	52.0	49.2	59.4	56.4	67.0	77.3	97.9	102.1	127.5	151.1	138.7	178.0
⅓	40	43.0	52.9	50.0	60.4	57.3	68.1	78.5	99.5	103.7	129.4	153.4	140.7	180.6
⅔	41	43.8	53.8	50.9	61.3	58.2	69.1	79.8	101.1	105.3	131.4	155.7	142.6	183.3
9	42	44.5	54.6	51.7	62.3	59.1	70.2	81.0	102.7	106.9	133.3	158.0	144.6	186.0
⅓	43	45.2	55.5	52.5	63.2	60.0	71.2	82.2	104.3	108.5	135.2	160.3	146.6	188.6
⅔	44	45.9	56.3	53.3	64.2	60.9	72.3	83.4	105.9	110.1	137.2	162.6	148.6	191.3
10	45	46.6	57.2	54.1	65.1	61.8	73.3	84.7	107.5	111.6	139.1	164.9	150.6	193.9
⅓	46	47.3	58.1	54.9	66.1	62.7	74.4	85.9	109.1	113.2	141.0	167.2	152.6	196.6
⅔	47	48.0	58.9	55.7	67.1	63.6	75.4	87.1	110.7	114.8	143.0	169.5	154.6	199.3
11	48	48.8	59.8	56.5	68.0	64.5	76.5	88.3	112.3	116.4	144.9	171.8	156.6	201.9
⅓	49	49.5	60.7	57.3	69.0	65.5	77.5	89.5	113.8	118.0	146.8	174.1	158.6	204.6
⅔	50	50.2	61.5	58.1	69.9	66.4	78.6	90.8	115.4	119.6	148.8	176.4	160.6	207.3
12	51	50.9	62.4	59.0	70.9	67.3	79.6	92.0	117.0	121.2	150.7	178.8	162.6	209.9
⅓	52	51.6	63.2	59.8	71.8	68.2	80.7	93.2	118.6	122.8	152.6	181.1	164.6	212.6
⅔	53	52.3	64.1	60.6	72.8	69.1	81.7	94.4	120.2	124.4	154.6	183.4	166.6	215.2
13	54	53.0	65.0	61.4	73.8	70.0	82.8	95.7	121.8	126.0	156.5	185.7	168.6	217.9
⅓	55	53.8	65.8	62.2	74.7	70.9	83.8	96.9	123.4	127.6	158.4	188.0	170.6	220.6
⅔	56	54.5	66.7	63.0	75.7	71.8	84.9	98.1	125.0	120.2	160.4	190.3	172.6	223.2
14	57	55.2	67.6	63.8	76.6	72.7	85.9	99.3	126.6	130.8	162.3	192.6	174.6	225.9
⅓	58	55.9	68.4	64.6	77.6	73.6	87.0	100.5	128.2	132.4	164.2	194.9	176.6	228.6
⅔	59	56.6	69.3	65.4	78.5	74.5	88.0	101.8	129.8	134.0	166.1	197.2	178.6	231.2
15	60	57.3	70.1	66.2	79.5	75.4	89.1	103.0	131.4	135.6	168.1	199.5	180.6	233.9
⅓	61	58.0	71.0	67.0	80.5	76.3	90.2	104.2	133.0	137.2	170.0	201.8	182.6	236.5
⅔	62	58.8	71.9	67.9	81.4	77.2	91.2	105.4	134.5	138.8	171.9	204.1	184.6	239.2
16	63	59.5	72.7	68.7	82.4	78.1	92.3	106.7	136.1	140.4	173.9	206.4	186.5	241.9
⅓	64	60.2	73.6	69.5	83.3	79.0	93.3	107.9	137.7	142.0	175.8	208.7	188.5	244.5
⅔	65	60.9	74.4	70.3	84.3	79.9	94.4	109.1	139.3	143.6	177.7	211.0	190.5	247.2
17	66	61.6	75.3	71.1	85.2	80.8	95.4	110.3	140.9	145.2	179.7	213.3	192.5	249.9
⅓	67	62.3	76.2	71.9	86.2	81.7	96.5	111.5	142.5	146.8	181.6	215.6	194.5	252.5
⅔	68	63.0	77.0	72.7	87.2	82.6	97.5	112.8	144.1	148.4	183.5	217.9	196.5	255.2
18	69	63.8	77.9	73.5	88.1	83.5	98.6	114.0	145.7	150.0	185.5	220.2	198.5	257.8
⅓	70	64.5	78.8	74.3	89.1	84.4	99.6	115.2	147.3	151.6	187.4	222.5	200.5	260.5
⅔	71	65.2	79.6	75.1	90.0	85.4	100.7	116.4	148.9	153.2	189.3	224.8	202.5	263.2
19	72	65.9	80.5	76.0	91.0	86.3	101.7	117.7	150.5	154.8	191.3	227.1	204.5	265.8
⅓	73	66.5	81.3	76.8	91.9	87.2	102.8	118.9	152.1	156.4	193.2	229.3	206.5	268.5
⅔	74	67.3	82.2	77.6	92.9	88.1	103.8	120.1	153.7	158.0	195.1	231.7	208.5	271.1
20	75	68.0	83.1	78.4	93.9	89.0	104.9	121.3	155.2	159.6	197.1	234.0	210.5	273.8
⅓	76	68.8	83.9	79.2	94.8	89.9	105.9	122.5	156.8	161.2	199.0	236.3	212.5	276.5
⅔	77	69.5	84.8	80.0	95.8	90.8	107.0	123.8	158.4	162.8	200.9	238.6	214.5	279.1
21	78	70.2	85.7	80.8	96.7	91.7	108.0	125.0	160.0	164.4	202.9	240.9	216.5	281.8
⅓	79	70.9	86.5	81.6	97.7	92.6	109.1	126.2	161.6	166.0	204.8	243.2	218.5	284.5
⅔	80	71.5	87.4	82.4	98.6	93.5	110.2	127.4	163.2	167.6	206.7	245.5	220.5	287.1

TABLE FOR ARCH CULVERTS.

CONTINUED.

No. of Culvert....		1	2	3	4	5	6	7	8	9	10	11	12	13
Size of Culvert....		3X1	3X2	4X1	4X2	5X1	5X2	6X2	6X3	8X2	8X3	8X4	10X2	10X4
Arch—Cub. Yds. per Lineal Foot........		0.2327	0.2327	0.2968	0.2968	0.3490	0.3490	0.4072	0.4072	0.6727	0.6727	0.6727	0.8815	0.8815
Walls—Cubic Yards per Lineal Foot........		0.1481	0.2962	0.1481	0.2962	0.1481	0.2962	0.3704	0.6666	0.3704	0.6666	0.9630	0.4411	1.0370
Paving—Cubic Yards per Lineal Foot........		0.3333	0.3333	0.3704	0.3704	0.4074	0.4074	0.4444	0.5185	0.5555	0.5926	0.6666	0.6666	0.7407
Coping—Cub. Yds....		2.2222	2.5926	2.5926	2.9700	2.9630	3.3333	3.7030	4.0740	4.4444	4.8148	5.1850	5.5555	5.9260
Wings—Cub. Yds....		8.2628	11.8148	10.6362	11.6360	13.2800	17.7240	20.6270	25.5760	28.6010	40.2383	48.1160	41.2735	49.3310
Sunk W'ls—Cub. Yds.		4.0000	4.0000	4.4444	4.4444	4.8888	4.8888	5.3333	6.2220	6.6666	7.1111	8.0000	8.0000	8.8888
Area Water-way..		6.334	9.334	10.283	14.283	14.817	19.817	26.137	32.137	41.133	49.133	57.133	58.270	79.270
Cub. Yds. in 1 Lineal Foot of Culvert........		0.7111	0.8622	0.8091	0.9373	0.9045	1.0526	1.2220	1.3923	1.5988	1.9318	2.2023	1.9935	2.6627
Base, Fifteen Feet. Height	Length of Culvert	Cub. Yds.	Cub. Yds.	Cub. Yds.	Cub. Yds.	Cub. Yds.	Cub. Yds.	Cub. Yds.	Cub. Yds.	Cub. Yds.	Cub. Yds.	Cub. Yds.	Cub. Yds.	Cub. Yds.
22	81	72.3	88.2	83.2	99.6	94.3	111.2	128.6	164.8	169.2	208.6	247.8	222.5	289.8
⅓	82	73.0	89.1	84.0	100.6	95.3	112.3	129.9	166.4	170.8	210.6	250.1	224.5	292.4
⅔	83	73.8	90.0	84.9	101.5	96.2	113.3	131.1	168.0	172.4	212.5	252.4	226.5	295.1
23	84	74.5	90.8	85.7	102.5	97.1	114.4	132.3	169.6	174.0	214.4	254.7	228.5	297.8
⅓	85	75.2	91.7	86.5	103.4	98.0	115.4	133.5	171.2	175.6	216.4	257.0	230.4	300.4
⅔	86	75.9	92.6	87.3	104.4	98.9	116.5	134.8	172.8	177.2	218.3	259.3	232.4	303.1
24	87	76.6	93.4	88.1	105.4	99.8	117.5	136.0	174.4	178.8	220.2	261.6	234.4	305.8
⅓	88	77.3	94.3	88.9	106.3	100.7	118.6	137.2	175.9	180.4	222.2	263.9	236.4	308.4
⅔	89	78.0	95.1	89.7	107.3	101.6	119.6	138.4	177.5	182.0	224.1	266.2	238.4	311.1
25	90	78.8	96.0	90.5	108.2	102.5	120.7	139.6	179.1	183.6	226.0	268.5	240.4	313.7
⅓	91	79.5	96.9	91.3	109.2	103.4	121.7	140.9	180.7	185.2	228.0	270.8	242.4	316.4
⅔	92	80.2	97.7	92.1	110.2	104.3	122.8	142.1	182.3	186.8	229.9	273.1	244.4	319.1
26	93	80.9	98.6	92.9	111.1	105.3	123.8	143.3	183.9	188.4	231.8	275.4	246.4	321.7
⅓	94	81.6	99.5	93.8	112.1	106.2	124.9	144.5	185.5	190.0	233.8	277.7	248.4	324.4
⅔	95	82.3	100.3	94.6	113.0	107.1	125.9	145.8	187.1	191.6	235.7	280.0	250.4	327.1
27	96	83.0	101.2	95.4	114.0	108.0	127.0	147.0	188.7	193.2	237.6	282.4	252.4	329.7
⅓	97	83.8	102.0	96.2	114.9	108.9	128.0	148.2	190.3	194.8	239.6	284.7	254.4	332.4
⅔	98	84.5	102.9	97.0	115.9	109.8	129.1	149.4	191.9	196.4	241.5	287.0	256.4	335.0
28	99	85.2	103.8	97.8	116.8	110.7	130.2	150.7	193.5	198.0	243.4	289.3	258.4	337.7
⅓	100	85.9	104.6	98.6	117.8	111.6	131.2	151.9	195.1	199.6	245.4	291.6	260.4	340.4
⅔	101	86.6	105.5	99.4	118.8	112.5	132.3	153.1	196.6	201.2	247.3	293.9	262.4	343.0
29	102	87.3	106.4	100.2	119.7	113.4	133.3	154.3	198.2	202.8	249.2	296.2	264.4	345.7
⅓	103	88.0	107.2	101.0	120.7	114.3	134.4	155.5	199.8	204.4	251.1	298.5	266.4	348.4
⅔	104	88.8	108.1	101.9	121.6	115.2	135.4	156.8	201.4	206.0	253.1	300.8	268.4	351.0
30	105	89.5	108.9	102.7	122.6	116.1	136.5	158.0	203.0	207.6	255.0	303.1	270.4	353.7
⅓	106	90.2	109.8	103.5	123.5	117.0	137.5	159.2	204.6	209.2	256.9	305.4	272.4	356.3
⅔	107	90.9	110.7	104.3	124.5	118.0	138.6	160.4	206.2	210.8	258.9	307.7	274.3	359.0
31	108	91.6	111.5	105.1	125.5	118.8	139.6	161.6	207.8	212.4	260.8	310.0	276.3	361.7
⅓	109	92.3	112.4	105.9	126.4	119.7	140.7	162.9	209.4	214.0	262.7	312.3	278.3	364.3
⅔	110	93.0	113.3	106.7	127.4	120.6	141.7	164.1	211.0	215.6	264.7	314.6	280.3	367.0
32	111	93.8	114.1	107.5	128.3	121.5	142.8	165.3	212.6	217.2	266.6	316.9	282.3	369.7
⅓	112	94.5	115.0	108.3	129.3	122.4	143.8	166.5	214.2	218.8	268.5	319.2	284.3	372.3
⅔	113	95.2	115.8	109.1	130.2	123.3	144.9	167.8	215.8	220.4	270.5	321.5	286.3	375.0
33	114	95.9	116.7	109.9	131.2	124.2	145.9	169.0	217.3	222.0	272.4	323.8	288.3	377.6
⅓	115	96.6	117.6	110.8	132.2	125.1	147.0	170.2	218.9	223.6	274.3	326.1	290.3	380.3
⅔	116	97.3	118.4	111.6	133.1	126.1	148.0	171.4	220.5	225.1	276.3	328.4	292.3	383.0
34	117	98.0	119.3	112.4	134.1	127.0	149.1	172.6	222.1	226.7	278.2	330.7	294.3	385.6
⅓	118	98.7	120.2	113.2	135.0	127.9	150.2	173.9	223.7	228.3	280.1	333.0	296.3	388.3
⅔	119	99.5	121.0	114.0	136.0	128.8	151.2	175.1	225.3	229.9	282.1	335.3	298.3	390.9
35	120	100.2	121.9	114.8	137.0	129.7	152.3	176.3	226.9	231.5	284.0	337.6	300.3	393.6
⅓	121	100.9	122.7	115.6	137.9	130.6	153.3	177.5	228.5	233.1	285.9	339.9	302.3	396.3
⅔	122	101.6	123.6	116.4	138.9	131.5	154.4	178.7	230.1	234.7	287.9	342.2	304.3	398.9
36	123	102.3	124.5	117.2	139.8	132.4	155.4	180.0	231.7	236.3	289.8	344.5	306.3	401.6
⅓	124	103.0	125.3	118.0	140.8	133.3	156.5	181.2	233.3	237.9	291.7	346.8	308.3	404.3
⅔	125	103.7	126.2	118.8	141.7	134.2	157.5	182.4	234.9	239.5	293.7	349.1	310.3	406.9
37	126	104.5	127.0	119.7	142.7	135.1	158.6	183.6	236.5	241.1	295.6	351.4	312.3	409.6
⅓	127	105.2	127.9	120.5	143.7	136.0	159.6	184.9	238.0	242.7	297.5	353.7	314.3	412.2
⅔	128	105.9	128.8	121.3	144.6	136.9	160.7	186.1	239.6	244.3	299.4	356.0	316.3	414.9
38	129	106.6	129.6	122.1	145.6	137.8	161.7	187.3	241.2	245.9	301.4	358.3	318.3	417.6
⅓	130	107.3	130.5	122.9	146.5	138.7	162.8	188.5	242.8	247.5	303.3	360.6	320.2	420.2
⅔	131	108.0	131.4	123.7	147.5	139.6	163.8	189.8	244.4	249.1	305.2	362.9	322.2	422.9
39	132	108.7	132.2	124.5	148.4	140.5	164.9	191.0	246.0	250.7	307.2	365.2	324.2	425.6
⅓	133	109.5	133.1	125.3	149.4	141.4	165.9	192.2	247.6	252.3	309.1	367.5	326.2	428.2
⅔	134	110.2	133.9	126.1	150.4	142.3	167.0	193.4	249.2	253.9	311.0	369.8	328.2	430.9
40	135	110.9	134.8	126.9	151.3	143.2	168.0	194.6	250.8	255.5	313.0	372.1	330.2	433.5
⅓	136	111.6	135.7	127.8	152.3	144.1	169.1	195.9	252.4	257.1	314.9	374.4	332.2	436.2
⅔	137	112.3	136.5	128.6	153.2	145.0	170.2	197.1	254.0	258.7	316.8	376.7	334.2	438.9
41	138	113.0	137.4	129.4	154.2	146.0	171.2	198.3	255.6	260.3	318.8	379.0	336.2	441.5
⅓	139	113.7	138.3	130.2	155.1	146.9	172.3	199.5	257.2	261.9	320.7	381.4	338.2	444.2
⅔	140	114.5	139.1	131.0	156.1	147.8	173.3	200.8	258.7	263.5	322.6	383.7	340.2	446.9
42	141	115.2	140.0	131.8	157.1	148.7	174.4	202.0	260.3	265.1	324.6	386.0	342.2	449.5
⅓	142	115.9	140.8	132.6	158.0	149.6	175.4	203.2	261.9	266.7	326.5	388.3	344.2	452.2
⅔	143	116.6	141.7	133.4	159.0	150.5	176.5	204.4	263.5	268.3	328.4	390.6	346.2	454.8
43	144	117.3	142.6	134.2	159.9	151.4	177.5	205.6	265.1	269.9	330.4	392.9	348.2	457.5
⅓	145	118.0	143.4	135.0	160.9	152.3	178.6	206.9	266.7	271.5	332.3	395.2	350.2	460.2
⅔	146	118.7	144.3	135.8	161.8	153.2	179.6	208.1	268.3	273.1	334.2	397.5	352.2	462.8

TABLE FOR ARCH CULVERTS.

CONTINUED.

No. of Culvert....		1	2	3	4	5	6	7	8	9	10	11	12	13
Size of Culvert....		3X1	3X2	4X1	4X2	5X1	5X2	6X2	6X3	8X2	8X3	8X4	10X2	10X4
Arch—Cub. Yds. per Lineal Foot........		0.2327	0.2327	0.2909	0.2909	0.3190	0.3490	0.4072	0.4072	0.6727	0.6727	0.6727	0.8843	0.8848
Walls—Cubic Yards per Lineal Foot...........		0.1481	0.2962	0.1481	0.2962	0.1481	0.2962	0.3701	0.6666	0.3704	0.6666	0.9630	0.4111	1.0370
Paving—Cubic Yards per Lineal Foot...........		0.3333	0.3333	0.3701	0.3701	0.4074	0.4074	0.4444	0.5185	0.5555	0.5926	0.6666	0.6666	0.7407
Coping—Cub. Yds...		2.2222	2.5926	2.5926	2.0700	2.0630	3.3333	3.7030	4.0740	4.4434	4.8148	5.1850	5.5555	5.9260
Wings—Cub. Yds....		8.2628	11.8183	10.6362	14.6360	13.2800	17.1240	20.6370	25.5200	28.6010	40.2383	48.1460	47.2153	59.3310
Sunk W'ls—Cub. Yds.		4.0000	4.0000	4.4444	4.4444	4.8888	4.8888	5.3333	6.2220	6.6666	7.1111	8.0000	8.0000	8.8888
Area Water-way..		6.534	9.534	10.283	14.283	14.817	18.817	26.137	32.137	41.133	49.133	57.133	59.270	79.270
Cub. Yds. in 1 Lineal Foot of Culvert...........		0.7144	0.8622	0.8094	0.9575	0.9045	1.0526	1.2220	1.5923	1.5986	1.8819	2.3023	1.9955	2.6622
Base, Fifteen Feet. HEIGHT	Length of Culvert	Cub. Yds.	Cub. Yds.	Cub. Yds.	Cub. Yds.	Cub. Yds.	Cub. Yds.	Cub. Yds.	Cub. Yds.	Cub. Yds.	Cub. Yds.	Cub. Yds.	Cub. Yds.	Cub. Yds.
44	147	119.5	145.2	136.7	162.8	154.1	180.7	209.3	269.9	274.7	336.2	399.8	354.2	465.5
⅓	148	120.2	146.0	137.5	163.8	155.0	181.7	210.5	271.5	276.3	338.1	402.1	356.2	468.2
⅔	149	120.9	146.9	138.3	164.7	155.9	182.8	211.8	273.1	277.9	340.0	404.4	358.2	470.8
45	150	121.6	147.7	139.1	165.7	156.8	183.8	213.0	274.7	279.5	341.9	406.7	360.2	473.5
⅓	151	122.3	148.6	139.9	166.6	157.7	184.9	214.2	276.3	281.1	343.9	409.0	362.2	476.1
⅔	152	123.0	149.5	140.7	167.6	158.6	185.9	215.4	277.9	282.7	345.8	411.3	364.1	478.8
46	153	123.7	150.3	141.5	168.5	159.5	187.0	216.6	279.4	284.3	347.7	413.6	366.1	481.5
⅓	154	124.5	151.2	142.3	169.5	160.4	188.0	217.9	281.0	285.9	349.7	415.9	368.1	484.1
⅔	155	125.2	152.1	143.1	170.5	161.3	189.1	219.1	282.6	287.5	351.6	418.2	370.1	486.8
47	156	125.9	152.9	143.9	171.4	162.2	190.2	220.3	284.2	289.1	353.5	420.5	372.1	489.4
⅓	157	126.6	153.8	144.7	172.4	163.1	191.2	221.5	285.8	290.7	355.5	422.8	374.1	492.1
⅔	158	127.3	154.6	145.6	173.3	164.0	192.3	222.7	287.4	292.3	357.4	425.1	376.1	494.8
48	159	128.0	155.5	146.4	174.3	164.9	193.3	224.0	289.0	293.9	359.3	427.4	378.1	497.4
⅓	160	128.7	156.4	147.2	175.3	165.9	194.4	225.2	290.6	295.5	361.3	429.7	380.1	500.1
⅔	161	129.5	157.2	148.0	176.2	166.8	195.4	226.4	292.2	297.1	363.2	432.0	382.1	502.8
49	162	130.2	158.1	148.8	177.2	167.7	196.5	227.6	293.8	298.7	365.2	434.3	384.1	505.4
⅓	163	130.9	158.9	149.6	178.1	168.6	197.5	228.9	295.4	300.3	367.1	436.6	386.1	508.1
⅔	164	131.6	159.8	150.4	179.1	169.5	198.6	230.1	297.0	301.9	369.0	438.9	388.1	510.7
50	165	132.3	160.7	151.2	180.0	170.4	199.6	231.3	298.6	303.5	370.9	441.2	390.1	513.4
⅓	166	133.0	161.5	152.0	181.0	171.3	200.7	232.5	300.1	305.1	372.9	443.5	392.1	516.1
⅔	167	133.7	162.4	152.8	182.0	172.2	201.7	233.7	301.7	306.7	374.8	445.8	394.1	518.7
51	168	134.5	163.3	153.7	182.9	173.1	202.8	235.0	303.3	308.3	376.7	448.1	396.1	521.4
⅓	169	135.2	164.1	154.5	183.9	174.0	203.8	236.2	304.9	309.9	378.7	450.4	398.1	524.1
⅔	170	135.9	165.0	155.3	184.8	174.9	204.9	237.4	306.5	311.5	380.6	452.7	400.1	526.7
52	171	136.6	165.8	156.1	185.8	175.8	205.9	238.6	308.1	313.1	382.5	455.0	402.1	529.4
⅓	172	137.3	166.7	156.9	186.7	176.7	207.0	239.9	309.7	314.7	384.5	457.3	404.1	532.0
⅔	173	138.0	167.6	157.7	187.7	177.6	208.0	241.1	311.3	316.3	386.4	459.6	406.1	534.7
53	174	138.7	168.4	158.5	188.7	178.5	209.1	242.3	312.9	317.9	388.3	461.9	408.0	537.4
⅓	175	139.5	169.3	159.3	189.6	179.4	210.2	243.5	314.5	319.5	390.2	464.2	410.0	540.0
⅔	176	140.2	170.2	160.1	190.6	180.3	211.2	244.7	316.1	321.1	392.2	466.5	412.0	542.7
54	177	140.9	171.0	160.9	191.5	181.2	212.3	246.0	317.7	322.7	394.1	468.8	414.0	545.4
⅓	178	141.6	171.9	161.7	192.5	182.1	213.3	247.2	319.3	324.3	396.0	471.1	416.0	548.0
⅔	179	142.3	172.7	162.6	193.4	183.0	214.4	248.4	320.8	325.9	398.0	473.4	418.0	550.7
55	180	143.0	173.6	163.4	194.4	183.9	215.4	249.6	322.4	327.5	399.9	475.7	420.0	553.3
⅓	181	143.7	174.5	164.2	195.4	184.8	216.5	250.9	324.0	329.1	401.8	478.0	422.0	556.0
⅔	182	144.5	175.3	165.0	196.3	185.8	217.5	252.1	325.6	330.7	403.8	480.3	424.0	558.7
56	183	145.2	176.2	165.8	197.3	186.7	218.6	253.3	327.2	332.3	405.7	482.7	426.0	561.3
⅓	184	145.9	177.1	166.6	198.2	187.6	219.6	254.5	328.8	333.9	407.6	485.0	428.0	564.0
⅔	185	146.6	177.9	167.4	199.2	188.5	220.7	255.7	330.4	335.5	409.6	487.3	430.0	566.7
57	186	147.3	178.8	168.2	200.1	189.4	221.7	257.0	332.0	337.1	411.5	489.6	432.0	569.3
⅓	187	148.0	179.6	169.0	201.1	190.3	222.8	258.2	333.6	338.7	413.4	491.9	434.0	572.0
⅔	188	148.7	180.5	169.8	202.1	191.2	223.8	259.4	335.2	340.2	415.4	494.2	436.0	574.6
58	189	149.4	181.4	170.6	203.0	192.1	224.9	260.6	336.8	341.8	417.3	496.5	438.0	577.3
⅓	190	150.2	182.2	171.5	204.0	193.0	225.9	261.9	338.4	343.4	419.2	498.8	440.0	580.0
⅔	191	150.9	183.1	172.3	204.9	193.9	227.0	263.1	340.0	345.0	421.2	501.1	442.0	582.6
59	192	151.6	184.0	173.1	205.9	194.8	228.0	264.3	341.5	346.6	423.1	503.4	444.0	585.3
⅓	193	152.3	184.8	173.9	206.8	195.7	229.1	265.5	343.1	348.2	425.0	505.7	446.0	588.0
⅔	194	153.0	185.7	174.7	207.8	196.6	230.2	266.7	344.7	349.8	427.0	508.0	448.0	590.6
60	195	153.7	186.5	175.5	208.8	197.5	231.2	268.0	346.3	351.4	428.9	510.3	450.0	593.3
⅓	196	154.4	187.4	176.3	209.7	198.4	232.3	269.2	347.9	353.0	430.8	512.6	451.9	595.9
⅔	197	155.2	188.3	177.1	210.7	199.3	233.3	270.4	349.5	354.6	432.7	514.9	453.9	598.6
61	198	155.9	189.1	177.9	211.6	200.2	234.4	271.6	351.1	356.2	434.7	517.2	455.9	601.3
⅓	199	156.6	190.0	178.7	212.6	201.1	235.4	272.9	352.7	357.8	436.6	519.5	457.9	603.9
⅔	200	157.3	190.9	179.6	213.6	202.0	236.5	274.1	354.3	359.4	438.5	521.8	459.9	606.6
62	201	158.0	191.7	180.4	214.5	202.9	237.5	275.3	355.9	361.0	440.5	524.1	461.9	609.2
⅓	202	158.7	192.6	181.2	215.5	203.8	238.6	276.5	357.5	362.6	442.4	526.4	463.9	611.9
⅔	203	159.4	193.4	182.0	216.4	204.7	239.6	277.7	359.1	364.2	444.3	528.7	465.9	614.6
63	204	160.2	194.3	182.8	217.4	205.6	240.7	279.0	360.7	365.8	446.3	531.0	467.9	617.2
⅓	205	160.9	195.2	183.6	218.3	206.6	241.7	280.2	362.2	367.4	448.2	533.3	469.9	619.9
⅔	206	161.6	196.0	184.4	219.3	207.5	242.8	281.4	363.8	369.0	450.1	535.6	471.9	622.6
64	207	162.3	196.9	185.2	220.3	208.4	243.8	282.6	365.4	370.6	452.1	537.9	473.9	625.2
⅓	208	163.0	197.7	186.0	221.2	209.3	244.9	283.8	367.0	372.2	454.0	540.2	475.9	627.9
⅔	209	163.7	198.6	186.8	222.2	210.2	245.9	285.1	368.6	373.8	455.9	542.5	477.9	630.5
65	210	164.4	199.5	187.6	223.1	211.1	247.0	286.3	370.2	375.4	457.9	544.8	479.9	633.2
⅓	211	165.2	200.3	188.5	224.1	212.0	248.0	287.5	371.8	377.0	459.8	547.1	481.9	635.9
⅔	212	165.9	201.2	189.3	225.0	212.9	249.1	288.7	373.4	378.6	461.7	549.4	483.9	638.5

TABLE FOR ARCH CULVERTS.

CONTINUED.

No. of Culvert....		1	2	3	4	5	6	7	8	9	10	11	12	13
Size of Culvert....		3X1	3X2	4X1	4X2	5X1	5X2	6X2	6X3	8X2	8X3	8X4	10X2	10X4
Arch—Cub. Yds. per Lineal Foot..........		0.2727	0.2727	0.2909	0.2909	0.3490	0.3190	0.4072	0.4072	0.6727	0.6727	0.6727	0.8843	0.8848
Walls—Cubic Yards per Lineal Foot		0.1481	0.2962	0.1481	0.2962	0.1481	0.2962	0.3704	0.6666	0.3704	0.6666	0.9630	0.4444	1.0370
Paving—Cubic Yards per Lineal Foot..........		0.3333	0.3333	0.3704	0.3704	0.4074	0.4074	0.4444	0.5185	0.5555	0.5926	0.6666	0.6666	0.7407
Coping—Cub. Yds....		2.2222	2.5926	2.5926	2.9700	2.9630	3.3333	3.7030	4.0740	4.4444	4.8148	5.1850	5.5555	5.9260
Wings—Cub. Yds....		8.2028	11.8153	10.6562	11.6360	13.2600	17.7240	20.6370	25.5260	28.6610	40.2382	48.1460	47.2753	59.3310
Sunk W'ls—Cub. Yds.		4.0000	4.0000	4.4444	4.4444	4.8888	4.8888	5.3333	6.2220	6.6666	7.1111	8.0000	8.0000	8.8888
Area Water-way..		6.334	9.531	10.283	11.283	14.617	19.817	26.137	32.137	41.133	49.133	57.133	58.270	78.270
Cub. Yds. in 1 Lineal Foot of Culvert		0.7141	0.8622	0.8084	0.9575	0.9045	1.0526	1.2220	1.5923	1.3966	1.8310	2.3093	1.9955	2.8627
Base, Fifteen Feet. Height.	Length of Culvert.	Cub. Yds.	Cub. Yds.	Cub. Yds.	Cub. Yds.	Cub. Yds.	Cub. Yds.	Cub. Yds.	Cub. Yds.	Cub. Yds.	Cub. Yds.	Cub. Yds.	Cub. Yds.	Cub. Yds.
66	213	166.6	202.1	190.1	226.0	213.8	250.1	290.0	375.0	380.2	463.7	551.7	485.9	641.2
⅓	214	167.3	202.9	190.9	227.0	214.7	251.2	291.2	376.6	381.8	465.6	554.0	487.9	643.9
⅔	215	168.0	203.8	191.7	227.9	215.6	252.3	292.4	378.2	383.4	467.5	556.3	489.9	646.5
67	216	168.7	204.6	192.5	228.9	216.5	253.3	293.6	379.8	385.0	469.5	558.6	491.9	649.2
⅓	217	169.4	205.5	193.3	229.8	217.4	254.4	294.8	381.4	386.6	471.4	560.9	493.9	651.8
⅔	218	170.2	206.4	194.1	230.8	218.3	255.4	296.1	382.9	388.2	473.3	563.2	495.8	654.5
68	219	170.9	207.2	194.9	231.7	219.2	256.5	297.3	384.5	389.8	475.3	565.5	497.8	657.2
⅓	220	171.6	208.1	195.7	232.7	220.1	257.5	298.5	386.1	391.4	477.2	567.8	499.8	659.8
⅔	221	172.3	209.0	196.6	233.7	221.0	258.6	299.7	387.7	393.0	479.1	570.1	501.8	662.5
69	222	173.0	209.8	197.4	234.6	221.9	259.6	301.0	389.3	394.6	481.0	572.4	503.8	665.2
⅓	223	173.7	210.7	198.2	235.6	222.8	260.7	302.2	390.9	396.2	483.0	574.7	505.8	667.8
⅔	224	174.4	211.5	199.0	236.5	223.7	261.7	303.4	392.5	397.8	484.9	577.0	507.8	670.5
70	225	175.2	212.4	199.8	237.5	224.6	262.8	304.6	394.1	399.4	486.8	579.3	509.8	673.1
⅓	226	175.9	213.3	200.5	238.4	225.5	263.8	305.8	395.7	401.0	488.8	581.7	511.8	675.8
⅔	227	176.6	214.1	201.4	239.4	226.5	264.9	307.1	397.3	402.6	490.7	584.0	513.8	678.5
71	228	177.3	215.0	202.2	240.4	227.4	265.9	308.3	398.9	404.2	492.6	586.3	515.8	681.1
⅓	229	178.0	215.9	203.0	241.3	228.3	267.0	309.5	400.5	405.8	494.6	588.6	517.8	683.8
⅔	230	178.7	216.7	203.8	242.3	229.2	268.0	310.7	402.1	407.4	496.5	590.9	519.8	686.5
72	231	179.4	217.6	204.6	243.2	230.1	269.1	312.0	403.6	409.0	498.4	593.2	521.8	689.1
⅓	232	180.2	218.4	205.5	244.2	231.0	270.1	313.2	405.2	410.6	500.4	595.5	523.8	691.8
⅔	233	180.9	219.3	206.3	245.1	231.9	271.2	314.4	406.8	412.2	502.3	597.8	525.8	694.4
73	234	181.6	220.2	207.1	246.1	232.8	272.3	315.6	408.4	413.8	504.2	600.1	527.8	697.1
⅓	235	182.3	221.0	207.9	247.1	233.7	273.3	316.8	410.0	415.4	506.2	602.4	529.8	699.8
⅔	236	183.0	221.9	208.7	248.0	234.6	274.4	318.1	411.6	417.0	508.1	604.7	531.8	702.4
74	237	183.7	222.8	209.5	249.0	235.5	275.4	319.3	413.2	418.6	510.0	607.0	533.8	705.1
⅓	238	184.4	223.6	210.3	249.9	236.4	276.5	320.5	414.8	420.2	512.0	609.3	535.8	707.7
⅔	239	185.2	224.5	211.1	250.9	237.3	277.5	321.7	416.4	421.8	513.9	611.6	537.8	710.4
75	240	185.9	225.3	211.9	251.9	238.2	278.6	323.0	418.0	423.4	515.8	613.9	539.8	713.1
⅓	241	186.6	226.2	212.7	252.8	239.1	279.6	324.2	419.6	425.0	517.8	616.2	541.7	715.7
⅔	242	187.3	227.1	213.5	253.8	240.0	280.7	325.4	421.2	426.6	519.7	618.5	543.7	718.4
76	243	188.0	227.9	214.4	254.7	240.9	281.7	326.6	422.8	428.2	521.6	620.8	545.7	721.1
⅓	244	188.7	228.8	215.2	255.7	241.8	282.8	327.8	424.3	429.8	523.5	623.1	547.7	723.7
⅔	245	189.4	229.6	216.0	256.6	242.7	283.8	329.1	425.9	431.4	525.5	625.4	549.7	726.4
77	246	190.2	230.5	216.8	257.6	243.6	284.9	330.3	427.5	433.0	527.4	627.7	551.7	729.0
⅓	247	190.9	231.4	217.6	258.6	244.5	285.9	331.5	429.1	434.6	529.3	630.0	553.7	731.7
⅔	248	191.6	232.2	218.4	259.5	245.4	287.0	332.7	430.7	436.2	531.3	632.3	555.7	734.4
78	249	192.3	233.1	219.2	260.5	246.4	288.0	334.0	432.3	437.8	533.2	634.6	557.7	737.0
⅓	250	193.0	234.0	220.0	261.4	247.3	289.1	335.2	433.9	439.4	535.1	636.9	559.7	739.7
⅔	251	193.7	234.8	220.8	262.4	248.2	290.1	336.4	435.5	441.0	537.1	639.2	561.7	742.4
79	252	194.4	235.7	221.6	263.3	249.1	291.2	337.6	437.1	442.6	539.0	641.5	563.7	745.0
⅓	253	195.2	236.5	222.5	264.3	250.0	292.3	338.8	438.7	444.2	540.9	643.8	565.7	747.7
⅔	254	195.9	237.4	223.3	265.3	250.9	293.3	340.1	440.3	445.8	542.9	646.1	567.7	750.3
80	255	196.6	238.3	224.1	266.2	251.8	294.4	341.3	441.9	447.4	544.8	648.4	569.7	753.0
⅓	256	197.3	239.1	224.9	267.2	252.7	295.4	342.5	443.5	449.0	546.7	650.7	571.7	755.7
⅔	257	198.0	240.0	225.7	268.1	253.6	296.5	343.7	445.0	450.6	548.7	653.0	573.7	758.3
81	258	198.7	240.9	226.5	269.1	254.5	297.5	344.9	446.6	452.2	550.6	655.3	575.7	761.0
⅓	259	199.4	241.7	227.3	270.0	255.4	298.6	346.2	448.2	453.7	552.5	657.6	577.7	763.7
⅔	260	200.2	242.6	228.1	271.0	256.3	299.6	347.4	449.8	455.3	554.5	659.9	579.7	766.3
82	261	200.9	243.4	228.9	272.0	257.2	300.7	348.6	451.4	456.9	556.4	662.2	581.7	769.0
⅓	262	201.6	244.3	229.7	272.9	258.1	301.7	349.8	453.0	458.5	558.3	664.5	583.7	771.6
⅔	263	202.3	245.2	230.5	273.9	259.0	302.8	351.1	454.6	460.1	560.3	666.8	585.6	774.3
83	264	203.0	246.0	231.4	274.8	259.9	303.8	352.3	456.2	461.7	562.2	669.1	587.6	777.0
⅓	265	203.7	246.9	232.2	275.8	260.8	304.9	353.5	457.8	463.3	564.1	671.4	589.6	779.6
⅔	266	204.4	247.8	233.0	276.7	261.7	305.9	354.7	459.4	464.9	566.0	673.7	591.6	782.3
84	267	205.1	248.6	233.8	277.7	262.6	307.0	355.9	461.0	466.5	568.0	676.0	593.6	785.0
⅓	268	205.9	249.5	234.6	278.7	263.5	308.0	357.2	462.6	468.1	569.9	678.3	595.6	787.6
⅔	269	206.6	250.3	235.4	279.6	264.4	309.1	358.4	464.2	469.7	571.8	680.6	597.6	790.3
85	270	207.3	251.2	236.2	280.6	265.3	310.1	359.6	465.7	471.3	573.8	683.0	599.6	792.9
⅓	271	208.0	252.1	237.0	281.5	266.3	311.2	360.8	467.3	472.9	575.7	685.3	601.6	795.6
⅔	272	208.7	252.9	237.8	282.5	267.2	312.3	362.1	468.9	474.5	577.6	687.6	603.6	798.3
86	273	209.4	253.8	238.6	283.4	268.1	313.3	363.3	470.5	476.1	579.6	689.9	605.6	800.9
⅓	274	210.1	254.7	239.4	284.4	269.0	314.4	364.5	472.1	477.7	581.5	692.2	607.6	803.6
⅔	275	210.9	255.5	240.3	285.4	269.9	315.4	365.7	473.7	479.3	583.4	694.5	609.6	806.3
87	276	211.6	256.4	241.1	286.3	270.8	316.5	366.9	475.3	480.9	585.4	696.8	611.6	808.9
⅓	277	212.3	257.2	241.9	287.3	271.7	317.5	368.2	476.9	482.5	587.3	699.1	613.6	811.6
⅔	278	213.0	258.1	242.7	288.2	272.6	318.6	369.4	478.5	484.1	589.2	701.4	615.6	814.2

www.ingramcontent.com/pod-product-compliance
Lightning Source LLC
Chambersburg PA
CBHW022105210326
41519CB00056B/1203